Atomic Weights of the Elements*

ELEMENT	SYMBOL	ATOMIC NUMBER	ATOMIC WEIGHT
Mercury	**Hg**	**80**	**200.59**
Molybdenum	Mo	42	95.94
Neodymium	Nd	60	144.24
Neon	**Ne**	**10**	**20.18**
Neptunium	**Np**	**93**	**237.0**
Nickel	**Ni**	**28**	**58.69**
Niobium	Nb	41	92.91
Nitrogen	**N**	**7**	**14.01**
Nobelium	No	102	253
Osmium	Os	76	190.2
Oxygen	**O**	**8**	**16.00**
Palladium	Pd	46	106.42
Phosphorus	**P**	**15**	**30.97**
Platinum	Pt	78	195.09
Plutonium	**Pu**	**94**	**244**
Polonium	**Po**	**84**	**209**
Potassium	**K**	**19**	**39.10**
Praseodymium	Pr	59	140.91
Promethium	Pm	61	145
Protactinium	**Pa**	**91**	**231.0**
Radium	Ra	88	226.05
Radon	Rn	86	222
Rhenium	Re	75	186.2
Rhodium	Rh	45	102.91
Rubidium	**Rb**	**37**	**85.47**
Ruthenium	Ru	44	101.07
Samarium	Sm	62	150.35
Scandium	Sc	21	44.96
Selenium	**Se**	**34**	**78.96**
Silicon	**Si**	**14**	**28.09**
Silver	**Ag**	**47**	**107.87**
Sodium	**Na**	**11**	**22.99**
Strontium	**Sr**	**38**	**87.62**
Sulfur	**S**	**16**	**32.06**
Tantalum	Ta	73	180.95
Technetium	Tc	43	99
Tellurium	Te	52	127.60
Terbium	Tb	65	158.92
Thallium	Tl	81	204.37
Thorium	**Th**	**90**	**232.04**
Thulium	Tm	69	168.93
Tin	**Sn**	**50**	**118.69**
Titanium	Ti	22	47.90
Tungsten	W	74	183.85
Uranium	**U**	**92**	**238.03**
Vanadium	**V**	**23**	**50.94**
Xenon	**Xe**	**54**	**131.29**
Ytterbium	Yb	70	173.04
Yttrium	Y	39	88.91
Zinc	**Zn**	**30**	**65.38**
Zirconium	Zr	40	91.22

*Elements discussed in *Chemistry in Perspective* appear in boldface.

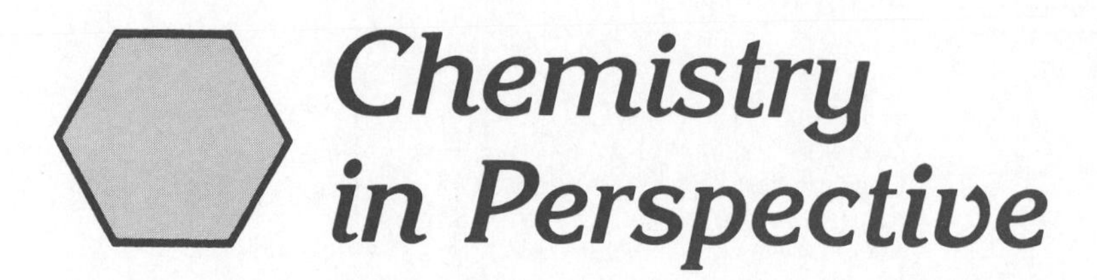

Chemistry in Perspective

Jerry R. Mohrig
Carleton College

William C. Child, Jr.
Carleton College

ALLYN AND BACON, INC.
Boston London Sydney Toronto

Managing editor: Ray Short
Composition buyer: Linda Cox
Cover administrator: Linda Dickinson
Cover designer: Design Ad Cetera
Production services: Lifland et al., Bookmakers
Production administrator: Elaine Ober

Library of Congress Cataloging-in-Publication Data

Mohrig, Jerry R.
Chemistry in perspective.

Includes index.
1. Chemistry. I. Child, William C. II. Title.
QD33.M72 1987 540 86–25893
ISBN 0–205–10270–0

Printed in the United States of America
10 9 8 7 6 5 4 3 2 1 90 89 88 87

CREDITS

Figure 1.1 © Peter L. Bloomer/Horizons West, 1984. All rights reserved.; *Figure 1.3* © Russ Kinne, National Audubon Society Collection/Photo Researchers, Inc.; *Figure 2.2* "Metric football" cartoon © Mike Witte.; *Figure 3.1* E. F. Smith Memorial Collection, Center for the History of Chemistry, University of Pennsylvania.; *Figure 3.8* North Wind Picture Archives.; *Figure 3.11* Courtesy of David J. Neihardt.; *B.C. cartoon* on page 50 reprinted by permission of Johnny Hart and News America Syndicate.; *Figure 4.3(a)* © Dr. Jeremy Burgess, Science Photo Library/Photo Researchers, Inc.; *Figure 4.3(b)* Science Photo Library/Photo Researchers, Inc.; *Figure 4.5* © 1978 National Geographic Society.; *Figure 5.2* University of Wisconsin Photographic Media Center.; *Figure 5.8* © M. Rangell/The Image Works, Inc.; *Figure 6.3* University of Wisconsin Photographic Media Center.; *Figure 6.4* E. F. Smith Memorial Collection, Center for the History of Chemistry, University of Pennsylvania.; *Figure 6.6* E. F. Smith Memorial Collection, Center for the History of Chemistry, University of Pennsylvania.; *Figure 7.5* With permission of the Hebrew University of Jerusalem, Israel.; *Figure 7.6* AIP Niels Bohr Library, W. F. Meggers Collection.; *Figure 7.11* University of Wisconsin Photographic Media Center.; *Figure 7.16(a) and (b)* © Dr. Jeremy Burgess, Science Photo Library/Photo Researchers, Inc.; *Figure 7.16(c)* © Prof. Stanley Cohen, Science Photo Library/Photo Researchers, Inc.; *Figure 7.16(d)* Courtesy of J. M. Huber Corporation.; *Cartoon* on page 158 © 1987 by Sidney Harris.; *Figure 9.6* Reproduced by permission of

(continued on page iv)

To Jean and Nancy with our love and thanks

CREDITS (continued from page ii)

the University Archivist, The Bancroft Library, University of California, Berkeley.; *Figure 9.14* Photographs by Tom Bochsler.; *Figure 10.1* © Tom McHugh, National Audubon Society Collection/Photo Researchers, Inc.; *Figure 10.3(a)* © George Whiteley, Science Photo Library/Photo Researchers, Inc.; *Figure 10.4* © National Audubon Society Collection/Photo Researchers, Inc.; *Figure 11.1* Courtesy of General Electric Research and Development Center.; *Figure 11.5(a)* © Muriel Orans/Horticultural Photography, Corvallis, Oregon.; *Figure 11.5(b)* © Jim Olive, 1986.; *Figure 11.7* © Ted Cordingley, 1986.; *Figure 11.11* University of Wisconsin Photographic Media Center.; *Figure 12.1* © Ted Cordingley, 1986.; *Figure 12.2* E. F. Smith Memorial Collection, Center for the History of Chemistry, University of Pennsylvania.; *Figure 12.4* Courtesy of Carleton College Archives.; *Figure 12.5* Cartoon © 1987 by Sidney Harris.; *Figure 12.7* Courtesy of Florida State News Bureau.; *Figure 12.12* © Mark Antman/The Image Works, Inc.; *Figure 13.4* Courtesy of IBM Corporation, Danbury, CT.; *Figure 14.9* North Wind Picture Archives.; *Figure 14.11* University of Wisconsin Photographic Media Center.; *Figures 14.18 and 14.23* © Ted Cordingley, 1986.; *Figure 15.4* E. F. Smith Memorial Collection, Center for the History of Chemistry, University of Pennsylvania.; *Figure 15.7* © Dr. Don Fawcett, Science Source/Photo Researchers, Inc.; *Figure 15.13* Illustration provided by M. F. Miller, Ph.D., Abbott Laboratories.; *Figure 15.19* Photo Researchers, Inc.; *Figure 15.31(b)* From J. D. Watson, *The Double Helix,* Atheneum, New York, 1968, p. 215, © 1968 by J. D. Watson. *Figure 16.2* Photo © Jim Whitmer.; *Cartoon* on page 419 © 1987 by Sidney Harris.; *Figure 16.15* © Science Source/Photo Researchers, Inc.; *Figure 16.17* Courtesy of Dr. William D. Kasimer.; *Figure 17.2* NASA photo from Science Source/Photo Researchers, Inc.; *Figure 18.6* AIP Niels Bohr Library. *Figure 18.9(a)* Painting by Gary Sheahan, courtesy of AIP Niels Bohr Library.; *Figure 18.9(b)* Courtesy of Argonne National Laboratory.; *Figures 18.11, 18.13, 18.14, and 18.16* © Tom Bochsler.; *Figures 18.17 and 18.18* © AP/Wide World Photos.; *Figure 19.5* Courtesy of South Coast Air Quality Management District, El Monte, CA 91731.; *Figure 19.9* © AP/Wide World Photos.

Contents

Preface

Science courses for nonscience majors are among the most important that are taught. They must be more than simplified versions of corresponding courses for science majors, which are often designed as part of a sequenced curriculum; they must be courses that stand by themselves. Today citizens and consumers face complex choices. Since chemistry is a central science with strong connections to all the other fields of natural science and to many fields of technology as well, it plays an important part in these choices. There is a danger that a whole generation of Americans may lack the understanding and the skills to participate fully in the technological world in which they work and live.

Chemistry in Perspective is designed for a one-semester course in college chemistry taken by students not majoring in the natural sciences. It is written to meet the needs of students with little or no background in science, but also to interest students who have studied chemistry in high school. The text aims for chemical literacy with a clear presentation of basic chemical concepts and a wide range of applications. *Chemistry in Perspective* talks about chemicals that are mentioned regularly in the news media—what they are, the many ways in which we use them, and how they can be misused. It links discussions of principles to examples of chemicals common in our lives, and it discusses technology in the areas of consumer chemistry, health, energy production and use, and the environment. Approximately 40% of the book is devoted to chapters that specifically discuss the applications of chemistry.

Science is an important part of the liberal arts. Students should come away from a chemistry course with the feeling that chemistry is a vital subject. Along with discussions of what chemistry tells about the world, there are glimpses in this book of its human dimension. Seeing the process of how scientists come to understand the world is an important part of the course. Showing the evidence, reasoning, and human intuition behind some important chemical insights reduces considerably the misperception that science cannot be understood by nonscientists.

Our interest in writing this book came from teaching a chemistry course for nonscience majors at Carleton College. The book was written over a six-year period. Initial versions of chapters were extensively rewritten, putting into effect some of the many and valuable comments of students who used those versions and of reviewers and colleagues, and making revisions to reflect the changing times. Our goal, however, has held constant: to write a chemistry textbook that treats the fundamentals of chemistry in a straightforward, interesting way and applies them to important issues in today's world as well as to everyday living.

The book's organization and approach provide a flexible and effective way to meet the many goals of the chemistry course for nonscience majors. In the thirteen chapters of *Part One—Principles,* we present the thinking and language of chemistry, or, in other words, the structure of chemical knowledge. When words will suffice, we avoid equations, which in many cases are not of fundamental importance to the argument and may stand in the way of understanding. There is no mathematics beyond elementary algebra, and not a great deal of that. We believe that *Chemistry in Perspective* is carefully written and readable. We have attempted to use a style of writing that is informal and engaging, yet clear and concise.

In the six chapters of *Part Two—Applications,* there are two separate sections of three chapters each. The first, *Chemistry and Your Life,* discusses consumer chemistry, biochemistry, and chemistry and health. The second, *Chemistry, Energy, and the Environment,* discusses the many forms of energy, radioactivity and nuclear power, and resources and the environment. Although energy and nuclear power issues are not solely a matter of chemistry, we include them because of their importance in our society.

Learning Aids and Teaching Options. The book has a number of features that are designed to meet student needs. In the chapters in *Part One—Principles,* there are over 75 examples and solutions immediately following important points in the text. Many of the almost 300 questions at the ends of chapters are modeled on these examples. At the end of each chapter in *Part Two—Applications,* there are additional questions on important kinds of decisions faced by citizens. These public-policy discussion questions follow examples that we develop within the text.

Every especially important scientific term first appears in boldface in the text and is redefined in the list of terms at the end of the chapter. There are over twenty short special interest features, ranging from radioactive nuclei in medicine to such lighter fare as hot air balloons and tracing water molecules through time. A short appendix provides conversions between different units of measurement. For easier teaching, the instructor's manual suggests simple lecture demonstrations that relate to individual topics and can enliven their presentation.

Instructors using *Chemistry in Perspective* have considerable freedom in their selection of topics, because it contains several optional chapters and sections. The availability of options in a textbook can be a great asset, but only if the understanding of later material does not depend on earlier optional matter. We have worked hard to do this. In *Part One—Principles,* six chapters contain optional parts and three entire chapters are optional. These three are on quantum theory (Chapter 7), the theory and experimental evidence for trends within the periodic table (Chapter 8), and the rates and mechanisms of chemical reactions (Chapter 13). We feel that these topics are too important to modern chemistry to omit from the book; however, the interests of some teachers and students will (in a one-semester course) lead to their spending more time on applications than on principles. Although the three optional chapters can enhance a student's appreciation of bonding and of metabolism, later chapters definitely do not require them. If Chapters 7 and 8 are read, they should be done in sequence.

In three of the chapters in *Part One—Principles,* there is optional material that goes into more detail than some instructors will want. However, other teachers will prefer to have these more detailed discussions available. This optional material includes background on how atomic and molecular weights were determined (Chapter 3), four sections on mole calculations (Chapter 5), and a section on reactions of organic compounds (Chapter 11). In three other chapters in the first part are optional sections on applications of the principles developed in those chapters. These are in Chapter 4 (a section on the escape velocity of hydrogen), Chapter 6 (a section on making aluminum), and Chapter 12 (three sections on acid rain). All optional chapters and topics are clearly indicated in the text (with an asterisk) as well as in the table of contents.

In *Part Two—Applications,* Chapters 15 *(Biochemistry)* and 16 *(Chemistry and Your Health)* can be used without Chapter 14 *(Chemistry and the Consumer),* but if omission is necessary we recommend skipping one of the later chapters rather than the first chapter in the sequence. Each of Chapters 17, 18, and 19 stands alone, although the three are related; any combination can be selected and read in any order. Also, the two sections on applications of chemistry are independent of one another. In addition, it is possible to use Chapter 18 *(Nuclear Reactions and Nuclear Power)* earlier in the course, directly after Chapter 6.

Acknowledgments

Many people deserve credit for their important contributions to *Chemistry in Perspective.* Jean Mohrig and Nancy Child made it all possible by their help and encouragement. Wendy Zimmerman has our sincere thanks for typing the original manuscript in its many versions. Charles Carlin helped to shape the original conception of the book and wrote first drafts of two chapters. Our colleagues Lynn Buffington and James Finholt class-tested various chapters in the chemistry course for nonscience majors at Carleton. Our students have provided many valuable comments and suggestions.

The capable staff at Allyn and Bacon were a pleasure to work with; Editor James Smith, Production Administrator Elaine Ober, and Developmental

Editor Jane Dahl all added their considerable professional expertise to improving the book's conception and providing high-quality production. Sally Lifland and Jane Hoover provided excellent editorial assistance. The following professors read portions of the manuscript and made numerous useful suggestions. We thank them all.

Wayne P. Anderson, Bloomsburg University; Thomas Dobbelstein, Youngstown State University; Frank Fazio, Indiana University of Pennsylvania; Elmer Foldvary, Youngstown State University; George Gorin, Oklahoma State University; C. David Gutsche, Washington University; Marcus E. Hobbs, Duke University; Kenneth Martin, Gordon College; Robert Parry, University of Utah; Nelson Sartoris, Wittenberg University; Wesley Smith, Ricks College; Judith Strong, Moorhead State University; Everett Turner, University of Massachusetts; Richard Wendt, Loyola University; Archie Wilson, University of Minnesota;

Jerry R. Mohrig
William C. Child, Jr.
Northfield, Minnesota

Science is nothing else than the search to discover unity in the wild variety of nature—or more exactly, in the variety of our experience. Poetry, painting, the arts are the same search for unity in variety. Each in its own way looks for likenesses under the variety of human experience.

Jacob Bronowski
Science and Human Values

PART ONE

Principles

1 Introduction

1.1 WHAT ARE CHEMICALS?

This book is about chemicals. According to one dictionary, a chemical is "any substance used in or obtained by a chemical process." This definition is all right as far as it goes. It says clearly that all chemicals are substances, or materials, but the meaning of "chemical process" is not obvious. Also, there is nothing in the definition to indicate whether the "chemical process" is initiated by human beings or occurs in nature without human action. Yet many people seem to feel that chemicals come mostly out of laboratories or factories. They do not realize that most chemicals are produced by natural biochemical and geochemical processes. All substances are chemicals. Our bodies are dynamic chemical systems, made up of chemicals, nourished by chemicals, and producing chemicals. Some chemicals are toxic and hazardous; others are necessary for life.

Is there any difference between a chemical made by an industrial process and the same chemical found naturally in the environment? You may have read in newspapers or magazines about the problem of acid rain in the

United States and Canada. Some lakes, particularly in New England, contain so much acid that all the fish have been killed. Most of this acid is produced when gases from coal-burning power plants dissolve in the water, clearly the result of human action. Some of the acid, however, results from the emission of gases by volcanoes, a natural phenomenon. The product is the same regardless of the source of the pollution. A chemical called sulfuric acid is formed in both cases, and sulfuric acid is a poison.

Many people believe that regular doses of vitamin C help prevent colds. A few argue that the vitamin C must come from natural sources in order to be effective. They say that synthetic vitamin C made by a pharmaceutical company is not as good. Most chemists disagree. According to all the evidence we have, vitamin C is a single chemical compound no matter where it comes from. Any pure substance has one set of properties, which do not depend on the source of the substance. Therefore, vitamin C that comes from a natural source, such as rose hips, must be identical with vitamin C made in the laboratory. Perhaps the vitamin C extracted from rose hips also contains tiny amounts of other chemicals that are beneficial; these trace chemicals are probably not present in the synthetic vitamin. In that case, any extra benefit from natural vitamin C compared with synthetic vitamin C results not from any difference in the vitamin itself but from extra chemicals included with the ''natural'' vitamin.

To a chemist, claiming that natural vitamin C is different from synthetic vitamin C is like arguing that table salt mined in Louisiana is different from table salt made in the lab by combining sodium and chlorine in a test tube. Each product tastes just as salty and dissolves to the same extent in water. As far as anyone can tell, the properties of the two are identical. They are the same substance, called sodium chloride by chemists.

A similar controversy surrounds ''natural'' food. Its supporters claim that food without chemical additives is safer and healthier. We won't take sides in this controversy. We do argue, though, that any differences between ''natural'' foods and those containing added substances result from the presence of different chemicals in the two, not from any inherent advantage of ''natural'' over ''artificial'' substances. In order to settle the controversy it will be necessary to determine what chemicals are present in each food. Then tests can be performed to determine which chemicals are most nutritious and tasty and least hazardous.

Another impression many people have is that all chemicals are harmful or at least unpleasant. What we have already said about the universal presence of chemicals should help dispel that feeling. Yet somehow built into our use of language is the idea that pollutants and toxic substances are chemicals whereas beneficial substances are something else. As a society we must learn which chemicals are harmful, regardless of their source, and then control their use. People can make good chemicals as well as bad ones.

1.2 CHEMICALS IN NATURE

Two types of chemicals are in especially great supply in nature. They are the well-known chemical water and a lesser-known class of chemical substances called silicates. Water is the main ingredient of rivers, lakes, oceans, and clouds. Invisible water vapor in the air around us is responsible for the rain and the snow. Water is made up of hydrogen and oxygen. The silicates are present in most rocks and contain two simpler substances—silicon and oxygen. Sand is perhaps the most familiar example of a silicate.

Without water, life in the forms we know would not exist on the earth. Many scientists believe that the first life appeared in the oceans. Millions

Figure 1.1. The Grand Canyon, Arizona.

of years later some of the ocean creatures moved onto land and adapted to this different environment. Today, water remains an essential part of the diets of animals and humans. People cannot survive without food and water, but a person can live longer without food than without water. Even people on hunger strikes drink a little water each day. Without water, plant life would be impossible, because plants need water along with carbon dioxide in order to grow.

Water has been called the "universal solvent" because it can dissolve so many substances. Water slowly changes the landscape as it dissolves materials in the rocks (Figure 1.1). The dissolved materials are then carried by rivers to the oceans, which contain over 97% of the water on the earth.

Silicates, the most common components of rocks, consist of fundamental substances called elements, which are the building blocks of all chemicals. All silicates contain the elements silicon and oxygen, and they may also contain several other elements. The widespread presence of silicates is indicated by the fact that silicon and oxygen are the most abundant elements in the earth's crust. Silicon makes up 28% of the crust; oxygen accounts for 46%. When silicon and oxygen are combined with each other and with other elements, they have properties that are very different from the properties of the separate elements.

The mineral quartz contains just silicon and oxygen. It is found in rocks as clear and often attractive crystals (Figure 1.2). The gemstones amethyst, tiger's-eye, agate, and onyx are quartz with colored impurities. Even more familiar are the tiny grains of quartz found on beaches; we know them as sand. One of the special properties of quartz is its hardness. Sandpaper derives its effectiveness from the ability of sand particles to wear away substances.

Mica is a silicate that contains additional elements, such as aluminum, potassium, sodium, and calcium (Figure 1.3). The most unusual property of mica is its tendency to split into thin layers. Each layer, however, is quite strong. It is very difficult to break a slice of mica by pulling from opposite ends. Something about its structure causes weak forces between layers but strong forces within layers. One of the chief objectives of chemists is to discover the relationship between the structure and the properties of substances—in this case what makes mica form strong, thin sheets. Mica finds important uses in such different products as electrical insulation, waterproof fabrics, and cosmetics.

We eat chemicals every day, since that is what our food is made up of. The proteins, carbohydrates, fats, vitamins, and minerals that we need for well-balanced diets come from plant and animal sources. For example, common sugar (the chemical sucrose) occurs in the fruits, seeds, flowers, and roots of many plants. It contains the elements carbon, hydrogen, and oxygen. Sucrose is isolated commercially mainly from sugar cane and sugar beets. The refining of sugar involves its separation from the other chemicals in these plants. In the final steps of the process the sucrose is crystallized,

Figure 1.2. Crystals of quartz.

Figure 1.3. Mica.

or granulated, from water solutions. White crystals of pure sucrose result. Each year in the United States the consumption of sucrose is more than 100 pounds per person.

1.3 MAN-MADE CHEMICALS

For centuries people have been converting the chemicals found in nature into other chemicals better suited to human needs. Some of the earliest examples involve metals; periods of early history are named for the metals produced—for example, the Bronze Age and the Iron Age. Early humans discovered, probably by accident, that heating certain rocks produces shiny metals, such as copper. The metal can then be shaped into useful and attractive tools and vessels. Later, people learned that adding tin to copper produces bronze, which is much harder than copper. The chemistry of producing iron came later. It is more difficult to make iron, but it was highly valued because it keeps a sharp edge much longer than bronze or copper does.

In the last century the number of man-made chemicals has expanded greatly. Modern plastics provide a good example of chemicals made by people and not by nature. One plastic is polyethylene, which is found in a variety of products—from plastic bags and children's toys to electrical insulation and pipes. It has made our material lives different from those of people a generation or two ago. For example, food can be kept longer without spoiling if it is placed in polyethylene bags rather than paper bags. Manufacture of this synthetic chemical is a very big business; over ten billion pounds per year are produced in the United States. The ultimate sources of polyethylene are petroleum and natural gas. A key property of polyethylene is its flexibility: it can bend and stretch without breaking. We have learned how to control the flexibility and toughness of polyethylene by understanding its chemical structure.

In the last three decades people have become more aware that most plastics simply do not go away. When plastic objects are discarded into garbage pails and then into sanitary landfills, they may disappear from view but they don't disappear from the environment. Concern about waste disposal has stimulated research on the production of biodegradable plastics that will decompose reasonably quickly upon exposure to weather and soil. The main problem in making a plastic biodegradable is to find an additive that makes the product appetizing to microorganisms but yet does not ruin the desirable properties of the plastic.

There are also, of course, man-made chemicals that are harmful. One example is the class of chemicals called polychlorinated biphenyls, or PCBs. These substances are excellent insulators and heat conductors and thus have been used in large electrical transformers. They do not deteriorate at high temperatures, and they are inexpensive. However, they are very toxic and do not break down in the environment, so PCB spills are dangerous. Unfortunately, PCBs from discarded transformers have been allowed to get into the environment because of improper disposal.

Other harmful man-made chemicals are unwanted by-products of processes that are beneficial to society. Oxides of nitrogen are an example. These substances, which are usual ingredients of the smog found in the air above Los Angeles, Denver, and other large cities, can make people susceptible to respiratory infections and can damage plant life. Such chemicals are formed from the two main components of air: life-giving oxygen and harmless nitrogen. At normal temperatures these two elements do not combine to form oxides of nitrogen. At the high temperatures found inside the cylinders of automobile engines or the furnaces of power plants, however, they can combine. Thus, oxides of nitrogen exit from the exhaust pipes of cars and from the smokestacks of power plants. This problem may be with us as long as society depends on the internal combustion engine for transportation and on the burning of coal, oil, and natural gas for the generation of electricity.

Nature's chemicals and man-made chemicals are all subject to the same scientific laws. Increased understanding of these laws should enable us to produce more useful chemicals and to control those that are less desirable.

1.4 CHEMISTRY, THE NATURAL SCIENCES, AND TECHNOLOGY

In our society you cannot avoid using and hearing about chemicals. They are everywhere. This book is about chemistry rather than physics or biology, but the boundaries between these different sciences are not as clear as you may think. In any academic or industrial laboratory you can find scientists whose fields of training are not obvious from their research. A person investigating the structure of a protein may have studied chemistry, biology, or physics. A new plastic may be developed by someone trained in materials science, chemical engineering, or chemistry.

These examples show the versatility of scientists and engineers, but they also suggest the overlap of the scientific disciplines. Fields such as biochemistry, geophysics, molecular biology, and chemical physics are used all the time, and many observers of science say that the most exciting research is taking place in these cross-disciplinary areas. Chemistry overlaps so many other disciplines that Joel Hildebrand—a chemist who taught at the University of California, Berkeley—defined it with the statement "Chemistry is what chemists do and how they do it."

We live in a technological era dominated by the chemical, electronic, and computer industries. Many people are fascinated by science but are suspicious of technology. Many cannot tell the difference between sense and nonsense in matters connected with science and technology. It isn't hard to find examples of nonsense in advertisements for some of the products that we use daily. Knowing about the character and limits of science is as important for non-scientists as understanding values and history is for scientists. In the following chapters you will learn about some basic chemistry, along with its applications and implications.

Chemists in Action

Consider the changing colors of leaves in the autumn. Just what causes the colors? The isolation and identification of chemicals in living things presents an exciting challenge to many chemists. Chief among the colored chemicals in plants is chlorophyll, which gives the plants their deep green color. With the onset of shorter days in the fall, chlorophyll production stops, and the yellow, orange, and red colors of other pigments appear. These pigments were there all along but were hidden by the intense chlorophyll color.

In addition to separating and identifying substances, chemists are interested in determining the relationships between the structures and properties of substances. For example, what makes skunk scent stink so badly? Chemists have analyzed it and found the chemical that is the culprit. However, we still do not understand enough about our sense of smell to know how this chemical produces such an intense odor. Something about the structure of skunk scent at the atomic level must lead to its recognition by our noses.

Many chemists are interested in the synthesis of new and useful substances. Synthetic chemicals have had a great effect on us. Many people owe their lives to new pharmaceutical products. Man-made fertilizers, insecticides,

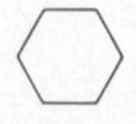

HONESTY IN SCIENCE

It has been said that you can't work in science without assuming the honesty of your co-workers. You can be skeptical, but you need to trust the integrity of your colleagues, since science is such a team effort. Occasional news reports, however, tell of scientists who have made up experimental data. Falsification may occur in fast-moving research fields where competition is strong and grant money and promotions are at stake.

In one case a graduate student had to withdraw his doctoral thesis on a unifying theory of the causes of cancer. Other labs were unable to repeat his work, and a colleague found that some of his samples had a different radioactive tracer from the one he claimed to have used.

In another case the falsification of an immunologist's research on skin grafts was exposed by a laboratory attendant. The immunologist had painted white mice with a black felt pen; he was forced to resign his job.

The scientific enterprise can succeed only if scientists publish accurate and honest reports. Trying to repeat faked data can be deeply frustrating for other scientists, and dishonesty is discovered sooner or later. In fact, cheating is rare. Those who cheat usually get no second chance.

---

and herbicides help to make U.S. agriculture the world's most productive. Synthetic fibers, construction materials, detergents, and materials for transistors and other electronic devices add to the list.

Once scientists have observed and organized some factual information, they attempt to understand, explain, and use it. Words such as *understand* and *explain,* however, are subject to many interpretations. Some people will be satisfied with the statement "Alcohol burns because it is combustible," whereas other people will want to know the general principles of combustion and how they apply to alcohol. One of the main goals of chemistry is to explain chemical phenomena.

Observation and experimental data are the foundation of any scientific explanation. Because each scientist can collect and verify only a tiny fraction of the data in even a narrow specialty, science is a community enterprise that depends on the contributions of many people. Chemists come in many varieties. Some are motivated by practical concerns, such as the need for new building materials or energy sources. Others are motivated by the hope of understanding more about our world. Still others hope to get rich or be admired by their colleagues. Some become deeply involved in the political issues of the day, whereas others couldn't care less about day-to-day concerns, wanting only to be left alone to pursue their research. Like all other scientists, chemists are human beings with all their faults and virtues.

2.1 SCIENTIFIC EXPLANATION

Almost as soon as observations are made, a search for similarities and connections between them begins. For example, water boils at a lower temperature when it is on a mountaintop than when it is at sea level. Do other liquids have lower boiling points as the elevation above sea level is increased? The answer is yes, as shown by measurements of the boiling points of alcohol, gasoline, and other liquids. These observations then permit a generalization or law, which in this case states that the boiling temperature of a liquid decreases with increasing elevation. In general, a **law** is a summary of many related experiences or experimental observations.

Once a body of laws has been accumulated, scientific explanations at the simplest level become possible. When scientists "explain" something, they first of all show that the key parts are consistent with one or more general laws. When you think about it carefully, you realize that most questions that ask "Why does that happen?" are answered by referring to general rules. For instance, to explain why a book falls to the floor rather than rising to the ceiling, you refer to the law of gravity (Figure 2.1). Even though scientists have much to learn about gravity, we feel comfortable with the general rule.

The law concerning the decrease in the boiling point of a liquid with increase in elevation is very specific and is a special case of a more general law: The boiling point of a liquid decreases with decreasing atmospheric pressure. This more general rule is justified by two kinds of observations. First, we know that atmospheric pressure decreases with increasing altitude. Second, we know that a liquid can be boiled in the laboratory at well below its normal boiling temperature merely by reducing the air pressure in its container.

Figure 2.1. The law of gravity?

Let's look at another example of scientific explanation. Some people might ask why a puddle of water on the street dries up much more rapidly on a windy day than on a quiet day. The general rule isn't obvious here. To start out, we need to make a tentative assumption, or hypothesis, which can be tested to find out if it is right or wrong. A **hypothesis** is usually just an educated guess. Let's assume initially that the wind heats the water to a higher temperature. If this turns out to be the case, we can refer to the law that says liquids evaporate more rapidly at higher temperatures. Some testing with a thermometer, however, shows that often a puddle of water is actually cooler on a windy day. Our hypothesis looks wrong, so let's try another one.

Perhaps the wind removes water vapor from the vicinity of the puddle, sweeping it away. This hypothesis is much harder to test by experimental means. One way is to collect samples of air just above the puddle on the two kinds of days. Then we need some kind of test for the amount of water vapor in the air. Although water vapor is colorless, it does absorb light that is not visible to our eyes. By measuring the amount of absorption, we find that the amount of water vapor in the air above puddles is usually lower on windy days. Therefore, the law that says evaporation of a liquid increases as the amount of vapor above it decreases appears to provide an explanation. More testing of alternative hypotheses would be necessary before one could have confidence in this explanation.

At this stage you may still have an uneasy feeling. "Sure," you say, "liquids evaporate more rapidly when the vapor is removed, but why is that true?" Eventually a chemist would develop a hypothesis relating the speed of evaporation to the properties of the tiny particles of which the liquid and the vapor are composed. Linking of a natural occurrence to characteristics of invisible particles is just one example of the use of the atomic theory in chemistry (see Chapter 3). Frequently an observation made on a macroscopic, or visible, system is explained by referring to the invisible, atomic level.

A very general law in science is usually called a **theory,** especially if it has withstood a number of attempts to disprove it. The atomic theory is an example. The most satisfying explanations in science involve reference to theories of great generality. In fact, a chief goal of any branch of science is to connect a large number of observations by a very few general theories. In chemistry these theories usually employ the language of molecules, atoms, and electrons.

Scientists continually test and revise theories. The results of new experiments sometimes can force outright rejection of the theory that inspired the experiments. The usefulness of any theory is often judged by the number and quality of new experiments that it stimulates.

Another characteristic of a good theory is the degree to which it allows scientists to predict and control future events. In 1939 a dramatic example of prediction from theory took place when Albert Einstein wrote President

Franklin Roosevelt to suggest that it might be possible to make an atomic bomb of unparalleled destructive power. Experiments had shown that when a certain kind of uranium is bombarded with neutrons some of the uranium atoms split apart and at the same time release energy. On the basis of their theories of nuclear reactions, Enrico Fermi, Leo Szilard, and others had calculated conditions under which it might be possible to produce a nuclear chain reaction in uranium. This work led to the letter to President Roosevelt. Everyone is now aware of the accuracy of this prediction.

Scientific explanation involves defining many new terms. Every field has its special vocabulary. Whenever possible, a scientist uses an **operational definition,** one that includes a description of an experiment that makes the concept distinctive. For example, in Chapter 3 we will define a chemical element as a substance that cannot be decomposed even after many tries. This is an operational definition because it implies that a chemist should test a substance by trying to decompose it with heat or other means. Many failures to do so indicate that the material is an element. Another definition of an element might be "a fundamental substance." Such a definition gives us a rough idea but is too imprecise to be an operational definition.

If a new concept has numerical values associated with it, an operational definition tells us how to measure the new quantity. Later in this chapter, density is defined as the mass of a sample of matter divided by its volume. Both the mass and the volume can be measured in the laboratory.

2.2 DISPROVING HYPOTHESES

To round out our look at the goals and methods of chemistry, let's look at one good method for developing new theories: *the process of elimination.* If you eliminate alternative possibilities, you are left with the most likely explanation. One feature of any science is that it deals only with hypotheses that are capable of being disproved. If there is no conceivable way in which a particular generalization could be disproved, then it lies outside the realm of the natural sciences. Take, for example, the belief that a person's horoscope predicts the kind of day he or she will have. This belief could be true, but it is very difficult to find experiments or observations for testing it. Whether one is having a "good" day or a "bad" day is very subjective. Since the hypothesis cannot be disproved, it must be left unresolved and is not useful in promoting understanding from a scientific standpoint.

Example Can this hypothesis be disproved: "The canals on Mars were constructed by intelligent life forms"?

Solution Astronomers debated this proposal for years after seeing lines that resembled canals on Mars. In principle the theory can be disproved because it would be possible to get more powerful telescopes or even to go to Mars and find out. Photographs from the recent Mars probe suggest that the surface "canals" are the result of natural processes. They do not seem to have any orderly pattern of construction.

It is difficult to prove any theory with certainty. The usual approach is to rule out, one by one, various alternative theories by doing experiments. After many alternatives have been rejected and one theory has survived numerous tests, confidence in it increases. However, in the future it too may be revised or replaced by another theory that agrees with even more facts. The theory that submicroscopic particles exist, so important to modern theories of chemistry, might be regarded as false someday. As a practical matter, however, most scientists behave as though these particles really do exist. Belief in the submicroscopic particles of nature is common to all the sciences and serves as a unifying force. References to the properties of these particles are frequently found in physics, chemistry, geology, and biology. The success of scientists in predicting precise numerical results of future quantitative experiments is the most convincing evidence of the validity of this atomic theory.

In summary, we might state the four stages of scientific explanation as follows:

1. Specific observations
2. Laws
3. Hypotheses
4. Theories

Systematic experiments are used to test each hypothesis and theory. Complex problems are broken down into a number of simpler steps. After each individual puzzle is solved, the scientist reconstructs the whole. Sometimes the combination and interplay of all these stages is called the "scientific method." By now, though, you have seen that there is no exact method for finding scientific explanations.

Scientific investigation is not a mechanical process. No set of observations leads automatically to a new hypothesis. More often a flash of insight or an inspired guess plays a role. Many of the major advances in science have depended on wild hunches that went far beyond the available facts. Progress in science requires a great deal of creativity.

2.3 MEASUREMENT AND UNITS

Because chemical experiments often involve measurements, we need to look at the units used to express such quantities as length, volume, and mass. By international agreement all countries of the world use or will adopt the **metric system,** which is now called Le Système International d'Unités, or SI for short. Although the United States is committed to voluntary conversion to SI units by the Metric Conversion Act of 1975, many of us seem to be clinging desperately to the clumsy English system of measurement. Even England has officially abandoned use of English units. Besides the problem of having several forms of a unit, such as the long and short ton, the system of measurement used in the United States suffers from an overabundance of units that are unrelated to each other in a simple, predictable manner. Consider length, for example. One mile has 1760 yards, 1 yard has 3 feet, and 1 foot is equivalent to 12 inches. We won't even consider the units of furlongs, chains, rods, and fathoms, all of which also measure length.

The metric system, on the other hand, utilizes our base-ten number system, just as our decimal monetary system does. Prefixes signify multiples or subdivisions on the basis of ten. The prefix milli- means one-thousandth of whatever unit it modifies, centi- means one-hundredth, and kilo- means one thousand times. To convert dollars into cents, one merely moves the decimal point to the right two places: $8.52 equals 852 cents. To convert kilometers into meters, one merely moves the decimal point to the right three places: 83.942 kilometers equals 83,942 meters.

A rectangle 5.6 meters by 7.1 meters has an area equal to 39.8 square meters:

$$5.6 \text{ meters} \cdot 7.1 \text{ meters} = 39.8 \text{ square meters}$$

Contrast the ease of this calculation with the annoyance of calculating the area of a rectangle 18 feet 4 inches by 23 feet 3 inches (the same rectangle). Even with a calculator, it is an awkward calculation. Perhaps pressure from around the globe will convince U.S. citizens of the advantages of the metric system. For some time the dimensions of an Olympic-size swimming pool have been expressed in meters. Recently it was agreed that for any track record to be internationally recognized the distance involved must be in the metric system. Even a football game has been based on the metric system (see Figure 2.2).

The fundamental unit of length in the metric system is the **meter (m).** As indicated in Figure 2.3, the meter is slightly larger than the yard, which originally was decreed by Henry I of England to be the distance from his nose to the fingertips of his outstretched arm. The meter is very close to one ten-millionth of the distance from the North Pole to the Equator. From great to small, distances are commonly measured in kilometers, meters, centimeters, and millimeters. Equivalences between these units and the familiar units of the English system are given in the Appendix.

Volume, which is closely related to length, might logically be measured in cubic meters, but this unit is too large for common use in the laboratory. Instead, the **liter (L)** and the milliliter (mL) are used. The milliliter is the

Figure 2.2. Metric meaning.

same as 1 **cubic centimeter (cc),** and this is the link to the metric system. Figure 2.3 shows that 1 liter (1000 mL) is about 6% larger than a quart. It takes 5 milliliters to fill a teaspoon.

Although both volume and mass seem to measure the amount of any substance, it is usually better to use mass. A volume of 5.00 milliliters of liquid water occupies 5.45 milliliters when frozen and 30,600 milliliters when completely converted into vapor. Clearly the volume depends on the temperature. The mass of this sample of water, on the other hand, is 5.00 grams regardless of whether it is a liquid, solid, or gas. The kilogram and the **gram (g)** are the two most commonly used units of mass in the metric system. One kilogram is equivalent to 2.2 pounds (see Figure 2.3). A paper clip weighs about 1 gram.

Mass is a tricky concept to define, and we shall settle for the statement that it is a measure of the resistance of a body to a change in motion. For example, consider a space vehicle far from any planet. It can change its course or speed by firing one of its rocket engines. The greater the mass of the spaceship, the longer the engine must be fired to bring about a given change in course or speed.

Closely related to mass (and more familiar to most people) is the idea of weight. The weight of an object is the force of gravity acting on it. Weight depends directly on the mass of the object but also on the mass of the earth and the distance between the object and the center of the earth. People weigh a little bit less on a mountaintop than at sea level and even

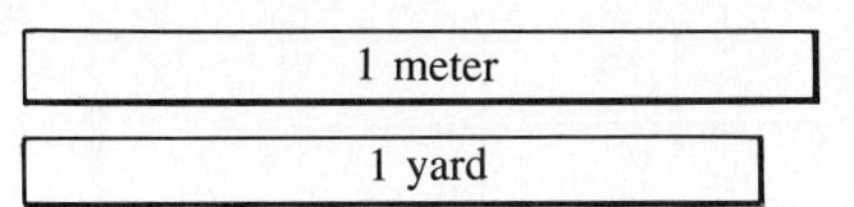

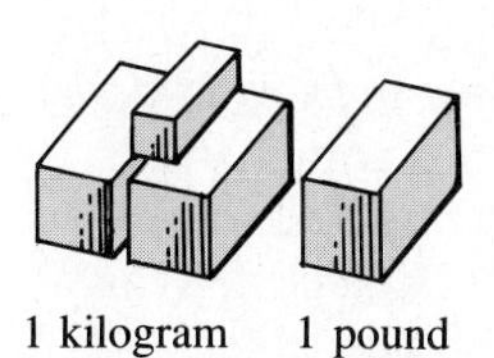

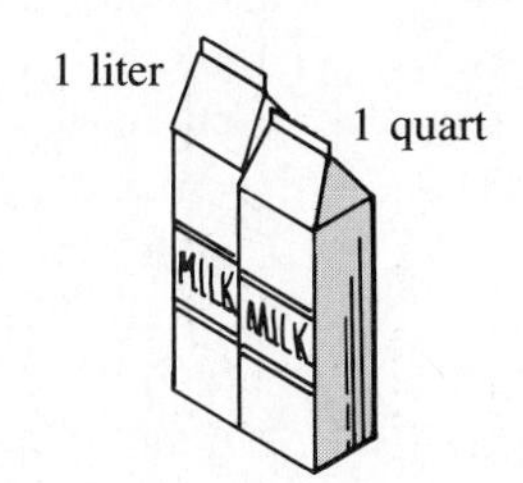

Figure 2.3. Comparison of some metric and English measures.

less if they happen to be on the surface of the moon. By definition, an object with a mass of 1 gram weighs 1 gram at sea level. The same unit, the gram, is used for mass and for weight. We will use the two words interchangeably in this book.

Fortunately, we can ignore the differences between mass and weight in an experiment if we use an equal-arm balance, shown in Figure 2.4. This device allows us to compare two masses directly. When the beam is exactly

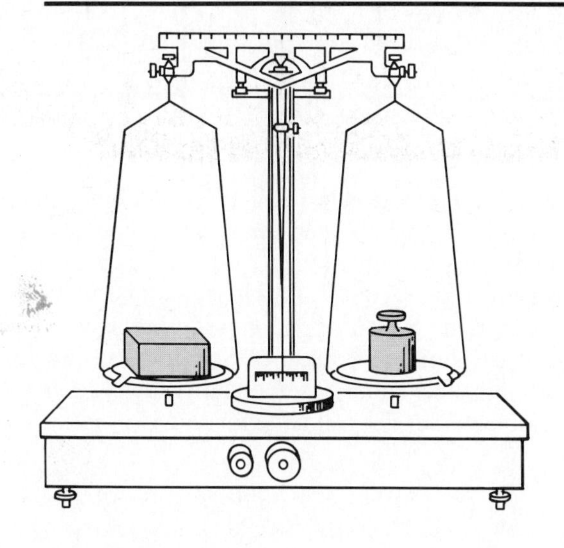

Figure 2.4. A simple equal-arm balance.

METRIC FOOTBALL

Northfield, Minn.—It was too warm for football, and many young men in the crowd took off their shirts as temperatures rose to 25 degrees.

Celsius, that is. It was about 78 degrees Fahrenheit.

In temperature, as in everything else inside Laird Stadium at Carleton College Saturday, the metric system prevailed.

The nation's first metric football game, billed as the Metric Bowl, was played here on a field 100 meters long and 53 meters wide. That means the playing field, not counting the end zones, was 109.36 yards long instead of the normal 100 yards, and 57.96 yards wide instead of 53.33.

Instead of yard lines every 10 yards there were meter lines every 10 meters. And instead of 4 downs to make 10 yards, the teams—Carleton and St. Olaf, two small colleges in the same town—had 4 downs to make 10 meters, or 10.9 yards.

Looking down from the press box, sportscaster Tom Albrecht inadvertently referred to a few "yard gains" rather than "meter gains" early in the game, but had little trouble thereafter.

He once called a pass a "15 yarder," and had difficulty changing that into a "15 meterer."

But he said during a break: "This is a lot easier than I thought it would be. The more I look at it, the less problem it is."

There was only one time in the game that the metric system had an obvious effect.

Quarterback Tim Bates of the white-shirted St. Olaf visiting team hurtled his 188-centimeter and 82-kilogram body (6 feet 2, 180 pounds) toward an onrushing 300-kilogram aggregate of blue-shirted Carleton defenders on a third-down play midway through the second quarter.

Although Bates went far enough for a first down under the 10-yard system, he was a few centimeters shy of a first down under the metric system.

Normally, Carleton has no cheerleaders, but a special student squad was hurriedly drafted for Saturday's game. Coeds donned "Cheer-liter" tee shirts for the occasion, while men wore "Drop back 10 meters and punt" shirts. (From "Town Gets Kick out of First 'Meterball' Game," Michael Hirsley, *Chicago Tribune,* September 18, 1977.)

horizontal, the weights and masses of the two objects are equal. All one needs is a set of standard, known masses for comparison. The ultimate standard is a carefully protected piece of platinum-iridium alloy kept at the International Bureau of Weights and Measures in Sevres, France. Its mass is exactly 1 kilogram. Secondary standards are located in many places, including the U.S. National Bureau of Standards in Washington, D.C.

Finally, we introduce density, a quantity derived from the mass and volume of an object. **Density** (D) is defined as the mass (M) divided by the volume (V):

$$D = \frac{M}{V}$$

Although both the mass and volume of a sample depend on the amount of the substance present, the density does not. It is a true characteristic of the material. For any substance at a definite temperature and pressure, the volume is directly proportional to the mass. Therefore the ratio of the mass to the volume is the same regardless of the size of the sample. The density of a substance is a useful property sometimes used for identification.

Example A small gold ingot weighs 965 grams and has a volume of 50 cubic centimeters. Using these data, calculate the density of gold.

Solution Density is usually given in units of grams (g) per cubic centimeter (cc).

$$D = \frac{M}{V} = \frac{\text{grams}}{\text{cubic centimeter}}$$

$$D = \frac{965 \text{ grams}}{50 \text{ cubic centimeters}} = \frac{19.3 \text{ grams}}{\text{cubic centimeter}}$$

Notice that the density of gold does not depend on the size of the ingot. If it had a volume of 500 cubic centimeters, its weight would be 9650 grams. The density would be the same.

2.4 VERY LARGE AND VERY SMALL NUMBERS

Whereas astronomers deal with extremely large objects, such as stars and galaxies, chemists and physicists talk of atoms and the nuclei of atoms. How much larger is the sun than an atom? Clearly, the sun is many, many, many times larger, but it is not easy to give an exact number. Scientists, however, need to express sizes on a quantitative basis.

Comparison of very large with very small things is best done using exponential notation. An **exponent** is a number placed to the right and slightly above another number; it indicates the power to which the number is to be raised. The larger the exponent, the larger the result.

The number 1000 can be expressed in exponential notation as a power of ten, or 10 with an exponent. The exponent tells the number of zeros or, more exactly, the number of places the decimal point must be moved. In the case of 1000, the decimal point must be moved three places to the left for the number to be written as a power of ten.

$$1000. = 10^3$$

or better,

$$1000. = 1 \times 10^3$$

The number 1×10^6 is the same as 1 with six zeros after it, or 1,000,000 (one million).

Example One large number that we read about from time to time is the U.S. national debt. It is over 1×10^{12} dollars and growing. Express this amount as a number without an exponent.

Soiution The number is 1 followed by 12 zeros, or $1,000,000,000,000 (one trillion dollars).

Example It is estimated that the United States has 1,500,000,000,000 tons of coal reserves. Give this number in exponential notation.

Solution It is usual to give the answer as a number between one and ten multiplied by ten raised to the appropriate power. The number between one and ten would be 1.5. The answer is 1.5×10^{12} tons of coal.

Very small numbers can be expressed using negative exponents. The mass of a DNA molecule is about 5×10^{-16} gram. In the case of a negative exponent, the actual number must have zeros to the right of the decimal point. The mass of a DNA particle would be

$$5 \times 10^{-16} \text{ gram} = 0.0000000000000005 \text{ gram}$$

When the decimal point is moved 16 places to the right, we get 5×10^{-16} again.

Figure 2.5 shows the relative sizes of some of the objects in the universe. The basis for these relative sizes is the size of an average person.

There are four prefixes commonly used with metric units. Each has an exponential relationship to the basic unit. From Table 2.1 you can see that a millimeter is a small unit of length equal to 1×10^{-3}, or 0.001, meter. A milliliter is a small unit of volume equal to 1×10^{-3} liter. A milligram is a small unit of mass equal to 1×10^{-3} gram. When you see the prefix milli-, you need to think small. On the contrary, when you see the prefix kilo-, you need to think big.

If you are dealing with small metric values, it is best to use the prefix micro-, milli-, or centi-. But it would not be convenient to use any of these

---

Figure 2.5. *The size of things.*

Galaxy	Sun	Earth	Person	Cell	Atom
10^{20}	10^{9}	10^{7}	1	10^{-4}	10^{-10}

Table 2.1 Common Metric Prefixes

PREFIX	SIZE
micro-	$\times 10^{-6}$ (one-millionth)
milli-	$\times 10^{-3}$ (one-thousandth)
centi-	$\times 10^{-2}$ (one-hundredth)
kilo-	$\times 10^{3}$ (one thousand times)

prefixes for a large amount. You measure a marathon in kilometers, the height of a person in meters, and the length of a cigarette in millimeters.

Example Radiation can damage our bodies. The unit of radiation commonly employed is the rem. Which dose is more dangerous, 0.2 rem or 100 millirems?

Solution The prefix milli- means 10^{-3}.

$$\begin{aligned} 100 \text{ millirems} = 100 \times 10^{-3} \text{ rem} &= 10^{2} \times 10^{-3} \text{ rem} \\ &= 10^{-1} \text{ rem} \\ &= 0.1 \text{ rem} \end{aligned}$$

So even though 100 of anything sounds like a lot, it may not be if the unit is very small. A radiation dose of 0.2 rem is more dangerous than one of 100 millirems.

2.5 CLASSIFYING MATTER

To set the stage for what will follow, we conclude this chapter with a useful way of organizing the kinds of matter that chemists study. Like any classification scheme, it is useful for orienting one to a new subject but presents difficulties with respect to borderline cases that do not clearly belong under any one heading.

Matter, which is anything that has mass, can first be divided into two categories, homogeneous and heterogeneous. The prefix homo- means the same or alike. A **homogeneous sample** is the same throughout. To homogenize something, you blend it so that it is the same all the way through. Hetero- means different. A **heterogeneous material** has distinctly different regions. The distinction is based on physical appearance. Most samples found in nature are composed of different solid phases having different colors and clear boundaries between them; they are heterogeneous. On examining a piece of granite we find at least three solid phases: colorless quartz crystals, white crystals of feldspar, and the grayish reflective layers of mica. A sample of oil floating on water has two liquid phases and so is heterogeneous. A handful of earth scooped at random might contain ants, bits of broken glass, a tab from a pop-top beverage can, and many colors of dirt particles; it clearly belongs in the heterogeneous category.

There is an uncertainty in the definition of a homogeneous sample as one that is uniform throughout and consists of one phase. Whether or not a substance appears uniform sometimes depends on the keenness of one's eyesight and the degree of magnification used. A sample is often considered to be homogeneous if it appears uniform under an ordinary optical microscope—the kind found in many biology and chemistry laboratories. Such a microscope might have a maximum magnification of a hundredfold.

Figure 2.6 shows the relationships among the various classifications of matter. For the chemist, the most important distinctions in the classification scheme are among solutions, compounds, and elements. It is also important to note that a sample of matter containing many particles of identical appearance and properties is homogeneous rather than heterogeneous. A pile of finely ground table salt is homogeneous even though it consists of many crystals. It constitutes a single solid phase, since every salt crystal has the same chemical composition. The state of subdivision is not relevant. However, if a single chemical has both solid and liquid phases present (as would be the case with ice water), the system is heterogeneous. Based on physical appearance, ice water has two very different regions, even though both consist of water. A sealed glass tube half-filled with red liquid bromine has a region above the surface of the liquid that has a less intense red color. This region consists of gaseous bromine, which is separated from the liquid by a definite boundary. The sample is heterogeneous because it has a gas phase and a liquid phase.

Figure 2.6. The classification of matter.

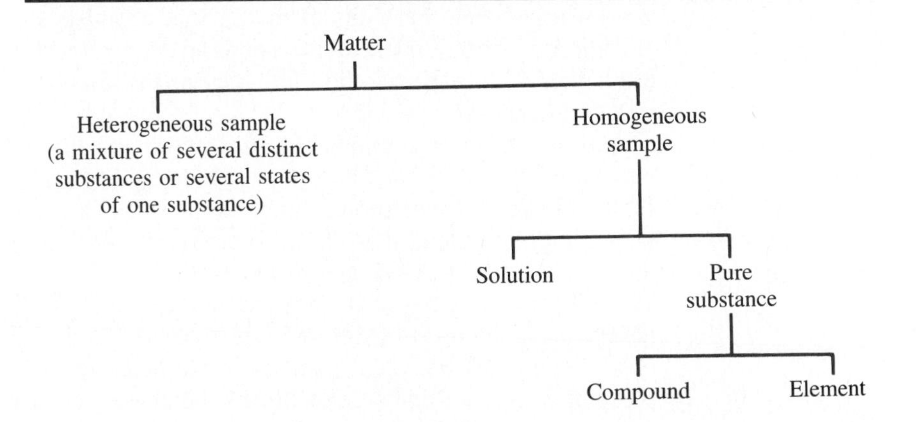

Two of the categories in Figure 2.6, heterogeneous sample and solution, are **mixtures:** that is, they contain two or more substances mixed together. Some of the complex mixtures found in nature may have hundreds of components. The distinguishing characteristic of a mixture is that its composition may vary. For instance, a solution of sugar in water may have 2% sugar or 10% sugar or any other amount up to the point where no more sugar will dissolve. The composition of the heterogeneous contents of a shopping cart can easily be altered by adding or removing items. For the chemist, the most important distinctions in the classification scheme are among solutions, compounds, and elements.

Example In which of the categories shown in Figure 2.6 would you place a sample of cranberry juice?

Solution Cranberry juice is a mixture of water and various substances obtained from crushed cranberries. Its appearance is that of a homogeneous material. Since it is a mixture and thus not a pure substance, it must be a solution.

2.6 SOLUTIONS AND PURE SUBSTANCES

A **solution** is a homogeneous mixture with a variable composition. A solution always contains more than one substance, but looks alike throughout. Most familiar solutions (such as salt water, wine, and gasoline) are liquid, but solid solutions (such as steel) and gaseous solutions (such as air) also exist.

The composition of any solution can be changed, and as the composition is varied, the properties of the solution gradually change. The boiling temperature of a salt-water solution increases slightly as more salt is dissolved, the taste of wine is altered if more alcohol is present, and the hardness of steel changes as the carbon content is varied. In contrast, the composition of a **pure substance** is fixed. Its composition can be changed only by causing the substance to react chemically to form a totally different substance or substances. In other words, the substance must be destroyed.

In order to determine whether a homogeneous sample is a solution or a pure substance, you can assume that the sample is a mixture and then test the hypothesis by seeing whether or not it can be separated into different substances. The separation can be achieved by any of a number of techniques, of which we will discuss only one—distillation. Distillation is useful only for liquid samples and is performed with an apparatus such as the one shown in Figure 2.7. After the liquid has been heated to its boiling temperature, it is slowly boiled away. The vapor is converted into a liquid in the water-cooled condenser, and the liquid is collected in a receiving flask.

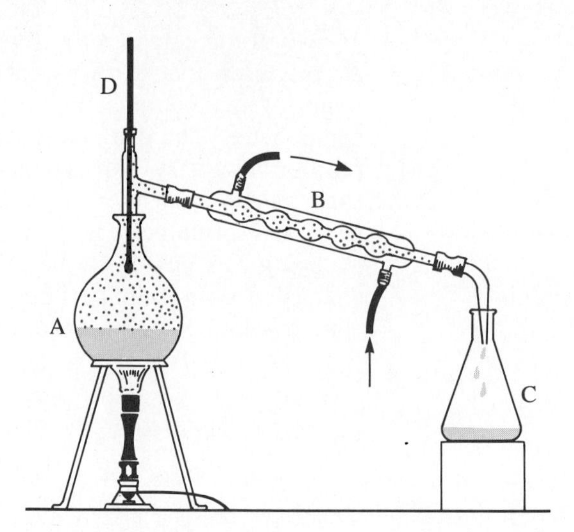

Figure 2.7. *Apparatus for the distillation of a liquid. A is the boiling flask, B the water-cooled condenser, C the receiving flask, and D the thermometer.*

The first clue as to whether the substance is pure or not comes from monitoring the boiling temperature during distillation. A pure substance boils at a nearly constant temperature, whereas a solution may boil over a range of temperatures. As a further check, the distillation can be halted before all the original liquid has boiled away, and some property of the liquid in the boiling flask and the liquid in the receiving flask can be measured and compared. If the two liquids differ in any property, the compositions are different. Agreement between the measurements for both samples suggests a pure substance. If toward the end of a distillation a solid appears as a residue in the boiling flask, then it is probable that the original sample is a solution of a solid in a liquid. The distillation of salt water produces this result.

Success in separating a homogeneous liquid into different substances clearly demonstrates that the sample is a mixture and a solution. Repeated failures strongly suggest, but do not prove, that the substance is pure. The hypothesis that a substance is pure is much easier to disprove than to prove.

It is not always necessary to attempt a separation in order to distinguish between a solution and a pure substance. If the sample in question is a solid and has a convenient melting temperature, then a careful study of the melting behavior can reveal whether or not it is pure. As the temperature is slowly raised, one carefully watches the crystals for any sign of melting. The temperature is noted when the very first liquid appears, and it is recorded again when the last bit of solid becomes liquid. The difference between these two temperatures indicates the purity of the sample. A small melting range of 1 to 2° means the substance is pure; a larger range suggests a mixture.

2.7 COMPOUNDS AND ELEMENTS

At the most fundamental level in the classification scheme (Figure 2.6), we have the distinction between a compound and an element. Any pure substance must be one or the other. A compound can be converted into simpler substances by a chemical reaction, whereas an element cannot be decomposed into more fundamental substances. A compound is therefore composed of elements, but, curiously, the properties of a compound usually bear little similarity to the properties of the elements of which it is made. Although consumption of a small amount of salt—the compound sodium chloride—is beneficial, eating either of its elements (sodium and chlorine) would produce results too horrible to describe. Alcohol, on the other hand, retains its properties when mixed with other substances in wine. Drinking a large amount of wine, which is a solution of water, alcohol, and other ingredients, will produce about the same level of intoxication as drinking a much smaller amount of pure alcohol.

The number of ways in which the 108 known elements can combine to create compounds is incredibly large. Currently known compounds number in the millions, and many others remain to be synthesized or isolated from the natural environment. The story of elements and compounds will be completed in Chapter 3.

IMPORTANT TERMS

law A statement that generalizes a group of observations.

hypothesis A temporary assumption or educated guess that can be tested by experiment.

theory A general explanation that can account for a wide range of different observations.

operational definition A definition that specifies a laboratory procedure capable of distinguishing the concept being defined from other concepts.

metric system The decimal system of physical measurement.

meter (m) A standard unit of length in the metric system, equal to 1.094 yards.

liter (L) A convenient unit of volume in the metric system, equal to 1.06 quarts.

cubic centimeter (cc) A unit of volume, equivalent to a cube 1 centimeter on each side. One cubic centimeter equals 1 milliliter (mL).

gram (g) A convenient unit of mass or weight. One thousand grams equals 2.2 pounds.

mass The measure of a body's resistance to a change in motion. Mass is closely related to weight and in this book is used interchangeably with weight.

density The mass of a sample of material divided by its volume. Chemists usually express the mass in grams and the volume in cubic centimeters or, equivalently, in milliliters.

exponent A number or symbol placed to the right and slightly above another number, signifying exponential notation. Exponents are commonly used to indicate powers of ten.

micro- Prefix meaning one-millionth.

milli- Prefix meaning one-thousandth.

centi- Prefix meaning one-hundredth.

kilo- Prefix meaning one thousand times.

homogeneous sample Matter that is the same throughout.

heterogeneous material Matter that has easily distinguished regions within it.

mixture A combination of two or more substances whose proportions can be varied.

solution A homogeneous mixture that can be separated into more than one pure substance.

pure substance A chemical that has a fixed composition, has a definite set of properties, and cannot be further purified.

QUESTIONS

1. Give an appropriate, convenient metric unit that might be used for stating each of the following physical quantities.

a. the weight of a football player
b. the volume of a glass of beer
c. the length of a cross-country ski
d. the elevation of Denver, Colorado, above sea level
e. the volume of an automobile gas tank
f. the mass of a penny
g. the volume of a thimble
h. the distance from Minneapolis to Chicago
i. the density of gasoline

2. Express each of the following numbers in exponential notation.

a. 3200 **d.** 0.1
b. 0.00024 **e.** 621.07
c. 4,000,000,000

3. Express each of the following numbers without exponential notation.

a. 4×10^2 **c.** 8×10^{-4}
b. 3.9×10^5 **d.** 2.95×10^{-6}

4. Which is larger?

a. 1 meter or 100 millimeters
b. 5 milliliters or 5 liters
c. 2 kilograms or 4 milligrams
d. 1 centimeter or 2 millimeters
e. 60 grams or 0.6 kilogram

5. Discuss whether each of the following hypotheses is capable in principle of being disproved.

a. All people over six feet tall are good basketball players.
b. There has never been a good basketball player who was less than six feet tall.
c. I had a particularly good day last July 17th because I was born under the sign of Virgo.
d. Carbon dioxide gas evaporates from the solid phase directly to the gas phase without ever becoming a liquid.
e. Sometime in the future all aspects of human existence will be explainable by science.

6. In a short paragraph summarize a science news article from a newspaper or magazine.

7. Carefully, but briefly, explain the different meanings of the word "law" as used in the federal law limiting the speed of automobiles to 55 m.p.h. and in the law that says liquids evaporate faster at higher temperatures.

8. Calculate the density in grams per cubic centimeter for each of the following (remember that 1 cubic centimeter = 1 milliliter).

a. A liter of water weighs 1 kilogram.
b. A sample of stainless steel has a mass of 1.5 kilograms and a volume of 183 cubic centimeters.
c. A sample of polyethylene weighs 262 milligrams and has a volume of 0.285 milliliter.
d. A volume of 0.5 liter of air has a mass of 600 milligrams.

9. Chromium, nickel, and lead are metals that look somewhat alike but have different densities:

METAL	DENSITY (in g/cc)
chromium	7.18
nickel	8.90
lead	11.35

A piece of metal thought to be either chromium, nickel, or lead was weighed and found to have a mass of 47.4 grams. Then it was dropped into some water in a graduated cylinder. Before the metal was added, the volume of water was read on the graduated cylinder as 10.3 milliliters. Afterward it was 16.8 milliliters. Identify the metal.

10. Decide whether each of the following samples is homogeneous or heterogeneous, and explain your choice.

a. powdered sugar
b. gasoline
c. a well-mixed sample of powdered sugar and powdered table salt
d. a carrot
e. a cross-section of the trunk of a 75-year-old tree

11. Classify each of the following specimens as heterogeneous or homogeneous. If homogeneous, is it a solution or a pure substance? Briefly give your reasoning.

a. tap water
b. a sample of water containing some ice cubes
c. smog
d. a chocolate-chip cookie
e. concrete
f. refined table salt
g. a piece of brass metal

12. For over 150 years chemists have believed that many substances consist of invisible particles called *molecules*. If the molecules are broken apart, the substance is changed into new substances. During this time period no experiment has contradicted this idea. Would you label the idea of the existence of molecules a hypothesis, a law, or a theory? Explain.

13. In each part below a description of some phenomenon is given, followed by an alleged explanation. Discuss whether the explanation adheres to the standards for a scientific explanation.

a. When water from any ocean in the world is distilled, the liquid that collects in the receiving flask is found always to have the same boiling point, melting point, density, and all other properties. Chemical analysis shows that every sample contains 11.2% hydrogen and 88.8% oxygen by weight. Why are the properties and composition always the same?

Explanation: The liquid appears to be a pure substance. Every pure chemical substance has a distinct set of properties and a definite composition.

b. In many parts of the world, including Europe and North America, people have observed that an ice skate glides readily over ice in wintertime at a temperature of 10° Fahrenheit. Yet at the same temperature an ice skate does not move readily over a smooth glass surface. Why is there so little resistance on ice?

Explanation: When a long, narrow metal surface is in contact with ice, the metal encounters very little resistance to gliding.

Elements and Compounds

Knowing whether a pure substance is a compound or an element is very important to a chemist. A **compound** can be decomposed—broken down into simpler things—whereas an **element** cannot be decomposed by a chemical reaction. To describe a compound, chemists must learn not only the identities of the elements combined within it but their proportions as well; otherwise, their knowledge of the substance is very meager. To describe an element, on the other hand, chemists need only identify that one element.

The correct classification of a pure substance as an element or a compound used to be far from easy. For example, as late as 1789 the famous French chemist Lavoisier published a list of 33 "elements" (see Figure 3.1). Some of these we now know to be compounds. Others, such as light and "caloric," we know as forms of energy. We have come a long way from Lavoisier's list; we now know of 108 chemical elements and millions of pure compounds.

Most people have had first-hand experience with only a few elements, because most elements are not encountered in the course of everyday life. Most everyone has seen and touched objects made of gold, silver, or aluminum. Anyone who has seen a diamond ring has seen the element carbon

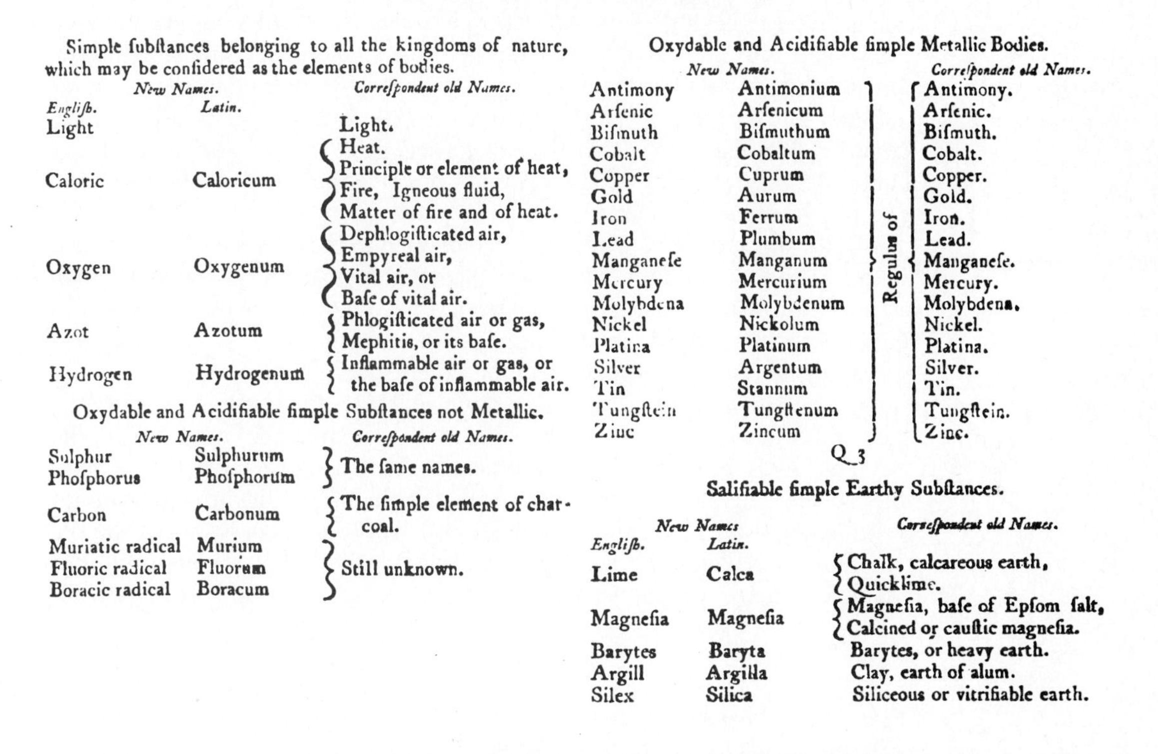

TABLE OF SIMPLE SUBSTANCES.

Simple ſubſtances belonging to all the kingdoms of nature, which may be conſidered as the elements of bodies.

New Names. English.	Latin.	Correſpondent old Names.
Light		Light.
Caloric	Caloricum	Heat. Principle or element of heat, Fire, Igneous fluid, Matter of fire and of heat.
Oxygen	Oxygenum	Dephlogiſticated air, Empyreal air, Vital air, or Baſe of vital air.
Azot	Azotum	Phlogiſticated air or gas, Mephitis, or its baſe.
Hydrogen	Hydrogenum	Inflammable air or gas, or the baſe of inflammable air.

Oxydable and Acidifiable ſimple Subſtances not Metallic.

New Names.		Correſpondent old Names.
Sulphur	Sulphurum	The ſame names.
Phoſphorus	Phoſphorum	
Carbon	Carbonum	The ſimple element of charcoal.
Muriatic radical	Murium	Still unknown.
Fluoric radical	Fluorum	
Boracic radical	Boracum	

Oxydable and Acidifiable ſimple Metallic Bodies.

New Names.		Correſpondent old Names.
Antimony	Antimonium	Regulus of Antimony.
Arſenic	Arſenicum	Arſenic.
Biſmuth	Biſmuthum	Biſmuth.
Cobalt	Cobaltum	Cobalt.
Copper	Cuprum	Copper.
Gold	Aurum	Gold.
Iron	Ferrum	Iron.
Lead	Plumbum	Lead.
Manganeſe	Manganum	Manganeſe.
Mercury	Mercurium	Mercury.
Molybdena	Molybdenum	Molybdena.
Nickel	Nickolum	Nickel.
Platina	Platinum	Platina.
Silver	Argentum	Silver.
Tin	Stannum	Tin.
Tungſtein	Tungſtenum	Tungſtein.
Zinc	Zincum	Zinc.

Q 3

Salifiable ſimple Earthy Subſtances.

New Names English.	Latin.	Correſpondent old Names.
Lime	Calca	Chalk, calcareous earth, Quicklime.
Magneſia	Magneſia	Magneſia, baſe of Epſom ſalt, Calcined or cauſtic magneſia.
Barytes	Baryta	Barytes, or heavy earth.
Argill	Argilla	Clay, earth of alum.
Silex	Silica	Siliceous or vitrifiable earth.

Figure 3.1. Lavoisier's table of elements.

in its sparkling form, and mercury is familiar as a slim silvery thread in many thermometers. If you have worked with copper wire, you have observed another element. Everyone comes in contact with the elements oxygen and nitrogen in the air we breathe, but these are invisible gases and hardly count. There aren't many more easily observed elements. Compounds, however, are everywhere.

3.1 DISTINGUISHING COMPOUNDS FROM ELEMENTS

How does a chemist know whether a particular sample is an element or a compound made up of two or more elements? Practical experience with chemical reactions is of primary importance. If a substance can be decomposed, it must be a compound. Some compounds decompose fairly easily when they are heated. This separation into simpler substances proves that the heated substance is a compound, because an element cannot be broken down by heat. After preparing a red powder by warming a mixture of mercury and oxygen, Lavoisier then heated the red powder to a high temperature. He discovered that it soon turned back into the original silvery liquid mercury and oxygen gas. This is pretty good evidence that the red powder, later called mercuric oxide, is a compound.

Electrical energy may also cause a compound to break down into elements. If two strips of platinum metal are dipped into water containing a small amount of salt and a battery is connected between the strips, as shown in Figure 3.2, bubbles of gas soon appear on each strip. Tests on each gas will show that one is hydrogen and the other oxygen. Since these two gases cannot possibly come from the sodium chloride dissolved in the water, they must come from the water. Water is a compound, made up of the elements hydrogen and oxygen. When a chemist can decompose a pure chemical into lighter, simpler substances, the chemical must be a compound.

Sometimes a chemical is involved in a reaction that is not a simple decomposition, and yet the results still show that it is a compound. For example, Figure 3.3 shows a weighed sample of a certain black powder

Figure 3.2. Using electricity for the decomposition of water containing some dissolved salt. Adding salt to the water makes it conduct electricity. Pure water is such a poor conductor that the decomposition would take a very long time.

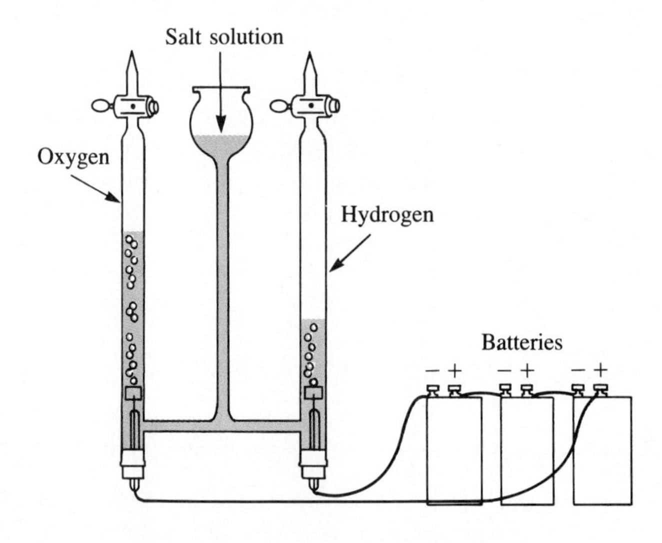

being heated in a tube under a stream of hydrogen gas. Another tube connected to the outlet is cooled in order to condense any vapor coming from the heated tube. After a while the black powder becomes a reddish-gold color, and drops of liquid collect in the cold tube. The properties of the liquid are those of water, and the reddish-gold solid (which weighs much less than the original black powder) proves to be copper metal. As in the two previous examples, a **chemical reaction** has taken place.

In a chemical reaction, elements and/or compounds change into other elements and/or compounds. For example, two elements may combine to form a compound, as when Lavoisier heated mercury and oxygen together to form a red powder, mercuric oxide. A compound can also be broken apart into elements, or more complicated rearrangements can occur.

In the reaction described in Figure 3.3, the **reactants,** or starting substances, are black powder and hydrogen; the **products,** or final substances, are copper and water. You can write the reaction as:

$$\text{black powder} + \text{hydrogen} \longrightarrow \text{copper} + \text{water}$$

Notice that the final products are written to the right of the arrowhead and the reactants are written on the left side.

From this observation, the black powder certainly seems to be a compound containing copper and at least one other element. What else could account for the loss in weight of the solid? Knowing that water—the other product

Figure 3.3. *Apparatus for reacting hydrogen and copper oxide. The hydrogen generator is at the left. Copper oxide is heated in the test tube in the center as hydrogen passes over it.*

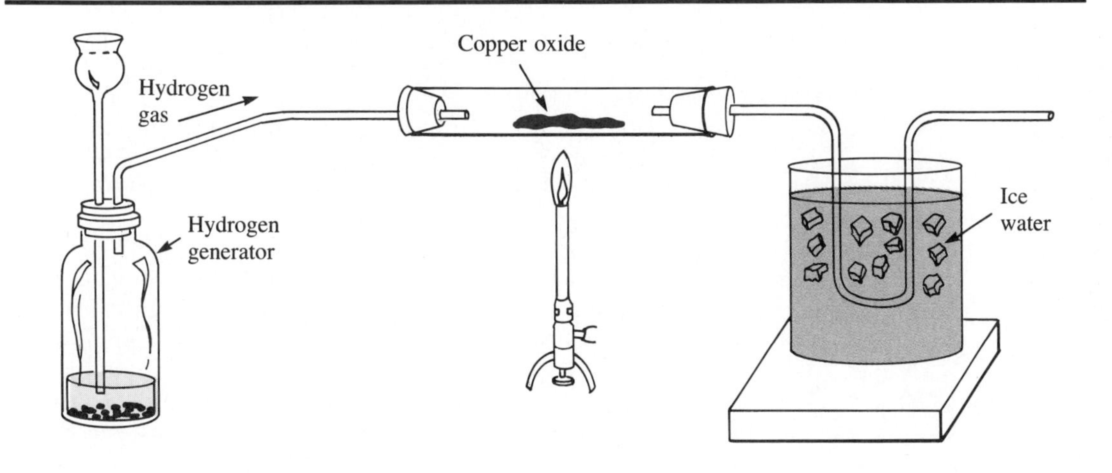

of the reaction—is composed of hydrogen and oxygen, we conclude that the black powder is a compound of copper and oxygen.

$$\underbrace{\text{copper oxide} + \text{hydrogen}}_{\text{reactants}} \longrightarrow \underbrace{\text{copper} + \text{water}}_{\text{products}}$$

The words *reactant* and *product* refer to chemical changes only. In a **chemical change** one or more chemical reactions occur. See Table 3.1 for some examples of chemical changes. A **physical change** involves no alteration of chemical compounds or elements. Mixing several substances to form a heterogeneous mixture, separating the components of a solution through distillation, and melting a solid into a liquid are all examples of physical changes. Sometimes a physical change occurs in the measurement of a physical property, such as the melting point. The words reactant and product are not used when there is only a simple physical change.

Elements, as we have said, are chemicals that cannot be decomposed in any chemical reaction. We have already mentioned several: sodium, chlorine, oxygen, nitrogen, copper, carbon, mercury, and hydrogen. Over 90 elements have been found in nature. More than a dozen others have been produced in the laboratory by means of high-energy nuclear reactions, raising the present total to 108 elements. Proving that a substance is an element rather than a compound is somewhat like proving that a liquid is a pure substance rather than a solution. Failure to decompose the substance may only mean that the chemist failed to run the correct chemical reactions. However, the use of sophisticated modern techniques makes identification of new elements far simpler.

Table 3.1 Examples of Reactants and Products in Chemical Changes

REACTANTS		PRODUCTS
water	⟶	oxygen + hydrogen
gasoline + oxygen	⟶	carbon dioxide + water
nitrogen + hydrogen	⟶	ammonia
iron + oxygen	⟶	iron oxide (rust)
carbon dioxide + water	⟶	sugar + oxygen

Example A student pours some sugar into water and stirs until the sugar completely dissolves. Is the change a physical change or a chemical change? If it is a chemical change, name the reactants and the products.

Solution It is a physical change, and there are no reactants or products in the chemical sense. No new compound forms when the sugar dissolves to form the homogeneous solution. If the water evaporates away, the sugar will remain behind. It will still be sugar—the same chemical it was at the beginning of the experiment. If the student were to drink the sugar water and convert the sugar to carbon dioxide and water by metabolic processes, *that* would be a chemical change.

Example When a piece of magnesium metal is strongly heated in air, it glows brightly as it changes into a white powder. Is the change physical or chemical? If chemical, what might the reactants and products be?

Solution The formation of a white powder from a metal looks very much like a chemical reaction. A reasonable hypothesis is that in air at a high temperature magnesium and oxygen (from the air) combine to form a compound of the two. If so, the reactants are magnesium and oxygen, and the product is a magnesium-oxygen compound. More evidence would be needed to prove that this is the reaction.

3.2 OCCURRENCE OF THE ELEMENTS

The elements are often referred to as the building blocks of nature. The evidence of astronomy, physics, and chemistry tells us that the same 90 or so elements exist everywhere in the universe. Because of the many nuclear reactions occurring in stars, the composition of the universe is evolving. Scientists estimate that our universe contains 76% hydrogen and 23% helium by weight. Oxygen is a poor third at 0.6%.

The earth and the other planets of our solar system have a richer variety of elements than many stars. Astronomers think that our solar system came from older stars that had already gone through several stages of development, so many of the less-common elements had a chance to form. Percentages for the most abundant elements in the crust (outermost layer) of the earth are given in Table 3.2 on page 37. The values given are far from exact because of the great difficulty of estimating the average amounts of various elements over the entire crust, which is roughly 10 miles thick.

None of these ten elements is found in the earth in the uncombined state. The elements are combined in clays and rocks, which are heterogeneous mixtures of compounds called silicates. These minerals consist of oxygen, silicon, and one or more metals such as aluminum, iron, and magnesium. A few of the less abundant elements (such as helium, sulfur, copper, and gold) have a very small tendency to react with other elements; they can exist in nature as pure elements. As far as we can tell from the rock samples brought back from the moon by the Apollo flights, the composition of the moon is not very different from that of the earth. Moon rocks contain more calcium, aluminum, and titanium but less sodium and potassium.

A look at Table 3.2 reveals that one of the important elements in living things, carbon, is not even among the ten most abundant elements in the crust of the earth. And yet carbon contributes almost 20% by weight to

COMPOSITION OF THE MOON AND THE EARTH

The overall compositions of the earth and the moon offer some interesting contrasts, but they are much harder to estimate than are the surface compositions. One method of finding the composition of the interior of the earth is to analyze sound waves that are directed downward and then bounce back from the various layers of matter beneath the surface. From this information scientists were able to estimate the densities of the different layers, and from the densities they determined the compositions. We have less knowledge of the individual layers of the moon. We must rely more on the overall density of the moon, which is 3.3 grams per milliliter. The earth is considerably more dense, at 5.5 grams per milliliter.

From this information geologists have concluded that iron is the most abundant element in the entire earth. It makes up about 30% of the mass. Iron accounts for only 9% of the mass of the moon, however. Much of the iron in the earth is located in the hot, compact, central core, accounting for the greater density of the earth. The composition of the moon has been likened to the composition of the earth's mantle, the layer about 1800 miles thick starting about 10 miles beneath the surface (see Figure 3.4). Any theory of the origin of the moon must take this observation into account.

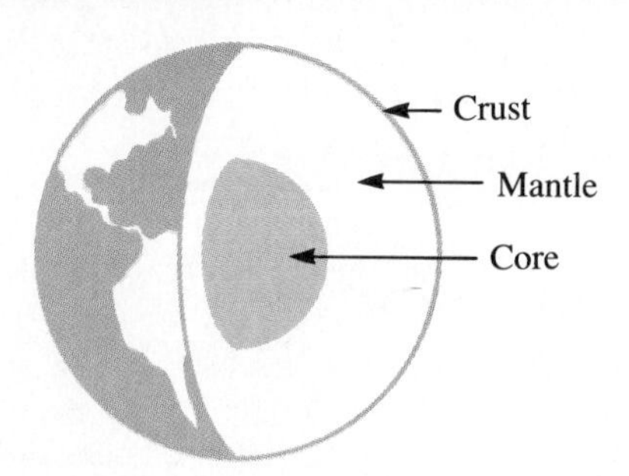

Figure 3.4. *The interior of the earth.*

There is an explanation for the smaller amount of sodium and potassium in moon rocks than in earth rocks. Geologists think of these elements as "volatile." Sodium and potassium evaporate easily from molten rocks at about 1500°C, which may have been the temperature when the moon and the earth were forming 4.5 billion years ago. Because the moon has much less mass than the earth, gravity on the moon is much weaker. Volatile elements such as sodium and potassium were able to overcome the force of gravity on the moon and escape into outer space. A similar loss of these two elements did not occur at the surface of the earth.

the mass of the human body. Clearly some unusual processes must have occurred during the development of plants and animals to account for their high concentration of carbon.

Table 3.2 gives the **symbol** as well as the name for each of the elements listed, again for the purpose of easy communication. Chemists have assigned a distinct one- or two-letter symbol to each of the 108 elements currently known. They are the chemists' shorthand way of writing the names of the elements. These symbols are used so often that you will come to know many of them simply through repeated use.

For some elements the symbol is just the first letter in the name; for many it consists of the first letter plus another from the remainder of the

Table 3.2 Average Composition of the Earth's Crust, Given as Percentages by Weight for the Ten Most Abundant Elements

ELEMENT	SYMBOL	PERCENT BY WEIGHT
oxygen	O	46.4
silicon	Si	28.2
aluminum	Al	8.2
iron	Fe	5.7
calcium	Ca	4.1
sodium	Na	2.4
magnesium	Mg	2.3
potassium	K	2.1
titanium	Ti	0.6
hydrogen	H	0.1

word. Mg and Mn are used to distinguish magnesium from manganese, for example. A few symbols are especially hard to remember because they were derived from the elements' Latin names. Thus, we have Na (natrium) for sodium and Fe (ferrum) for iron. Others in this category are Cu for copper, Au for gold, Ag for silver, Hg for mercury, Pb for lead, and Sn for tin. The symbol for potassium is K for kalium, which is Latin and German. You will find a complete list of the elements and their symbols inside the front cover.

3.3 ANALYSIS OF COMPOUNDS—TWO KEY EXPERIMENTS

One of the basic facts about any compound is the percent by weight of each element in it. Unfortunately, there is no instrument that gives such readings directly. Instead we must carry out a chemical reaction involving the compound in order to analyze it. Such a procedure depends on an important generalization—the **Law of Conservation of Mass:**

> In any chemical reaction the sum of the masses of the reactants equals the sum of the masses of the products.

In other words, during a reaction there is no change in mass. If you decompose a compound, separate the elements from each other, and weigh each one, you can be confident that the weights are those of the elements in the original compound.

Breaking apart a compound is not the only means by which its composition can be determined. Sometimes the weights of elements that combine to form the compound can be measured. We can find the composition of lead sulfide, for example, by carrying out the reaction

$$\underset{\text{reactants}}{\text{lead + sulfur}} \longrightarrow \underset{\text{product}}{\text{lead sulfide}}$$

The strategy is similar to what you would do if you wanted to find the weight of sand in a bucket filled just level with the top (see Figure 3.5). First, weigh the empty bucket. Next, fill the bucket heaping full with sand and scrape off the excess with a board. Then weigh the filled bucket. Finally, subtract the weight of the empty bucket; the difference is the weight of the sand. You have determined the weights of materials involved in the process

$$\text{bucket + sand} \longrightarrow \text{filled bucket}$$

without ever weighing the sand separately.

The goal of our experiment with lead and sulfur is to find two values: the weight of the lead that reacts and the weight of the lead sulfide that forms. The amount of sulfur present in the compound is found by subtracting the weight of the lead from the weight of the compound. To carry out the analysis, one first places a weighed amount of shiny, gray lead metal in a crucible, as shown in Figure 3.6. Then enough yellow powdered sulfur is added so that the lead is completely covered. This mixture is heated for some time to cause the reaction to take place. The success of the procedure depends on the presence of more than enough sulfur to react with all the lead. When the heating is continued, the excess sulfur combines with oxygen in the air and escapes as a gas from around the crucible's lid. Finally, only blue-gray lead sulfide remains, and this is cooled and weighed. The result might be as follows:

Weight of lead metal = 5.37 g
Weight of lead sulfide = 6.20 g
Weight of combined sulfur = 6.20 g − 5.37 g = 0.83 g

Figure 3.5. The steps in determining the weight of sand in a bucket filled just to the top. The weight of the sand equals the weight of the filled bucket minus the weight of the empty bucket.

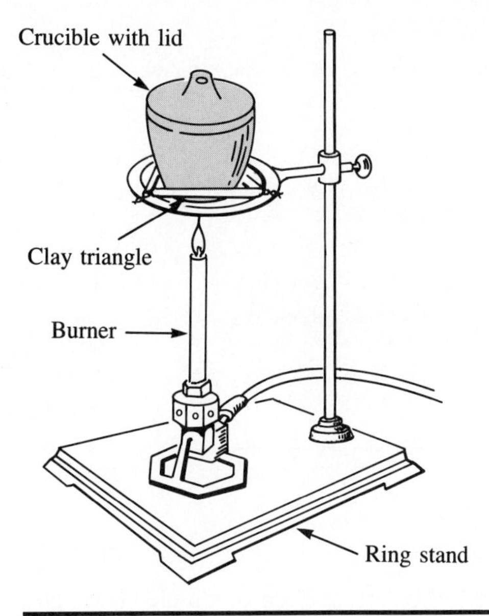

Figure 3.6. Heating a crucible that has a thin sheet of lead and sulfur powder inside.

Thus

$$\%\text{ sulfur in lead sulfide} = \frac{0.83\text{ g sulfur}}{6.20\text{ g lead sulfide}} \cdot 100$$
$$= 13.4\%\text{ sulfur}$$

The percentage of combined lead can be calculated similarly.

$$\%\text{ lead in lead sulfide} = \frac{5.37\text{ g lead}}{6.20\text{ g lead sulfide}} \cdot 100$$
$$= 86.6\%\text{ lead}$$

The two percentages total 100%, as expected. The results are illustrated in Figure 3.7.

As many high school and college students who have done this experiment can testify, it can go wrong in many ways. First of all, one must cope with the choking gas that is given off. (Some people quit before they even weigh the product.) Then, there are potential errors and uncertainties in the measurements of this or any experiment. For example, it is possible that not all the lead reacted because too little sulfur was present or sulfur did not come into contact with every bit of lead. This possibility can be checked by grinding the solid residue and then adding more sulfur and reheating. If the new weight of the contents of the crucible is greater than the old,

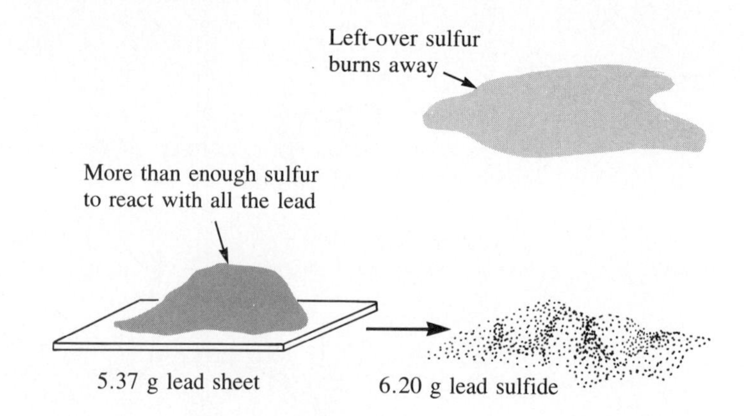

Figure 3.7. A schematic representation of the lead sulfide experiment.

the first reaction may not have been complete. If a nearly constant weight is found after several additions of sulfur, the results are probably reliable.

Whenever the experiment just described is carried out, the composition of lead sulfide is found to be close to 86.6% lead, 13.4% sulfur. Even when samples of lead and sulfur from different sources are used, the results agree. It is possible to make lead sulfide by a variety of reactions. Its composition by weight is always constant within the limits of experimental uncertainty.

By the late eighteenth century the compositions of many compounds had been determined and their unchanging proportions observed. This generalization was named the **Law of Constant Composition:**

> The proportions by weight of the elements in a given compound are always the same.

Pure sodium chloride, regardless of its source, is 39.3% sodium and 60.7% chlorine. Methane, the chief constituent of natural gas, is 25.1% hydrogen and 74.9% carbon. Nowadays chemists take this law for granted. Constant composition is part of the modern definition of any compound.

Example Liquefied petroleum (LP) gas is used as a fuel in many rural areas. Analysis of a compound in a sample of LP gas showed it to be 18.3% hydrogen and 81.7% carbon. Would you identify this compound as methane?

Solution No. If it were methane, it would contain 25.1% hydrogen and 74.9% carbon, since the percentages of the elements in a given compound are always the same. Actually, the LP gas component is another compound called propane.

3.4 ATOMIC THEORY

The atomic theory states that all matter consists of extremely small particles. This theory rests on the observations of thousands of chemists and physicists over many decades. John Dalton (Figure 3.8), an English chemist and philosopher, usually gets the credit for having made the first clear statement of modern atomic theory.

When elements and compounds are broken up into smaller and smaller particles (so small that they can barely be seen even with a powerful microscope), they still have the same properties. These tiny particles, which give elements and compounds their characteristic properties, are called atoms and molecules; the belief that they exist is basic to the thinking of chemists. Even though many, many trillions of them wouldn't even begin to fill a teaspoon, atoms and molecules do have definite, measurable masses. The smallest part of an element that retains its identity during a *chemical reaction* is the **atom.** The smallest part of an element or of most compounds that retains its identity during a *physical change* is the **molecule.** Although the history of chemistry is not a major theme of this book, we will show how just two laws, the Law of Conservation of Mass and the Law of

Figure 3.8. *John Dalton, chemist and philosopher.*

Constant Composition, have helped support the major ideas of the atomic theory.

To see the relevance of the conservation of mass to the atomic theory, consider the formation of water from hydrogen and oxygen. This reaction can be brought about by mixing hydrogen and oxygen gas in a heavy-walled steel container (called a combustion bomb) and then passing a spark through the mixture. A rapid reaction occurs with explosive force, and liquid water condenses on the walls of the bomb. The weight of the bomb and its contents is the same before and after the reaction.

$$\text{hydrogen} + \text{oxygen} \longrightarrow \text{water}$$

What microscopic, invisible structure of matter is consistent with this constancy of weight during a chemical reaction? Dalton's explanation was that each element consists of many identical, invisible particles, or atoms. *These atoms are permanent and indestructible units of matter having constant masses.* Furthermore, Dalton proposed that a chemical reaction, such as the formation of water from hydrogen and oxygen, involves rearrangements of atoms. Any bonds between hydrogen atoms in hydrogen gas and oxygen atoms in oxygen gas are broken. New bonds between the two kinds of atoms form to produce water. Because no atoms are created or destroyed, the total mass remains constant. Of course we cannot say that the conservation of mass proves the atomic theory of matter. We can only say that they are consistent. Dalton's way of visualizing atoms in chemical reactions is shown in Figure 3.9.

Atomic theory also provides a satisfying way to understand the Law of Constant Composition. Pure water from anywhere in the world contains 11.2% hydrogen and 88.8% oxygen by weight. The atomic theory says that water, like any other compound, has a fixed ratio of the numbers of atoms of each element. Although it took many years to determine this ratio, chemists are now confident that water has twice as many hydrogen atoms as oxygen atoms. This conclusion is expressed by the **formula** H_2O, in which the subscripts give the relative numbers of atoms of each element. The subscript 1 is never actually used but is understood to be present if no other subscript is present. Given that all hydrogen atoms have the same mass and all oxygen atoms have a different but constant mass, the proportions by weight of hydrogen and oxygen in water must always be the same.

This is not the only possible explanation of constant composition. One can imagine other sets of assumptions that agree with the facts, but they are all considerably more complicated. When presented with two theories that fit all the facts, scientists will nearly always choose the simpler of the two.

As already suggested, many elements and compounds consist of submicroscopic identical particles, called molecules, each of which contains a small number of atoms. Although some elements consist of separate atoms,

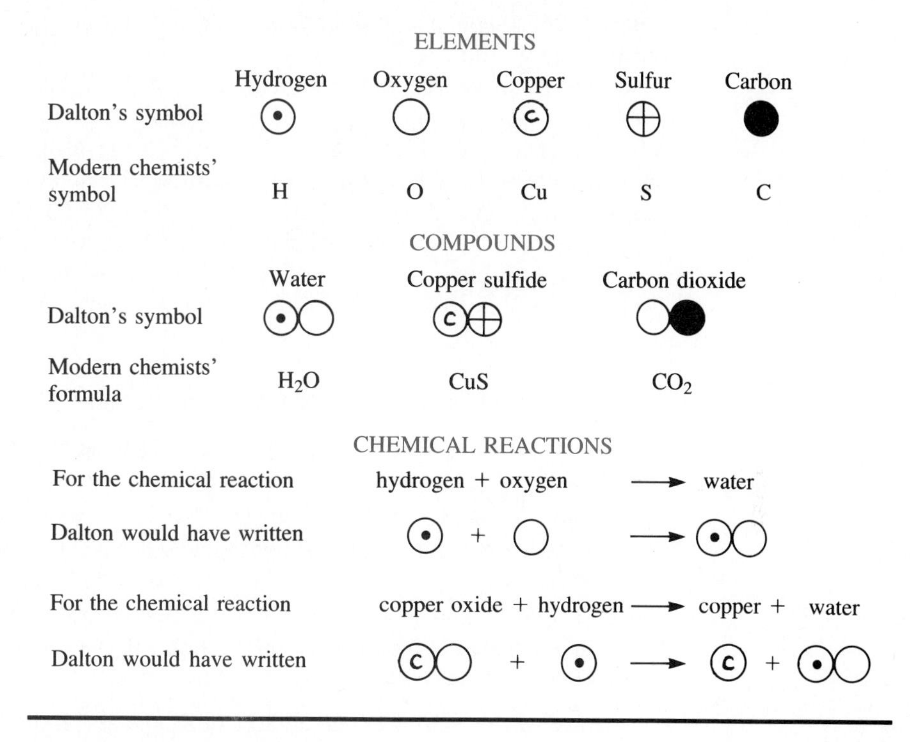

Figure 3.9. Atomic symbols help chemists to visualize atoms, molecules, and chemical reactions.

other elements consist of molecules composed of two or more identical atoms. In a molecular compound each molecule consists of at least two different kinds of atoms. The atoms within a molecule are linked together by chemical bonds. If the bonds in a compound are ruptured, the compound decomposes, because its basic unit, the molecule, has been destroyed. Water is a molecular compound. The formula H_2O not only gives the atomic ratio of hydrogen to oxygen but also says that each molecule of water contains two atoms of hydrogen and one of oxygen linked together in some fashion. These linkages, or bonds, are represented by lines in Figure 3.10.

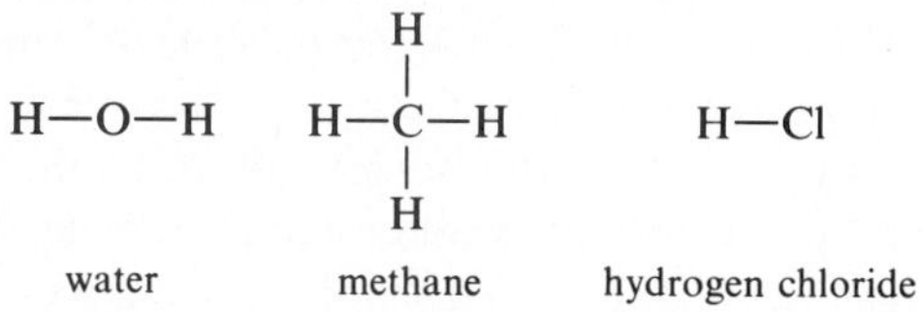

```
                   H
                   |
H—O—H           H—C—H           H—Cl
                   |
                   H

water            methane        hydrogen chloride
```

Figure 3.10. Representations of three common molecular compounds.

Some other molecular compounds are carbon dioxide, CO_2; methane, CH_4; and hydrogen chloride, HCl.

Sodium chloride, on the other hand, is one of a large number of ionic compounds that do not consist of molecules. Its formula, NaCl, tells us only that the ratio of sodium atoms to chlorine atoms is 1:1. Calcium chloride, or $CaCl_2$, is a salt in which the ratio of calcium atoms to chlorine atoms is 1:2. The distinction between molecular compounds and ionic compounds will be clarified in Chapters 9 and 10. For now you need only note that many but not all formulas of compounds describe molecules.

The assumptions of Dalton's atomic theory can be summarized in modern terms as follows.

Assumptions of the Modern Atomic Theory

1. Each element consists of tiny particles, called atoms. Atoms are neither destroyed nor created in ordinary chemical reactions.
2. In any sample of an element the atoms usually have the same average mass. The atoms of different elements have different properties, including different masses.
3. The several kinds of atoms in a given compound are present in fixed ratios.
4. Many compounds consist of groups of different atoms bound together. These particles are called molecules.

Because of the extremely small size of atoms and molecules, any visible amount of a substance contains an amazingly large number of them. The number of people on the earth is insignificant in comparison with the number of molecules in one drop of water. The smallest grain of table salt that can be seen under a microscope with one-hundredfold magnification has approximately ten billion sodium atoms and an equal number of chlorine atoms. A straight line of water molecules, in contact with each other and running from here to the moon (239,000 miles), would weigh 0.0001 gram, which is barely detectable on a balance used for accurate weighing of chemicals.

3.5 ATOMIC AND MOLECULAR WEIGHTS

According to atomic theory, one difference between atoms of different elements is their different masses. Therefore, a knowledge of these masses is necessary in order to apply atomic theory to chemical phenomena. There are two problems involved in finding the masses of atoms. The first is the difficulty of counting the number of atoms in a sample of any element. If one could somehow determine the number of atoms in a lump of sulfur, a person could obtain the weight of one sulfur atom by weighing the lump and then dividing the weight by the number of atoms.

$$\frac{\text{Weight of lump}}{\text{Number of atoms in lump}} = \text{weight of one atom}$$

But since the atoms and molecules are invisible to all but the most powerful microscopes, counting must be done by indirect means. As strange as it may seem, we count by weighing samples of elements and compounds. Chemists have adopted an atomic weight scale that gives the relative masses of different kinds of atoms. In other words, an arbitrary number is assigned to the mass of one kind of atom—a hydrogen atom, for example. Then the atomic weight of a second element, say helium, is set by nature at four times the number given to hydrogen, because one atom of helium weighs four times as much as one atom of hydrogen, according to the evidence. From the atomic weight scale we learn that the mass of a sulfur atom is twice the mass of an oxygen atom. A carbon atom weighs twelve times as much as a hydrogen atom. The actual masses of single atoms are seldom used, since the use of relative masses is easier.

Determining the relative masses of atoms is the second problem that frustrated scientists for many years. At the time Dalton proposed his theory, the compositions by weight for many compounds were known, but neither the formulas of the compounds nor the relative masses of the atoms in them were known. And it is impossible to calculate one without the other.

* To see the kind of trap nineteenth-century chemists were caught in, consider some data for the compound we now call carbon dioxide. This gas is a major product of the metabolism of our food; it also causes the fizz in soda pop and champagne. In Dalton's time carbon dioxide was called carbonic oxide. Assume that we know only the composition of carbonic oxide: 27.3% carbon and 72.7% oxygen by weight. The weight ratio of oxygen to carbon is

$$\frac{\text{Weight of oxygen in carbonic oxide}}{\text{Weight of carbon in carbonic oxide}} = \frac{72.7}{27.3} = 2.66$$

Still we can say nothing about the relative numbers of oxygen atoms and carbon atoms in carbonic oxide. Suppose that the formula of this compound was proposed to be CO_2 (now known to be correct). Then there would be twice as many oxygen atoms as carbon atoms in any quantity of the compound. One oxygen atom would weigh 2.66/2, or 1.33 times as much as one carbon atom.

$$\frac{\text{Weight of oxygen}}{\text{Weight of carbon}} \cdot \frac{\text{number of carbon atoms}}{\text{number of oxygen atoms}} = 2.66 \cdot \frac{1}{2} = 1.33$$

$$1.33 = \frac{\text{Weight of one oxygen atom}}{\text{Weight of one carbon atom}}$$

*The material between the asterisks is optional.

However, suppose instead that the formula was CO. Then there would be equal numbers of carbon and oxygen atoms in any sample. One oxygen atom would weigh 2.66 times as much as one carbon atom. The point is that a ratio of atomic weights can be found to go with any assumed formula. Additional information is needed to determine the correct formula. *

After Dalton announced his atomic theory, it took chemists about 50 years to learn how to determine relative atomic weights. Rather than tell that lengthy story, we will show how a modern chemist might obtain evidence for the formula of a compound from the x-ray diffraction method. In brief, the method involves directing x-rays toward a crystal and observing the pattern made by the rays that bounce off the solid surface. From this pattern (see Figure 3.11) it is possible to determine the location of the atoms in the crystal. A rough kind of molecular picture is obtained. This method was used with spectacular success in proving the existence of the double-helix structure of DNA, which directs the expression of our genetic information.

When x-ray diffraction is done with the solid form of carbonic oxide, commonly called dry ice, it is found that two atoms of oxygen are linked with each atom of carbon. In other words, the correct formula of carbonic oxide is CO_2. We have already seen that this formula leads to an atomic weight for oxygen that is 1.33 times that of carbon. With such data chemists were able to develop an accurate atomic weight scale for all the known elements.

There is one further complication though. It has been found that nearly every element occurring in nature has more than one kind of atom. The various **isotopes** differ only in mass. Otherwise, the atoms are the same.

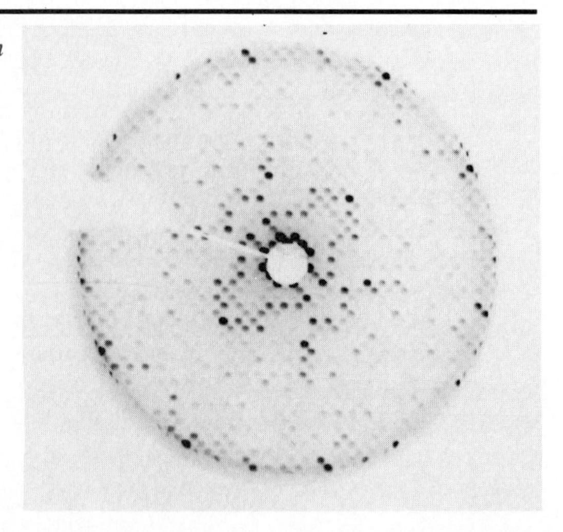

Figure 3.11. An x-ray diffraction pattern of an enzyme, alanine racemase.

In spite of the fact that an element can have many isotopes, for most elements the average mass of the atoms in any sample is the same. This is why we said in the second assumption of modern atomic theory (page 44), "In any sample of an element the atoms usually have the same average mass."

Isotopes of a given element have nearly identical chemical properties, and one can tell them apart only with the help of expensive instruments. Any sample of an element is a mixture of isotopes, or atoms of slightly different mass. This is true whether the element is combined in a compound or by itself. With few exceptions the proportions of the several isotopes of an element are the same everywhere on earth where that element is found.

After a long period during which chemists and physicists used slightly different atomic weight scales, all scientists finally agreed to assign an atomic weight of exactly 12.000 to the most common isotope of carbon. In any naturally occurring sample of carbon, a small amount of a heavier carbon isotope is also present. The average **atomic weight** for the carbon we deal with turns out to be 12.01. Having already decided that the atomic weight of oxygen must be 1.33 (1.332, to be more exact) times the atomic weight of carbon, we find the average atomic weight of oxygen to be 16.00 $(12.01 \cdot 1.332)$.

It is a simple step to represent the relative masses of molecules on the same scale. Since a molecule of carbon dioxide has one carbon atom and two oxygen atoms, the **molecular weight** of the compound is

$$\begin{aligned} 1 \cdot \text{atomic weight of C} &= 1 \cdot 12.01 = 12.01 \\ 2 \cdot \text{atomic weight of O} &= 2 \cdot 16.00 = \underline{32.00} \\ \text{Molecular weight of } CO_2 &= 44.01 \end{aligned}$$

This means that the mass of a molecule of CO_2 is 44.01/12.01 times the mass of an average carbon atom. Similarly, the molecular weight of methane, CH_4, is $12.01 + (4)(1.008) = 16.04$.

* The percent by weight of an element in a compound can now be calculated directly from atomic and molecular weights and the chemical formula. For example, the percent by weight of carbon, C, in carbon dioxide, CO_2, is given by

$$\begin{aligned} \%\,C &= \frac{\text{number of C atoms} \cdot \text{atomic weight of C}}{\text{molecular weight of } CO_2} \cdot 100 \\ &= \frac{12.01}{12.01 + (2)(16.00)} \cdot 100 = \frac{12.01}{44.01} \cdot 100 = 27.29\%\,C \end{aligned}$$

*The material between the asterisks is optional.

Example Find the percent by weight of each element in the compound sodium sulfate, Na_2SO_4.

Solution First, you need to find the relevant atomic weights on the inside front cover of this book.

Element:	Na	S	O
Atomic weight:	23.00	32.06	16.00

Next, you need to calculate the compound's weight.

$$\begin{aligned}\text{Atomic weight of 2 sodium atoms} &= 2 \cdot 23.00 = 46.00 \\ \text{Atomic weight of 1 sulfur atom} &= 1 \cdot 32.06 = 32.06 \\ \text{Atomic weight of 4 oxygen atoms} &= 4 \cdot 16.00 = \underline{64.00} \\ & \qquad\qquad\quad 142.06\end{aligned}$$

Now you can find the percent of Na in Na_2SO_4. Since there are two sodium atoms in one unit of Na_2SO_4, you need to multiply the relative weight of sodium by 2 and then divide by the weight of Na_2SO_4. Multiplying by 100 gives the percent of sodium.

$$\%\ \text{Na} = \frac{\text{number of Na atoms} \cdot \text{atomic weight of Na}}{\text{weight of Na}_2\text{SO}_4} \cdot 100$$

$$= \frac{2 \cdot 23.00}{142.06} \cdot 100 = 32.38\%\ \text{Na}$$

In the same way, you can calculate the percent of sulfur. This calculation is even simpler, because there is only one sulfur atom in Na_2SO_4.

$$\frac{32.06}{142.06} \cdot 100 = 22.57\%\ \text{S}$$

With four oxygen atoms in one unit of Na_2SO_4, the percent oxygen can be found using this same kind of calculation.

$$\frac{4 \cdot 16.00}{142.06} \cdot 100 = 45.05\%\ \text{O}$$

Notice that the percentages add up to 100%.

*

3.6 EQUATIONS FOR CHEMICAL REACTIONS

For several examples of chemical reactions we have used word equations such as

$$\text{copper oxide} + \text{hydrogen} \longrightarrow \text{copper} + \text{water}$$

This reaction, described at the beginning of this chapter, occurs when hydrogen gas is passed over heated copper oxide. Much more informative and useful to chemists is an equation that gives the formulas of the compounds involved.

$$CuO + H_2 \longrightarrow Cu + H_2O$$

The correct formula of each element and compound is determined by experimental evidence.

The formulas of the elements require comment, for we have shown two hydrogen atoms linked together in hydrogen gas but only a single atom in the formula for solid copper metal. Many of the gaseous elements, such as hydrogen, nitrogen, oxygen, and chlorine, occur not as single atoms but as molecules in which two identical atoms are bonded together. The individual particles that move in all directions inside a container of gas are these diatomic molecules. Two elements—bromine and iodine—are liquid and solid, respectively, at normal laboratory temperatures. Yet they also consist of diatomic molecules that keep their identities in solid, liquid, and gaseous states. The common elements that occur as diatomic molecules are listed in Table 3.3.

Metals such as copper, on the other hand, consist of orderly arrangements of atoms. As measured by x-ray diffraction, the distances between neighboring copper atoms are all the same. No molecules can be identified. Thus we use the symbol Cu to indicate the copper metal.

A very important requirement for any chemical equation is that every single atom of the reactants on the left side of the equation must show up in the products on the right side. This simply means that *the chemical equation must be balanced.* In a **balanced chemical equation,** the number of atoms of each kind must be the same on each side of the equation, since atoms are not destroyed and mass is conserved in chemical reactions. In the equation for the copper oxide–hydrogen reaction, we have one copper atom, one oxygen atom, and two hydrogen atoms on each side. The arrow pointing toward the right separates the reactants from the products and shows the direction of the chemical action. Often an equal sign is used for the same purpose. The symbols *s*, *l*, and *g* in parentheses indicate whether the substance is a solid, a liquid, or a gas, respectively.

Table 3.3 Diatomic Elements

ELEMENT	FORMULA	ELEMENT	FORMULA
hydrogen	H_2	fluorine	F_2
nitrogen	N_2	chlorine	Cl_2
oxygen	O_2	bromine	Br_2
		iodine	I_2

Consider the burning of hydrogen in oxygen to produce water. Knowing the correct formula for each reactant and product, we can write a preliminary, unbalanced version of the reaction:

$$\underset{\text{hydrogen}}{H_2(g)} + \underset{\text{oxygen}}{O_2(g)} \longrightarrow \underset{\text{water}}{H_2O(l)} \qquad \textit{Incomplete}$$

Because there are two oxygen atoms on the left and only one on the right, one might be tempted to write

$$H_2(g) + O(g) \longrightarrow H_2O(l) \qquad \textit{Incorrect}$$

Although this equation is balanced, an incorrect formula for the oxygen molecule is used. The correct equation is

$$2\,H_2(g) + O_2(g) \longrightarrow 2\,H_2O(l) \qquad \textit{Correct}$$

Notice that when only one molecule of a compound reacts or is produced, no number appears before its formula.

To balance the equation for a simple reaction, a trial-and-error procedure is sufficient.

Example The combustion of methane provides a large part of the energy that heats and cools our buildings, since methane is the chief compound in natural gas. What is the balanced chemical equation for this combustion?

Solution First you need to know the correct formula of every reactant and every product. For this reaction they are

$$\underset{\text{methane}}{CH_4(g)} + \underset{\text{oxygen}}{O_2(g)} \longrightarrow \underset{\text{carbon dioxide}}{CO_2(g)} + \underset{\text{water}}{H_2O(l)} \qquad \textit{Incomplete}$$

Carbon atoms are already balanced, as there is one on each side of the equation. Let's begin by balancing atoms in the other molecules without changing the number of carbon atoms on either side. It is easiest to ignore any element that appears on either side uncombined with any other element, such as O_2. It can be balanced last. To balance hydrogen atoms, place a 2 before H_2O:

$$CH_4(g) + O_2(g) \longrightarrow CO_2(g) + 2\,H_2O(l)$$

The atoms of carbon and hydrogen are now balanced, but we have four oxygen atoms on the right side and only two on the left. Putting a 2 before O_2 balances the entire equation:

$$CH_4(g) + 2\,O_2(g) \longrightarrow CO_2(g) + 2\,H_2O(l) \qquad \textit{Correct}$$

It is always a good idea to check your arithmetic at this point to make sure that a simple error hasn't crept in. We have one carbon atom on each side, four hydrogen atoms on each side, and four oxygen atoms on each side; therefore, the equation is balanced.

3.7 THE PERIODIC TABLE

In Table 3.4 we have taken the lightest 20 elements and arranged them in order of increasing atomic weight—except for argon and potassium, which have been interchanged. Two physical properties—physical state at room temperature and electrical conductivity—and one chemical property—the formula of the simplest compound that the element forms in combination with hydrogen—are given for each element. The latter is classed as a chemical property because a chemical reaction is required to form each compound. The elements selected are either solids or gases. As a matter of fact, of the 108 elements known, only two—bromine and mercury—are liquid at room temperature.

It should be clear from the table that groups of solids and gases alternate with one another. The sequence is certainly not random. The other physical property, electrical conductivity, refers to the ease with which the pure element conducts electricity. More obvious than the sequence of physical states is the cyclic variation in conductivity. The change from negligible to large and back to negligible is repeated once and seems to be starting again with the very last elements listed.

Cyclic variations in the formulas of the hydrides (compounds containing hydrogen and one other element) are also evident. To emphasize trends, we have written the formula of the boron compound, diborane, as $(BH_3)_2$ rather than B_2H_6; the $(BH_3)_2$ notation reveals immediately the 3:1 ratio of hydrogen to boron. Between interruptions at He, Ne, and Ar, which form no compound with hydrogen, we see a regular change in the ratio of hydrogen to the other element from 1:1 to 4:1 and back to 1:1. From Table 3.4 it seems that He, Ne, and Ar have very similar properties, as do Li, Na, and K. Other pairs and triads show similar patterns.

Table 3.4 Physical and Chemical Properties of the First 20 Elements

ELEMENT	SYMBOL	ATOMIC WEIGHT	PHYSICAL STATE AT ROOM TEMPERATURE	ELECTRICAL CONDUCTIVITY	FORMULA OF SIMPLEST HYDRIDE
hydrogen	H	1.008	gas	negligible	H_2
helium	He	4.00	gas	negligible	none
lithium	Li	6.94	solid	large	LiH
beryllium	Be	9.01	solid	large	BeH_2
boron	B	10.81	solid	small	$(BH_3)_2$
carbon	C	12.01	solid	very small	CH_4
nitrogen	N	14.01	gas	negligible	NH_3
oxygen	O	16.00	gas	negligible	H_2O
fluorine	F	19.00	gas	negligible	HF
neon	Ne	20.18	gas	negligible	none
sodium	Na	23.00	solid	large	NaH
magnesium	Mg	24.30	solid	large	MgH_2
aluminum	Al	26.98	solid	large	$(AlH_3)_x$
silicon	Si	28.09	solid	small	SiH_4
phosphorus	P	30.97	solid	very small	PH_3
sulfur	S	32.06	solid	very small	H_2S
chlorine	Cl	35.45	gas	negligible	HCl
argon	Ar	39.95	gas	negligible	none
potassium	K	39.10	solid	large	KH
calcium	Ca	40.08	solid	large	CaH_2

To underscore these similarities and the cyclic variations in properties, we can place the symbols in a pattern, as shown in Figure 3.12. In this arrangement, which is a portion of the modern **periodic table,** the elements are still in the same order as in Table 3.4 if one starts at the top and reads from left to right along each row. The beginning of each row marks the beginning of a new cycle, and elements with similar properties, called **families of elements,** are in vertical columns. The end of one cycle and the beginning of the next are defined somewhat arbitrarily.

Notice that if we put K and Ar in order of increasing atomic weight, K would be under Ne and Ar would be under Na. Such an arrangement would not fit with the similar properties of the two families (vertical columns). Atomic weight is not precisely the correct criterion for ordering the elements in the periodic table.

The idea of organizing the elements in a periodic arrangement originated

						H	He
Li	Be	B	C	N	O	F	Ne
Na	Mg	Al	Si	P	S	Cl	Ar
K	Ca						

Figure 3.12. A portion of the periodic table.

more than 100 years ago, even though at that time the noble gases (the helium group) and many of the elements past the first 20 had not yet been discovered. The periodic table is really an astounding idea—one that has given us unexpected powers to understand and predict chemical phenomena. During the decade of the 1860s, three chemists in three European countries produced versions of the periodic table independently of one another. The first was J. A. R. Newlands, a British chemist, who noticed regularities of the kind we have discussed. When he presented his arrangement of elements in rows of seven to the Chemical Society in 1864, the members thought the idea amusing and unworthy of serious consideration. In their defense it must be said that Newlands had some elements in the wrong families. The result was several serious mismatches in properties.

In 1869 a Russian chemist, Dmitri Mendeleev, presented his periodic table in a paper to the Russian Chemical Society, by whom it was very warmly received. The next year Mendeleev included his periodic table in a textbook of elementary chemistry, which later became popular in many countries. Finally, unaware of Mendeleev's work, the German chemist Lothar Meyer published a periodic table in an 1870 revision of his textbook. Mendeleev is given most of the credit for the periodic table, perhaps because he left empty spaces for undiscovered elements when properties did not agree. He then correctly predicted the properties of several undiscovered elements from the properties of their neighbors in the periodic table. Furthermore, Mendeleev recognized several exceptions to the order of increasing atomic weight. The periodic theory of elemental properties, summed up in the modern periodic table, is the single most valuable predictive theory used by chemists. It is amazing that such a simple arrangement of the elements could have such great scientific value.

A current periodic table of all the elements is included here as Figure 3.13 and can also be found inside the back cover. In this table the elements are ordered according to atomic number, which turns out to be the correct property to replace atomic weight. A definition of atomic number must be left until after we have studied the structure of the atom. After calcium (element number 20), there are ten elements whose properties do not match those of any of the preceding elements. Then, beginning with gallium (number 31), properties again correspond in vertical groups, or families. For this

GROUPS I	II											III	IV	GROUPS V	VI	VII	NOBLE GASES
1 H 1.01																	2 He 4.00
3 Li 6.94	4 Be 9.01											5 B 10.81	6 C 12.01	7 N 14.01	8 O 16.00	9 F 19.00	10 Ne 20.18
11 Na 22.99	12 Mg 24.31											13 Al 26.98	14 Si 28.09	15 P 30.97	16 S 32.06	17 Cl 35.45	18 Ar 39.95
19 K 39.10	20 Ca 40.08	21 Sc 44.96	22 Ti 47.88	23 V 50.94	24 Cr 52.00	25 Mn 54.94	26 Fe 55.85	27 Co 58.93	28 Ni 58.69	29 Cu 63.55	30 Zn 65.38	31 Ga 69.72	32 Ge 72.59	33 As 74.92	34 Se 78.96	35 Br 79.90	36 Kr 83.80
37 Rb 85.47	38 Sr 87.62	39 Y 88.91	40 Zr 91.22	41 Nb 92.91	42 Mo 95.94	43 Tc (98)	44 Ru 101.07	45 Rh 102.90	46 Pd 106.42	47 Ag 107.87	48 Cd 112.41	49 In 114.82	50 Sn 118.69	51 Sb 121.75	52 Te 127.60	53 I 126.90	54 Xe 131.29
55 Cs 132.91	56 Ba 137.33	57 La* 138.91	72 Hf 178.49	73 Ta 180.95	74 W 183.85	75 Re 186.2	76 Os 190.2	77 Ir 192.22	78 Pt 195.08	79 Au 196.97	80 Hg 200.59	81 Tl 204.38	82 Pb 207.2	83 Bi 208.98	84 Po (209)	85 At (210)	86 Rn (222)
87 Fr (223)	88 Ra 226.03	89 Ac† 227.03	104 Unq (261)	105 Unp (262)	106 Unh (263)												

*Lanthanides

58 Ce 140.12	59 Pr 140.91	60 Nd 144.24	61 Pm (145)	62 Sm 150.36	63 Eu 151.96	64 Gd 157.25	65 Tb 158.93	66 Dy 162.50	67 Ho 164.93	68 Er 167.26	69 Tm 168.93	70 Yb 173.04	71 Lu 174.97

†Actinides

90 Th 232.04	91 Pa 231.0	92 U 238.03	93 Np 237.0	94 Pu (244)	95 Am (243)	96 Cm (247)	97 Bk (247)	98 Cf (251)	99 Es (252)	100 Fm (257)	101 Md (258)	102 No (259)	103 Lr (260)

Figure 3.13. *The modern periodic table. (Hydrogen has a few properties that place it in group I and a few that place it in group VII. It is shown here in group I.)*

ARTIFICIAL ELEMENTS

Before 1940 the periodic table consisted of just the first 92 elements. In that year, while studying the "nuclear fission," or splitting of uranium atoms, two physicists—Edwin McMillan and Philip Abelson—identified an isotope of element 93, which had been produced unexpectedly by a different kind of process. This new element was quite unstable and disappeared after an hour or so. It was given the name neptunium (Np) because it follows uranium, just as the planet Neptune is next beyond Uranus in our solar system. Because the preparation of new actinides was closely linked with the development of the atomic bomb, new discoveries were kept secret during the years of World War II, from 1940 through 1945.

Since the war, about a half-dozen laboratories around the world have been competing with one another to produce and identify new elements. Most of these elements were first made and conclusively identified by scientists at the Lawrence Berkeley Laboratory in Berkeley, California. Sometimes two research groups have made competing claims for the discovery of a new element. In such cases a dispute has often arisen over its name. This is one reason why recent periodic tables have a location and number (such as 108) for an element but list no name or symbol.

The lifetimes of the artificial elements farther along in the periodic table become shorter and shorter. There is considerable challenge in identifying an element that exists for only a few seconds. Furthermore, the amounts produced are submicroscopic. Amazingly, for the most recent discoveries only a few atoms of the element were present. These atoms were identified by their distinctive radioactivity.

Some theories predict that element number 114 ought to be more stable than those on either side. This region has been called an "island of stability." The prediction has encouraged some scientists to play leapfrog; instead of attempting to make all the elements between the most recently prepared one and number 114, they are trying to make 114 first.

reason the earlier part of the table is split, and the third row is longer by ten elements, as shown. The fourth and fifth rows are also longer. Moreover, the fifth row is increased by still more elements, the lanthanides (58–71), as is the sixth row by the actinides (90–103). These two series have been removed and are shown separately to keep the periodic table from being awkwardly long. Beyond uranium (92), the actinides are produced artificially; they are highly radioactive.

The stepwise colored line on the periodic table separates the metals on the left from the nonmetals on the right. Metals are almost always solids. More importantly, **metals** conduct electricity very well, whereas **nonmetals** are poorer conductors. You can also see from Table 3.4 (page 52) that the nonmetals closer to the stepwise colored line in Figure 3.13 are better electrical conductors than those farther to the right.

In the second row of the periodic table, the nonmetals are B, C, N, O, and F. Of these nonmetals, boron has the most metallic character, with carbon next, and so on. Fluorine is the least metallic of all. The general pattern is that elements closer to the metals are more metallic than those farther away. For example, the elements in Group IV are more metallic than those in Group V, etc.

3.8 NAMING SIMPLE COMPOUNDS

Because we want to refer to compounds or classes of compounds by name, we need a few rules for naming them. If a compound consists of a metal (for example, sodium, or Na) and a nonmetal (for example, chlorine, or Cl), the name of the metal is given first, followed by the name of the nonmetal. The name is then altered by replacing the last several letters with the ending -ide. Thus, the compound NaCl is called sodium chloride. Notice also that the more metallic element comes first in the formula. Names for the most common nonmetals are shown in Table 3.5.

When simple compounds that combine two nonmetals are named, the more metallic element is given first. Only the element named second receives the -ide ending. Thus, the compound between carbon and chlorine is written CCl_4 rather than Cl_4C. It is known as carbon tetrachloride.

If more than one atom of a particular kind is present in a compound, a prefix is often used to indicate the number. Examples are sulfur dioxide (SO_2), boron trifluoride (BF_3), and carbon tetrachloride (CCl_4). The standard prefixes, which come from Greek, are as follows.

Number:	1	2	3	4
Prefix:	mono-	di-	tri-	tetra-

The prefix mono- is almost always omitted before the first element in the name, and the prefixes are rarely used for compounds that contain metals. Thus, $AlCl_3$ is simply called aluminum chloride.

Table 3.5 The Most Common Nonmetals

SYMBOL	NAME OF ELEMENT	NAME OF COMPOUND
H	hydrogen	hydride
F	fluorine	fluoride
Cl	chlorine	chloride
Br	bromine	bromide
I	iodine	iodide
O	oxygen	oxide
S	sulfur	sulfide

Unfortunately, these rules are not always followed for very common compounds. Common compounds often have common nicknames. Strictly speaking, water should be called ''dihydrogen oxide,'' but, of course, it isn't. ''Carbon tetrahydride'' (CH_4) is always called methane. ''Nitrogen trihydride'' (NH_3) goes by the name ammonia. These and some other important compounds retain the common names that have been used for generations.

Example Write the formula for phosphorus trichloride.

Solution First, we need to find phosphorus (P) and chlorine (Cl) in the periodic table (see Figure 3.13). Phosphorus (number 15) is in Group V, whereas chlorine (number 17) is in Group VII. Therefore, phosphorus is more metallic than chlorine is. Knowing this, we know that the symbol for phosphorus will come first in the formula. The prefix tri- means that there are three chlorine atoms per phosphorus atom in phosphorus trichloride. The formula is PCl_3.

Because the periodic table is so important as a correlating tool for chemists, it is the subject of much of Chapter 8. The application we stress now is the correlation and prediction of formulas for compounds. We have already seen the similarities in the formulas of the hydrides of any particular group. Comparable patterns can be made for many other classes of compounds. Consider, for example, the chlorides of the elements in the family of Group I, which begins with lithium (number 3) and ends with francium (number 87). (Hydrogen, often shown in Group I, is distinctive and doesn't fit well in any group.) The formulas are LiCl, NaCl, KCl, and so on, all with identical ratios of atoms. Thus, a knowledge of one chloride leads to a knowledge of all chlorides in Group I. In fact, the formulas of the chlorides parallel those of the hydrides fairly closely. Group II forms dichlorides (for example, $MgCl_2$); Group III, trichlorides; Group IV, tetrachlorides; Group V, trichlorides; and Group VI, dichlorides. We hasten to add that nature is seldom as simple as we would like. Most of these generalizations have exceptions. For example, lead (Group IV) forms a dichloride, $PbCl_2$, as well as $PbCl_4$. Despite the exceptions, the periodic table reduces what might be a huge, random collection of formulas to a manageable scheme.

Example Predict the formula of calcium fluoride. This chemical in its mineral form is known as fluorspar. It is the source of much of the fluoride that is added to drinking water to protect our teeth.

Solution Using Figure 3.13 we can see that calcium (number 20) is in Group II and fluorine (number 9) is in Group VII. Knowing that elements from the same group in the periodic table form compounds of similar formula with other elements, we can now proceed to predict the formula of calcium fluoride.

According to Table 3.4, calcium hydride has the formula CaH_2. Calcium is found in Group II. We have also said that Group II elements form dichlorides. Notice that hydrogen, chlorine, and fluorine are in the same family (Group VII). We would predict that if calcium combines with two chlorine atoms, it will also combine with two fluorine atoms. The predicted formula for calcium fluoride is therefore CaF_2.

IMPORTANT TERMS

compound A pure substance that can be decomposed by a chemical reaction.

element A pure substance that cannot be decomposed by any chemical reaction.

chemical reaction A process in which atoms are rearranged so that the substances originally present are transformed into new substances.

reactant A chemical that is transformed in a chemical reaction.

product A substance produced in a chemical reaction.

chemical change A change in which one or more chemical reactions occur.

physical change A change in which one or more elements or compounds undergo a change of physical state. All of the elements or compounds present initially are also present after the change.

symbol One or two letters used as the shorthand notation for the name of an element.

Law of Conservation of Mass The law stating that in any chemical reaction the sum of the masses of the reactants equals the sum of the masses of the products.

Law of Constant Composition The law stating that the proportions by weight of the elements in a given compound are always the same.

atom The smallest particle of matter that retains its identity in a chemical reaction. Each element has a different kind of atom.

molecule A discrete tiny particle made up of atoms. Molecules of different compounds contain different kinds and numbers of atoms.

formula A combination of chemical symbols and numbers that gives the kinds of atoms and their relative numbers in a compound or element.

isotopes Atoms of the same element that differ only in mass.

atomic weight The relative mass of an atom on a scale in which the mass of the most common isotope of carbon is given a value of 12.000.

molecular weight The relative mass of a molecule, found by adding together the atomic weights of all atoms making up the molecule.

balanced chemical equation A chemical equation that has not only the correct formula for each reactant and product but also the same number of atoms of each element on both sides.

periodic table An arrangement of the elements in rows approximately in order of increasing atomic weight. This arrangement emphasizes similarities within certain groups of elements.

family of elements Elements having similar properties and appearing in a vertical column of the periodic table. Also called a *group*.

metal An element that is a good conductor of electricity. Metals are also usually shiny and can be hammered into various shapes without breaking.

nonmetal Any element that is not a metal.

QUESTIONS

1. Calculate the molecular weight of the following:

a. CO   d. C_2H_6O g. $C_2H_2ClF_3$
b. F_2 e. CS_2 h. SiF_4
c. PCl_3 f. H_3PO_4

2. State whether each of the following describes a chemical or a physical change, and explain your answer.

a. Water from the Atlantic Ocean evaporates and becomes water vapor.
b. In a junk yard a strong magnet pulls iron away from other metals.
c. An egg is fried in a frying pan.
d. When a piece of iron is placed in hydrochloric acid, it slowly disappears as hydrogen gas bubbles from the solution.

3. In a steam engine, steam expands in a cylinder and pushes a piston as it cools down. In an automobile engine, gasoline burns in a cylinder, giving off carbon dioxide and water vapor, which push a piston. For each kind of engine, explain whether a chemical change or a physical change occurs in the cylinder.

4. Tell which chemicals are the reactants and which are the products in the following reactions:

a. A student drops a copper penny into nitric acid, HNO_3, and notices that copper nitrate, $Cu(NO_3)_2$, and nitrogen dioxide, NO_2, form.
b. Heating potassium chlorate, $KClO_3$, produces a white salt, KCl, and vigorous bubbling. A gas is evolved that makes a glowing wooden splint burst into flame. The gas is identified as molecular oxygen.
c. Silver metal tarnishes easily in the presence of H_2S, the gas that causes the bad smell of rotten eggs. The black tarnish is silver sulfide, Ag_2S.

5. Balance the following equations:

a. $H_2\ (g) + Cl_2\ (g) \longrightarrow HCl\ (g)$
b. $SO_2\ (g) + O_2\ (g) \longrightarrow SO_3\ (g)$
c. $C\ (s) + H_2O\ (g) \longrightarrow CO\ (g) + H_2\ (g)$
d. $PbO_2\ (s) \longrightarrow PbO\ (s) + O_2\ (g)$
e. $PCl_5 + H_2O \longrightarrow POCl_3 + HCl$
f. $MgO + Si \longrightarrow Mg + SiO_2$
g. $KClO_3 \longrightarrow KCl + KClO_4$

6. Write the formula for each of the following compounds:

a. water
b. carbon disulfide
c. boron trichloride
d. silicon tetrafluoride
e. hydrogen iodide
f. iodine

7. We know that 22.0 grams of carbon dioxide contains a definite number of CO_2 molecules. A sample of carbon monoxide, CO, contains exactly the same number of molecules. What is the weight of this sample of carbon monoxide? (The atomic weights of C and O are 12.0 and 16.0, respectively.)

8. As discussed in the text, when elements from the same group in the periodic table form compounds with other elements, the compounds have similar formulas. For example, in Table 3.4 you can see that oxygen and sulfur, both from Group VI, combine with hydrogen to give H_2O and H_2S, respectively. Predict the formula of each of these compounds:

a. potassium chloride   d. boron chloride
b. hydrogen selenide   e. potassium sulfide
c. aluminum chloride   f. calcium oxide

9. A homogeneous liquid A in a beaker is heated over a Bunsen burner. After a while the liquid level in the beaker slowly drops, and a solid, B, appears. Eventually only solid B remains at a very high temperature. On cooling to room temperature, there is no visible change in solid B. Two interpretations of this experiment come to mind. (1) A is a liquid solution. When it is heated, one of the components of the solution evaporates away.

During the evaporation, the other component, B, starts to come out of solution, and finally only B remains. (2) A is a pure compound that, on heating, decomposes into a gas and another substance, B.

a. What additional experiment or experiments could you do in order to decide between these two interpretations?

b. Do you believe from the evidence given that B must be an element?

First Optional Section

10. For some time John Dalton thought (incorrectly, as it turns out) that the formula of ammonia was NH, since he knew neither its true formula nor the correct atomic weights of nitrogen and hydrogen. He knew only that ammonia is 82% nitrogen and 18% hydrogen by weight. Given these data, explain how Dalton could have assumed an incorrect formula for ammonia.

Second Optional Section

11. Find the percent by weight of each element in the following compounds:

a. MgH_2 **e.** Fe_2O_3 **i.** H_2SO_4
b. SO_2 **f.** $PbCl_2$ **j.** NH_4NO_3
c. KCl **g.** AgBr
d. PCl_3 **h.** Br_2

12. Two common ores of copper are CuS and $CuFeS_2$. Which one has the higher percentage of sulfur? What is the percentage of sulfur in that ore?

13. Cocaine, $C_{17}H_{21}NO_4$, is a narcotic drug much discussed in newspapers and magazines. A chemist analyzed a sample suspected to be cocaine. She found the following percent composition by weight: 67.30% C, 7.00% H, 4.60% N, 21.10% O. Could the sample be cocaine?

Molecules and Energy

Energy plays many roles in our lives. On the grandest scale, our ultimate source of energy is the sun. The sunlight striking the earth keeps the land and oceans warm, evaporates water, and causes plants to grow. Even the energy we extract from fossil fuels—coal, petroleum, and natural gas—came from the sun long ago. On the smallest scale, the molecules and atoms of which all matter is composed have several kinds of energy. These molecular energies are at the heart of the physical and chemical changes which make our world such a dynamic place.

4.1 THE NATURE OF ENERGY

All of us are aware of the need for energy when there is work to be done. **Energy,** in fact, is defined as the capacity for doing work. We are all accustomed to relating these two terms. We use muscular (biochemical) energy to do the physical work necessary to lift an object. We convert chemical energy into mechanical energy in internal combustion engines to make our cars move. Electrical energy was used to run the presses that printed this book.

Most familiar is the energy we generate internally by the oxidation of the food we eat. An adult consuming and metabolizing a normal diet is functioning at about the same energy level as a 250-watt light bulb. In both cases, energy is eventually transferred to the environment, mostly as heat.

Seldom can one read a newspaper without seeing a story about energy or the fossil fuels with which we generate useful energy. Every one of us uses the various forms of energy—light, heat, chemical, mechanical, and electrical—in our daily lives. The conversion of energy from one form to another is essential to us and occurs repeatedly in homes and manufacturing plants. Every product we buy was made using energy.

Scientists use the **joule (J)** as the standard unit of energy. The joule is approximately the energy spent in lifting a 1-kilogram weight a distance of $\frac{1}{10}$ meter. One thousand joules of electrical energy will keep a 100-watt light bulb lit for 10 seconds. Somewhat under 12 joules of heat will raise the temperature of 1 teaspoon of water by 1°F. So you can see that a joule is a rather small unit.

Another energy unit of historical importance is the **calorie.** It was originally defined as the amount of energy required to increase the temperature of 1 gram of water by 1°C (see temperature scales, next section). One calorie is equivalent to 4.18 joules. Dietitians have chosen to use the **dietary calorie,** equal to 1000 ordinary calories, as their standard measure of food energy. A 25-gram bar of milk chocolate with almonds provides 133 dietary calories—in theory, that is enough energy to allow a 100-pound person to leap over New York City's World Trade Center three times!

4.2 TEMPERATURE AND TOUCH

Unfortunately, our senses cannot measure energy directly. Instead we use our sense of touch to gain a rough feeling for hotness, or the **temperature** of an object. Temperature and energy are not the same thing. We know that a glass of milk is cold and a cup of coffee is hot, but we cannot feel how much internal energy there is in a cup of hot coffee or a bathtub of hot water. They both feel hot, but considerably more energy is used in heating a tubful of bathwater than in making a cup of coffee. Also, our touch can sense only a limited range of temperatures, and none too reliably at that. Yet even in ancient times it was known that a person has a higher body temperature when he or she is sick. It was a common practice then, as it is today, for a physician to place a hand on a patient's forehead in order to judge how sick the person is.

A quantitative measure of temperature depends on the use of thermometers and accurate temperature scales. The literal meaning of thermometer is a meter for measuring temperature. Temperature scales provide a numerical value to replace the subjective sense of touch. The Fahrenheit scale is of historical significance and is still in use in the United States (see Figure 4.1). The story of its development provides some insight into the inner workings of science and scientists.

Gabriel Fahrenheit was a German instrument maker who became interested in temperature measurement. He published most of his experimental work in 1724–1725, shortly before his death. He perfected the art of sealing mercury in glass capillary tubes so that the expansion and contraction were clearly visible. Fahrenheit then turned to the problem of marking these tubes so that temperatures could be given numerical values. For this he needed a standard, constant, and reproducible temperature in an appropriate range for his as yet unmarked thermometers. As fate would have it, he chose the temperature found under the tongue of a human being as the standard for the upper end of his scale. He called this temperature 96°. He chose as his lowest temperature the coldest system he knew—a mixture of sea salt and ice. Its temperature became 0° on his now nearly complete thermometer. He divided his scale into 96 equal parts.

After this 96° thermometer was used to measure the temperature at which ice melts or water freezes, the 32° mark became another standard fixed point on the temperature scale. Later work indicated that the boiling point

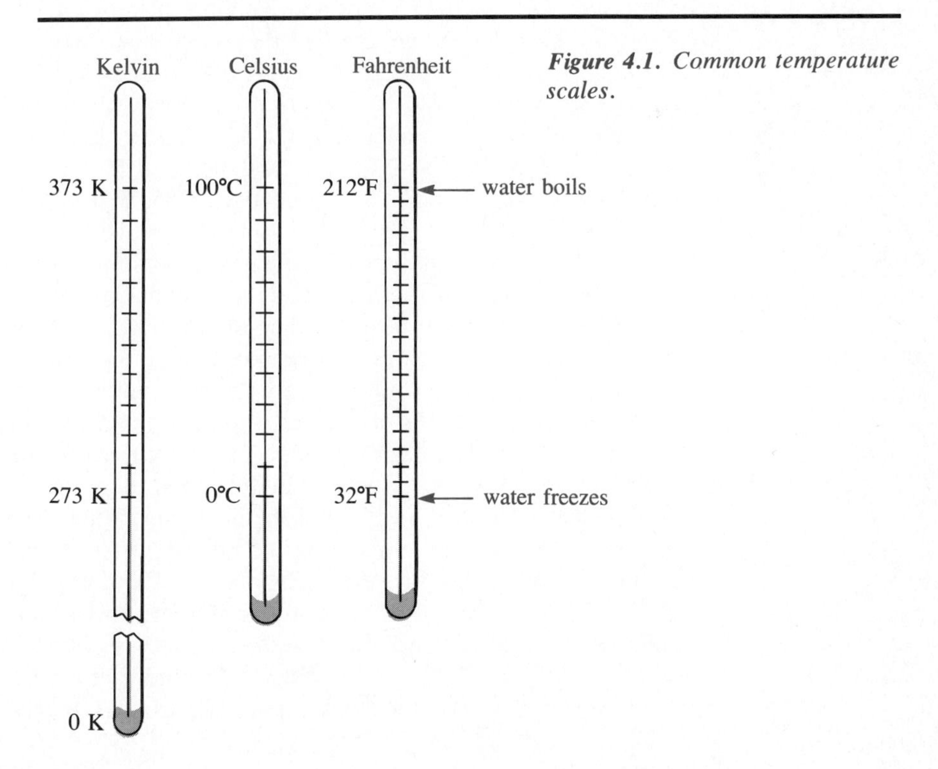

Figure 4.1. *Common temperature scales.*

of water is 212°. This figure, as it turned out, was in error (as is to be expected given the assumption that the body temperature of humans is 96°). Fahrenheit's successors, equally determined to establish a firm temperature scale, decided to use the value of 212° for the boiling point of water as a standard even though it was in error. They divided the scale from 32° to 212° into 180 equal parts. The body temperature of a healthy human being was revised upwards to 98.6°. Thus, our "quantitative" temperature scale was born.

The **Celsius temperature scale** (°C), developed in 1742 and widely adopted at the time of the French Revolution, is now used by much of the world. It takes its name from the Swedish astronomer, Anders Celsius. The Celsius scale uses the freezing and boiling of water as its fixed reference points: 0°C for the freezing of water or melting of ice and 100°C for the boiling of water (see Figure 4.1). These two points are divided by 100 equal Celsius degrees. This scale is very convenient, because pure water is readily available and its freezing and boiling points can readily be reproduced accurately. The scale used to be called the centigrade scale in some countries because of the 100-degree difference between the freezing point and the boiling point of water. (The Latin prefix centi- means 100.) By international agreement, however, it is now known everywhere as the Celsius temperature scale.

The international language of science virtually ignores the Fahrenheit scale. We will use the Celsius and Kelvin scales throughout this book. The **Kelvin temperature scale** (K) has the same size "degree" as the Celsius scale, but its zero point is the lowest temperature possible. This temperature is called *absolute zero*. No experimenter has ever produced the absolute zero temperature, although several have come very close to it. On this scale, water freezes at 273.15 K and boils at 373.15 K (Figure 4.1). By international agreement, the degree sign is not written when the symbol K is used.

Two liquids commonly used in thermometers today are alcohol and mercury. (Some medical thermometers use alcohol mixed with a red dye for easier reading.) When these two liquids are placed first in a cool bath and then in a warm bath, the percentage increase in the volume of each is about the same. Calibration based on reference points such as the freezing and boiling of water results in the creation of a thermometer. Alcohol and mercury work well in thermometers but are strictly limited in their useful temperature range. A "normal" body temperature of 98.6°F, or 37°C, is easily measured with an alcohol medical thermometer, but you couldn't use it to measure the temperature of boiling water, since alcohol is a gas at 100°C under usual pressures. The standard boiling point of ethyl alcohol is 78°C. The lustrous and toxic metal mercury has a much higher boiling point and is regularly used to measure temperatures of up to 360°C. Mercury's melting point is −39°C, so it could not be used to measure the temperature in Antarctica, where the winter temperature has been known to reach −87°C.

OTHER WAYS TO MEASURE TEMPERATURE

Scientists and engineers have been ingenious in devising new ways to measure temperature. They have used several physical properties of substances in addition to the volumes of liquids.

One device used is the thermocouple, which consists of two wires of different metals connected by twisting an end of one to an end of the other. At the contact between the two metals a small voltage develops, just as if a tiny battery were located there. The size of the voltage depends on the temperature. A measurement of this voltage indirectly gives the temperature of whatever is in contact with the connected wires.

Thermocouples are used in many gas furnaces to provide protection against leaking gas from a pilot light that has blown out. The thermocouple is placed above the flame of the pilot light. Should the pilot light go out, the temperature of the thermocouple will drop and the voltage will change. This change in voltage causes a valve to shut off the gas.

A novel thermometer has recently been developed to assist physicians who want to measure the skin temperature of patients. By sensing skin temperature at various places, a doctor can tell something about the pattern of blood circulation. A substance that forms liquid crystals is used in this thermometer. A liquid crystal is somewhat fluid, like a liquid, but is cloudy and has a fairly ordered arrangement of molecules, like a solid. Some liquid crystals change color when the temperature changes. All the doctor has to do is look at the color to tell the patient's temperature.

4.3 ENERGY AND MOTION

How are the properties of a substance (element or compound) affected by putting energy into it? The answer to this question lies in understanding how the energy level affects the molecules that make up matter. In this section we will not attempt to give all of the evidence for the theories presented but will simply describe a physical picture of the behavior of one substance—a collection of water molecules. Each molecule is made up of two atoms of hydrogen and one atom of oxygen bound together to form H_2O. The water molecule has incredibly small dimensions, with a roughly spherical diameter of 3×10^{-8} centimeters. A glass of water contains more water molecules than there are stars in the sky.

When we pump energy into the collection of water molecules by heating it, the temperature of the water rises. Eventually the water boils. If we heat it long enough, all of the liquid water is converted to gaseous water vapor or steam. In the same way, the earth's water cycle involves the evaporation of water from the oceans, rivers, and lakes, using energy from the sun. The water vapor condenses when cooled and falls back to the surface as rain or snow. Several trillion metric tons of water undergo this process daily.

When energy is absorbed by a sample of water, the H_2O molecules are thought to move faster. This relationship between energy and motion is a **model** (a simplified picture of physical reality) used by all scientists. Molecules are always moving, often at incredible speeds. This very second, you are being bombarded by many billions of nitrogen and oxygen molecules from the air around you. Each one of them is traveling thousands of miles per hour. Yet, because they are so small and have so little mass, you aren't aware of them, even if you have super-sensitive skin. Heating up a collection of molecules makes them move even faster; cooling slows them down. At a temperature of absolute zero (0 K) their motion would be at a minimum.

Water vapor is still composed of water molecules, each having the formula H_2O, but it is different from liquid water or ice. The gaseous vapor has more energy than the other forms, and the properties of substances depend very much on how much energy they have. The molecules absorb energy when they break away from the liquid and become relatively free of one another in the gas. Adding energy to molecules allows them to overcome the forces that hold them together in a liquid or a solid.

4.4 ICE–WATER

A collection of molecules can exist in three **states,** or phases: **solid, liquid,** or **gas** (see Figure 4.2). Water molecules in the liquid phase have far more energy than those in the solid phase. The behavior of water isn't at all unusual. Any solid has less energy than the corresponding liquid, and the gas has the most energy of all.

Like other crystals, solid ice has an orderly three-dimensional arrangement (Figure 4.3). The molecules are locked into a pattern, and their movements are restricted to vibrations about the positions they occupy. As the solid

Figure 4.2. *Solid, liquid, and gaseous states.*

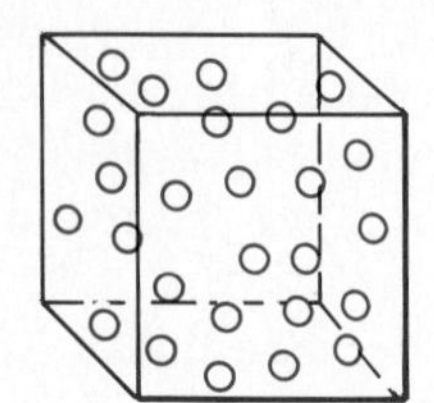

Some particles in the gaseous state (not drawn to scale). Particles are far apart and are bouncing off all sides of the container.

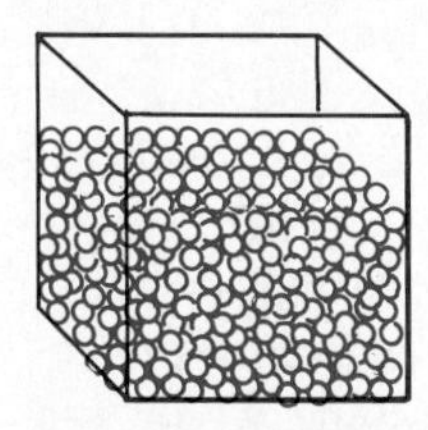

Some particles in the liquid state. Particles are not in fixed positions (liquids flow); usually the volume is somewhat larger for the liquid state than for the solid state.

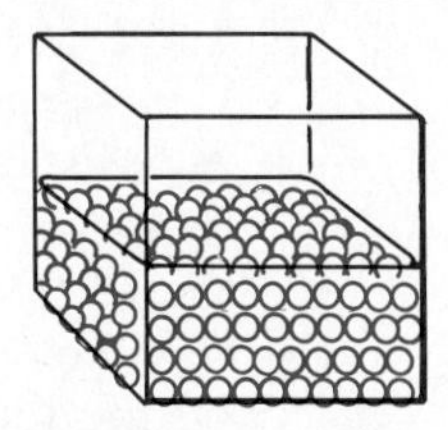

Particles in the solid state. The particles are in fixed positions.

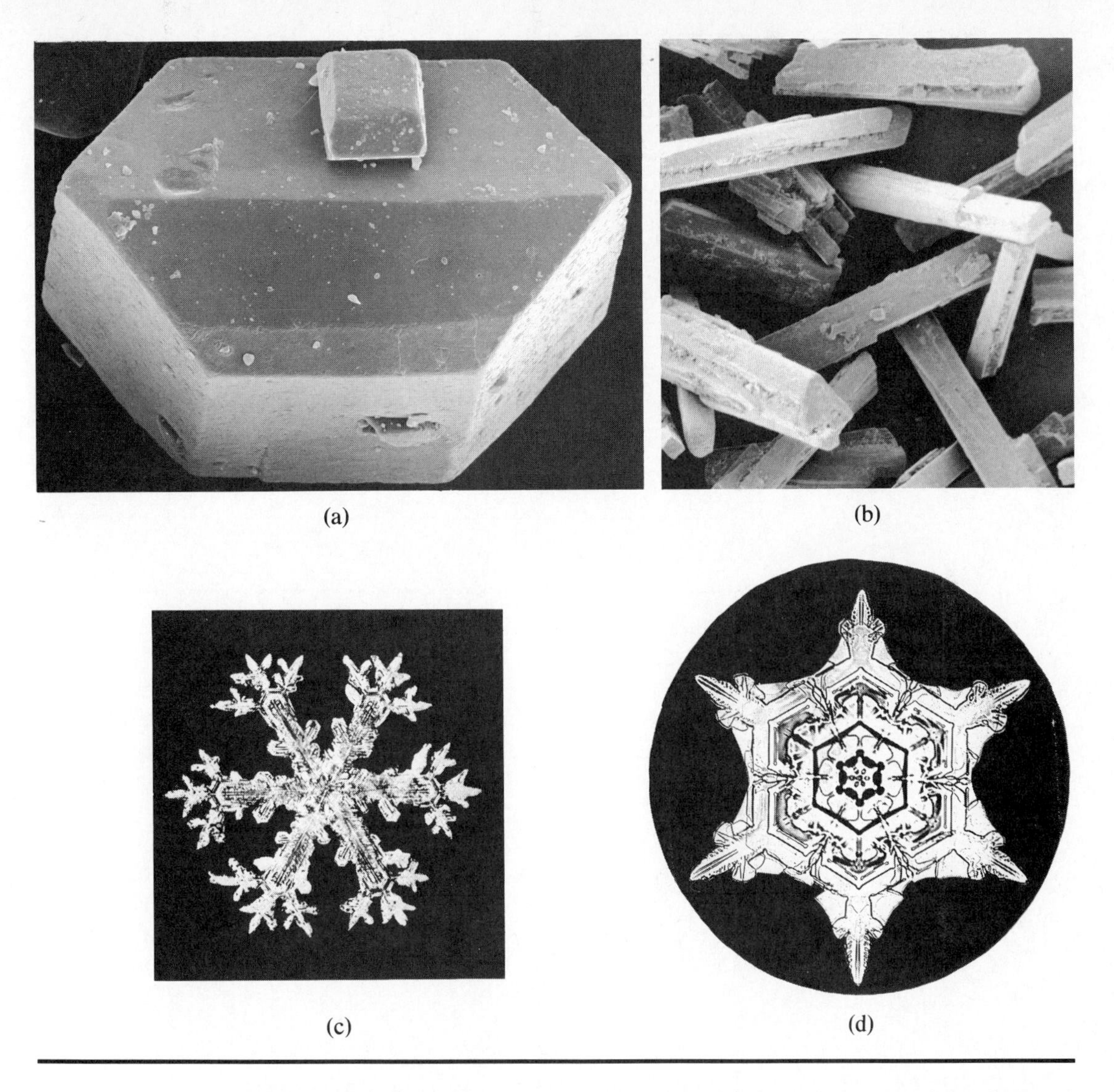

Figure 4.3. Some different crystal shapes. (a) Sucrose. (b) Monosodium glutamate (MSG). (c) and (d) Snowflakes.

absorbs energy and the temperature rises, the water molecules begin to bounce around more and more, back and forth, faster and faster. When the melting temperature is reached, the molecules bounce so vigorously that some of them move out of their positions in the solid and travel more freely. They form a new state—a liquid. The addition of more energy causes more of the solid to break down and become a mobile liquid.

Figure 4.4 shows how the temperature of water changes as energy is added. In the laboratory one would obtain this graph by plotting temperature against time, while heating the sample at a constant rate. Notice that the temperature stays constant during melting, even though energy is being added. It only rises again after all the solid has melted. As energy is added

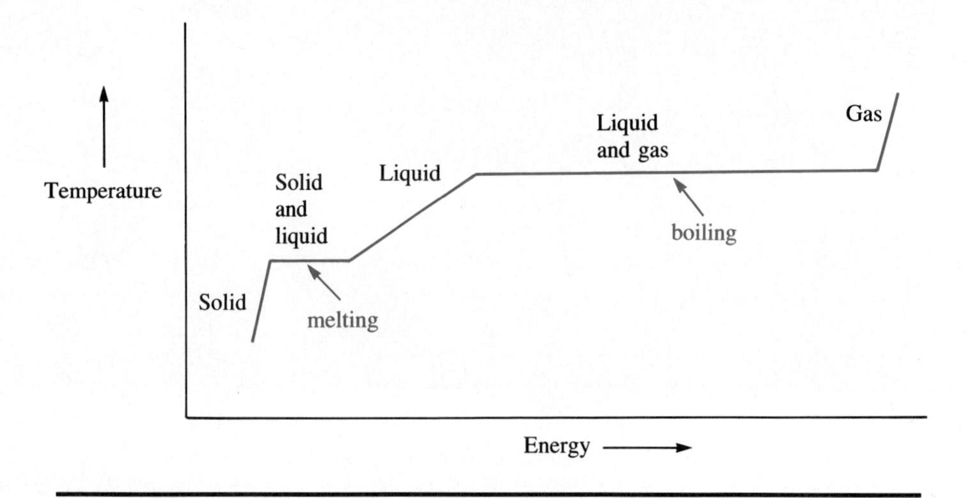

Figure 4.4. Temperature-energy diagram for water.

to an ice cube, the temperature of the ice rises uniformly. Then at a crucial temperature for water, 0°C or 273 K, all of the additional energy goes to break up the rigid three-dimensional crystal structure. This means that at 0°C liquid water has more energy per gram than ice does.

From 0°C to 100°C the temperature of the liquid water again rises uniformly as energy is added. Then during boiling the temperature stays constant. At 100°C a gram of steam has quite a bit more energy than a gram of water does. It takes energy to change the physical state of any chemical. Almost every substance has a temperature-energy diagram similar to the one shown in Figure 4.4.

Example Arrange the following substances in order of decreasing energy: liquid mercury at 150°C, liquid mercury at 200°C, liquid mercury at 357°C, solid mercury at −40°C, mercury vapor at 357°C. (Mercury freezes at −40°C and boils at 357°C.)

Solution Figure 4.4 shows that a solid has the least energy and a gas has the most energy. Also, mercury vapor at the boiling point has more energy than liquid mercury does at the same temperature. Liquid mercury gains energy as the temperature increases. So, the answer is as follows: mercury vapor at 357°C, liquid mercury at 357°C, liquid mercury at 200°C, liquid mercury at 150°C, solid mercury at −40°C.

The amount of energy necessary to melt an ice cube weighing 1 gram is 334 joules. Five times as much energy, 1668 joules, is necessary to melt a 5-gram ice cube. An enormous amount of energy is needed to melt an iceberg; it's no wonder they float so far from the North and South Poles into warmer waters. With the iceberg or the ice cube the speed of melting depends on the surface area and the rate of energy exchange with the environment.

Most materials increase in volume upon melting. Liquids usually have more open arrangements, with more holes in them. Thus most liquids are less dense than the corresponding solids, and the solids sink in their liquid phase. There are exceptions; the most common is water. Solid ice is about 8% less dense than liquid water at 0°C; therefore, ice floats on water. Ice has an unusually open three-dimensional crystal structure. This fact is very important for aquatic life, because it means that ice forms on the top of water, insulating the lower levels. Plant and animal life can continue to exist at 0°C below the frozen surface of lakes and streams. If ice were more dense than water, it would sink to the bottom and a lake would freeze from the bottom up. During the next spring the sun's energy might never penetrate much below the surface layers and our lakes might never thaw completely!

4.5 WATER–STEAM

Even though liquid water has no rigid three-dimensional crystal structure, there is still a lot of attraction between nearby water molecules. A great deal of energy is necessary to pull them away from one another and produce the gas phase. Each gram of water needs 2260 joules for conversion to vapor. As we add energy to liquid water, random motions of the molecules increase and the temperature rises. The average speed of the molecules increases with the added energy. Some molecules gain enough energy to go bursting past their neighbors and out from the surface into the air. This loss of highly energetic molecules from the liquid to the gaseous state is called *evaporation*. We know that as we heat water by adding energy, the rate of evaporation increases.

Example When water evaporates from your skin, it feels cool. What is the reason for this cooling effect?

Solution Energy must be added to water molecules when they change from liquid to vapor. Some of that energy comes from molecules in your skin, which then move more slowly than before. The temperature of your skin consequently falls.

You are also familiar with what happens when you take a cold bottle from the refrigerator on a hot humid day. Soon drops of water appear on the glass and begin to run down its side. This happens because fast-moving

HOT-AIR BALLOONS

In some parts of the United States people are fond of the sport of ballooning. They generally use hot-air balloons, with a propane burner as the energy source for heating the air. When the air is heated, its volume increases, and some of the air escapes from the opening at the bottom of the balloon.

Adding heat energy makes the molecules move faster, and they tend to fly away from one another. Molecules of the hotter gas are farther apart on the average, so there are fewer of them in a given volume. Hot gases are therefore less dense and tend to rise, which gives lift to the balloon. Balloons filled with hot, light air can rise into the skies and float along until their air cools. The cool air has a smaller volume, and when more air is sucked back into the balloon, it becomes heavier. A heater is used to keep the air hot and to prevent the balloon from sinking.

Hot air does not have enough lifting power for the heavy balloons used on long voyages. A lighter gas must be used. Helium is the lightest gas next to hydrogen, and helium is far safer because it is nonexplosive. The first successful balloon trip across the Atlantic Ocean (without a motor) took place in 1978, when the Double Eagle II (see Figure 4.5) flew from Maine to France. It used 4500 cubic meters of helium gas in its huge bag, made of neoprene-coated nylon. The trip took 7 days, and the balloon flew at altitudes of over 10 kilometers. Success came on the fifteenth attempt to cross the Atlantic in a balloon.

Figure 4.5. The Double Eagle II.

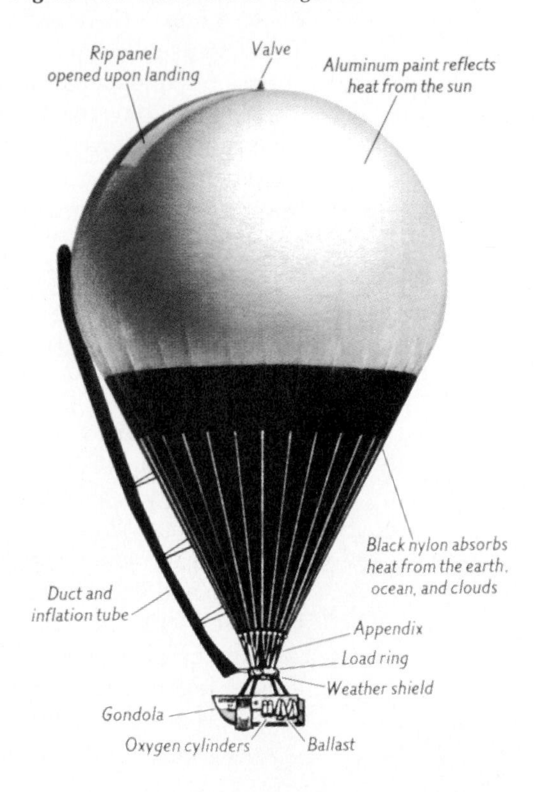

water molecules in the air collide with the cold glass, lose energy, and condense back into the liquid state.

Imagine now that some liquid water is put into an empty, closed vessel. The liquid immediately begins to evaporate, because there are always some molecules with large enough energies to escape into the empty space above the liquid. As water molecules accumulate in the vapor, some of them strike the surface of the liquid, lose energy, and condense. Thus, two opposing processes—evaporation and condensation—occur simultaneously. At first, the rate of evaporation exceeds the rate of condensation, and the number of molecules in the gas phase continues to increase. Finally, the number of vapor molecules is large enough that the two opposing processes take place at equal speeds. The same number of water molecules return to the liquid state as go into the vapor above it. In effect, they cancel each other. At the visible level there is no further change, although at the molecular level there is still frantic activity. This is one example of an equilibrium process.

The number of vapor molecules required to produce equilibrium in the vessel depends on the temperature. The rate of evaporation of the liquid is greater at a higher temperature, and the rate of condensation must also be greater, since the two rates must be equal at equilibrium. A greater rate of condensation can only be achieved if the number of vapor molecules is greater. So the concentration of vapor molecules at equilibrium must be larger at higher temperatures. This law is often stated in a different way:

The vapor pressure of a liquid increases with increasing temperature.

The term **vapor pressure** means the pressure exerted by the molecules in the vapor when equilibrium exists.

When water is in an open bottle in contact with air, evaporation takes place. Equilibrium cannot be achieved because many of the vapor molecules escape entirely from the neighborhood of the bottle. The rate of condensation is always less than the rate of evaporation, and the liquid slowly evaporates away. At 100°C the vapor pressure of water reaches 1 atmosphere, which is also the pressure of the air above the liquid at sea level. Under these conditions bubbles of vapor can form in the interior of the liquid and rise to the surface without collapsing. This is the process of boiling. A boiling liquid sends molecules into the gas phase much faster than the process of evaporation can.

4.6 KINETIC ENERGY AND GASES

Earlier in this chapter we said that adding more energy to a substance makes its molecules move faster. This energy of motion is called **kinetic energy.** In order to understand how energy affects the molecules that make up matter, we must examine kinetic energy. The kinetic energy of a nonmoving hard body is assumed to be zero. A crumpled, crushed automobile is dramatic testimony that a speeding body has lots of kinetic energy, which can be lost in an abrupt collision. The car's energy of motion is transferred into the energies of another car or a tree or whatever. The energy transfer can result in heat or work. Metal can be twisted and glass shattered.

The kinetic energy that an object has when it moves from one place to another is called **translational energy.** High speed and large mass give a body high translational energy. The formula that relates them is

$$\text{Translational kinetic energy} = \tfrac{1}{2} mv^2 \qquad \textbf{(Equation 4.1)}$$

where m = mass of the moving body

v = velocity, or speed, at which it is traveling

All molecules, whether part of a gas, liquid, or solid, have kinetic energy. Even though the molecular motion is highly restricted in a solid, there is kinetic energy associated with the vibrations of the molecules.

The concept that energy is related to molecular motion is a simple idea with enormous practical utility. This idea is given the reasonable name of the **kinetic-molecular theory.** Like all important theories in science, it is firmly tied to experimental observations. In this case many of the observations were made on gases, and some are listed here:

1. A gas will expand to fill any container. *Example:* The air in a room reaches all the way to the ceiling.
2. A gas can be compressed when the pressure on it increases. *Example:* When the piston in a bicycle pump is pushed down, the air within the pump is compressed to a fraction of its original volume.
3. Gases exert **pressure,** or force, on each square centimeter of the container. *Example:* The gas inside a balloon keeps it from shrinking.
4. If more gas is added to a container, the volume or the pressure increases. *Example:* Pumping air into bicycle and automobile tires keeps them from collapsing even under heavy weight.
5. In a closed container pressure increases with higher temperature. *Example:* The recommended pressure is always lower for a cold tire than for a warm one since the pressure will increase when friction from the automobile tire's moving on the road heats it up.
6. Gases diffuse into each other readily. *Examples:* Smokestacks have been built ever higher to carry away local air-pollution problems. A skunk need not be very close for the smell to reach you.

These observations and others were made over several centuries. During the nineteenth century in particular, natural scientists struggled to make some sense of them. It was a difficult puzzle, but there were possible solutions and there were experiments to be done with the equipment that was available. Scientists are always ready to tackle problems with possible solutions; those that offer no hope of solution must often wait for a new generation.

Example When the air is sucked out of an empty metal can with a vacuum pump, the can collapses. Which one of the general experimental observations in the list above does this specific observation fit?

Solution Observation 3 fits best: gases exert pressure. The pressure of the atmosphere crushes the can when there is no opposing pressure from the inside.

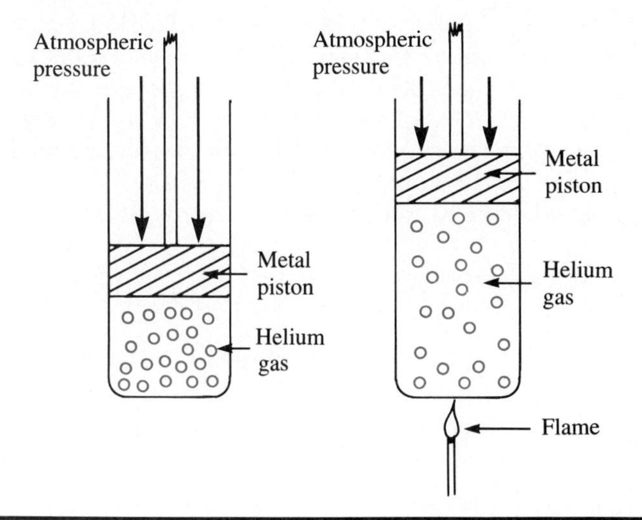

Figure 4.6. Measurement of gas volume at different temperatures.

It is possible to measure the change in volume of a gas, such as air, when the temperature is increased or decreased (see Figure 4.6). A container with a piston at one end is a convenient vessel. Figure 4.7 shows the result of three such experiments, each done at constant pressure and with a constant quantity of gas. Not only does the volume of the gas increase with temperature, but the graph of volume versus temperature is a straight line. With the assumption that straight lines stay straight, one can project

Figure 4.7. Charles's Law.

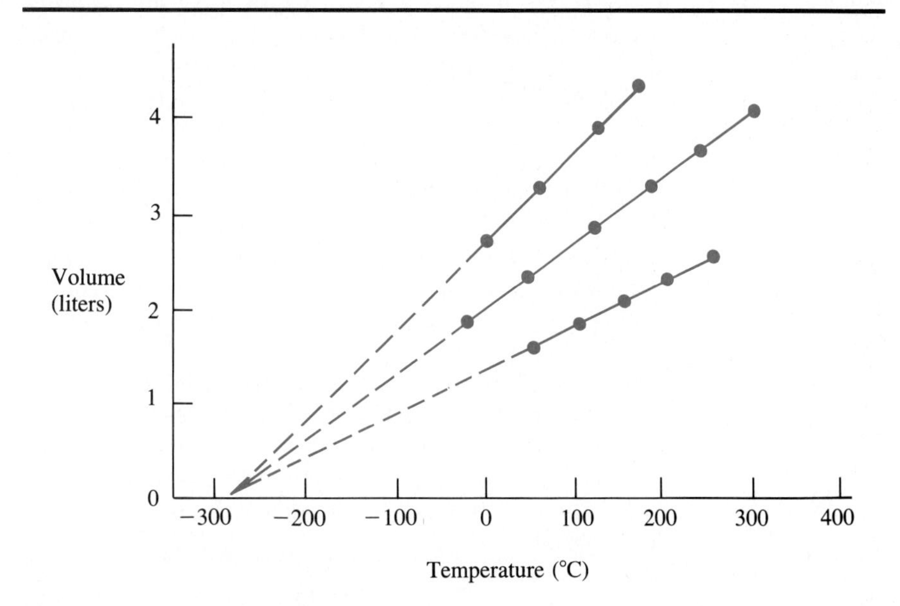

the lines to zero volume. Zero volume would occur at −273.15°C. Lord Kelvin's absolute temperature scale, in which 0°C = 273.15 K, came out of these kinds of observations. The straight-line relationship of gas volume and temperature was named Charles's Law in honor of J. A. C. Charles, the French physicist who first made these quantitative measurements in the late eighteenth century.

Charles's Law can be stated in two other ways. In words, it is

> At any constant pressure, a constant quantity of a gas occupies a volume directly proportional to the absolute temperature (Kelvin scale).

The most compact way to give Charles's Law is in the form of this mathematical equation:

$$V = CT$$

where V = volume
C = a constant
T = temperature in degrees Kelvin (K)

It is common for chemists to express the relationship as an equation rather than in words.

In addition to Charles's Law, other quantitative relationships were found between pressure and volume, as well as between pressure and temperature. These were of crucial importance in the development of the kinetic-molecular theory.

The kinetic-molecular theory uses the model of an **ideal gas.** No such gas exists, although some gases (including H_2 and He) come close. Yet the model gives a valid and useful physical picture of gases over a wide range of conditions. More important, it allows us to combine the idea of energy with the particle theory of matter.

The model pictures a gas as a space, very thinly populated by minute particles (molecules). These molecules are in continuous, rapid, erratic motion. Collisions with the walls of the container produce the pressure of the gas. In the ideal-gas model the molecules are completely independent. Their entire energy is kinetic. This means that no attracting or repelling forces act between the gas molecules. All collisions that the molecules make with each other or with the walls of their container are frictionless or elastic. No kinetic energy is lost in collisions, and the molecules do not settle out after a time.

Since the ideal-gas model accounts for all of the energy of the molecules as kinetic energy, the addition of heat to the system can only increase the kinetic energies of the molecules. Therefore, an increase in temperature will only lead to faster movement. Scientists have expressed these ideas in quantitative equations. The calculations can predict the properties of many gases.

In Figure 4.7 we assumed that the straight lines could be projected to zero volume. In fact, real gases do not follow this straight-line behavior at low temperatures. For example, the oxygen molecules in air are not in the gaseous state below −183°C at atmospheric pressure. The boiling point of O_2 is −183°C. Surely liquid oxygen wouldn't be expected to show the same kind of properties that we discussed for gases. Even through the ideal-gas model does not account for the properties of all gases under all conditions, it is predictive within a few percent error for many gases. Chemists accept the limitations of the ideal-gas model and continue to use it where it is helpful in understanding nature.

* 4.7 THE ESCAPE VELOCITY OF HYDROGEN *(optional)*

Examination of a chart showing the composition of dry air shows several interesting facts. It doesn't make too much difference where the sample is taken as long as it isn't in a city with air pollution. Because of gaseous diffusion, the composition of the air at the earth's surface is much the same the world around. Molecular nitrogen (N_2) is by far the most abundant chemical in the air (78%). The oxygen that we need to breathe comes next at nearly 21%. Argon is present in almost 1% abundance. But where are the hydrogen (H_2) and helium (He) that you might expect to find? They are present in very small amounts. Yet all of our speculations about the origin of the earth suggest that they were once here in abundance, and the two of them are far and away the most common elements in the universe. Where did the H_2 and He go?

To answer this question, we need to estimate the velocities of these small molecules. Then we can see whether they might have escaped into outer space. Using the ideas of the kinetic-molecular theory, especially the idea that an increase in the temperature of a gas only leads to faster movement, we obtain the equation

$$\text{Average translational kinetic energy} = \frac{mv^2}{2} = \frac{3kT}{2} \qquad \textbf{(Equation 4.2)}$$

where m = the mass of a molecule of the ideal gas

v = the average velocity of the molecules

k = a numerical constant = 1.38×10^{-23} when the mass is in kilograms and the velocity is in meters per second

T = temperature (Kelvin) of the gas sample

Calculation of the average speed of molecular hydrogen (H_2) at 0°C using this equation gives a figure of 6600 kilometers per hour. It is done as follows:

$$\frac{mv^2}{2} = \frac{3kT}{2}$$

$$v^2 = \frac{3kT}{m}$$

$$v = \sqrt{\frac{3kT}{m}}$$

The absolute temperature, T, is 273 K, and the mass of a hydrogen molecule is 3.35×10^{-27} kilograms. When these values, along with the value of k given above, are substituted into the equation, the result is

$$v = 1840 \text{ meters/second, or } 6600 \text{ kilometers/hour}$$

Remember that this is only the average speed (more precisely, the root mean square speed) of the molecules. At any instant the velocities range from zero to many times the average value.

Even on a cold winter day, some hydrogen molecules have the necessary velocity to slowly make their way upward from the earth. The reason that their progress is so slow is that the hydrogen molecules bump into other molecules in the air billions of times every second. These collisions occur from all possible directions, and a molecule will bounce one way and then another. Sometimes head-on collisions occur, and the molecules will bounce straight backward. When a molecule reaches the escape layer at an altitude of about 600 kilometers, it will be traveling about 18,000 kilometers per hour, since the temperature there is 1700°C. The molecules at that altitude make many fewer collisions because of the low pressure. It takes a long time for any single H_2 molecule to reach the escape layer, but it has been estimated that many trillions of H_2 molecules escape from the earth's gravitational field each second.

The world's supply of helium (which is useful in scientific research and industrial applications, as well as for blowing up balloons with real lift) is limited for the most part to the gas fields of the southern United States. The helium can be recovered when natural gas is taken from the wells. The helium was trapped underground long ago when it was formed, probably through radioactivity in the surrounding rocks. Like hydrogen, helium can readily escape from earth, so there is far too small an amount to recover economically from the atmosphere.

Hydrogen and helium escape from the earth's atmosphere, but calculations show that very few oxygen, nitrogen, or heavier molecules escape. Their speeds are simply not great enough. We saw in Equation 4.2 that the average translational kinetic energy of gas molecules is proportional to the absolute temperature of the gas. Since many common constituents of the atmosphere act like ideal gases, why should hydrogen and helium escape but not oxygen and nitrogen? The answer to this question lies in the expression for kinetic energy:

$$\frac{mv^2}{2} = \frac{3kT}{2}$$

It is the energy—not just the speed—that is proportional to temperature. Thus the mass of a gas molecule is also an important factor. At a given temperature, the smaller the mass, the greater the speed. One can calculate that H_2 has four times the average velocity of O_2 at the same temperature. The heavier molecules simply don't have the necessary escape velocity.

Example Which have the higher average velocity in the gas phase at the same temperature—oxygen molecules or nitrogen molecules?

Solution First we need to assume that both nitrogen and oxygen act like ideal gases. At the same temperature, then, nitrogen and oxygen would have the same average translational kinetic energy. Since kinetic energy equals $\frac{1}{2}mv^2$, the question can be answered if we know which gas has the smaller mass (m). The product of a small mass and a high velocity squared will equal the product of a larger mass and a smaller velocity squared.

Oxygen gas, O_2, has a molecular weight equal to 32. Nitrogen, N_2, has a molecular weight equal to 28. Therefore, the nitrogen molecules have the higher average velocity.

*

4.8 POTENTIAL ENERGY

Up to this point, the energy of molecules has been portrayed primarily in terms of movement. This idea works well for gases, but it is only part of the story for liquids and solids. To introduce the second important kind of energy, let us consider a stationary rock on a cliff, high above a valley. Because of its elevation, the rock has **potential energy,** or energy of position. Potential energy has nothing to do with movement.

A stationary rock on a cliff has potential energy first because of its position and secondarily because of its mass. The same is true for an apple on its branch. The apple is attracted to earth by the force of gravity. Since it isn't on the ground, it has a higher potential energy because of its position above the earth. This apple has the potential to fall to the ground. As it falls, the apple's potential energy will change into kinetic energy, as you can see from Figure 4.8 (on page 78).

Just before reaching ground level, the apple has kinetic energy only. When the apple strikes the ground, this kinetic energy becomes zero. The apple might break apart or make the earth vibrate a little bit—these processes involve energy exchange. Energy may flow from system to system in various forms, but it is never destroyed. We may impart energy of motion to an object, only to see it gradually settle to rest as the forces of friction slow

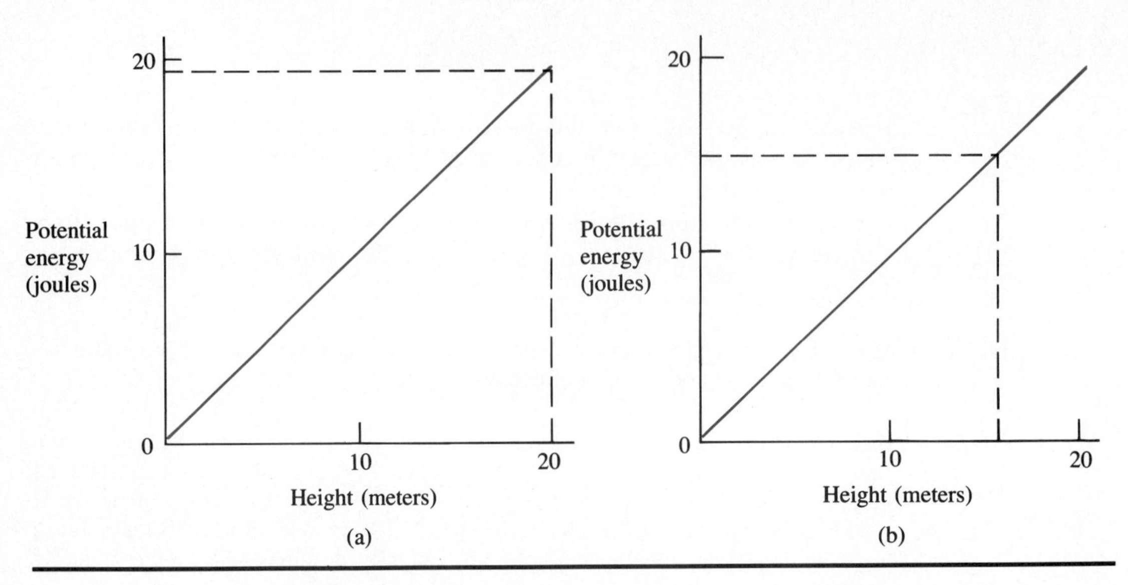

Figure 4.8. (a) Potential energy of a 100-gram apple at various heights. At a height of 20 meters the potential energy is 19.6 joules. (b) A short time after release of the same apple from a height of 20 meters, the potential energy has decreased to 15 joules. The remainder of the original potential energy (4.6 joules) has become kinetic energy.

it down. The energy of motion is not lost; it merely appears in the altered form of heat, as the friction heats up the surface with which it is in contact.

The potential energy of a group of particles results from attractive or repulsive interactions. Electrical, gravitational, and magnetic forces can cause potential energy differences. In general, the potential energy of any collection of particles is defined as the work done to bring them very slowly from reference positions to their present positions. An example on a large scale is the work that has to be done in moving a rock from the bottom of a hill (reference position) to the top (present position). If molecules attract one another, they naturally come closer together from their initial positions. This natural attraction can actually do work. When the molecules move closer together, they are capable (in principle) of doing work on other objects. Their final potential energy is lower than before. This is also the situation when a positive and a negative electrical charge are attracted to each other. As they move together, they reach a lower and lower potential energy and a more stable arrangement, as shown in Figure 4.9(a).

When particles are very far apart, they do not interact with each other at all. Scientists define the potential energy for this reference position as zero. The choice of a reference position and corresponding zero potential energy is always arbitrary. It is often convenient to have the reference position be that for which there is no interaction. If particles attract one

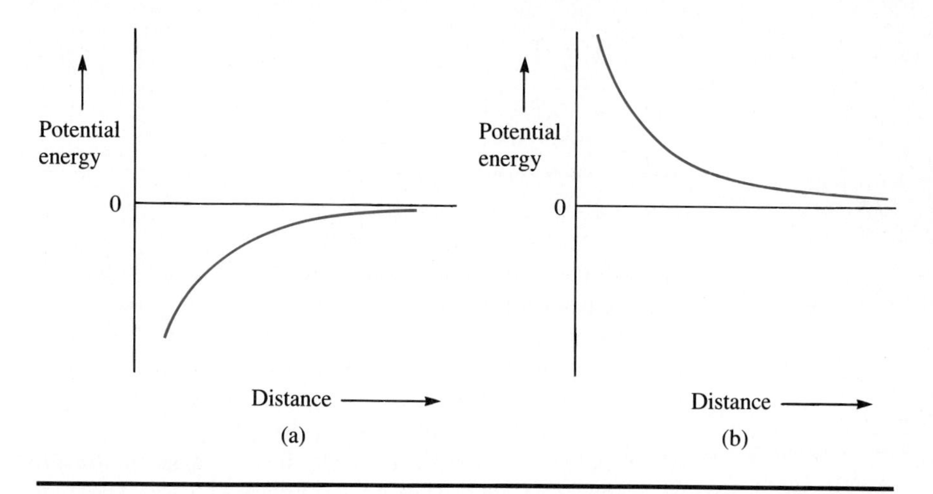

Figure 4.9. Potential energy curves. (a) A positive and a negative charge. (b) Two positive charges.

another at shorter distances, their potential energy drops as they move closer together [Figure 4.9(a)]. Since an energy lower than zero must be negative, the potential energy of such particles will be negative at short-to-moderate separations.

Now consider the case in which two positive electrical charges come into contact. When they are an infinite distance apart [far to the right in Figure 4.9(b)], their interaction is zero. The potential energy of this situation is again considered to be zero. If the two positive charges have molecular dimensions, an "infinite" distance can be much less than a centimeter. Work must be done to bring these two positive charges together, since they repel each other. The necessary work gives the system higher energy. As the particles are pushed closer, the potential energy increases. The positive charges are apt to fly apart, which would again give a lower potential energy and a more stable system.

The potential energy for both situations is given very simply by the equation

$$\text{Potential energy} = \frac{q_1 q_2}{d}$$

where q_1 and q_2 are the charges (including their signs) on particles 1 and 2, respectively, and d is the distance between them. As the separation d becomes very large, the potential energy approaches zero, in agreement with previous statements. The smaller the separation, the more the potential energy differs from zero. When one particle is positive and the other negative, the product $q_1 q_2$ is negative, yielding a negative potential energy, as we saw earlier. On the other hand, for two positive or two negative particles,

$$\text{Solid} \xrightarrow{\text{melting}} \text{Liquid} \xrightarrow{\text{boiling}} \text{Gas}$$

Increase in potential energy (melting) — Large increase in potential energy (boiling)

Figure 4.10. Energy changes that accompany phase changes.

the product in the numerator is positive, and the potential energy is positive. So, the equation agrees with all the qualitative statements made previously.

Molecules tend to attract each other, so substances have lower potential energy when their molecules are close together. Usually they are closest together in the solid phase, and so solids have lower potential energy than liquids. Melting a solid breaks up the orderly arrangement of its molecules. Molecules in the liquid are usually farther apart on the average. The process of melting increases the potential energy of the substance, but the kinetic energy stays about the same.

Boiling a liquid produces a gas, in which the molecules are much farther apart than they are in the liquid. The work required to separate the molecules from one another results in a large increase in potential energy (see Figure 4.10). At the constant temperature of boiling, the kinetic energy remains about the same. Remember that for an ideal gas the potential energy is zero; the molecules no longer have any attraction for each other. Chemists have found the concept of potential energy to be crucial in analyzing the energy changes that occur in chemical reactions as well as in changes of state.

Example Which has the greatest potential energy—solid mercury at −40°C, liquid mercury at −40°C, liquid mercury at 357°C, or mercury vapor at 357°C?

Solution Solids generally have low potential energy. Gases have the highest potential energy (close to zero), since the particles are far apart. So mercury vapor at 357°C has the highest potential energy.

4.9 CAN ENERGY BE CREATED?

We have seen that molecules possess both kinetic and potential energy. All molecules are in motion, even in solids, and they therefore have some kinetic energy. Their potential energy depends in part on whether they are in a liquid, a solid, or a gas. For example, molecules of a gas have a much larger potential energy than they do when the same substance is a liquid or a solid. In every case, though, the addition of energy to a substance results in an increase in the energy of its molecules. In an overall sense, no energy has disappeared, nor has any been created. The sum of the original energy in the substance and the added energy equals the new amount of energy in the substance.

A chemical reaction produces more drastic changes. Atoms that were bound together in the molecules of the reactants are separated, and new combinations of atoms are present in the products. Almost always the strength of bonding in the reactant molecules differs from that in the product molecules. Strong bonding means that much work is required at the molecular level to break apart the atoms. Tightly bound molecules have low potential energy compared with molecules in which the atoms are loosely held together.

Many chemical reactions, such as the burning of fuels, are exothermic. An **exothermic reaction** produces heat. Some reactions need heat added from the outside in order to occur; these are classified as **endothermic reactions.** In an exothermic reaction the heat energy released is exactly equal to the energy lost by the reacting substances. The energy of the products is less than the energy of the reactants by an amount equal to the heat given off.

Figure 4.11 shows the energy relationship for the combustion of methane.

$$\underset{\text{methane}}{CH_4(g)} + 2\,O_2(g) \longrightarrow \underset{\text{carbon dioxide}}{CO_2(g)} + 2\,H_2O(g)$$

High and low energies are indicated in the vertical direction of the diagram. Consistent with the stronger bonds between atoms in the H_2O and CO_2 molecules, the energy of these compounds is shown at a lower level than the energy of the reactant molecules.

For the diagram in Figure 4.11 we assume that the products are at the same temperature as the reactants. All the heat released is transferred to anything in contact with the products. If the reaction takes place in the furnace of a house, the heat is transferred to air or water which is circulated through the house. If they are at the same temperature, the products of the reaction have about the same kinetic energy as the reactants. The energy

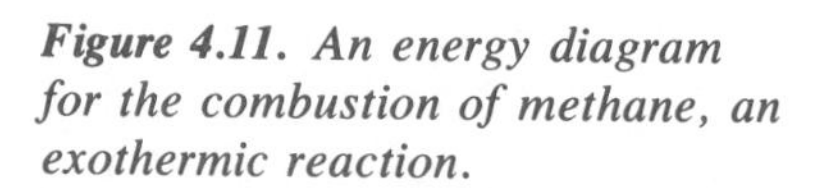

Figure 4.11. An energy diagram for the combustion of methane, an exothermic reaction.

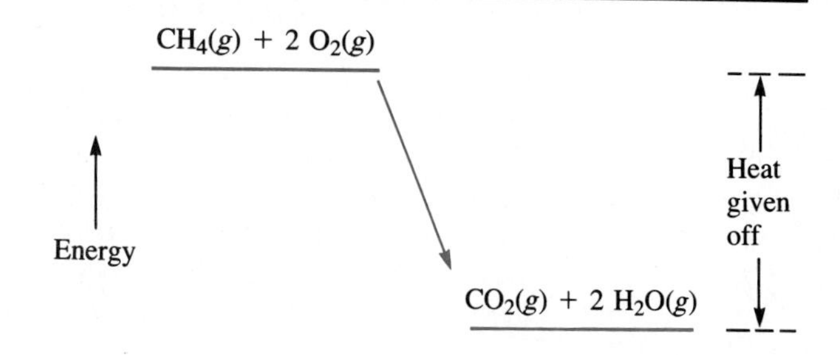

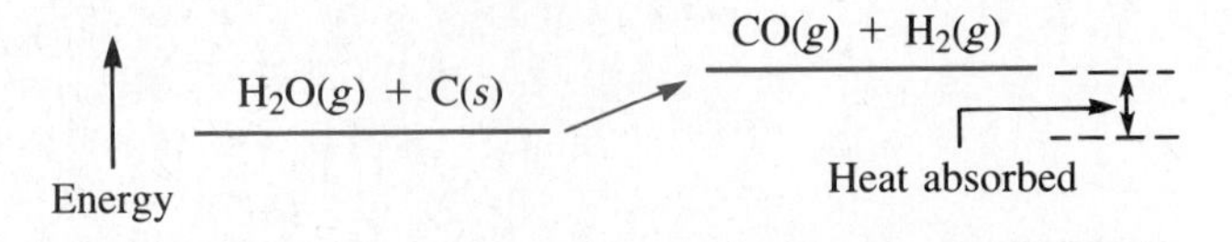

Figure 4.12. Energy diagram for an endothermic reaction.

difference between reactants and products is primarily a difference in potential energies.

An endothermic reaction is diagrammed with just the opposite energy relationship. The heat absorbed during the reaction equals the increase in the energy of the molecules. In this case the products have a greater energy than the reactants. The reaction between steam and carbon is shown in Figure 4.12.

$$H_2O(g) + C(s) \longrightarrow \underset{\text{carbon monoxide}}{CO(g)} + H_2(g)$$

During at least a century and a half of scientific investigation, there have been no exceptions to the **Law of Conservation of Energy** in physical and chemical changes:

Energy is neither created nor destroyed in physical and chemical changes.

One kind of energy can be converted into another, but the amounts are always equal; there is no net gain or loss. This idea led to another statement of the conservation of energy: The total amount of energy in the universe is constant. This seems reasonable, but it may be too sweeping since scientists' experiments have not extended to the farthest reaches of the universe. Conservation of energy is often called *the First Law of Thermodynamics*. Thermodynamics is the study of the transfer of heat (thermo-) and work from one body to another.

4.10 THE DRIVING FORCE FOR CHEMICAL REACTIONS

A fundamental question in chemistry is why some reactions have a strong tendency to take place and others do not. Fuels, such as coal (carbon) and natural gas (methane), burn completely once they are ignited. The complex reactions involved in the cooking of food occur with modest heating. Medicines cure our ailments by reacting in our bodies, and the chemical industry has developed many reactions for the synthesis of useful products.

Other reactions, however, have little tendency to occur. Even at extreme summer and winter temperatures, water has little tendency to decompose into hydrogen and oxygen. In fact, no products can be detected. It is only at very high temperatures that significant amounts of hydrogen and oxygen form from water vapor. It would be nice if ethylene, C_2H_4, could be produced directly by a reaction between water and carbon from coal.

$$2\,C(s) + 2\,H_2O(l) \longrightarrow C_2H_4(g) + O_2(g)$$

This reaction, however, simply does not occur. The ethylene utilized to make polyethylene for packaging and other uses must be synthesized in another way. Three separate reactions are used, with special catalysts.

One feature of many reactions having a strong tendency to take place is that they are exothermic. Certainly this is true for the burning of fuels, and there are lots of other examples. In an exothermic reaction the products have a lower potential energy than the reactants. In the world of physics there are many examples of natural changes accompanied by decreases in potential energy. For example, a raindrop that falls to earth has a lower potential energy at the end of the process. The release of heat is one of the driving forces for both chemical and physical changes.

The release of heat energy favors a process, but this is not the only driving force in chemistry. The endothermic reaction shown in Figure 4.12 produces moderate amounts of products despite the fact that the reaction absorbs heat from the surroundings. A more familiar example is the melting of ice. When a chunk of ice is placed in an environment warmer than 0°C, it melts with an absorption of heat from the environment. The melting of ice is endothermic, and yet it has a strong tendency to occur. We must consider a second driving force.

4.11 MOLECULAR DISORDER

Imagine an experiment done with the apparatus shown in Figure 4.13. A container is divided into two parts by a glass plate at the center. This plate fits into a slot so that it can be removed at the appropriate time. We pour clear water into the left side of the apparatus and some black ink into the right side. Very carefully we remove the glass plate separating the two liquids. The boundary between the black and the clear regions is relatively sharp, and the black color is still observed in the right half of the container. Several hours later we look again at our experiment. Now the black color has spread into the left half, and there is no longer a distinct boundary between the black and the clear regions. A simple way to explain this change is that some of the molecules responsible for the black color have

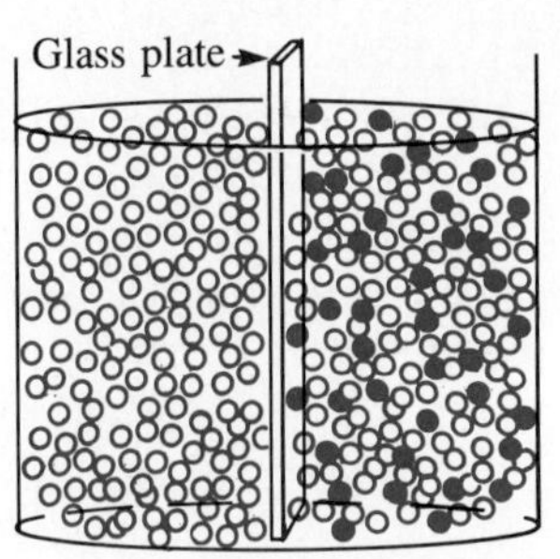

Clear water (left) and ink (right) before mixing

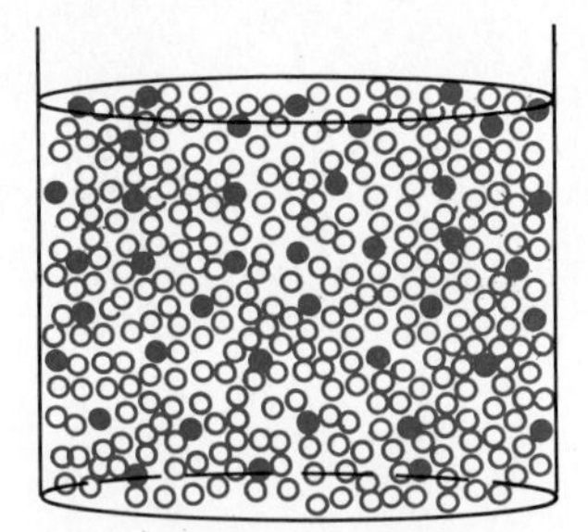

Distribution of molecules several hours after removal of glass plate

Figure 4.13. A demonstration of spontaneous mixing. (Ink molecules are shown as solid dots.)

migrated into the left half as a result of their random motions. After a very long time we could expect the liquid to be a uniform gray color throughout.

The driving force for this change cannot be a lowering of potential energy with the release of heat, because no heat can be detected. Instead, the degree of randomness in the arrangement of molecules provides the explanation. Before removal of the glass plate, all the "black" molecules were confined to the right side. Immediately after the removal of the partition, every part of the liquid is, in principle, available to these molecules. From this point of view, the arrangement of black molecules is rather organized: all are in the right half. A much more random arrangement is an even distribution of black molecules over the entire volume of liquid.

The mixing process that occurs naturally is toward an even distribution of molecules. Everyone has observed spontaneous mixing of substances to form homogeneous solutions, as when cream is added to coffee, for example. Without stirring, the process may be very slow, but eventually a uniform mixture results. In every case an ordered arrangement of molecules changes into a more disordered or "mixed up" arrangement. Spontaneous mixing is an example of a universal tendency in nature.

> Overall, change moves in the direction of a more random or disordered distribution of molecules.

This is a statement of the **Second Law of Thermodynamics.**

There is a great difference between the degree of disorder in a gas, a liquid, and a solid. The molecules of a gas in, say, a 1-liter flask have access to the entire container. The molecules are moving in erratic paths at high speeds, and at any moment the arrangement of molecules fits no pattern. If at one instant you knew where a particular molecule was, there is no way you could know where it would be a fraction of a second later. All of this adds up to a large measure of disorder in a gas.

The molecules of a liquid, on the other hand, are in a more ordered arrangement. Because they are confined to a much smaller volume, their freedom of motion is restricted. They are in constant motion, but each molecule bumps against the same neighbors for a while before it moves to an adjacent site in the liquid. Over short distances there is a regular arrangement of molecules, but this order is not evident over longer distances.

Finally, in a solid the molecules are highly ordered. An organized structure exists over many molecular diameters. Each molecule vibrates back and forth but rarely moves to a new position in the crystal.

The tendency for change to occur in the direction of greater disorder explains why ice in a warm environment melts despite the fact that the process is endothermic. An increase in disorder accompanies the breakdown of the ice crystal into the more random molecular arrangement of the liquid. Apparently this effect more than makes up for the increase in potential energy of the molecules, which by itself does not favor melting.

Scientists have invented a mathematical quantity called **entropy** to measure the amount of disorder in any substance. The term entropy is used a good deal in current writing, and it is applied to a variety of fields. Some social scientists refer to an increase in entropy in certain kinds of social change. For our purposes entropy simply means the degree of molecular disorder.

Example Which has a higher entropy—10 grams of ice or 10 grams of steam?

Solution The substance that has the greater molecular disorder has the greater entropy. Water vapor (steam) has much greater disorder than ice does, since ice crystals are highly ordered and a gas has almost no pattern. Therefore, 10 grams of steam has the higher entropy.

Let's return to a few of the exothermic and endothermic reactions we discussed earlier and look again at the driving force behind them. To summarize, there are two kinds of driving forces for chemical and physical change:

1. Changes tend to occur in the direction of lower potential energy with the evolution of heat energy. In other words, exothermic processes are favored.
2. Changes tend to occur in the direction of greater molecular disorder.

In short, low potential energy and high entropy are the favored states. Often only one of these two driving forces favors a particular process. In such cases it is not possible to predict which driving force will dominate without using a mathematical form of the Second Law of Thermodynamics. It is clear, though, that any endothermic process that occurs naturally must involve a substantial increase in disorder.

Thus, for the endothermic reaction shown in Figure 4.12, the products must have more disorder, or entropy, than the reactants do. On the other hand, the exothermic reaction of Figure 4.11 clearly has in its favor the lower potential energy of the products. Any change in molecular disorder occurring in this reaction is less important.

IMPORTANT TERMS

energy The capacity for doing work.

joule (J) A standard unit of energy, equivalent to the power of 1 watt operating for 1 second.

calorie A secondary unit of energy, equal to 4.18 joules.

dietary calorie 1000 calories. Sometimes called a "large" calorie, a dietary calorie is used to specify energies derived from foods.

temperature Degree of hotness or coldness of a body. Temperature is usually measured by observing the volume of a substance such as mercury in a calibrated tube.

Celsius temperature scale The temperature scale used by most of the world. The freezing point of water (0°C) and the boiling point of water at 1 atmosphere pressure (100°C) are its reference points.

Kelvin temperature scale Scientists' temperature scale that gives the absolute temperature of a body; 273 K equals 0°C. The Kelvin scale has the same size "degree" as the Celsius scale.

model A simplified picture of physical reality. It is an approximate theory, often dealing with atomic or molecular behavior.

state The designation of a substance as either a gas, a liquid, or a solid. Also called *phase*.

solid The state of matter in which the substance retains its shape and the particles have fixed positions.

liquid The state of matter in which the substance takes the shape of its container. The particles are mobile rather than in fixed positions.

gas The state of matter in which the particles completely fill their container. Also called a *vapor* if it is easily condensed.

vapor pressure Pressure exerted by the molecules in the vapor state above a liquid when the vapor is confined within a closed vessel.

kinetic energy The energy of motion, which depends on the mass and the speed of an object. Translational kinetic energy is the kinetic energy of motion from one location to another.

translational energy The kinetic energy that an object has when it moves from one place to another.

kinetic-molecular theory The theory that relates properties of a gas, such as pressure and speed of diffusion, to molecular behavior.

pressure Force per unit area of surface. A gas exerts pressure on the walls of the container that confines it.

ideal gas A gas in which the molecules have only kinetic energy. In the ideal-gas model the molecules are completely independent, and their potential energy is zero.

potential energy The energy of position relative to that of other positions. It has nothing to do with motion.

exothermic reaction A reaction in which heat is released.

endothermic reaction A reaction in which heat must be added.

Law of Conservation of Energy The law stating that energy is neither created nor destroyed in a chemical or physical change. Also called the *First Law of Thermodynamics*.

Second Law of Thermodynamics The law stating that there is a tendency for the molecular disorder to increase in any process that naturally occurs in the world.

entropy The degree of molecular disorder.

QUESTIONS

1. Which has more energy?

a. 1 gram of iron at 273 K or 1 gram of iron at 373 K

b. 1 gram of solid iron at 1535°C (its melting point) or 1 gram of liquid iron at 1535°C

c. 1 gram of liquid iron at 2800°C or 1 gram of iron vapor at 3200°C

2. Which of the six general experimental observations on gases (page 72) do the following specific observations fit?

a. If there is a natural gas leak inside a house, the smell of gas can soon be detected everywhere in the house.

b. A pressurized aerosol can says on the label "Danger—do not incinerate."

c. A warm can of cola spurts liquid all over when it is opened.

d. The person next to you in the movie theater is wearing perfume.

e. A basketball must be pumped up before use.

f. A large amount of LP gas can be stored in a small tank.

3. When a bottle of ammonia water is uncapped in a closed room, the odor of ammonia gas can soon be detected several feet from the bottle. Explain why gases diffuse readily and do not settle to the bottom of a container.

4. a. Where does water go when it evaporates from your skin?

b. Does the water have more or less energy after it evaporates?

c. Is the energy change during evaporation primarily kinetic or potential?

5. a. When both are at 25°C, do either the molecules in liquid alcohol or the molecules in alcohol vapor have greater potential energy, or are the potential energies about equal? Explain.

b. Does a stretched or an unstretched rubberband have greater potential energy? Explain.

6. When sulfuric acid is mixed with water, the solution that forms becomes so hot that touching even its container is uncomfortable. Is this process exothermic or endothermic? Explain.

7. When ammonium nitrate, $NH_4NO_3(s)$, dissolves in water, the solution feels cold to the touch. Is the dissolving process endothermic or exothermic? Explain.

8. Predict whether there is an increase or a decrease in entropy when each of the following physical changes occurs.

a. Water boils.

b. $Al(s) \longrightarrow Al(l)$

c. Dry ice (solid CO_2) evaporates.

d. Water freezes on a cold winter day.

9. A sugar cube is dropped into a glass of water and allowed to stand. Three days later the sugar cube has disappeared, and the water tastes sweet. Explain whether an increase or a decrease in entropy has occurred.

10. An English scientist, James Joule, predicted and then verified in the middle of the nineteenth century that the water in the pool at the bottom of a high waterfall is warmer than the water at the top of the waterfall. Explain why this is so.

11. a. Two decks of cards are side by side on a table. In the first deck, the four suits are segregated; that is, all the spades are together, and so on. The second deck has been thoroughly shuffled. Which deck has the greater entropy? Discuss.

b. A row of five pennies shows all five heads. Another row of five pennies shows three heads and two tails. Which group of five pennies has the greater entropy? Explain.

12. When 5 milliliters of liquid water is placed in a 1-liter container which is then closed, the amount of liquid decreases slightly and then remains constant. When the same amount of water is put into a 10-liter container, the decrease in the amount of liquid is greater. Why does more liquid evaporate in the 10-liter container?

13. Make a sketch of Figure 4.4, and add the temperatures in degrees Celsius corresponding to the two horizontal portions of the graph (melting and boiling). For each straight line segment, state whether the molecules are primarily undergoing an increase in kinetic energy or an increase in potential energy as thermal energy is added to the system.

14. At 100°C liquid water in a pan boils away and turns into water vapor or steam. Which of the two driving forces for physical change must dominate in this process? Explain.

15. It was recently reported that production of 1 pound of white potatoes requires the following amounts of energy, if packaging energy costs are ignored.

1 POUND OF POTATOES	ENERGY USED IN PRODUCTION (in kJ)
canned	9,450
frozen	15,700
dehydrated	28,100

In the canning process heat is used to cook the potatoes. Frozen potatoes require some preliminary cooking, and then the water in them must be frozen. Dehydrated potatoes are first partially cooked, and then the water in them is evaporated away. Provide short answers to the following questions:

a. Why is more energy required to produce frozen potatoes than canned potatoes?

b. Why is more energy required to produce dehydrated potatoes than frozen potatoes?

16. Ethanol, C_2H_5OH, is being mixed with gasoline and sold as "gasohol" for automobiles in order to extend our supply of petroleum.

a. Write the equation for the combustion of liquid ethanol to give carbon dioxide gas and liquid water.

b. Judging by what you know about combustion, is the reaction endothermic or exothermic?

c. Does the driving force toward lower potential energy operate in this reaction? Explain.

d. Does the driving force toward higher entropy operate in this reaction? Explain.

Optional Section

17. Suggest why the most abundant element in the universe, hydrogen, is not the most abundant element on earth.

18. According to a statement on page 76, hydrogen molecules at an altitude of 600 kilometers, where the temperature is 1700°C, have an average speed of about 18,000 kilometers per hour. By means of calculations, verify this speed.

Moles and Chemical Reactions

Knowing something about the number of molecules in a piece of matter is very important to a chemist, because the properties of substances often depend on how many molecules are present. For example, the intensity of color in a dyed fabric or a colored liquid depends on the number of molecules that are reflecting colored light. The toxicity of soil contaminated with polychlorinated biphenyls (PCBs) from discarded transformer oil increases as the number of PCB molecules increases. In a chemical reaction the amount of product formed depends on how many molecules of each reactant were present initially. Some properties of solutions, such as their freezing and boiling points, are determined by the relative numbers of molecules of the several components.

Chemists don't try to count individual molecules simply because there are so many in any visible sample of matter. Instead, they use a unit like a dozen, only much, much larger. It is somewhat like dealing with bunches of carrots rather than individual carrots. Chemists determine the number of "bunches" of molecules in a sample by weighing it.

5.1 COMPARING NUMBERS OF MOLECULES

Usually it is necessary to know only whether the number of molecules in one substance is the same as the number of molecules in a second substance, or by what factor they differ. One doesn't need the actual numbers. A chemist might ask: How does the number of molecules of water in a beaker compare with the number of molecules of sugar in a test tube? (See Figure 5.1.) Are there 100 times as many water molecules as sugar molecules, or what exactly is the ratio?

There are two methods for comparing numbers of molecules. Unfortunately, the first one applies only to gases, but it is still useful. More than 170 years ago the Italian physicist Amadeo Avogadro (Figure 5.2) proposed a hypothesis that was ignored by most other scientists of the time. We now know it to be true. This hypothesis, more than any other, has helped chemists to compare numbers of molecules.

> **Avogadro's Hypothesis:** Equal volumes of different gases contain the same number of molecules if they are at the same temperature and pressure.

For instance, 1 liter of oxygen gas at a pressure of 0.4 atmosphere (atm) and a temperature of 350 K has the same number of molecules as does 1 liter of argon gas at 0.4 atmosphere and 350 K. In other words, the number of O_2 molecules per liter equals the number of Ar molecules per liter. If at the same temperature and pressure the volumes of O_2 and Ar are 2 liters and 1 liter, respectively, then there are twice as many O_2 molecules as Ar molecules. This is an easy method for comparing numbers of molecules.

Example Which sample has more molecules at 20°C and 1 atmosphere pressure—450 milliliters of methane gas (CH_4) or 0.6 liter of hydrogen gas (H_2)?

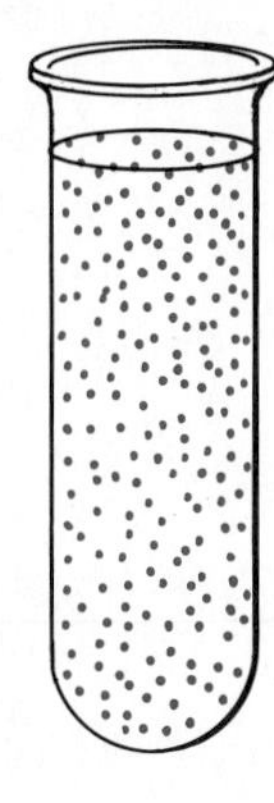

Figure 5.1. A figurative comparison of the number of water and sugar molecules in two samples.

Figure 5.2. Amadeo Avogadro.

Solution By Avogadro's Hypothesis, equal volumes of gases under the same conditions will have the same number of molecules. Therefore, we need to see which of the two gases has the larger volume. The larger volume will represent more molecules. There are 1000 milliliters in a liter, so 450 milliliters is equal to 0.450 liter. The volume of H_2 gas is larger, and thus the sample of hydrogen gas contains more molecules.

An interesting fact emerges if we *weigh* equal volumes of two gases at the same temperature and pressure. Figure 5.3 illustrates the procedure. In effect, we are weighing equal numbers of molecules. The flask can be evacuated, weighed, and then filled with hydrogen gas at room temperature (say, 296 K) until the pressure reaches 1 atmosphere. Then the mass of the hydrogen gas can be found by measuring the difference in the weight of the flask when evacuated and when filled with hydrogen. This weight is 0.0231 gram in our example. If the same flask is re-evacuated and filled with oxygen gas to a pressure of 1 atmosphere and at the same temperature as before, the mass of the oxygen is then found to be 0.3667 gram. This careful comparison of the weights of equal volumes of hydrogen and oxygen indicates that a volume of oxygen is approximately 16 times heavier than an equal volume of hydrogen.

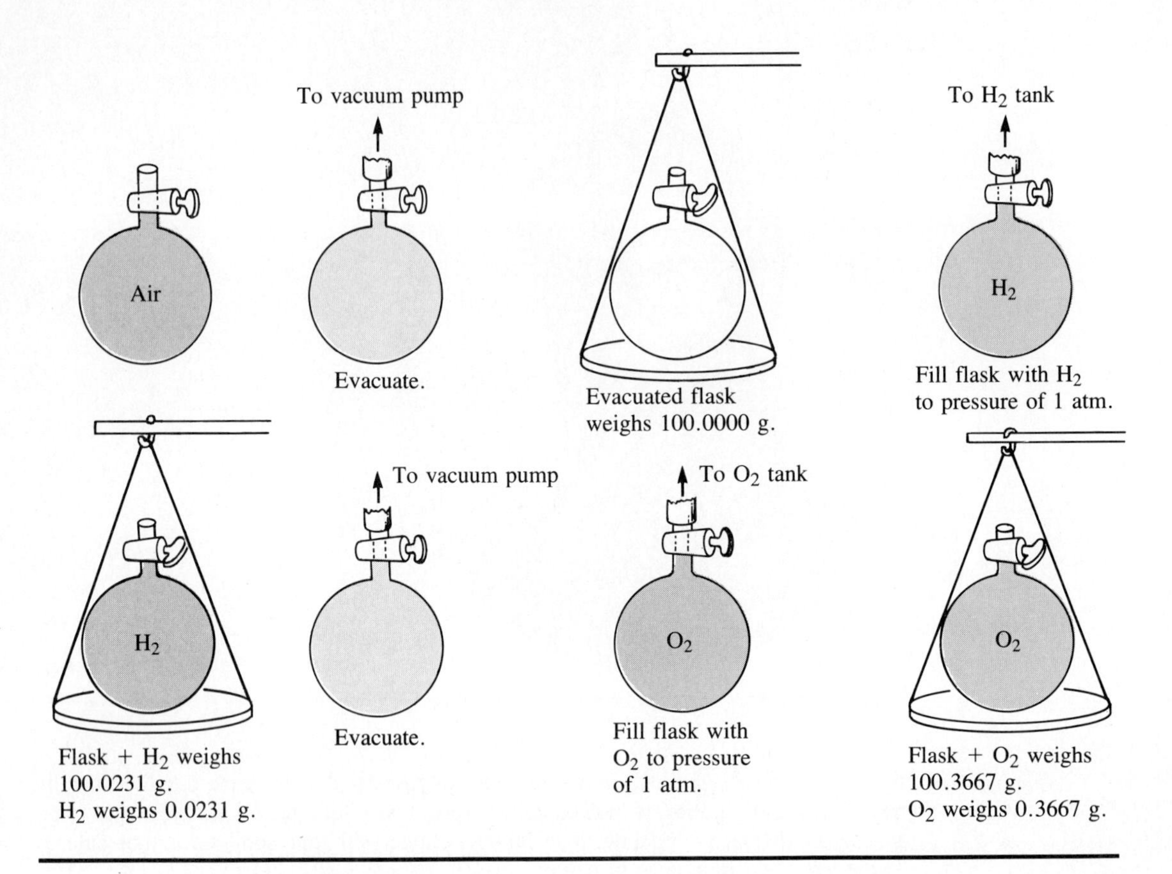

Figure 5.3. *Weighing equal volumes of hydrogen gas and oxygen gas. Since the gases have equal volumes, the number of H_2 molecules equals the number of O_2 molecules.*

$$\frac{0.3667 \text{ gram } O_2}{0.0231 \text{ gram } H_2} = 15.9$$

The temperature, pressure, and volume are the same in both cases, and therefore the number of hydrogen molecules is identical to the number of oxygen molecules.

The experiment illustrated in Figure 5.3 shows that a certain very large number of oxygen molecules weighs more than an equal number of hydrogen molecules—or one oxygen molecule weighs more than one hydrogen molecule. In fact, since the ratio of weights is 16:1, we can conclude that an oxygen molecule weighs 16 times as much as a hydrogen molecule. This should come as no surprise. The molecular weights of O_2 and H_2 are 32 and 2, respectively, for a ratio of 16:1. In fact, Avogadro's Hypothesis was used originally to calculate these molecular weights.

This brings us to the second method for comparing numbers of molecules or obtaining equal numbers of molecules. It is based on molecular weights and applies not only to gases but also to liquids and solids. How could we measure out equal numbers of molecules of liquid water (H_2O) and dry ice (a solid oxide of carbon, CO_2)? Their molecular weights are 18 and 44, respectively. In other words, the ratio of the weights of the two kinds of molecules is

$$\frac{\text{Mass of a } CO_2 \text{ molecule}}{\text{Mass of an } H_2O \text{ molecule}} = \frac{44}{18} = 2.44$$

Suppose that we weigh 25 grams of water on a balance. How much CO_2 should we weigh to get an equal number of molecules? Because a molecule of CO_2 weighs 2.44 times as much as a molecule of H_2O, we need 2.44×25 grams = 61 grams of CO_2. This method, then, involves weighing amounts of substances that are proportional to their molecular weights.

Example A sample of neon gas, Ne, weighs 47 grams. What is the weight of sulfuric acid, H_2SO_4, having the same number of molecules as there are in the 47-gram sample of neon? Some useful atomic weights are Ne, 20.2; H, 1.0; S, 32.1; and O, 16.0.

Solution First we must calculate the molecular weights of the two substances. The molecular weight of Ne is the same as the atomic weight (20.2), because the neon atoms do not combine together into larger molecules. The molecular weight of H_2SO_4 is

$$\begin{aligned}(2 \cdot \text{atomic weight of H}) + (1 \cdot \text{atomic weight of S}) + (4 \cdot \text{atomic weight of O})\\ = (2 \cdot 1.0) + (1 \cdot 32.1) + (4 \cdot 16.0)\\ = 98.1\end{aligned}$$

Note that the molecular weight of sulfuric acid is greater than that of neon. Therefore, for the same number of molecules, the weight of sulfuric acid is greater than that of the neon. We must multiply the weight of the neon, 47 grams, by a ratio of molecular weights that is greater than 1.

$$\frac{\text{Molecular weight of } H_2SO_4}{\text{Molecular weight of Ne}} = \frac{98.1}{20.2} = 4.86$$

Then

$$\text{Weight of } H_2SO_4 = 47 \text{ grams} \cdot 4.86 = 228 \text{ grams}$$

The number of H_2SO_4 molecules in 228 grams of sulfuric acid equals the number of Ne molecules in 47 grams of neon.

Because the method is so general, chemists usually measure numbers of molecules by weighing them. Although counting by relative weights is not common in everyday life, there are situations in which weighing objects can be handy.

Example Imagine that you are in a supermarket shopping for a detergent. Your budget is strained, and you are after the best bargain. Two brands look good, but which is the better buy? Brand A requires 30 grams of detergent per washing-machine load to do the job. Brand X requires 45 grams per load of clothes. A 750-gram box of A costs $3.00; a 1000-gram box of X also costs $3.00. Which detergent is the better bargain?

Solution You want to find which brand washes more loads of clothes for the same price. First, you can determine the relative weights of the detergents necessary for one washing load.

$$\frac{\text{Weight of Brand A}}{\text{Weight of Brand X}} = \frac{30 \text{ grams}}{45 \text{ grams}} = 0.67$$

This means that to wash any particular number of loads the weight of A needed is 0.67 times the weight of X needed. Now, how much Brand A do you need in order to wash the number of loads you can wash with a 1000-gram box of Brand X? You must multiply the 1000 grams by the ratio of washing efficiency, a number less than 1.

$$\text{Weight of Brand A} = 1000 \text{ grams} \cdot 0.67 = 670 \text{ grams}$$

Since a 750-gram box of Brand A gives more washes and yet costs the same as a 1000-gram box of Brand X, Brand A is the better bargain. You can wash more loads of clothes (over 10% more) with Brand A for the same amount of money.

5.2 THE MOLE

It is not convenient to calculate ratios of molecular weights repeatedly in order to obtain samples of substances containing equal numbers of molecules. What we need is a chemical "dozen," a standard measure of a constant number of molecules. Such a measure has been defined, and it is commonly used by chemists. Its name smacks of neither romance nor poetry, nor does it honor any of the distinguished scientists of the past. The measure is called, simply, the *mole*. An association with a small animal burrowing underground is unavoidable. A careful reading of the dictionary, however, reveals a Latin origin—"moles," meaning a mass or pile of stones.

In general, a **mole** of any substance contains the same number of particles as the number of carbon atoms in 12.0 grams of carbon (atomic weight 12.0). According to this definition, 1 mole of any molecular substance has the same number of molecules as 1 mole of any other substance. Furthermore, 3 moles of one substance has three times as many molecules as 1 mole of

Table 5.1 Some Gram Molecular Weights

SUBSTANCE	MOLECULAR WEIGHT	GRAM MOLECULAR WEIGHT (or grams per mole)
H_2O	18	18 g
O_2	32	32 g
CO_2	44	44 g
H_2SO_4	98.1	98.1 g

a second substance. Moles and molecules are in direct proportion to each other.

To define a mole only as a certain number of molecules, however, is of little practical value, because we have no simple way of counting molecules. Instead, in order to make the definition of the mole more practical, we define a mole of a molecular substance in terms of a weight: *A mole of molecules of a substance is 1 gram molecular weight of the substance.* A **gram molecular weight,** in turn, is the number of grams of a substance equal to its molecular weight. This definition provides a practical, working relationship between the weight of chemical compounds and the number of molecules they contain. Table 5.1 gives some examples.

It is important to realize that the first and second definitions above are totally consistent with each other. According to the first, 1 mole of H_2O contains the same number of particles (in this case, molecules) as the number of carbon atoms in 12 grams of carbon. According to the second definition, 1 mole of H_2O weighs 18 grams. Because one molecule of H_2O weighs 18/12 times one atom of C, the number of molecules in 18 grams of H_2O equals the number of atoms in 12 grams of C.

Example Which has the greater number of molecules—17 grams of ammonia (NH_3, molecular weight 17), which is responsible for the odor of some smelling salts, or 30 grams of hydrogen sulfide (H_2S, molecular weight 34), which has the smell of rotten eggs?

Solution The 17 grams of ammonia is 1 mole of ammonia, but 30 grams of hydrogen sulfide is less than 1 mole, because 1 mole of hydrogen sulfide weighs 34 grams. Therefore, 17 grams of ammonia has the greater number of molecules.

Some substances don't consist of molecules. The largest class of such substances is the metals, which make up the bulk of the periodic table. A metal—such as iron (Fe), nickel (Ni), or sodium (Na)—consists of atoms packed closely together in a regular pattern. There are no clusters of atoms that we could call molecules. The noble gases (in group VIII) also exist as single atoms, although you could refer to the independent gas particles as molecules consisting of one atom each. An operational definition of a mole of atoms is as follows: *A mole of atoms of a substance is 1 gram atomic weight of the substance.* In parallel with gram molecular weight, a **gram atomic weight** is the weight in grams equal to the atomic weight of the element. The gram atomic weight of the metal Fe (atomic weight 55.8) is 55.8 grams, and the gram atomic weight of the noble gas Ar (atomic weight 40.0) is 40.0 grams. The number of atoms in 55.8 grams of iron equals the number of atoms in 40.0 grams of argon. This number also equals the number of atoms in 12.0 grams of carbon. In addition, this number is the same as the number of molecules in 17 grams of ammonia, for example. A mole stands for a fixed number of particles—either atoms or molecules, depending on the context.

A final case to consider is that of a compound that does not consist of molecules. As we said in Chapter 3, ionic compounds, such as sodium chloride (NaCl), are not molecular. For these we have this definition: *A mole of a nonmolecular compound is 1 gram formula weight of the compound.* A **gram formula weight** is the number of grams equal to the sum of the atomic weights in the formula of the compound. Thus 1 mole of NaCl equals 58.5 grams (23.0 + 35.5).

Example Determine the weight in grams of each of the following: 1 mole of N_2 molecules, 1 mole of Zn atoms, 1 mole of KI (an ionic compound). Use the table of atomic weights inside the front cover to calculate whatever molecular or formula weights you need.

Solution By appropriate addition of atomic weights, we find the following weights of 1 mole: N_2, 28 grams; Zn, 65.4 grams; KI, 166 grams.

We come now to that oft-quoted number, the number of particles in 1 mole of a substance. This number was not easy to determine. Methods such as measurement of the emissions from a radioactive element were finally used. In recognition of the scientific contributions of Amedeo Avogadro, the number has been named **Avogadro's number,** and it has the value

602,200,000,000,000,000,000,000 atoms per mole of an element

or

602,200,000,000,000,000,000,000 molecules per mole of a molecular substance

Clearly this is an awkward way of expressing such a huge number. A much more practical way is to write Avogadro's number using exponential notation:

$$6.022 \times 10^{23} \text{ particles per mole}$$

This value, then, is the chemist's standard measure of molecules or atoms. As you have seen, however, a mole of anything is always determined by weighing a gram atomic weight, a gram molecular weight, or a gram formula weight.

5.3 MOLES, CHEMICAL REACTIONS, AND HEAT

The balanced chemical equation for a reaction gives more information than we indicated earlier. Not only does it tell us how many molecules of each reactant and product are involved, but it also gives us the relative numbers of moles. Take, for example, the reaction between hydrogen and chlorine to give hydrogen chloride.

$H_2(g)$	+	$Cl_2(g)$	$\longrightarrow$	$2\ HCl(g)$
hydrogen		chlorine		hydrogen chloride
1 molecule		1 molecule		2 molecules
1 mole		1 mole		2 moles

The equation says that from one molecule of H_2 and one molecule of Cl_2 we get two molecules of HCl. Or, putting it differently, twice as many molecules of HCl are produced as there were molecules of H_2 or of Cl_2 originally. Now, a similar statement can be made about the numbers of moles, because 1 mole of anything contains the same number of molecules. The ratio of moles of HCl to moles of either H_2 or Cl_2 is 2:1. One way of thinking about a chemical equation is that the numbers before each chemical formula give the relative numbers of moles.

As we have seen, the great majority of chemical reactions are exothermic and give off heat when they occur. Only a small fraction are endothermic. For either kind of reaction, we can specify just how much heat is associated with the reaction with the help of the mole concept. To illustrate, let's examine the burning of two major fuels: coal (which is mainly carbon) and natural gas (which is primarily methane, CH_4). Figure 5.4 shows the kind of apparatus used for the combustion of a sample of coal. The equations for the combustion reactions are

$$C(s) + O_2(g) \longrightarrow CO_2(g)$$

$$CH_4(g) + 2\ O_2(g) \longrightarrow CO_2(g) + 2\ H_2O(g)$$

From careful measurements chemists have learned that 393,000 joules (J), or 393 kilojoules (kJ), of heat is released when 1 mole of carbon burns. From the burning of 1 mole of methane, 802 kilojoules is evolved. In each of these exothermic reactions, the heat given off can be thought of as an additional product of the reaction. For example, whenever 1 mole of CO_2 and 2 moles of H_2O form in the combustion of CH_4, 802 kilojoules of heat

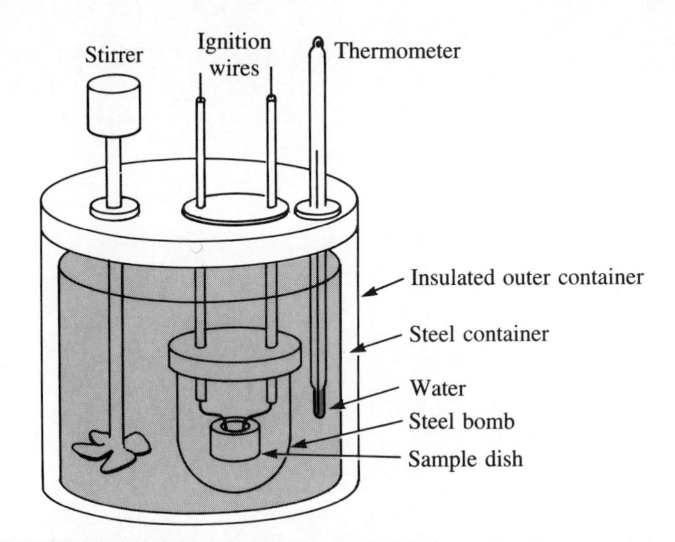

Figure 5.4. A bomb calorimeter for measuring heats of combustion of a liquid or solid fuel. The temperature increases when the reaction occurs. The amount of heat is calculated from the temperature rise.

accompanies these chemical products. The above two equations can be rewritten as thermochemical equations to show the heat effects.

$$C(s) + O_2(g) \longrightarrow CO_2(g) + 393\ \text{kJ}$$

$$CH_4(g) + 2\ O_2(g) \longrightarrow CO_2(g) + 2\ H_2O(g) + 802\ \text{kJ}$$

A **thermochemical equation** shows the quantity of heat involved when the indicated numbers of moles of reactants are consumed and the indicated numbers of moles of products are formed. In other words, in a thermochemical equation the coefficient before each substance always stands for a number of moles. Also, in a thermochemical equation it is important to show the state (gas, liquid, or solid) of each substance, because the quantity of heat depends on these states.

The release of 802 kilojoules of heat in the combustion of 1 mole of methane means that the energy of the products is less than the energy of the reactants by 802 kilojoules. The energy diagram in Figure 4.11 can now be improved to show the amount of energy released—see Figure 5.5.

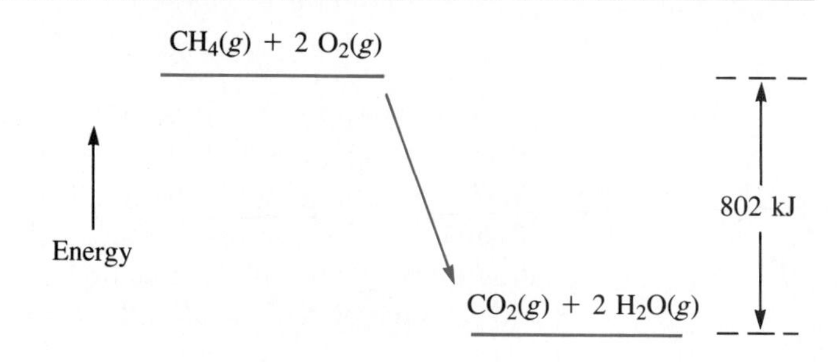

Figure 5.5. A quantitative energy diagram for the combustion of 1 mole of methane.

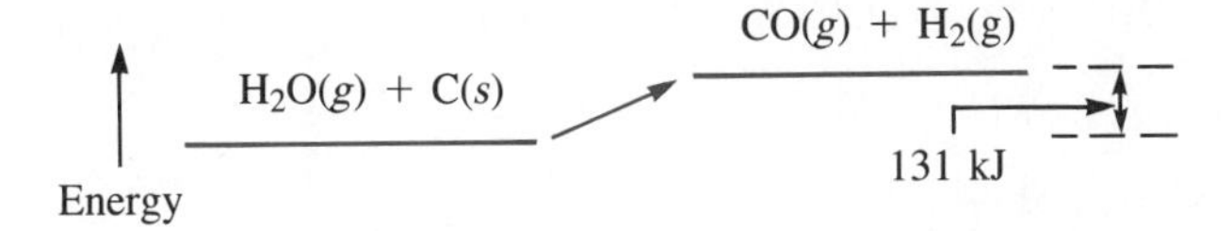

Figure 5.6. A quantitative energy diagram for an endothermic reaction.

When a reaction is endothermic, the quantity of heat is shown on the left side of the thermochemical equation, because the heat is analogous to a reactant:

$$H_2O(g) + C(s) + 131\ kJ \longrightarrow CO(g) + H_2(g)$$

A total of 131 kilojoules of heat is absorbed from the surroundings and fed into the reaction for each mole of $H_2O(g)$ and $C(s)$ consumed (see Figure 5.6). For exothermic reactions, then, the heat is shown on the right side of the equation, and for endothermic reactions it is on the left.

The reaction shown in Figure 5.6 is important in industry because the products, CO and H_2, are both fuels. To make the reaction go, steam at a high temperature is passed over hot coke (a form of carbon). The gaseous mixture that is produced can be transported through a pipe to a place where heat is needed. We express the combustion of this gaseous mixture with two thermochemical equations:

$$2\ CO(g) + O_2(g) \longrightarrow 2\ CO_2(g) + 566\ kJ$$

$$2\ H_2(g) + O_2(g) \longrightarrow 2\ H_2O(g) + 484\ kJ$$

Example In a process often used by the steel industry to obtain iron from iron ore, iron oxide, Fe_2O_3 (s), reacts with carbon monoxide gas to give carbon dioxide gas and solid iron metal. For every 2 moles of iron metal formed, 384 kilojoules of heat is absorbed by the reaction. Write the thermochemical equation for this reaction.

Solution First, we write a balanced equation for the reaction that shows the formation of 2 moles of Fe. The state (*s*, *l*, or *g*) must also be given for each substance.

$$Fe_2O_3(s) + 3\ CO(g) \longrightarrow 2\ Fe(s) + 3\ CO_2(g)$$

Because the reaction is endothermic, 384 kilojoules must be written on the left side. The thermochemical equation is

$$Fe_2O_3(s) + 3\ CO(g) + 384\ kJ \longrightarrow 2\ Fe(s) + 3\ CO_2(g)$$

The remaining sections of this chapter are optional.

* 5.4 THE ARITHMETIC OF THE ATOMS *(optional)*

In order to use the mole in chemistry, we must perform one of two kinds of arithmetic operations. The first is to calculate the weight in grams of a given number of moles of a substance. For example, what is the weight of 5.78 moles of water? The second arithmetic operation is to find the number of moles present in a given weight of a substance. For instance, how many moles are there in 14.3 grams of water? In doing arithmetic of this kind, it is easy to become confused over whether to multiply or divide the given number by the weight of 1 mole. In the two examples just given, is the gram molecular weight of water (18 grams per mole) multiplied by the given value or divided into the given value?

To make this question easy to answer, we use the **labeled-factor method.** According to this method, each calculation involves multiplying the given value, labeled with its units, by a fraction consisting of a numerator and a denominator, each of which also has a label. When the calculation is set up properly, some labels will cancel to give the desired label in the answer.

This is a rather abstract description. To illustrate the method, let's digress from chemistry and talk about the calculations of a grocery shopper. Say that on the shelf of your local grocery store are coffee cans of two sizes. A 1-pound can costs $2.91, whereas a 3-pound can costs $7.69. Which is the better buy—three 1-pound cans or one 3-pound can (see Figure 5.7)?

Although this question is simple and most of us would answer it intuitively, we'll use the labeled-factor method to calculate the cost of three 1-pound cans. The number we operate on is 3 pounds. This must be multiplied by a factor containing both a numerator and a denominator so as to obtain a total cost.

$$(3 \text{ pounds}) \left(\frac{2.91 \text{ dollars}}{1 \text{ pound}} \right) = 8.73 \text{ dollars}$$

We set up the factor so that the unit *pound* is in the denominator. Then this unit cancels the *pounds* in the number we are operating on (3 pounds), leaving *dollars* as the remaining unit, which is what we want. The 3-pound

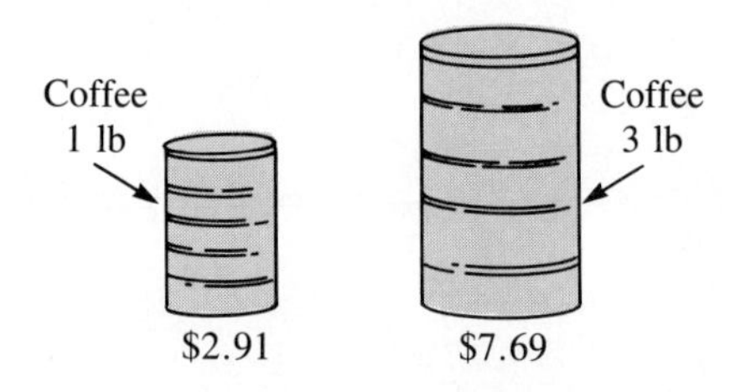

Figure 5.7. *Coffee cans of two sizes. Which is the better deal?*

can is cheaper than three 1-pound cans. Note that in the factor the numerator and the denominator give two properties of the same thing. The thing is a sample of coffee that weighs 1 pound and costs $2.91.

Example You are in a grocery store and have picked up a 5-pound bag of oranges. A clerk tells you that 1 dozen oranges weighs about 4 pounds. How many dozen oranges are in the bag?

Solution To use the labeled-factor method, you multiply the quantity 5 pounds by a fraction that will allow you to obtain a number of dozens in the answer. The factor will contain two properties of this particular sample of oranges—1 dozen and 4 pounds. One of these units must be in the numerator and the other in the denominator. By placing the 1 dozen in the numerator and the 4 pounds in the denominator, you can make the pound units cancel and end up with a number of dozens.

$$(5 \text{ pounds}) \left(\frac{1 \text{ dozen}}{4 \text{ pounds}}\right) = 1.25 \text{ dozen}$$

The 5-pound bag contains about 1.25 dozen oranges.

Now let's apply the labeled-factor method to two kinds of calculations involving numbers of moles. First, we will calculate the weight of a given number of moles of a substance. For example, what is the weight of 0.26 mole of Br_2 molecules? The given value (0.26 mole of Br_2) is multiplied by a labeled factor that comes directly from the definition of a mole: 1 gram molecular weight of the substance. In this case, the gram molecular weight of Br_2 is 160 grams. The labeled factor is either

$$\frac{1 \text{ mole } Br_2}{160 \text{ grams } Br_2}$$

or its reciprocal,

$$\frac{160 \text{ grams } Br_2}{1 \text{ mole } Br_2}$$

Because the desired answer is the weight of Br_2 in grams, the unit *moles* must cancel. The second labeled factor is what we want.

$$\text{Number of grams of } Br_2 = (0.26 \text{ mole } Br_2) \left(\frac{160 \text{ grams } Br_2}{1 \text{ mole } Br_2}\right)$$
$$= 42 \text{ grams } Br_2$$

The second kind of calculation is just the reverse of the first. Given a weight in grams, what is the corresponding number of moles? This time the labeled factor must have 1 mole in the numerator and the gram molecular weight or gram atomic weight in the denominator. We can calculate the number of moles in 63 grams of ammonia (NH_3, molecular weight 17) as follows:

$$\text{Number of moles of } NH_3 = (63 \text{ grams } NH_3)\left(\frac{1 \text{ mole } NH_3}{17 \text{ grams } NH_3}\right)$$
$$= 3.7 \text{ moles } NH_3$$

The secret of the method lies in setting up the labeled factor so that the units of the quantity given (grams, in the last example) cancel and the desired unit (in this case, moles) appears in the answer.

Example Calculate (a) the number of moles in 4 grams of O_2 and (b) the weight in grams of 0.15 mole of iron metal, Fe. Find the necessary atomic weights inside the front cover.

Solution (a)

$$\text{Number of moles of } O_2 = (4 \text{ grams } O_2)\left(\frac{1 \text{ mole } O_2}{32 \text{ grams } O_2}\right)$$
$$= 0.12 \text{ mole } O_2$$

(b)

$$\text{Weight of Fe} = (0.15 \text{ mole Fe})\left(\frac{56 \text{ grams Fe}}{1 \text{ mole Fe}}\right)$$
$$= 8.4 \text{ grams Fe}$$

*

* 5.5 WORKING WITH AVOGADRO'S NUMBER *(optional)*

Usually, chemists only need to know the number of moles of a substance. Sometimes, however, it is interesting to find the actual number of molecules or atoms in a sample. For example, even after a 1-liter flask has been evacuated by the best vacuum pump available, it still contains about 3×10^{10}, or 30 billion, air molecules. The weight of these air molecules, though, is only 1.5×10^{-12} gram, much less than can be weighed on the best analytical balance.

One additional labeled factor allows us to determine the number of molecules in a sample. It involves Avogadro's number and can be written as

$$\frac{6.02 \times 10^{23} \text{ particles (molecules, atoms)}}{1 \text{ mole of substance}}$$

To find the number of molecules in a given number of moles of substance,

OLD WATER MOLECULES

Each of us may be a descendant of the first prehistoric woman in a way that never occurred to us. Avogadro's number provides the key.

The human body is about two-thirds water. Assume that this early cavewoman contained the same fraction of water. When she died her body returned to the soil, and during the millions of years since then, there has been ample time for all the water molecules in her body to become uniformly distributed throughout the rivers, lakes, oceans, and atmosphere of the earth. In short, wherever there is water there is some chance that it may contain a water molecule from the first cavewoman. Is it likely that your body contains some of these water molecules? If so, how many?

For simplicity let's say that the early cavewoman had the same weight as a modern person—about 60 kilograms. The weight of water in her body was two-thirds of this, or 40 kilograms. Let's calculate the number of water molecules in the cavewoman's body. The 40 kilograms of water is the same as 40,000 grams.

$$\text{Number of water molecules} = (40{,}000 \text{ grams } H_2O)\left(\frac{1 \text{ mole } H_2O}{18 \text{ grams } H_2O}\right) \times \left(\frac{6 \times 10^{23} \text{ molecules } H_2O}{1 \text{ mole } H_2O}\right) = 1.3 \times 10^{27} \text{ molecules } H_2O$$

This is the number of water molecules released from the cavewoman's body and spread throughout all the water-containing objects of the world.

How many of these water molecules are in any one of our bodies? Because the present-day human also weighs about 60 kilograms and consists of two-thirds water, his or her body also contains 1.3×10^{27} molecules of water. Now we need to calculate the fraction of these molecules that came from the original cavewoman. Geologists estimate that the total amount of water on the earth is 1.3×10^{21} kilograms. Therefore, the fraction of this total that originally was in the cavewoman's body is

$$\frac{40 \text{ kilograms}}{1.3 \times 10^{21} \text{ kilograms}} = 3.0 \times 10^{-20}$$

This fraction multiplied by the number of water molecules in any of our bodies gives the number of water molecules from the cavewoman that now belong to each of us.

$$\text{Number of molecules from original cavewoman} = (3.0 \times 10^{-20})(1.3 \times 10^{27} \text{ molecules}) = 4 \times 10^{7}, \text{ or 40 million, molecules}$$

Each of our bodies contains *40 million* water molecules from the first prehistoric woman. Furthermore, whenever anyone drinks a soda or a beer, he or she ingests 360,000 more of her molecules!

we simply multiply the number of moles by this labeled factor. If the amount of substance is given as a weight, an extra step is involved.

Example It has been estimated that the total amount of oxygen in the earth's atmosphere weighs 1.2×10^{21} grams. How many molecules are in this weight of O_2?

Solution We begin with the given information, 1.2×10^{21} grams, and multiply by a labeled factor to change this value into a number of moles.

$$\text{Number of moles of } O_2 = (1.2 \times 10^{21} \text{ grams } O_2)\left(\frac{1 \text{ mole } O_2}{32 \text{ grams } O_2}\right)$$
$$= 3.7 \times 10^{19} \text{ moles } O_2$$

Next, we convert the number of moles into the number of molecules through use of a labeled factor involving Avogadro's number.

$$\text{Number of molecules of } O_2 = (3.7 \times 10^{19} \text{ moles } O_2)\left(\frac{6.02 \times 10^{23} \text{ molecules } O_2}{1 \text{ mole } O_2}\right)$$
$$= 2.2 \times 10^{43} \text{ molecules } O_2$$

It is not really necessary to calculate the number of moles of O_2, because this value is an intermediate rather than a final answer. Instead, we can combine the two steps above into one.

$$\text{Number of molecules of } O_2$$
$$= (1.2 \times 10^{21} \text{ grams } O_2)\left(\frac{1 \text{ mole } O_2}{32 \text{ grams } O_2}\right)\left(\frac{6.02 \times 10^{23} \text{ molecules } O_2}{1 \text{ mole } O_2}\right)$$
$$= 2.2 \times 10^{43} \text{ molecules } O_2$$

*

* 5.6 HOW MUCH PRODUCT IS FORMED? *(optional)*

Many chemical reactions are constantly occurring in nature, in industry, and in the laboratory. The amount of product that can form in any of these reactions depends on how many molecules of the various reactants were originally present. Rather than deal directly with numbers of molecules, we again use the concept of the mole.

The labeled-factor method makes it easy to calculate the number of moles of product produced from a given number of moles of reactant. For example, how many moles of HCl will form from 2.4 moles of H_2 and an excess of Cl_2 in the reaction

$$\underset{1 \text{ mole}}{H_2(g)} + \underset{1 \text{ mole}}{Cl_2(g)} \longrightarrow \underset{2 \text{ moles}}{2\,HCl(g)}$$

The factor that is multiplied by 2.4 moles of H_2 comes from the chemical equation

$$\text{Number of moles of HCl formed} = (2.4 \text{ moles } H_2)\left(\frac{2 \text{ moles HCl}}{1 \text{ mole } H_2}\right)$$
$$= 4.8 \text{ moles HCl}$$

The chemical equation tells us that 2 moles of HCl will form from 1 mole of H_2. From these two we select for the numerator of the labeled factor the one containing *moles HCl*, which is what we want in the result. The denominator then has *moles* H_2, which cancels the other *moles* H_2. The method is the same as the one we used before.

Example Under suitable conditions nitrogen and hydrogen react to form ammonia according to the equation

$$N_2(g) + 3\ H_2(g) = 2\ NH_3(g)$$

Much of the chemical fertilizer used on farms in the United States is synthesized through this reaction. How many moles of NH_3 will be obtained from 0.73 mole of H_2 if there is more than enough N_2 present to react with all of the H_2?

Solution The amount of NH_3 formed is determined by the amount of H_2 originally present, because there is an excess of N_2 available. The labeled factor involves the relationship between moles of NH_3 and moles of H_2.

$$\text{Number of moles of } NH_3 \text{ formed} = (0.73 \text{ mole } H_2)\left(\frac{2 \text{ moles } NH_3}{3 \text{ moles } H_2}\right)$$

$$= 0.49 \text{ mole } NH_3$$

In this kind of problem we treat the labeled factor containing numbers of moles of two different substances in the same way as we handled the factor containing dozens and pounds in an earlier problem. The unit we want in the answer is in the numerator, and the unit we want to cancel is in the denominator.

Three steps are necessary to calculate the weight in grams of product formed from the weight of reactant initially present. First, the weight of reactant must be converted into the corresponding number of moles of reactant. Then the relationship between the moles of reactant and moles of product (given by the chemical equation) can be used. Finally, the calculated number of moles of product formed must be changed into a weight. Each of these steps has been done in the preceding problems. We will show the overall procedure in the next section with an example based on the burning of coal in a power plant. *

* 5.7 FUELS AND THE ATMOSPHERE *(optional)*

The element sulfur is a bright yellow powder. When sulfur burns, the product is a colorless, pungent, and toxic gas called sulfur dioxide. The chemical equation for this reaction is

$$\underset{\substack{\text{sulfur} \\ \text{1 mole}}}{S} + \underset{\substack{\text{oxygen} \\ \text{1 mole}}}{O_2} \longrightarrow \underset{\substack{\text{sulfur dioxide} \\ \text{1 mole}}}{SO_2}$$

One of our most important fuels is coal, from which we produce much of the electricity in the United States. The heat from burning coal converts water into steam, which drives a turbine, which in turn operates an electrical generator. Unfortunately, however, the coal we use contains small amounts of contaminants, one of which is sulfur. Sulfur is present in amounts ranging from 1% to 7% of the weight of the coal itself, depending on the source of the coal.

The fuel requirements of a coal-burning electrical generating plant are seldom described in terms of moles of coal but rather in terms of tons of coal. A modern 1000-megawatt power plant requires about 400 tons of coal per hour of full-capacity operation. At this level of consumption even small percentages of undesirable sulfur impurities can cause a serious air-pollution problem (see Figure 5.8). Let's determine the amount of sulfur dioxide coming from the smokestack of the 400-ton-per-hour plant when it burns coal containing 1% sulfur, which is often referred to as "clean" coal.

How much sulfur is contained in the coal burned in this plant in one hour? If the sulfur is 1% of the coal by weight,

$$400 \text{ tons coal} \times 0.01 = 4 \text{ tons sulfur}$$

Figure 5.8. Polluted air over a city.

SMOKESTACK EMISSIONS

In order to comply with government anti-pollution regulations, a power plant must either remove sulfur from the coal it uses or decrease the amount of sulfur dioxide gas escaping from the smokestack. Processes are under development both for washing the sulfur out of the coal before it's burned and for scrubbing the gases leaving the stack. Either process adds to the cost of the fuel consumed and increases the cost of the electricity generated.

Power plants in the eastern United States are at a particular disadvantage. The cleanest coal mined in the United States comes from the states just east of the Rocky Mountains; it contains about 1% sulfur. If the power companies in the East use this coal, they must transport it at great cost. It is cheaper for them to burn the more easily available high-sulfur coal from the eastern mines. Unfortunately, this coal contains as much as six times more sulfur than is found in western coal. Therefore the cost of controlling sulfur dioxide emissions is higher for the eastern coal. It is no wonder that electricity is especially expensive in the East.

So, every hour 4 tons of sulfur will be fed into the fireboxes of the boilers along with 396 tons of coal (carbon). In grams this amount of sulfur is

$$(\text{4 tons sulfur})\left(\frac{9.1 \times 10^5 \text{ grams}}{1 \text{ ton}}\right) = 3.6 \times 10^6 \text{ grams sulfur}$$

The strategy for calculating the amount of sulfur dioxide produced involves three parts:

$$\text{grams S} \longrightarrow \text{moles S} \longrightarrow \text{moles } SO_2 \longrightarrow \text{grams } SO_2$$

Each arrow stands for a single calculation with the use of one labeled factor. We could do each calculation separately, but it is simpler to combine them.

$$\begin{aligned}&\text{Weight of } SO_2 \text{ formed}\\ &\quad = (3.6 \times 10^6 \text{ grams S})\left(\frac{1 \text{ mole S}}{32 \text{ grams S}}\right)\left(\frac{1 \text{ mole } SO_2}{1 \text{ mole S}}\right)\left(\frac{64 \text{ grams } SO_2}{1 \text{ mole } SO_2}\right)\\ &\quad = 7.2 \times 10^6 \text{ grams } SO_2\end{aligned}$$

Here we have used the atomic weight of S, 32, and the molecular weight of SO_2, 64. After use of the first labeled factor, cancellation of *grams S* against *grams S* gives *moles S*. Similarly, after using the second labeled factor we get *moles* SO_2, and after the third, *grams* SO_2. With practice, it becomes easy to set up the appropriate factors. The calculation shows that the power plant emits 7.2×10^6 grams of SO_2 each hour.

It may be informative to calculate the volume of this weight of SO_2. We can do so by using a fact that is based on some experimental measurements and Avogadro's Hypothesis. The measurements involve determining the volume of 1 mole of several common gases at a temperature of 0°C (273 K) and a pressure of 1 atmosphere. For each gas the volume is found to be approximately **22.4 liters** under these conditions. The volume should be the same for each gas, because 1 mole of any gas has the same number of molecules, and equal numbers of molecules are in equal volumes, according to Avogadro's Hypothesis. The combination of temperature and pressure specified above, 273 K and 1 atmosphere, is used often enough to be given a special name—standard temperature and pressure, abbreviated **STP.** Because of Avogadro's Hypothesis, we can be confident that 1 mole of any gas at STP has a volume close to 22.4 liters, whether or not its volume has been measured.

Suppose that the amount of SO_2 produced in 1 hour could be collected in a large vessel at 1 atmosphere pressure and a temperature of 273 K. What would its volume be?

$$\text{Number of moles of } SO_2 \text{ formed} = (7.2 \times 10^6 \text{ grams } SO_2)\left(\frac{1 \text{ mole } SO_2}{64 \text{ grams } SO_2}\right)$$

$$= 1.12 \times 10^5 \text{ moles } SO_2$$

$$\text{Volume of } SO_2 \text{ at STP} = (1.12 \times 10^5 \text{ moles } SO_2)\left(\frac{22.4 \text{ liters } SO_2}{1 \text{ mole } SO_2}\right)$$

$$= 2.5 \times 10^6 \text{ liters } SO_2$$

In other words, the volume of SO_2 coming out of the smokestack in 1 hour would be 2,500,000 liters at STP if the SO_2 could be segregated from the other gases with which it is mixed.

Sulfur dioxide pollution has corrosive effects on stone and metal surfaces, damages plants, and affects human health. Furthermore, it is a major cause of acid rain, discussed in Chapter 12.

Example In a coal-burning power plant some of the nitrogen in the air used for combustion reacts with the oxygen present to form nitric oxide, NO. The reaction is

$$N_2 + O_2 \longrightarrow 2\,NO$$

The nitric oxide is another agent responsible for acid rain. According to one estimate, 1×10^{10} grams of N_2 reacts this way in a 1000-megawatt power plant in 1 year. What weight of NO results?

Solution The molecular weights of N_2 and NO are 28 and 30, respectively. Using the labeled-factor method, we find

Number of grams of NO formed

$$= (1 \times 10^{10} \text{ grams } N_2)\left(\frac{1 \text{ mole } N_2}{28 \text{ grams } N_2}\right)\left(\frac{2 \text{ moles NO}}{1 \text{ mole } N_2}\right)\left(\frac{30 \text{ grams NO}}{1 \text{ mole NO}}\right)$$

$$= 2 \times 10^{10} \text{ grams NO}$$

*

IMPORTANT TERMS

Avogadro's Hypothesis Equal volumes of gases, measured at the same temperature and pressure, contain equal numbers of molecules.

mole A mole of any substance contains the same number of particles, $6.02 \cdot 10^{23}$, as the number of carbon atoms in 12.00 grams of carbon. In more practical terms, a mole of molecules, atoms, or formulas is 1 gram molecular weight, 1 gram atomic weight, or 1 gram formula weight, respectively.

gram molecular weight The weight in grams equal to the molecular weight of a substance.

gram atomic weight The weight in grams equal to the atomic weight of an element.

gram formula weight The weight in grams equal to the formula weight of an ionic compound.

Avogadro's number The number of particles (molecules, atoms, or formula units) in a mole of a substance. Its numerical value is $6.02 \cdot 10^{23}$.

thermochemical equation An equation for a reaction that shows the quantity of heat either absorbed or evolved for the stated amount of reaction.

Optional Sections

labeled-factor method A method of calculation in which an initial value is multiplied by a factor having a numerator and a denominator. The numerator is labeled with the unit desired in the answer. The denominator is labeled with the unit to be canceled. Both the numerator and the denominator refer to the same quantity.

22.4 liters The volume of 1 mole of gas (not liquid or solid) at STP.

STP Standard temperature and pressure, or 0°C (273 K) and 1 atmosphere.

QUESTIONS

1. The following volumes of gases were measured at 0°C and 0.58 atmosphere pressure. Arrange these volumes in order of increasing number of molecules: 450 milliliters CH_4, 25 milliliters N_2, and 0.039 liter CO_2.

2. For the following pairs of samples, determine which sample has the greater number of molecules. Show how you arrive at your answers.

a. 50 grams of chlorine gas (Cl_2) or 0.5 mole of water

b. 100 grams of copper or 100 grams of chromium. (Compare the number of atoms.)

c. 71 grams of NO_2 or 2 grams of H_2

3. Calculate the weight in grams of 1 mole of each of the following:

a. Na atoms
b. I_2 molecules
c. N_2O_4 molecules
d. NaCl (ionic compound)
e. SF_6 molecules
f. Ar atoms

4. a. When 1 mole of octane, C_8H_{18}, a component of gasoline, burns inside an automobile cylinder,

5450 kilojoules of heat is evolved. Carbon dioxide and water are the products. Write a balanced thermochemical equation for the combustion of 1 mole of liquid octane to give carbon dioxide gas and liquid water.

b. Glucose, $C_6H_{12}O_6$, "burns" inside the human body by reacting with oxygen. Again, carbon dioxide and water are the products. We know that 2800 kilojoules of heat is released for each mole of glucose that reacts. Write the thermochemical equation for the reaction of 1 mole of solid glucose with oxygen to give carbon dioxide gas and liquid water.

c. Lime (which is solid CaO) is manufactured from limestone (solid $CaCO_3$) by heating the limestone. Carbon dioxide gas is the other product. A total of 178 kilojoules of heat is absorbed by the reaction per mole of lime produced. Write the thermochemical equation for the reaction.

5. For each of the following reactions, tell (1) whether the reaction is endothermic or exothermic and (2) whether the products have a greater or a smaller potential energy than the reactants.

a. $3\ O_2(g) + 285\ kJ \longrightarrow 2\ O_3(g)$

oxygen → ozone

b. $2\ CO(g) + O_2(g) \longrightarrow 2\ CO_2(g) + 566\ kJ$

Optional Sections

6. a. How many moles are there in 15.0 grams of helium, He?

b. How many moles of I_2 molecules are there in 100 grams of iodine?

c. Calculate the number of moles of methane, CH_4, in 47.5 grams of that compound.

7. a. What is the weight of 2.5 moles of tin, Sn?

b. What is the weight of 0.61 mole of carbon tetrachloride?

8. The human body is approximately 70% water by weight. A typical person weighs 75 kilograms. How many molecules of water are there in a typical person?

9. The elements of highest atomic number can only be made artificially. When californium was first made, only ten atoms were detected. If the atomic weight of californium is 246, what is the weight of ten atoms?

10. When gasoline is consumed in an internal combustion engine, most of the carbon atoms from the gasoline end up in carbon dioxide molecules. A few form carbon monoxide, an undesirable pollutant. Although gasoline is a mixture of organic compounds, one important component is heptane, C_7H_{16}. The following equation approximates the desired complete combustion of heptane to carbon dioxide and water.

$$C_7H_{16}(l) + 11\ O_2(g) \longrightarrow 7\ CO_2(g) + 8\ H_2O(g)$$

Calculate the mass in grams of carbon dioxide that would be formed from 700 grams (0.26 gallon) of C_7H_{16}.

11. One of the ores of copper is $CuFeS_2$. It reacts with oxygen in a furnace to give copper metal according to the equation

$$2\ CuFeS_2 + 5\ O_2 \longrightarrow 2\ Cu + 2\ FeO + 4\ SO_2$$

Sulfur dioxide (SO_2) is a prime producer of acid rain, which may destroy our wilderness lakes. If only 10% of the SO_2 gas is allowed to escape into the environment, how many grams of SO_2 will go out of the chimney for every kilogram of $CuFeS_2$ processed?

12. The smallest mass that can possibly be detected on most analytical balances used in the laboratory is about 0.0002 gram. How many atoms are there in this mass of iron?

13. a. What is the volume of 0.071 mole of neon gas when that volume is measured at STP?

b. How many moles are there in 1084 milliliters of nitrogen gas measured at 273 K and 1 atmosphere pressure?

14. Astronomers have discovered in interstellar space both "diffuse" and "dark" clouds, which consist of tiny solid particles and gas—mainly hydrogen. In a dark cloud there might be 10,000 H_2 molecules per milliliter. If the smallest amount of matter that can be weighed is $1 \cdot 10^{-6}$ gram, could the amount of hydrogen in 1.0 liter of a dark cloud be weighed?

Atomic Structure—Ions and Electrons

The chief difference John Dalton saw between the atoms of each element was in their masses. Why these masses are different and why one element combines with another in a compound were questions he could only wonder about. Many years later scientists began to realize that the properties of atoms might be related to their internal structures. In other words, the visible properties of elements depend on the structures of those incredibly small atoms. Several observations suggest that atoms are not the indivisible bodies that Dalton and his friends believed in.

Anyone who has lived in a cold, dry climate knows the unpleasant tingle of the spark that leaps between one's hand and a doorknob after one has shuffled across a carpet inside a warm house. Even more painful is the shock of 110-volt house current when one accidentally touches a metal contact in an electrical outlet. And we all know that high voltages can kill. A dramatic example of electricity occurs in nature whenever a thunderstorm causes lightning to flash through the sky. In one of his more famous experiments, conducted during a storm in 1752, Benjamin Franklin flew a kite to which was attached a metal wire. At the lower end of the kite string he fastened a key. Whenever a storm cloud passed overhead, Franklin was able to draw sparks from the key, showing that electricity was conducted from the wire through the wet kite string to the key. These phenomena, as well as the simple process of turning on a lamp by flipping a switch, are examples of the flow of charged particles through an electrical circuit. As we shall see, the study of electrical phenomena was important in understanding the structure of the atom.

During a change of state, such as from liquid to gas, atoms seem to retain their identities. It is in connection with phenomena such as electrical conduction or the extraordinary events of radioactive decay that atomic structure is revealed. For example, the element radium, which has been used in medical radiation treatments and in the luminous dials of watches, slowly emits highly energetic particles and radiation. This radioactive decay involves the breaking apart of atoms, which suggests that atoms are composed of simpler units.

6.1 ENTER THE ELECTRON

Over 2300 years ago Plato wrote of the strange power of amber—the hard, yellow, fossilized resin that the Greeks called "elektron." Amber is able to attract bits of hair and dust after it has been rubbed with fur. Many substances become electrically charged when rubbed. We now call this property *static electricity*. Sometimes they attract each other; in other cases there is a mutual repulsion. From such observations has arisen the idea that there are two kinds of charge: positive and negative. Unlike charges attract; like charges repel. We made use of this fact in Chapter 4 when discussing potential energy. Furthermore, when charges move between the poles of a magnet, a force acts on the charges and causes changes in their

Figure 6.1. *A gas discharge tube.*

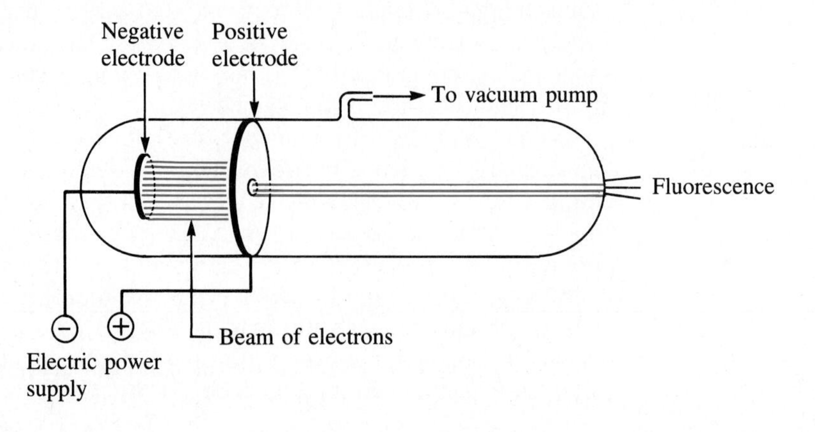

ELECTRON BEAMS IN SCIENCE FICTION

The interplay of science and fantasy in science fiction attracts a host of fans. Some of the fantasy even seems to be authentic. An electron-beam weapon was described so realistically in 1947 in George O. Smith's "Venus Equilateral" that a good deal of scientific interest in particle beams resulted.

This scene from "Venus Equilateral," in which the electron-beam gun is introduced, has two characters—engineer Walter Franks and his secretary, Jeanne.

Walter: Now I've been trying to devise a space gun that will blast meteors directly instead of avoiding them by coupling the meteor detector to the autopilot.

Jeanne: Gonna shoot 'em out of existence?

Walter: Not exactly. Popping at them with any kind of rifle would be like trying to hit a flying bird with a spitball. Look, Jeanne, speed on the run from Mars to Terra at major opposition is up among the thousands of miles per second at the turnover. A meteor itself may be blatting along at fifty miles per second. Now a rifle, shooting a projectile at a few thousand feet per second, would be useless. You'd have the meteor in your lap and out of the other side while the projectile is making up its mind to move forward and relieve the pressure that is building up behind it due to the exploding powder.

I've designed an electron gun. It is a superpowered, over-sized edition of the kind they used to use in kinescope tubes, oscilloscope tubes, and electron microscopes. Since the dingbat is to be used in space, we can leave the works of the gun open and project a healthy stream of electrons at the offending object without their being slowed and dispersed by an impeding atmosphere.

Jeanne: But that sounds like shooting battleships with a toy gun.

Walter: Not so fast on the objections, gal. I've seen a simple oscilloscope tube with a hole in the business end. It was burned right through a quarter inch of glass because the fellows were taking pix and had the intensity turned up high.

paths. A beam of positive charges is bent in one direction, and a beam of negative charges is bent in the opposite direction.

Over the course of the nineteenth century, experiments with gas discharge tubes revealed more and more properties of electrical charges, particularly the negative ones. A simplified diagram of a discharge tube appears in Figure 6.1. Whether the gas inside the tube is air, hydrogen, neon, or any other gas, nothing seems to happen until the pressure is lowered substantially. Then the high voltage between the **electrodes** (or conductors) causes a current to flow, and the tube glows, much like the neon signs that light up our cities.

In addition, the glass near the positive electrode emits a different kind of light called *fluorescence;* the same process is at work in the modern television tube. If the end of the tube is coated with a chemical such as zinc sulfide, the chemical also emits light. Careful investigation of fluorescence shows that it consists of separate flashes of light, as if a beam of individual particles were hitting the chemical. After some controversy as to what was going on in gas discharge experiments, scientists agreed that the beam is a stream of particles that travels from the negative to the positive electrode. From the direction in which the beam is bent by a magnet, it became evident that the beam consists of negative particles. We now call these negative particles **electrons**.

Credit for the discovery of the electron is usually given to J. J. Thomson, a physicist at Cambridge University, who published an important description of his work in 1897. By subjecting electron beams to both electric and magnetic fields, Thomson was able to determine the ratio of the charge to the mass of the particles in the beam. The value he came up with for the ratio was disbelieved at first by many other scientists because it suggested either a very large electric charge or else a surprisingly small mass. Later, when the charge alone was accurately measured, the mass could be calculated from the ratio, and it proved to be about 1/1840 of the mass of the hydrogen atom, the lightest atom known. So electrons are very, very light particles.

The fact that the properties of the negatively charged particles are always the same, regardless of the identity of the gas in the discharge tube, indicates that electrons must be present in all atoms. Nowadays chemists think that many of the key properties of an atom are related to its internal arrangement of electrons. In chemistry, the twentieth-century might well be called the century of the electron.

6.2 THE CASE FOR CHARGED ATOMS—IONS

We picture the atom as being electrically neutral, having a mass thousands of times greater than the mass of an electron, and containing negatively charged electrons. But there must be something else within atoms besides electrons. There is a great deal of mass to be accounted for, and a positive charge must be present to balance the negative charge. To find out about these positively charged particles, we return to a study of electricity.

Since the earliest days of chemical science it has been known that pure water is a poor conductor of electricity; yet a water solution of a salt such as sodium chloride, NaCl, or potassium sulfate, K_2SO_4, is a good conductor. Other solutions, such as sugar dissolved in water, are poor conductors of electricity. By the late nineteenth century most chemists were agreed that solutions conduct electricity because positively and negatively charged particles move between the electrodes and carry the electrical current. Figure 6.2 shows how this happens.

In his 1884 Ph.D. dissertation, Svante Arrhenius (Figure 6.3), a Swedish chemist, presented the hypothesis that the charged atoms, which he called **ions,** form in the solution when the salt dissolves in water. Many of his

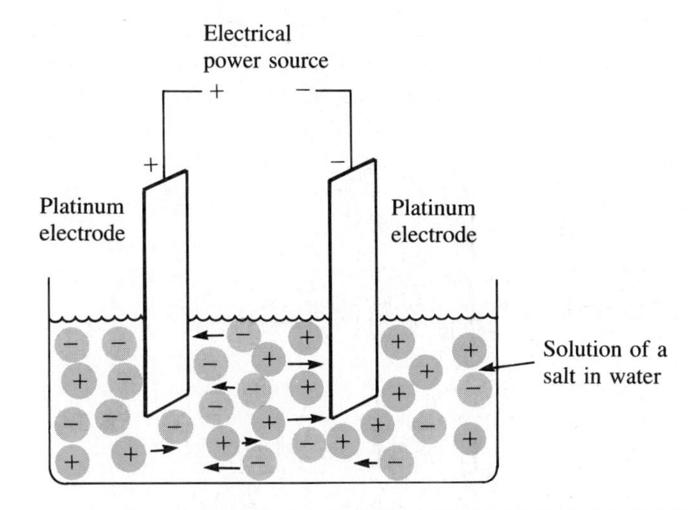

Figure 6.2. Electrical conduction in solution. Charged atoms between the electrodes move toward the electrode of opposite charge.

Figure 6.3. Svante Arrhenius.

contemporaries believed that charges resided in atoms only when current was flowing. The idea that oppositely charged ions would separate from each other in a solution of a salt was revolutionary. Because of the attraction between these charges, much work would be required to separate them, as we saw in Chapter 4. Arrhenius's professors thought so little of his hypothesis that they delayed granting him a degree. Arrhenius described his professor's reception of his idea as follows:

> I came to my professor, Cleve, whom I admire very much, and I said: "I have a new theory of electrical conductivity as a cause of chemical reactions." He said: "That is very interesting," and then said, "Good-bye." He explained to me later, when he had to pronounce the reason for my receiving the Nobel prize, that he knew that there are many different theories formed which are all almost certain to be wrong, for after a short time they disappear. Therefore, by using the statistical manner of forming his ideas, he concluded that my theory also would not exist very long.

After graduation Arrhenius moved on to other laboratories where the scientific community was more open-minded. Nineteen years after his Ph.D. dissertation was finally accepted, Arrhenius was given the Nobel Prize in Chemistry for his revolutionary and powerful view of nature. By then even his professors had to admit that they had been wrong.

Arrhenius found support for his ideas about ions in data on the freezing points of salt-water solutions. Pure water, of course, freezes at 0°C. A salt or sugar solution, on the other hand, always freezes at a lower temperature. For example, a solution containing 0.1 mole of sugar in 1000 grams of water freezes at −0.19°C. If the amount of sugar is doubled, to 0.2 mole in 1000 grams of water, the freezing point becomes −0.38°C (Table 6.1). Doubling the amount of sugar just about doubles the depression in the freezing point. Other experiments verify that the reduction in the freezing point is directly proportional to the number of moles of sugar dissolved in a fixed amount of water. Since a mole measures a definite number of molecules, it follows that the reduction in the freezing point is proportional to the relative number of sugar molecules in solution.

Sodium chloride solutions present a somewhat different picture. A solution containing 0.1 mole of NaCl in 1000 grams of water freezes at −0.35°C; 0.2 mole of NaCl in the same amount of water freezes at approximately −0.69°C. The reductions in the freezing point of salt solutions are nearly twice as great as those for nonionic solutions, such as sugar or antifreeze (see Table 6.1). Arrhenius argued that when 0.1 mole (0.1 gram formula weight) of NaCl is dissolved in 1000 grams of water, the NaCl splits apart, or **ionizes,** into 0.1 mole of Na^+ ions and 0.1 mole of Cl^- ions. (Naturally, there must be an equal number of Na^+ ions and Cl^- ions, since NaCl has no overall electrical charge.) The total number of ions from 0.1 mole of NaCl is 0.2 mole, and the freezing point is about the same as when 0.2 mole of sugar or antifreeze is present. It again appears that the reduction

Table 6.1 Freezing Points of Aqueous Solutions

SUBSTANCE (moles per 1000 grams water)	FREEZING POINT OF SOLUTION (°C)
Nonionic	
sucrose (sugar), 0.1 mole	−0.19
sucrose, 0.2 mole	−0.38
ethylene glycol (antifreeze), 0.1 mole	−0.19
ethylene glycol, 0.2 mole	−0.37
ethylene glycol, 0.3 mole	−0.56
Ionic	
NaCl, 0.1 mole	−0.35
NaCl, 0.2 mole	−0.69
NaCl, 0.3 mole	−1.05
KI, 0.1 mole	−0.35
$CoCl_2$, 0.1 mole	−0.50

in the freezing point is proportional to the relative number of particles (molecules or ions) dissolved in the water.

$$\begin{array}{lcl} \text{NaCl} & \longrightarrow & \text{Na}^+ \quad + \quad \text{Cl}^- \\ 0.1\text{ mole} & \longrightarrow & 0.1\text{ mole} + 0.1\text{ mole} \\ 0.1\text{ mole} & \longrightarrow & 0.2\text{ mole of particles} \end{array}$$

You may have noticed that the freezing point depressions for sodium chloride solutions are not exactly twice as great as those for nonionic solutions. The reason for this is that not all Na^+ and Cl^- ions are completely independent particles in solution. A very small percentage of them are paired, and each pair behaves like a single particle. So the total number of particles is less than it would be if dissociation were complete.

Example (a) According to Table 6.1, a solution of 0.1 mole of $CoCl_2$ (cobalt chloride) in 1000 grams of water freezes at −0.50°C. Predict the freezing point of a solution of 0.2 mole of $CoCl_2$ in 1000 grams of water.

(b) Write a chemical equation, based on the observed freezing point of the $CoCl_2$ solution, for the ionization of $CoCl_2$ in water.

Solution (a) If 0.1 mole of $CoCl_2$ in 1000 grams of water produces a freezing-point depression of −0.50°C, you would expect 0.2 mole of $CoCl_2$ to produce a freezing-point depression twice as large:

$$2 \cdot (-0.50°C) = -1.0°C$$

The measured value for the freezing point is $-0.98°C$. ($CoCl_2$, by the way, dissolves in water to give a pretty pink solution.)

(b) Since 0.1 mole of sugar, which does not ionize, produces a freezing point of $-0.19°C$, we can see immediately that $CoCl_2$ (0.1 mole), with its freezing point of $-0.50°C$, must ionize. But how many moles of ions are formed? A ratio of the freezing points will provide the answer:

$$\frac{-0.50}{-0.19} = 2.7$$

It looks as if almost three times as many particles are formed, so the answer is

$$CoCl_2 \longrightarrow Co^{2+} + 2\,Cl^-$$

$$0.1 \text{ mole} \longrightarrow 0.1 \text{ mole} + 0.2 \text{ mole}$$

$$0.1 \text{ mole} \longrightarrow 0.3 \text{ mole of ions}$$

By the early years of the twentieth century most chemists and physicists accepted the existence of ions. They also agreed that a positive ion results from the loss of one or more electrons from an atom and a negative ion results from the addition of one or more electrons to an atom. Although experiments on the electrical conduction and freezing points of solutions are very different from those involving gas discharge tubes, they also reveal the complex nature of atoms. Once more we see that there must be some positive particles in atoms, along with the negative electrons. See Table 6.2 for some examples of positive and negative ions.

The presence of ions in certain kinds of compounds has very practical consequences. Many metals on the surface of the earth are in the form of oxides or sulfides, which consist of positive metal ions and negative oxygen or sulfur ions. For instance, aluminum oxide, Al_2O_3, contains positive aluminum ions and negative oxygen ions. Most of the aluminum in this country is produced by passing an electric current through a molten mixture of

Table 6.2 Some Ions Commonly Encountered in Chemistry

POSITIVE IONS		NEGATIVE IONS
Na^+	Co^{2+}	F^-
K^+	Ni^{2+}	Cl^-
Mg^{2+}	Zn^{2+}	Br^-
Ca^{2+}	Ag^+	I^-
Cr^{3+}	Sn^{2+}	S^{2-}
Fe^{2+}	Pb^{2+}	OH^-
Fe^{3+}	Al^{3+}	O^{2-}

bauxite, an ore rich in aluminum oxide, and another aluminum ore. Positive aluminum ions receive electrons at the negative electrode to become aluminum metal.

6.3 RADIOACTIVE ELEMENTS

The existence of **radioactivity** gives strong support to the idea that atoms are not the smallest particles of nature. In 1896, quite by accident, Henri Becquerel, Professor of Physics at the Museum of Natural History in Paris, discovered that photographic plates that were covered with paper and left next to crystals of a uranium compound had dark spots when developed later. It was as though the plates had been placed near a source of x-rays—a then recently discovered, highly penetrating type of radiation. Yet the uranium compound did not seem to have anything in common with an x-ray tube. After more experiments Becquerel found that uranium in any form (a solid compound, a compound dissolved in water, or the pure metal) gives off those penetrating rays. The presence of uranium atoms in any substance causes the appearance of these rays.

Figure 6.4. *Marie Sklodowska Curie and her daughter Irène.*

USES OF RADIOACTIVE NUCLEI IN MEDICINE

The field of nuclear medicine has become an important specialty within the last few years. Radioactive nuclei have many applications in our hospitals for diagnosis and for treatment. Injection of a radioactive tracer in the bloodstream makes it possible for physicians to see a patient's heart almost as if they were watching a motion picture of it. By monitoring the blood flow through the chambers of the heart, they can detect congenital abnormalities. Emitters of radioactive gamma rays are used, and the emission pattern is read by a scintillation counter outside the body, with computer enhancement of the signals. The scintillation method involves chemicals that give off light when struck by gamma rays.

A radioactive isotope of iodine is applied in a routine diagnostic test for abnormalities in the thyroid gland and in treating cancer of the thyroid. Moreover, radium is used in treating cancer of the uterus and radioactive phosphorus in treating leukemia.

Other scientists became interested in Becquerel's discovery, and after a few years other new radioactive elements, such as polonium and radium, were isolated. Both polonium and radium were discovered in Paris by Marie (see Figure 6.4) and Pierre Curie in 1898. Starting with a large amount of an ore called *pitchblende*, the Curies labored many months in order to separate small amounts of polonium and radium compounds. Each element was identified by its characteristic radioactivity.

It soon became evident that not all the rays from these elements were the same. On the basis of the ways in which the rays were bent or not bent by a magnet and the thickness of metal required to stop them, **alpha** (α) **rays, beta** (β) **rays,** and **gamma** (γ) **rays** were identified (see Table 6.3). Figure 6.5(a) shows that there are two kinds of rays emitted from ^{238}U (an atom of uranium having a mass of 238 on the atomic weight scale). One kind of ray is deflected in a magnetic field, which suggests that it is electrically charged. The other kind is not bent. In Figure 6.5(b) we see the result of a similar experiment performed with radioactive thorium, ^{234}Th, which is the element left after ^{238}U decays. In this case one kind of ray is bent a great deal and in the direction opposite to that observed for the ray from ^{238}U. We now know that the alpha rays (from U) are positively charged and the beta rays (from Th) are negatively charged.

Table 6.3 Particles and Rays Given Off by Radioactive Atoms

PARTICLE OR RAY	CHARGE	MASS (on atomic weight scale)
alpha (α)	+2	4
beta (β)	−1	1/1840
gamma (γ)	0	0

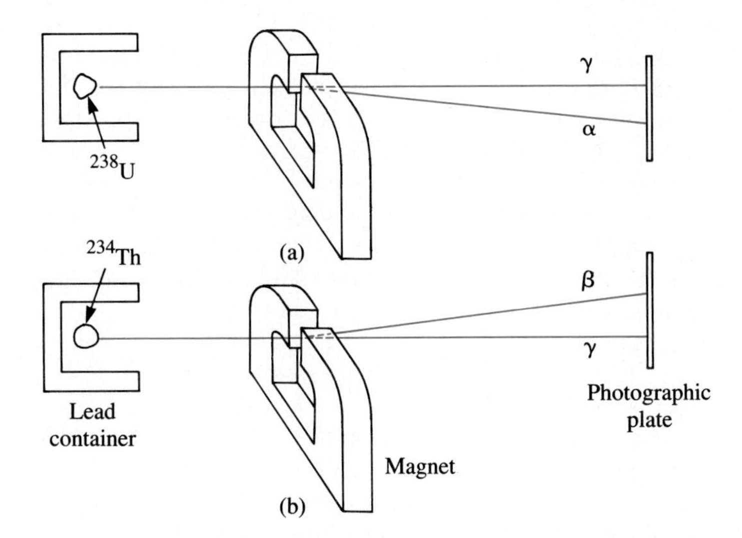

Figure 6.5. Effect of a magnet on radioactive emissions.

Later investigations revealed that the alpha-ray particles are positive helium ions (He^{2+}) that have been expelled from radioactive atoms with large kinetic energies. Their charge is twice that of a single electron but opposite in sign. Therefore, an alpha particle is a helium atom that is lacking two electrons. A beta particle is nothing more than a highly energetic electron; its path is bent by a magnet much more than that of an alpha particle because of the much smaller mass of the electron. The alpha particle, being much heavier, resists changes in its motion. Finally, the gamma rays are not particles at all but rather a form of radiation somewhat like visible light or x-rays. The energy associated with gamma rays, however, is much greater than the energy of visible light and considerably greater than that of x-rays. It is hard to see how alpha or beta particles could break away from the atoms of radioactive elements if the atoms were not made from smaller units. Table 6.3 summarizes the properties of alpha and beta particles and gamma rays.

6.4 THE NUCLEAR ATOM

How are the electrons and other particles arranged within an atom? During the early 1900s this question brought forth various ideas from scientists. One person suggested than an atom might resemble the planet Saturn, with a sphere of positive charge and a ring of negative charge. Other people believed the atom to be a uniform sphere of positive charge with negative charges embedded at regular intervals.

In 1911, at the University of Manchester in England, Ernest Rutherford (see Figure 6.6) ended these speculations by giving a convincing interpretation to experiments he and his colleagues had done. In one experiment, alpha particles were directed toward a very thin gold leaf, and the paths taken by the particles emerging from this foil were studied. Figure 6.7 shows a beam of particles from a radioactive source coming from the left toward the gold leaf. The eyepiece on the right contains a zinc sulfide screen, which can be placed at any angle. Remember that zinc sulfide will give off light if it is struck by high-energy rays. The alpha particles that make it through the gold leaf will make the zinc sulfide fluoresce. This allows the experimenter to obtain information on the number of particles coming from the foil at various angles with respect to the direction of the beam.

Most of the heavy, positively charged alpha particles pass straight through the foil with no change in direction. A few undergo small deflections in direction, and a very few are deflected through large angles. Occasionally a particle bounces back in the general direction of the source. These results astounded the experimenters. They worked to verify and extend them by trying several different metal foils and other means for tracking the paths of the alpha particles. The nature of the observations was always the same:

Figure 6.6. Ernest Rutherford.

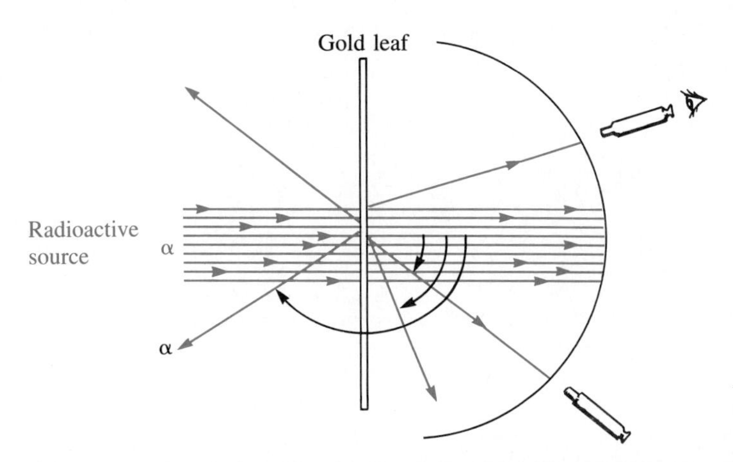

Figure 6.7. The Rutherford scattering experiment. Most particles pass straight through the gold foil, but a few are deflected through large angles.

a few of the high-energy alpha particles always bounced back toward the radioactive source. As Rutherford said, "It was almost as incredible as if you fired a 15-inch shell at a piece of tissue paper and it came back and hit you."

The passage of most of the alpha particles straight through the foil convinced Rutherford that a large fraction of the volume of the atoms in the metal foil contains very little mass. Therefore, most of the mass of an atom must be concentrated in a very small volume. Such a high concentration of mass would also account for the occasional huge deflection—such a deflection would occur when an alpha particle made a direct hit on this concentrated mass and bounced backward. Rutherford later called this concentrated mass the **nucleus** of the atom. Rutherford's discovery of the atomic nucleus solved the problem of where most of the mass of an atom resides, but the question of the electrical charge remained unresolved. He wondered, could the nucleus be positively charged, just enough to balance the negative charges of electrons?

Figure 6.8 compares the predicted path of a positively charged alpha particle traveling near a neutral nucleus with the expected path of such a particle traveling near a positive nucleus. Because of the repulsion between like charges, the alpha particle veers away from a positive nucleus but is unaffected by a neutral nucleus. So the number of small deflections should be greater for a charged nucleus than for a neutral nucleus.

Rutherford's careful mathematical analysis of the number of small deflections persuaded him and others that the nucleus is indeed positively charged. His picture of an atom consisted of a positive nucleus occupying a very small volume but containing most of the mass, surrounded by enough negative electrons to fill over 99% of the volume (Figure 6.9). If the nucleus were the size of a baseball, the whole atom would be a sphere 1 kilometer in diameter.

Chemists and physicists now accept Rutherford's model of the nuclear atom as truth. A very, very large amount of experimental data is consistent

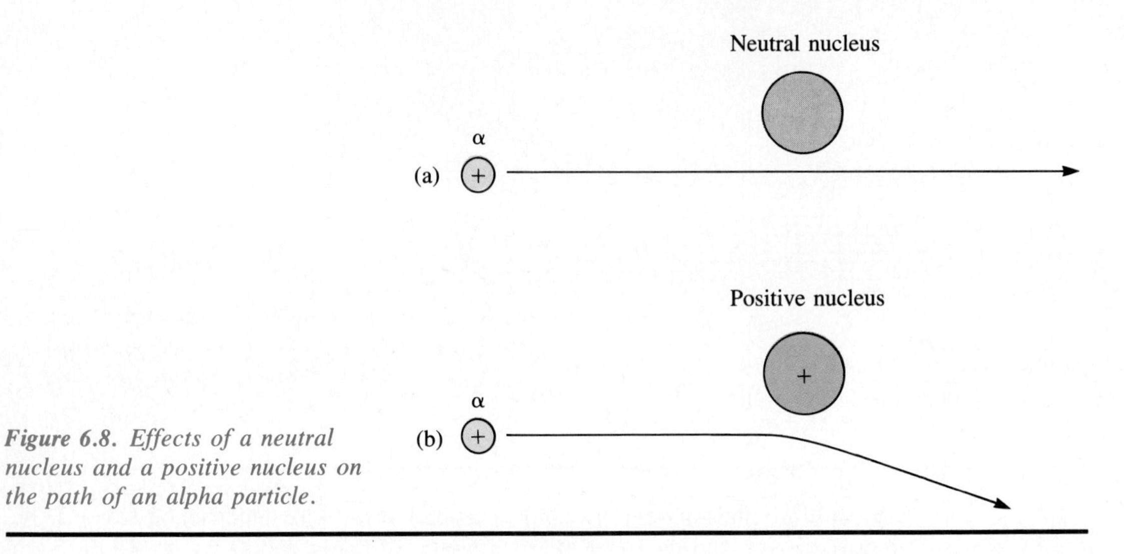

Figure 6.8. Effects of a neutral nucleus and a positive nucleus on the path of an alpha particle.

with it. In a neutral atom the total negative charge of the electrons is exactly balanced by the positive charge of the nucleus. In positive ions one or more electrons have been stripped away; in negative ions extra electrons have been added.

It was not until the **neutron** was discovered in 1932 that our present picture of the basic structure of the atom was completed. The main question remaining after Rutherford's contribution was the nature of the particles in the nucleus. Despite the large number of nuclear particles that have been discovered by contemporary physicists, the model of the nucleus most often used by chemists includes just two types of particles: protons and neutrons. A **proton** has a relative mass very close to 1 on the atomic weight scale and a positive charge equal in magnitude to the negative charge on the

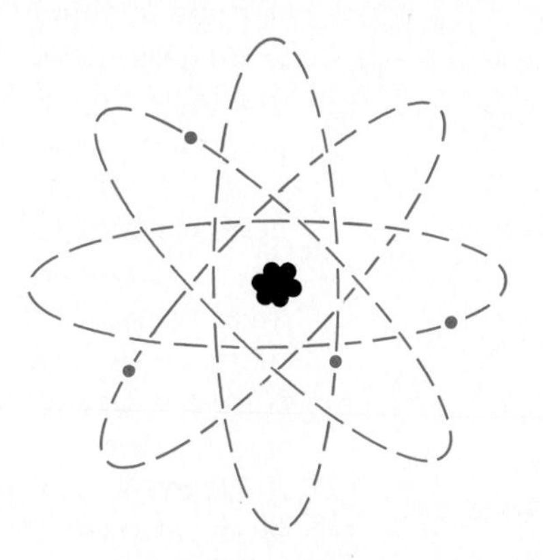

Figure 6.9. Simulated view of a nuclear atom.

Table 6.4 Atomic Particles

PARTICLE	CHARGE	MASS
proton	+1	1
neutron	0	1
electron	−1	0.0005

electron. As the name implies, a neutron has no electrical charge. Its relative mass is also almost exactly 1.

In a neutral atom the number of protons in the nucleus equals the number of electrons outside the nucleus. This number of protons is defined as the **atomic number** of the element. Each element is characterized by its atomic number, which ranges from 1 for hydrogen to over 100 for the artificially produced radioactive elements. Because most elements have about the same number of neutrons and protons in their nuclei, the atomic weights of most elements are approximately twice their atomic numbers.

Now it is possible to connect our modern view of atomic structure to the atomic weight scale of the elements. One of the assumptions of modern atomic theory (p. 44, Chapter 3) is that in any sample of an element the atoms have the same average mass. The atomic weight scale is based on average atomic weights because each chemical element can have a number of isotopes. The several **isotopes** of an element differ in mass but not in the number of protons and electrons. Thus they must have a different number of neutrons. By contrast, if two atoms have a different number of protons in their nuclei, they are atoms of two different elements. For most elements, samples from anywhere in the world have the same proportions of isotopes.

Example In the analysis of some moon rocks, samples of two nuclides were isolated and purified. A **nuclide** is a specific atom, identified by its atomic number and atomic mass. Nuclide X has an atomic weight of 40 and an atomic number of 19. Nuclide Y has an atomic weight of 40 and an atomic number of 20. Are X and Y isotopes of the same element?

Solution To be isotopes of the same element, the two nuclides must have the same number of protons, which is given by the atomic number. Since the atomic numbers of the two nuclides (19 and 20) are different, they must be nuclides of *different* elements. The element with atomic number 19 is potassium, and the element with atomic number 20 is calcium.

To obtain a good approximation to the atomic weight of an isotope, you simply add the number of protons and the number of neutrons. This sum is called the *mass number*. Recall that the atomic weight of each proton and each neutron is close to 1. The total mass of the electrons is very small because the mass of a single electron is only 0.05% of the mass of a proton or neutron. Since all of the isotopes of an element have nearly identical chemical and physical properties, these properties must not depend heavily on the number of neutrons. The chemistry of an element depends mainly on the number of protons in the nucleus and on the number and arrangement of electrons surrounding the nucleus.

Example (a) How many neutrons are in the nucleus of potassium-40 (the isotope of potassium with a mass number of 40) and of calcium-40?

(b) How many electrons does a neutral atom of potassium-40 have? How many electrons does a neutral atom of calcium-40 have?

Solution (a) The number of neutrons can be found by subtracting the atomic number (number of protons) from the mass number of each isotope. For potassium-40, we have $40 - 19 = 21$ neutrons. For calcium-40, we have $40 - 20 = 20$ neutrons.

(b) In any neutral atom the number of protons must be equal to the number of electrons. The atomic number gives the number of protons. So, an atom of potassium-40 has 19 electrons, and an atom of calcium-40 has 20 electrons.

6.5 TRANSFERRING ELECTRONS—OXIDATION AND REDUCTION

In every chemical reaction the distribution of electrons around the nuclei of the reacting atoms is changed in some fashion. The change may be subtle, or it may be drastic. Electron transfer reactions are best described as drastic, because electrons are physically removed from some atoms and given to others. The driving force for these reactions comes either from the potential energy stored within the reacting atoms or from some outside source, such as an electrical generator.

Figure 6.10 shows a copy of a letter written in 1907 to the manager of an amusement park in Grand Rapids, Michigan. The subject is "lighter than air" flying machines. It seems that the amusement park owner wanted to arrange some flights for his customers, so he had inquired about what chemicals would be needed. The answer was cast-iron borings and sulphuric (sulfuric) acid.

What would you make from small pieces of iron and sulfuric acid? Hydrogen gas! This lightest of all gases, H_2, was used to provide lift to balloons and blimps. The chemical reaction that produces hydrogen is

$$\underset{\text{iron}}{Fe} + \underset{\text{sulfuric acid}}{H_2SO_4} \longrightarrow \underset{\text{hydrogen}}{H_2} + \underset{\text{ferrous sulfate}}{FeSO_4}$$

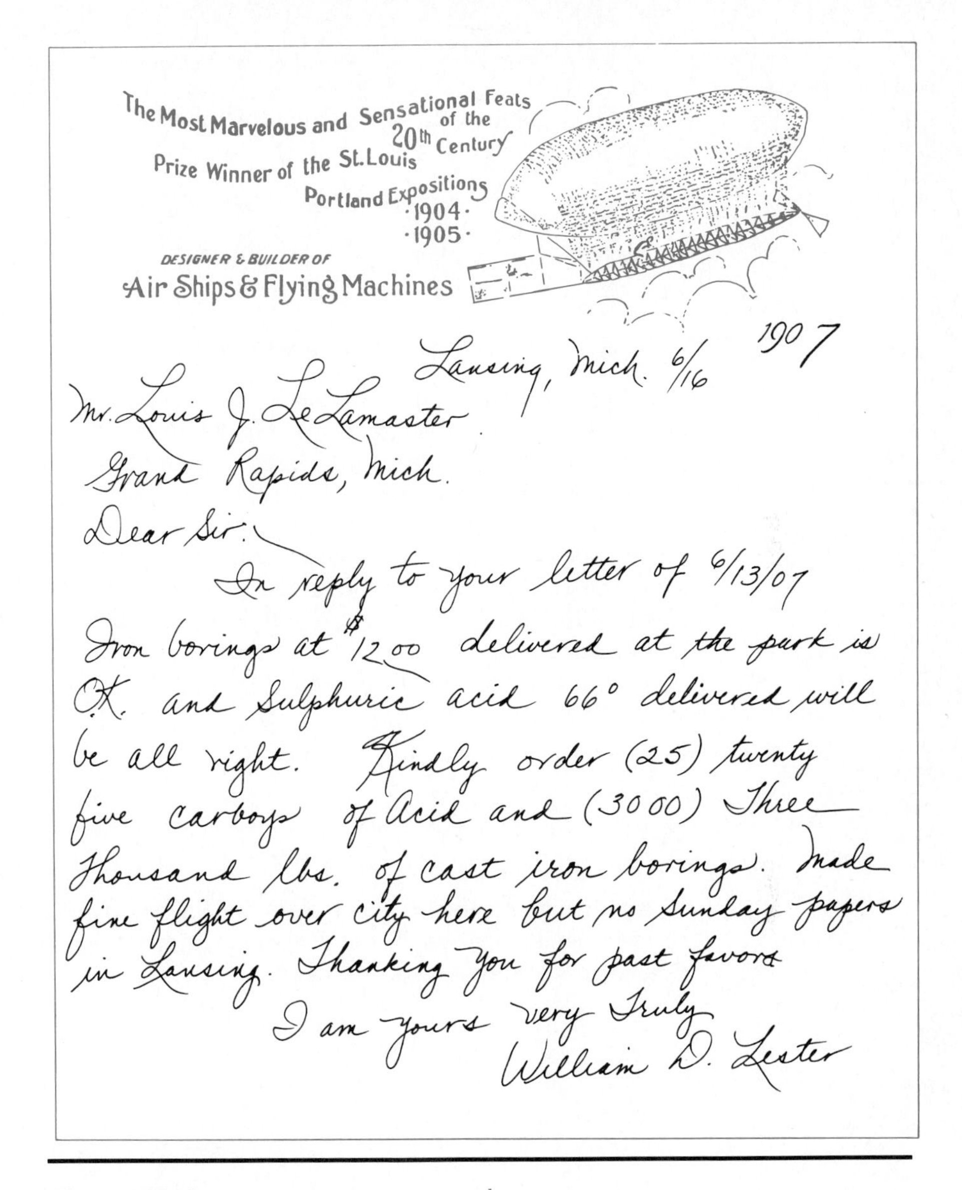

The Most Marvelous and Sensational Feats of the 20th Century
Prize Winner of the St. Louis Portland Expositions ·1904· ·1905·
DESIGNER & BUILDER OF
Air Ships & Flying Machines

Lansing, Mich. 6/16 1907

Mr. Louis J. LeLamaster
Grand Rapids, Mich.

Dear Sir:

In reply to your letter of 6/13/07 Iron borings at $12.00 delivered at the park is O.K. and Sulphuric acid 66° delivered will be all right. Kindly order (25) twenty five carboys of Acid and (3000) Three Thousand lbs. of cast iron borings. Made fine flight over City here but no Sunday papers in Lansing. Thanking you for past favors

I am yours very Truly
William D. Lester

Figure 6.10. Letter to an amusement park manager.

This reaction is an example of a large class of chemical reactions called *oxidation-reduction reactions*. Every oxidation is accompanied by a reduction. The iron metal is oxidized in this reaction, and the hydrogen atoms are reduced. A neutral iron atom has 26 electrons. After it has been oxidized (forming Fe^{2+}, or the ferrous ion), the iron atom has two electrons less, or 24. Each iron atom has lost two electrons and is said to be **oxidized**. A general definition of **oxidation** is the loss of electrons.

$$Fe \longrightarrow Fe^{2+} + 2 \text{ electrons}$$

In water solution the sulfuric acid will ionize as shown in the following equation:

$$H_2SO_4 \longrightarrow 2\,H^+ + SO_4^{2-}$$

In order for the two hydrogen ions to become a molecule of hydrogen gas, each of them must gain one electron:

$$2\,H^+ + 2 \text{ electrons} \longrightarrow H_2$$

The hydrogen ions are then said to be **reduced.** A general definition of **reduction** is the gain of electrons. Oxidation and reduction always occur together in a reaction because electrons cannot be created from nothing, nor can they be destroyed.

Anytime a neutral element reacts to form a positive ion, the element undergoes the process of oxidation. Anytime a negative ion reacts to form a neutral element, the ion undergoes oxidation. Each of these processes involves the loss of electrons. The following partial reactions are examples of oxidation:

$$Na \longrightarrow Na^+$$
$$Mg \longrightarrow Mg^{2+}$$
$$Al \longrightarrow Al^{3+}$$
$$2\,Cl^- \longrightarrow Cl_2$$
$$2\,I^- \longrightarrow I_2$$

Anytime an element reacts to form a negative ion or a positive ion reacts to form the element itself, the process is a reduction. The following partial reactions are examples of reduction:

$$Na^+ \longrightarrow Na$$
$$Mg^{2+} \longrightarrow Mg$$
$$Al^{3+} \longrightarrow Al$$
$$Cl_2 \longrightarrow 2\,Cl^-$$
$$I_2 \longrightarrow 2\,I^-$$

Example In the following reaction, what becomes oxidized and what gets reduced?

$$2\,K + Br_2 \longrightarrow 2\,K^+Br^-$$

Solution It is the reactants that are changed in the reaction, so we must look at the reactants to see what is being oxidized and reduced. The potassium metal becomes a positive ion and therefore becomes oxidized. A potassium ion has one less electron than does a neutral potassium atom; loss of electrons is oxidation.

The bromine molecule turns into negative bromide ions. Each bromine atom gains an electron in the process and thus is reduced.

In some oxidation-reduction reactions, no independent ions are present as reactants or products. This makes it difficult to apply the definitions given for oxidation and reduction. In most of these cases either H_2 or O_2 is among the reactants. An example, of course, is the burning of a fuel, such as methane:

$$CH_4 + 2\,O_2 \longrightarrow CO_2 + 2\,H_2O$$

Like all combustion reactions, this one is an oxidation-reduction reaction. Whenever a chemical reacts with oxygen molecules, it can be said to be oxidized. In the above example the methane (CH_4) is oxidized. Naturally, the O_2 is reduced here, since reduction goes hand in hand with oxidation.

When a chemical reacts with a hydrogen molecule (H_2), the chemical becomes reduced in the process. In the following reaction, the carbon dioxide is reduced and the H_2 is oxidized.

$$CO_2 + H_2 \longrightarrow CO + H_2O$$

Example In the following reaction, what undergoes oxidation and what is reduced?

$$2\,H_2 + O_2 \longrightarrow 2\,H_2O$$

Solution The chemical that reacts with O_2 is oxidized; therefore, H_2 is oxidized. The chemical that reacts with H_2 is reduced; thus, O_2 is reduced.

* 6.6 MAKING ALUMINUM *(optional)*

Ever since 1825, when aluminum metal was first prepared, the importance of this metal to industrialized societies has been increasing. It is lightweight and is a good conductor of electricity. Its alloys have great strength. World production of aluminum from ore is over 10 million tons per year. Aluminum is used in the construction of buildings, airplanes, electrical conductors, cooking utensils, beverage containers, and in many other ways. In the United States over 20 kilograms of aluminum per person is produced each year. Among the metals, aluminum is surpassed only by iron with respect to the amount used.

Aluminum appeared rather late in the chronology of metal production not because of scarcity (it makes up 8.2% of the crust of the earth) but because of the great stability of aluminum ores. Chief among these is bauxite, which has a high content of aluminum oxide (Al_2O_3). Pure Al_2O_3 is a white solid that melts at 2072°C. It consists of aluminum ions, Al^{3+}, and oxide ions, O^{2-}. Having lost three electrons, the aluminum in Al_2O_3 is present in an oxidized form. An oxide ion is an oxygen atom that has gained two electrons. The ratio of two Al^{3+} ions for every three O^{2-} ions in solid Al_2O_3 causes the compound to be electrically neutral.

To prepare elemental aluminum from Al_2O_3, one must force the Al^{3+} ions to gain electrons. The problem is that this process has very little tendency to occur. Only a few reactive metals such as sodium and potassium can transfer electrons to aluminum ions in Al_2O_3. Historically, the first aluminum was prepared not from Al_2O_3 but from $AlCl_3$. The reaction employed was

$$\underset{\text{aluminum chloride}}{AlCl_3} + \underset{\text{potassium}}{3\,K} \longrightarrow \underset{\text{aluminum}}{Al} + \underset{\text{potassium chloride}}{3\,KCl}$$

In simplified form this reaction can be written

$$Al^{3+} + 3\,K \longrightarrow Al + 3\,K^+$$

The transfer of electrons is now apparent. Each Al^{3+} ion gains a total of three electrons—one from each of the three potassium atoms, which in turn become K^+ ions. The aluminum ions are reduced, and the potassium metal is oxidized in the reaction. The chloride ions, Cl^-, do not participate in the electron transfer, and so they need not be written in the simplified form of the reaction. Only because potassium metal loses electrons easily does the reaction occur spontaneously.

This method of preparing aluminum was very expensive simply because metallic potassium was expensive (and still is). Later, a slight reduction in cost was achieved by using sodium in place of potassium, but still aluminum could not compete with other commonly used metals.

Often electrical energy can decompose a compound that otherwise is quite stable. This process is called **electrolysis.** You have already seen an example in Chapter 3, where the decomposition of water into hydrogen and oxygen by an electrical current was described. When a compound is electrolyzed, it must be liquid and must be able to conduct electricity. The melting point of Al_2O_3 is so high as to make the electrolysis of pure Al_2O_3 impractical. During the latter half of the nineteenth century, a number of chemists tried to find a substance capable of dissolving Al_2O_3 and forming a liquid solution at a reasonable temperature.

Charles Martin Hall, a student at Oberlin College in the early 1880s, was fascinated by a professor's remark that a fortune awaited the person who could find a suitable solvent for aluminum oxide, and after graduation Hall began an intensive search. In 1886 he discovered that a mineral called cryolite, Na_3AlF_6, had a melting point well below that of Al_2O_3 and could dissolve Al_2O_3. The electrolysis could be accomplished at approximately 1000°C. Surprisingly, one week later a Frenchman, Paul L. T. Héroult, independently discovered the same process. Credit for the method, which is used today almost in its original form, is given to both men.

Figure 6.11 shows a simplified version of the electrolysis cell. Because the liquid solution of cryolite and Al_2O_3 is reactive toward iron, the entire tank is lined with carbon. Through its connection to a source of electrical energy, the iron tank is negatively charged; the carbon rods that dip into the solution are positively charged. Within the solution, Al^{3+} ions move toward the tank, where they become reduced and form neutral aluminum atoms. These atoms form a layer of liquid aluminum at the bottom of the tank. This change can be written as a partial reaction:

$$Al^{3+} + 3 \text{ electrons} \longrightarrow Al$$

At the same time, oxide ions (O^{2-}) move toward the positively charged carbon rods and become oxidized, forming O_2 gas.

$$2\,O^{2-} \longrightarrow O_2 + 4 \text{ electrons}$$

At the high temperature of the solution the O_2 gas immediately oxidizes the carbon rod to produce carbon dioxide, which bubbles away. The O_2 is reduced in the process:

$$C + O_2 \longrightarrow CO_2$$

The carbon electrode is gradually consumed and must be replaced from time to time.

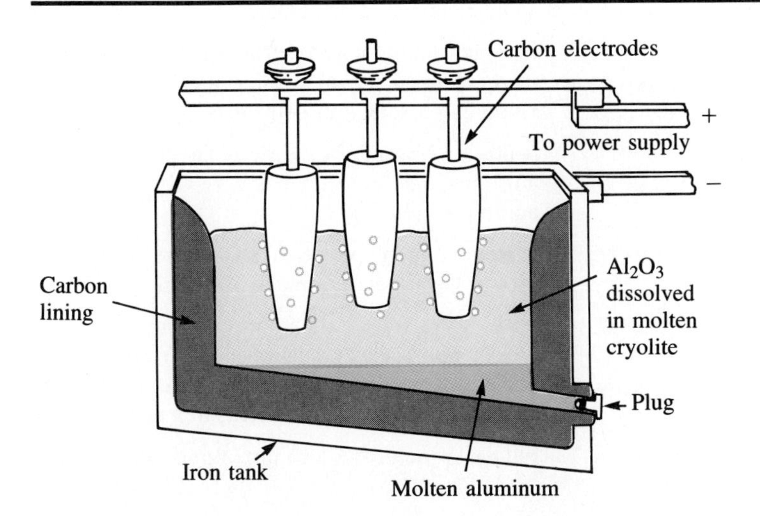

Figure 6.11. Electrolysis cell for the production of aluminum.

The principal features of this process are typical of all electrolyses. Positive ions in the liquid travel to the negative electrode to accept electrons provided by an external source of electrical power. These positive ions are reduced. At the same time negative ions move to the positive electrode, where they give up electrons to the power source and are oxidized. In the Hall-Heroult process for the production of aluminum, the final products of the electrolysis are aluminum metal and carbon dioxide. The Al^{3+} ions are reduced to aluminum atoms, and carbon atoms are oxidized:

$$2\ Al_2O_3 + 3\ C \longrightarrow 4\ Al + 3\ CO_2$$

A key aspect is that a nonspontaneous chemical reaction is forced to occur by the application of electrical energy.

A huge amount of electricity is used in the production of aluminum. For each kilogram of aluminum produced, about 16 kilowatt-hours of electricity is consumed. About one-half of the final price of aluminum comes from the cost of the electricity needed to produce it. Since energy is also used in extracting aluminum oxide from bauxite, aluminum production requires great amounts of energy. As the price of energy changes, so does the price of aluminum. Both the huge energy consumption and the fact that discarded cans and other objects made of aluminum contribute to our solid-waste problem make the recycling of aluminum look especially attractive. *

IMPORTANT TERMS

electrode A solid conductor through which electricity enters or leaves a liquid or a gas.

electron A component of atoms that has a negative charge (-1) and a very small mass (1/1840 of the mass of a hydrogen atom).

ion An atom carrying an electrical charge. A positive ion results when one or more electrons are removed from an atom. A negative ion has one or more extra electrons added to a neutral atom.

ionize To form ions that can be separated from each other.

radioactivity Spontaneous emission of alpha, beta, or gamma rays from a sample.

alpha rays Heavy, fast moving, positively charged ions produced by some radioactive elements. Alpha rays are composed of helium ions (He^{2+}).

beta rays Highly energetic electrons produced by some radioactive elements.

gamma rays High-energy radiation produced by some radioactive elements.

nucleus (plural: **nuclei**) The small, dense, positively charged region at the center of an atom. It contains protons and neutrons.

neutron One of the particles in the nucleus of an atom. The neutron has a relative mass of 1 and a charge of 0.

proton A component of an atomic nucleus. A proton has a relative mass of 1 and a charge of $+1$.

atomic number The number of protons in the nucleus of an atom of an element; also the number of electrons around the nucleus of a neutral atom.

isotope One of two or more atoms whose nuclei

have the same number of protons but a different number of neutrons.

nuclide Any neutral atom specified by its atomic number and atomic mass. Examples are carbon-14 (atomic number 6) and oxygen-16 (atomic number 8).

oxidized Having undergone the process of oxidation.

oxidation The loss of one or more electrons by an atom or ion; alternatively, the reaction of a molecule with O_2.

reduced Having undergone the process of reduction.

reduction The gain of one or more electrons by an atom or ion; alternatively, the reaction of a molecule with H_2.

electrolysis A chemical reaction brought about by the application of electricity to a liquid sample.

QUESTIONS

1. When amber is rubbed with fur, static charges are produced on both materials, but they have opposite signs. Suggest how these static charges come about; base your discussion on your understanding of the atom and its constituent particles.

2. Give the number of protons, neutrons, and electrons in each of the following nuclides:

a. oxygen: atomic number, 8; atomic weight, 16
b. hydrogen: atomic number, 1; atomic weight, 1
c. aluminum: atomic number, 13; atomic weight, 27
d. gold: atomic number, 79; atomic weight, 197
e. uranium: atomic number, 92; atomic weight, 238
f. uranium: atomic number, 92; atomic weight, 235

3. Give the number of protons, neutrons, and electrons in each of the following ions:

a. Ca^{2+}: atomic number, 20; atomic weight, 40
b. I^{-}: atomic number, 53; atomic weight, 127
c. Cr^{3+}: atomic number, 24; atomic weight, 53
d. S^{2-}: atomic number, 16; atomic weight, 32

4. Based on information given in this chapter, estimate the freezing point of a solution containing 0.1 mole of the salt $MgCl_2$ and 1000 grams of water.

5. Most brands of antifreeze that people add to their automobile radiators use the chemical ethylene glycol (CH_2OHCH_2OH) to lower the freezing point enough to protect the engine from damage in cold weather. A colored dye is added to the antifreeze, and various rust inhibitors are used to protect the metal surfaces. A 50% solution of ethylene glycol will keep radiator water from freezing in all but the most bitter winter temperatures. Such a solution has 16.0 moles of ethylene glycol, which is nonionic, in every 1000 grams of water. Using the data found in Table 6.1, calculate the freezing point of a 50% ethylene glycol solution in water.

6. Dalton said that all the atoms of a given element are identical. How should this statement be revised in view of the ideas presented in this chapter?

7. In the electrolysis of molten aluminum oxide the following half-reactions occur at the electrodes:

$$Al^{3+} + 3 \text{ electrons} \longrightarrow Al$$

and

$$2\ O^{2-} \longrightarrow O_2 + 4 \text{ electrons}$$

In these reactions, is Al^{3+} oxidized or reduced? Is O^{2-} oxidized or reduced?

8. Which chemical substance undergoes oxidation and which undergoes reduction in each of the following chemical reactions?

a. $Al^{3+} + 3\ Na \longrightarrow Al + 3\ Na^+$
b. $Ca + I_2 \longrightarrow Ca^{2+} + 2\ I^-$
c. $2\ Fe^{2+} + Cl_2 \longrightarrow 2\ Fe^{3+} + 2\ Cl^-$
d. $2\ C + O_2 \longrightarrow 2\ CO$
e. $Cr^{3+} + 3\ K \longrightarrow Cr + 3\ K^+$
f. $2\ Mg + O_2 \longrightarrow 2\ MgO$
g. $NiO + H_2 \longrightarrow Ni + H_2O$

9. Some man-made radioactive isotopes emit a particle called a *positron*. In a magnetic field a beam of positrons bends in the direction opposite to that followed by a beam of beta particles. The extent of bending (i.e., the shape of the path), however, is the same for the beta particles and the positrons. What are the charge and the mass (on the atomic weight scale) of the positron?

10. Some contemporaries of Arrhenius thought that ions exist in a solution of a salt in water only when an electric current flows. This hypothesis might be stated as follows: "In a sodium chloride solution, the solute particles are NaCl molecules. When the solution is electrolyzed and a current flows, however, the NaCl molecules dissociate into Na^+ and Cl^- ions." Give the evidence for and against this hypothesis.

11. Shortly after the electron and the proton were discovered, some scientists believed that atoms consisted of uniform mixtures of electrons and protons. For example, they might have argued that the silver atom contained 107 protons and 107 electrons (no neutrons), evenly distributed throughout the atom. If this were true, what results would you expect in a Rutherford scattering experiment with silver foil? How do these results differ from the actual results?

12. It is conceivable that the negative particles in a gas discharge tube are negative ions rather than electrons. Take a position for or against this proposition, and support your case.

7 Quanta—Light and Matter *

One of the most exciting scientific adventures of this generation is the exploration of our solar system with space probes. We now know far more about the atmospheres and conditions on other planets than we did even 10 or 15 years ago. Yet the only heavenly body on which human beings have set foot is our own moon. In every other case, with the Mariner, Pioneer, Venera, and Voyager space probes, we have used complex instruments to send back messages. Most of the messages have been in the form of radio waves. Although we usually think of **light** as energy that enters the eye and causes a visual image in the brain, these radio waves that we cannot see are actually a kind of light wave.

Receiving messages from space is not new—people have been looking up at the stars to obtain information with their eyes for many thousands of years. But for the past 130 years we have had an advantage: the ability to analyze light by resolving it into a spectrum of colors, with an instrument called a **spectroscope** (Figure 7.1). This **spectrum** is the same kind you see in a rainbow: a separation of light into its different colors. With a spectroscope, however, clearer distinctions can be made between the colors.

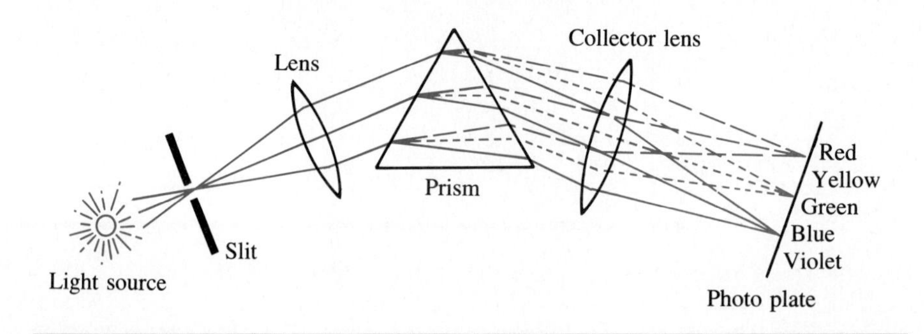

Figure 7.1. Diagram of a simple spectroscope. Rays from a light source enter a narrow slit and are made parallel by a lens. Next, they are beamed onto a prism, which separates rays of different colors. A second lens collects the rays and beams them onto photographic film.

In 1868, just a few years after the invention of a practical spectroscope, there was a total eclipse of the sun in India. French astronomer Pierre Janssen decided to make the first spectroscopic study of the sun's chromosphere (or lower atmosphere), which could only be studied during an eclipse. He noticed that the yellow light that came through the spectroscope attached to his telescope did not coincide with the well-known yellow light from the common element sodium. An English astronomer, Norman Lockyer, later found that the exact spectrum of colors couldn't be accounted for by any element known on earth. He proposed the presence of a new element in the sun and named it *helium,* after the Greek sun god, Helios.

Were scientists fascinated by the idea? Not exactly. Most doubted the existence of Lockyer's helium. Some laughed at the idea. It was only after a quarter of a century and the isolation of helium from uranium ores (it is a product of alpha decay) that Lockyer's claim was verified.

Modern astronomers study the heavens with a large number of spectroscopic telescopes. They study the whole range of light radiation that bombards the earth, from high-energy x-rays to low-energy radio waves. These astronomers look at the stars and the planets as well as the molecular clouds between the stars. Their many, many observations are consistent with the idea of a universal chemistry—that the same elements exist throughout the universe, though some are more common in stars and others in planets.

7.1 WAVES

As in the case of outer space, much of what we know about the structure of atoms comes from our understanding of light and how it interacts with matter.

When compounds of metals are sprinkled on a flame, characteristic colors are produced depending on the metal used. Brilliant yellow is associated with sodium chloride, sodium sulfate, and sodium nitrate; bright red comes from strontium compounds; and green is produced by nickel compounds. Recordings of these emitted colors are called *emission spectra,* because particular colors of light are emitted, or sent out, from the flame. Somehow the flame's energy makes the metal ions glow in different colors.

Another pattern of colors is seen if one flies over a city at night. Strings of bluish-white light are emitted from mercury vapor lamps, and patches of yellowish-orange light come from the newer sodium lamps. As time goes on, we will see the energy-efficient sodium lamps more and more. In either case, the energy that produces the lamplight comes from electricity rather than from a flame. However, before we can make any sense of these observations, we must look into the nature of light itself.

Until the end of the nineteenth century, most scientists thought of a light beam as being made up of waves traveling through space. Crudely speaking, a light **wave** is an electrical force that varies in intensity from point to point. At one point the force acts in a certain direction; a little farther along the wave, the force acts in the opposite direction. This changing back and forth in space is an important feature of a wave. The entire wave also moves through space, so the electrical force at any fixed position changes with time. In any wave motion, energy moves from one place to another without the transfer of matter. This picture accounts quantitatively for many optical phenomena. It accounts nicely for the different colors of light found in a rainbow or produced by passing white light through a prism: light waves of different colors have different wavelengths and frequencies (see Figure 7.2).

Experiments have shown that in a vacuum all light travels at the same speed, regardless of wavelength. The **wavelength** is the distance between two repeating, adjacent points in a wave. As a light wave travels through space, it takes a certain amount of time for a wavelength to pass by a fixed point. The number of wavelengths, or cycles, passing the point each second is called the **frequency** of the light wave. Frequency is expressed in cycles per second, where **cycle** refers to the smallest pattern (one wavelength long) which repeats many times to make the entire wave. The longer the wavelength, the smaller the frequency, since more time is needed for a wave cycle to pass a fixed point. Conversely, light with a relatively short wavelength will have a greater frequency. This relationship is shown by the following equation:

$$c = \lambda\nu$$

where c = the speed of light, 3.0×10^8 meters per second
λ = wavelength, in meters per cycle
ν = frequency, in cycles per second

The Greek letters λ (lambda) and ν (nu) are commonly used for wavelength and frequency, respectively.

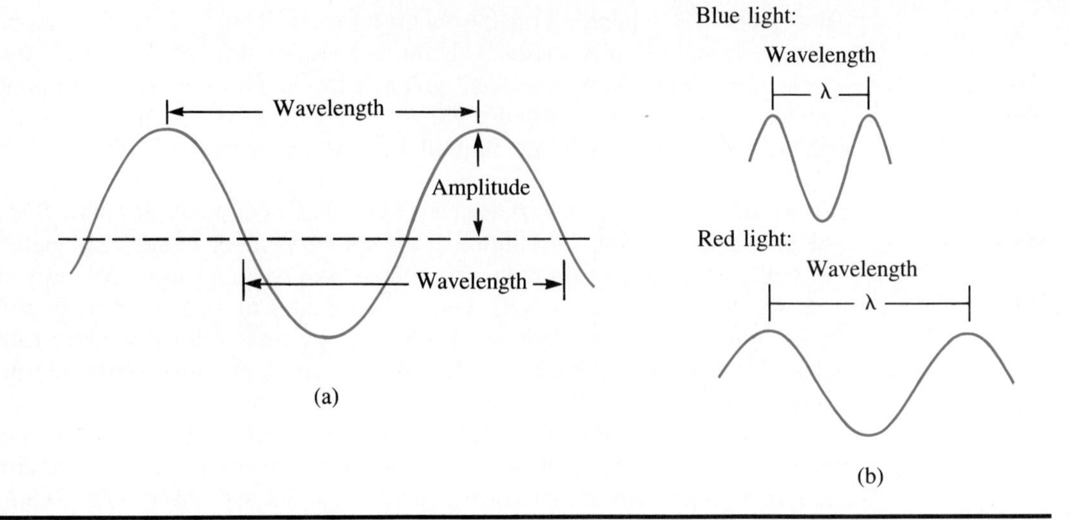

Figure 7.2. (a) Portion of a light wave. (It would be traveling through space along the dashed line.) (b) Portions of light waves that have different wavelengths and frequencies.

Example By looking at Figure 7.2(b) and the relationship $c = \lambda\nu$, decide whether blue light or red light has the greater frequency.

Solution According to Figure 7.2(b), blue light has a smaller wavelength than does red light. From the equation $c = \lambda\nu$, we see that a smaller wavelength means a larger frequency, because the speed of light is the same for all kinds of light. Therefore, the frequency of blue light is greater.

Many other kinds of light are known in addition to the light that human eyes can see. They range from radio waves, which can have a wavelength of over 100 meters, to gamma rays, which have a wavelength of less than a billionth of a meter. The general term for all kinds of light is **electromagnetic radiation.** The "magnetic" part of the name refers to the fact that these waves have magnetic as well as electrical properties. Figure 7.3 shows the full range of frequencies of electromagnetic radiation and the names of different parts of the spectrum. (In practice most scientists use the word "light" only for the visible region.)

At the end of the nineteenth century, the wave theory of light seemed to be completely satisfactory. There were a few puzzling phenomena remaining, but no one suspected that their explanation would turn the orderly world of physics upside down.

INFRARED AND MICROWAVE RADIATION

Although our eyes don't respond to electromagnetic radiation outside the visible region, it has benefited society in many ways. Everyone knows about the use of x-rays in medicine to observe broken bones. Also important are the applications of waves of longer wavelengths (lower frequencies), such as infrared waves and microwaves.

All objects emit **infrared** radiation, which has frequencies just below those of the visible region of the spectrum. The intensity of this radiation at different wavelengths depends on the temperature. This property gives rise to a unique form of photography in which anything warm produces a strong image. Infrared photography in total darkness reveals heat sources, such as people and cars. It is even possible to view an infrared image directly. In either the photography or the direct viewing of infrared images, a sophisticated instrument converts infrared radiation into visible light so that we can see the results. Infrared photographs of commercial buildings and houses taken in the winter show heat leaks and guide owners in improving insulation. Military gadgets that use infrared radiation include heat-seeking missiles and aircraft and missile trackers.

The microwave oven, used in many homes and restaurants, takes advantage of waves even longer than infrared waves. The wavelength of the radiation in a typical microwave oven is 12 centimeters, which is equivalent to a frequency of 2.5×10^9 cycles per second. In an ordinary oven, heat flows from the source (gas flame or electric heating element) to the outside of the food, where it is absorbed. Then the heat energy must be slowly transmitted to the interior of the food before the center can be cooked. As a result, the cooking time can be quite long.

In a microwave oven, the electromagnetic energy penetrates the food and heats the inside and the outside at the same time. Water molecules, which are present in most foods, are especially efficient at absorbing microwaves. As the water molecules absorb energy, they become warmer and transfer some of their excess energy to neighboring food molecules. The temperature of the entire morsel rises quickly, and so the cooking time is relatively short. It is best to use a glass or plastic container to hold the food in a microwave oven, because these substances absorb very little microwave radiation, allowing the food to receive most of the energy.

Figure 7.3. The electromagnetic spectrum.

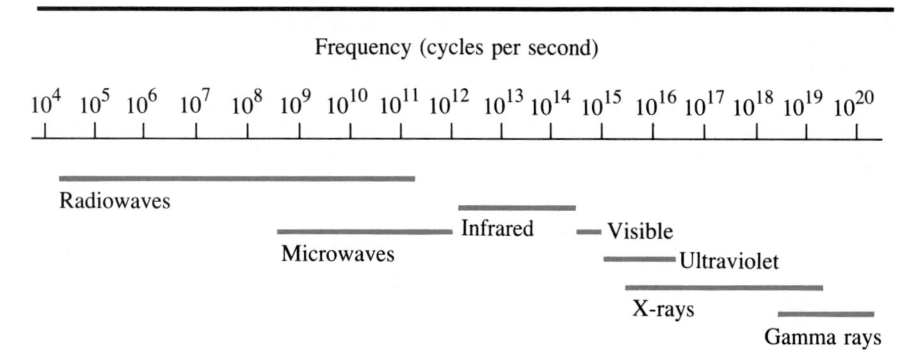

7.2 THE PHOTOELECTRIC EFFECT

One unexplained phenomenon was the **photoelectric effect,** in which light shining on the surface of a polished metal causes a stream of negative electrons to be emitted from the surface [see Figure 7.4(a)]. This effect is the principle behind the "electric eye." For most metals, electrons are ejected only if the light used is high-frequency **ultraviolet** light, which has frequencies beyond those of visible light. If visible light shines on zinc metal, for example, no current flows no matter how great the intensity of the light is. On the other hand, ultraviolet light of very weak intensity causes a small current of electrons to flow. A few metals, such as sodium and potassium, show the photoelectric effect with visible as well as with ultraviolet light. It appears that for each metal there is a different threshold frequency. When light with a frequency below the threshold is used, no electrons are ejected. The measurement apparatus is shown in Figure 7.4(b).

The wave theory of light did a very poor job of explaining these results. According to this theory the energy contained in a beam of light increases with the intensity of the beam. Therefore, if light of a certain intensity causes no electrons to be ejected, it should be possible to cause electrons to be ejected simply by increasing the intensity of the light. According to the wave theory, the frequency of the light should have little to do with its ability to remove electrons. These predictions are obviously contrary to the observations.

There was another part to the photoelectric experiment that was also puzzling. When the maximum energies of the ejected electrons were measured, they were found to increase directly with the frequency of the light. Typical

Figure 7.4. (a) The photoelectric effect. (b) Apparatus used for measuring it. (c) A sample result.

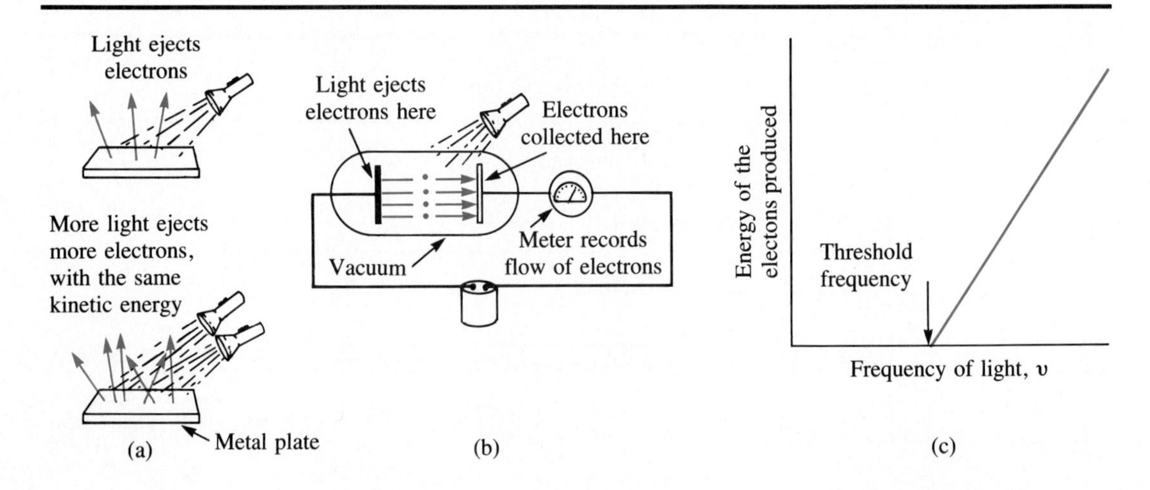

behavior is shown in Figure 7.4(c). Below the threshold frequency no electrons are released and no current flows. Above this cut-off value a straight line describes the way the energy of the ejected electrons varies with the frequency of the applied light.

This extra significance of the frequency only added to the mystery. In all these experiments the only effect of increasing the intensity was to increase the number of electrons ejected in a given time. The maximum energies of the individual electrons were unaffected by light intensity.

7.3 ENTER THE QUANTUM

An explanation of the photoelectric effect was published by Albert Einstein (Figure 7.5) in one of his famous papers of 1905. He assumed that a certain critical energy is necessary to remove an electron from the metal surface. This seemed reasonable, because electrons are attracted to the nuclei of metal atoms and work must be done on an electron to pull it from the metal. Since each metal is different in nuclear charge and in the arrangement of its electrons, the threshold energy for each metal is different.

Now comes the really radical part of Einstein's explanation. He said that light comes in packets of energy called **photons** or **quanta** (singular: **quantum**). A light beam doesn't fill the entire space through which it travels. Instead, it is "grainy" or "lumpy." Furthermore, the energy carried by each photon or quantum of light depends not on the intensity but on the

Figure 7.5. *Albert Einstein at his desk in the patent office in Bern, Switzerland, 1905.*

frequency of the light. Five years earlier, another German physicist by the name of Max Planck (Figure 7.6) had coined the term "quantum of light" to account for the color and intensity of light given off by very hot objects. Einstein built on this earlier idea.

The landmark equation for the energy of a photon is

$$E = h\nu$$

where E = energy of the photon
h = Planck's constant

As before, ν is the frequency of the light. The value of h must be found from experimental measurements. According to this equation, when the frequency of the light is doubled, the energy of a photon also doubles. The equation contains a paradox, however. The left side refers to the lumpy or particle-like quality of light, whereby energy is transferred in discrete packets. The right side features the frequency, which is based on the wavelike nature of light. In other words, the energy of a photon can be given only in terms of the frequency of a wave.

With this drastic statement about the existence and energies of photons, Einstein removed the mystery from the photoelectric effect. When the metal absorbs light, the entire energy of one photon is transferred to one ejected electron. Some of the energy, or work, is required to yank the electron away from the atomic nucleus, and the remainder becomes the kinetic energy of the liberated electron. Einstein's famous photoelectric equation

Figure 7.6. Max Planck.

is

$$h\nu = \varepsilon + \frac{1}{2}mv^2$$

where ε = work needed to free the electron from the metal

$\frac{1}{2}mv^2$ = kinetic energy of the liberated electron

Figure 7.7 may help to clarify Einstein's equation. It is the same as Figure 7.4(c) except that the horizontal axis now shows the photon energy rather than the frequency. Because the two are directly proportional, the graph has exactly the same form as before. Parts (a) and (b) of Figure 7.7 represent two metals, A and B, with two values of ε, ε_A and ε_B, respectively, which are indicated on the figures. These values are the threshold energies. A photon with less energy than the threshold value does not produce any current.

Suppose that blue light of a particular frequency is directed at metals A and B. The energy of a photon of this light is $h\nu$, which is shown on the horizontal axis in parts (a) and (b) of Figure 7.7. For each metal, this photon energy is greater than the threshold energy, and the excess becomes the kinetic energy of the ejected electrons. Because metal A has a lower threshold

Figure 7.7. Plots of the energy of ejected electrons against the photon energy for the photoelectric experiment. (a) Metal A has a threshold energy of ε_A. (b) Metal B has a larger threshold energy, ε_B.

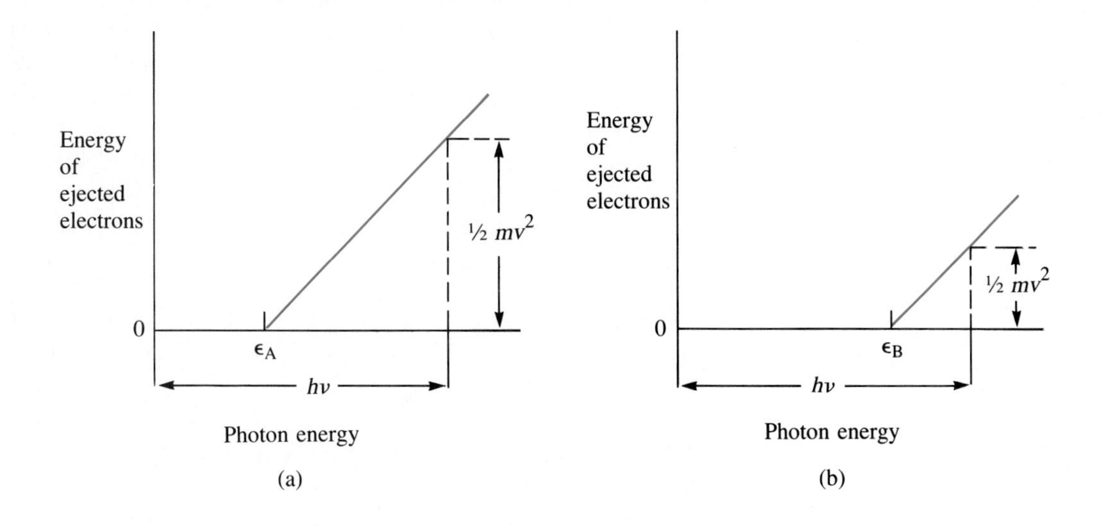

energy, more energy from each of the incident photons goes into kinetic energy. Once quantitative data, such as those presented in Figure 7.4(c) and in Figure 7.7, became available, Einstein's and Planck's idea of light beams made up of packets of light was warmly received by scientists.

To describe light, we imagine that each wave is divided into many parts. Each part, or photon, contains light energy. Between photons there is no energy. When a light beam interacts with a large object, such as a lens or a prism, a great many photons come into contact with the object at the same instant. Then we aren't aware of individual photons. So the wave theory of light explains very well the passage of light beams through lenses and prisms. On the other hand, when light causes electrons to be ejected from the surface of a metal, the light must interact with individual atoms. Only a tiny fraction of a wave hits an atom; this fraction is a single photon. Now the photon or particle-like nature of light becomes very important. The quantum notion caused a revolution in modern physics, and both Planck and Einstein received Nobel prizes for their work on it. Nowadays the word "photon" is probably used more often than the term "quantum" to signify a packet of light energy.

The equation $E = h\nu$ has been crucial in our understanding of the detailed structures of atoms and molecules, as we shall see. From it, one can determine that photons of red light are less energetic than photons of blue light and that photons of radio waves have much less energy than photons of x-rays. Gamma rays, x-rays, and ultraviolet radiation have biological dangers because of their high-energy photons; infrared and radio waves seem to be less dangerous. In the quantum model of light, the intensity of light received by a certain object depends on the number of photons per second bombarding the object. The greater the number of photons per second, the greater the intensity of the light.

Let us compare the energy of a photon of infrared radiation ($\nu = 1 \times 10^{14}$ cycles per second) to a photon of blue visible light ($\nu = 7.9 \times 10^{14}$ cycles per second). A handy unit of energy to use in this calculation is the *electron-volt*—the energy that must be transferred to an electron to move it from a positive electrode to an electrode that is more negative by a potential of 1 volt. Because the electron is attracted to the positive electrode and repelled by the negative electrode, work must be done to move it in this direction. One electron-volt equals 1.6×10^{-19} joule. It is an extremely small unit of energy convenient for measuring the energy of a photon or an electron in an atom.

To calculate the photon energy in electron-volts, we must have the value of Planck's constant in consistent units: 4.1×10^{-15} electron-volt second.

$$E = h\nu$$

$$E = (4.1 \times 10^{-15} \text{ electron-volt second}) \cdot (1 \times 10^{14} \text{ second}^{-1})$$

$$E = 0.41 \text{ electron-volt}$$

The units following the value of Planck's constant must include *second* because a frequency has the unit $second^{-1}$. ("Cycles" doesn't count.) Multiplying h by $second^{-1}$ gets rid of *second* and leaves us with electron-volts, as desired. We find that 0.41 electron-volt is the energy of 1 photon of infrared radiation having a frequency of 1×10^{14} cycles per second.

By contrast, a photon of blue light ($\nu = 7.9 \times 10^{14}$ cycles per second) has 7.9 times as much energy. The energy of a photon of blue light is

$$0.41 \cdot 7.9 = 3.2 \text{ electron-volts}$$

Using this type of calculation, we can add an energy scale to Figure 7.3, producing Figure 7.8.

Example What is the energy of 1 photon of red light, which has a frequency of 3.8×10^{14} cycles per second?

Solution Photon energy is directly proportional to frequency. We saw above that light with a frequency of 1×10^{14} cycles per second has a photon energy of 0.41 electron-volt. Therefore, red light must have a photon energy 3.8 times as great, or

$$3.8 \cdot 0.41 = 1.6 \text{ electron-volts}$$

To capture the essence of light, one must use both the wave concept and the photon concept. Unfortunately, scientists have found no single concept that includes all the attributes of light. Some experiments bring out the wavelike properties of light. An example is using a prism to spread

Figure 7.8. *The electromagnetic spectrum, with an energy scale.*

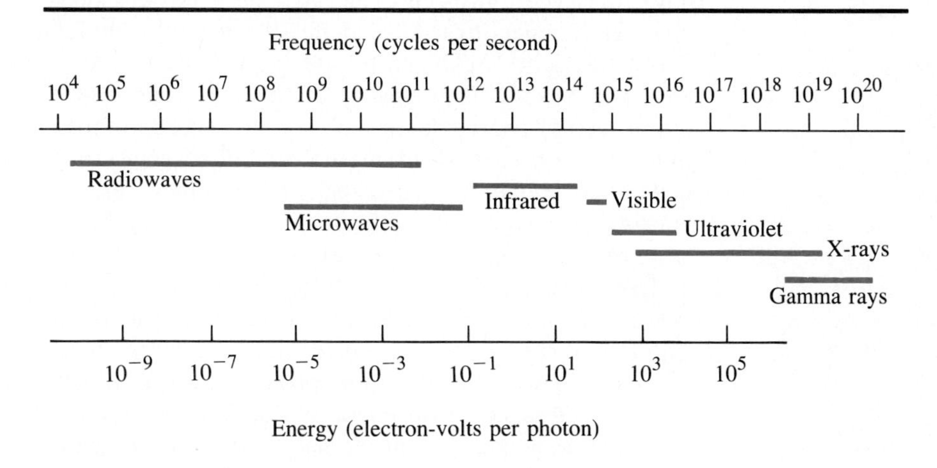

out the colors hidden in a sunbeam. In different circumstances, such as the photoelectric experiment, the particulate nature of light is the key to understanding. This is not to imply that light changes back and forth between wave and photon—it is a single entity. We give it a dual nature because the familiar ideas of waves and particles are all that we have to work with. These ideas might seem inadequate for dealing with the microscopic world of molecules, atoms, and electrons, but scientists have learned to work with this dual nature of light.

7.4 LIGHT AND ATOMS

Long before Planck proposed the quantum idea and Einstein used it to explain the photoelectric effect, scientists had been using the spectroscope to identify "chemical fingerprints." Each element was known to give rise to a different emission spectrum when it was excited by heat or electricity. Lockyer used such an analysis when he discovered helium on the sun. Spectral analysis is simplified by the fact that spectra of the elements do not contain all colors in a continuous "rainbow." On the contrary, the spectra have individual lines of color at exact frequencies—a strange, even astounding, fact. No light at all appears over the frequencies between the bright lines; these areas appear black.

The set of lines with their associated frequencies or wavelengths is called a **line spectrum.** It was not difficult for scientists to catalog the exact frequency and brightness of each colored line in the spectra of neon, sodium, hydrogen, and the other elements. After that, newly discovered samples of gas and rock could be analyzed to see which elements they contained. The process is much like solving a puzzle.

Take the line spectrum of the element neon, Ne, for example. The neon lights that add color to many commercial buildings in the United States have a bright orange-red color. A neon light consists of a glass tube with electrical connections, filled with neon gas at a moderately low pressure. When the light is plugged in and the electricity turned on, the electrical energy makes the neon glow red. Our eyes cannot resolve the spectral lines of the neon discharge; instead we see a composite of them all. Neon happens to have twenty-one bright lines in its visible spectrum; ten intense lines are at the red end of the spectrum, and six more are in the yellow-orange portion. Our eyes see mainly the red light, with a hint of orange. The blue and green lines are swamped by the intensity of the red and orange.

The spectrum of sodium is quite different. It has only seven bright lines (see Figure 7.9). Four of them are much more intense than the other three. These four lines are in the yellow and orange region of the light spectrum. So what do we see when we sprinkle sodium chloride onto a flame? A brilliant yellow-colored flame. As von Goethe wrote in the poem *Faust,* "in the yellow flame of an ordinary lamp, whose wick was sprinkled with salt, he saw the possibility of analyzing the most distant stars."

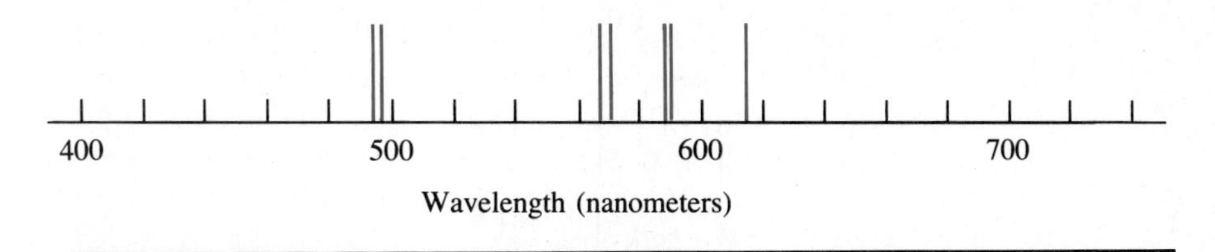

Figure 7.9. The line spectrum of sodium, Na.

7.5 SPECTRUM OF THE HYDROGEN ATOM

Hydrogen, the lightest element, emits rose-magenta light when it is excited in a glass tube by an electric current. The input of electricity breaks apart the H_2 molecules into H atoms, and it is the excited H atoms that give off the visible light. Human eyes cannot see ultraviolet light, although bees can, it seems. At any rate, it is possible to analyze ultraviolet light in a simple manner, using an ultraviolet spectroscope. These instruments show that the atomic spectrum of hydrogen contains a series of lines in the ultraviolet region in addition to the series in the visible region. There is also a third set of emission lines in the infrared portion of the electromagnetic spectrum. The line spectrum of atomic hydrogen is shown in Figure 7.10.

The middle group of lines in Figure 7.10 is what you would see if the light from a hydrogen discharge tube was passed through a prism and then was focused on a screen, as shown in Figure 7.1. Most of the screen is blank. At particular locations there are narrow bands of light, corresponding to the special frequencies emitted by hydrogen atoms after they have been energized by electrical energy. These lines are generally identified by their wavelengths rather than by their frequencies. [Remember the reciprocal relation between these two terms ($\nu = c/\lambda$); a short wavelength means a high frequency.]

Figure 7.10. Major lines in the spectrum of the hydrogen atom. There are 1 × 10^9 nanometers in 1 meter. The separations between the three groups of lines are actually much greater than those shown here.

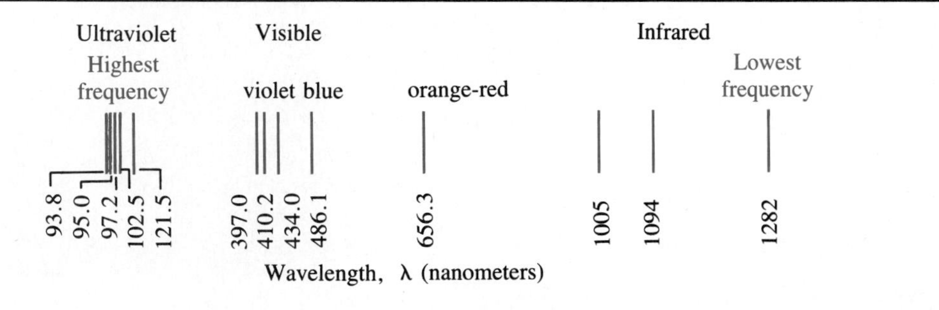

The emission of light from atoms that have been energized by electricity or heat is a general phenomenon. But where does the light come from? Scientists suspected that the lines in an atomic spectrum arise when excited electrons return to more stable arrangements. Remember the photoelectric effect, in which the energy of a photon can be used to make an electron go to a higher energy level and leave the atom altogether. It seems reasonable for light to be associated with the process of an electron's changing energy states. After all, light is a form of energy.

7.6 STRUCTURE OF THE HYDROGEN ATOM

The exact explanation of the line spectrum of the hydrogen atom was the work of the Danish physicist Niels Bohr (Figure 7.11). Basing his approach on the nuclear atom hypothesized by Rutherford, Bohr assumed that the hydrogen atom consists of a central proton around which an electron moves. In keeping with the quantum ideas of Planck and Einstein, Bohr reasoned that an atom can attain only certain electronic energy states and that electrons can stay in these states indefinitely. Only when an electron changes from one energy state to another will an energy exchange take place. If an excited electron returns to a state of lower energy, closer to the positive nucleus, energy must be given off. If the change in energy is large enough, visible or even ultraviolet light is emitted.

These energy changes must agree with the Law of Conservation of Energy,

Figure 7.11. The Danish physicist Niels Bohr.

which says that energy is neither created nor destroyed. When an electron in a hydrogen atom is energized by electricity, the increase in energy of the electron exactly equals the consumption of electrical energy (Figure 7.12). When the energized electron drops to a lower energy state with the emission of a photon of light, the decrease in the energy of the electron exactly equals the energy of the photon created. It is like the collisions of balls on a pool table. In each collision, the gain in energy of one ball equals the loss in energy of the ball that strikes it.

Bohr assumed that in each energy state the electron moves in a circular orbit about the nucleus. The greater the energy, the greater is the radius of the orbit. The unique feature of Bohr's model was the assumption that only certain orbits and energies are possible. By contrast, a satellite circling the earth can have any orbit and energy of motion, provided that the satellite is high enough that air resistance does not slow it down. Using the ideas of classical physics and the quantum ideas of Planck and Einstein, Bohr arrived at the following equation for the energy of the electron:

$$E = -\frac{B}{n^2}$$

The constant B contains a number of fundamental physical constants, including the mass and the electrical charge of the electron; Planck's constant, h; and the atomic number of hydrogen. Bohr put limits on the energy values possible for the hydrogen atom's electron by assuming that n can have only integral values; that is, $n = 1, 2, 3, 4, 5, \ldots$, etc. The lowest energy state for the hydrogen atom is $n = 1$. Higher energy states are $n = 2$, $n = 3$, etc. The state in which $n = 1$ is called the **ground state,** because the

Figure 7.12. An energy diagram showing the changes that occur when an electron is excited by electrical energy and then loses energy by emitting a photon of light. Each horizontal line stands for one of the possible energy levels of an electron in a hydrogen atom.

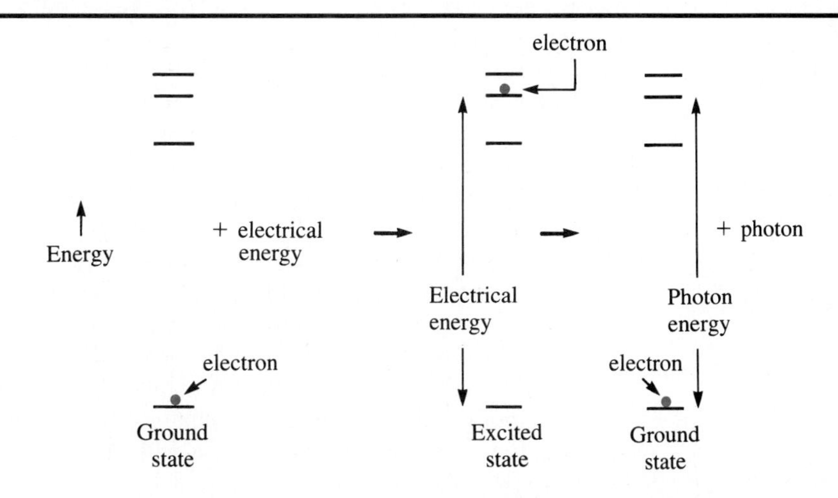

electron has the lowest possible energy, as if it were resting on the ground. Higher-energy electronic states are called **excited states.** The number of the energy state, which we have designated by the symbol n, is referred to as the **quantum number.**

Because the opposite charges of the nucleus and the electron attract each other, the potential energy of each allowed state is negative (see page 79). The sum of the kinetic energy of the electron, which is always positive, and this potential energy gives the total energy of the hydrogen atom. Although the kinetic energy cancels some of the negative contribution from the potential energy, the potential energy is always larger in magnitude. Therefore, the total energy, given by Bohr's equation, is negative.

Example In the equation $E = -B/n^2$, why does the state in which $n = 1$ correspond to the lowest energy?

Solution The equation may be somewhat confusing because of the minus sign. The magnitude of B/n^2 is greatest when n is smallest—that is, when $n = 1$. Because of the minus sign, the value with the largest magnitude is the most negative. On the number line the most negative value is the smallest, or lowest.

Using his equation, Bohr calculated the energies of electronic energy states with different values of n. Then he made a dramatic correlation. He calculated the amounts of energy that should be given off when electrons move from high energy states to lower ones. The answers fit almost exactly with the wavelengths of visible light in hydrogen's atomic spectrum! Bohr's calculations also predicted a series of spectral lines in the ultraviolet range, which were discovered later in almost the exact positions Bohr had predicted (Figure 7.10). Light does indeed provide a message as to the position and energy of the electron in a hydrogen atom. The relationship between hydrogen's electronic energy states and the wavelengths of emitted light in its spectrum is depicted in Figure 7.13.

The four arrows pointing down in Figure 7.13(a) show changes that may occur after the electron has been excited to the $n = 2$, 3, or 4 level by heat or electricity. The excited states won't last very long, as the electron "prefers" to be in the ground state at usual temperatures. It can return to the ground state by releasing energy as a photon of light. If the quantum number in the excited state is 3 or greater, the electron need not go directly to the ground state. Two of the arrows show the electron dropping to the $n = 2$ state, which has an energy between those of the $n = 3$ and $n = 1$ states. Because the photons emitted are of smaller energy, these arrows are shorter than the others. After a brief time, the electron in the intermediate ($n = 2$) state will drop to the innermost and lowest energy orbit ($n = 1$), releasing another photon. Of course, many other arrows, corresponding to many other possible energy changes, could also be drawn.

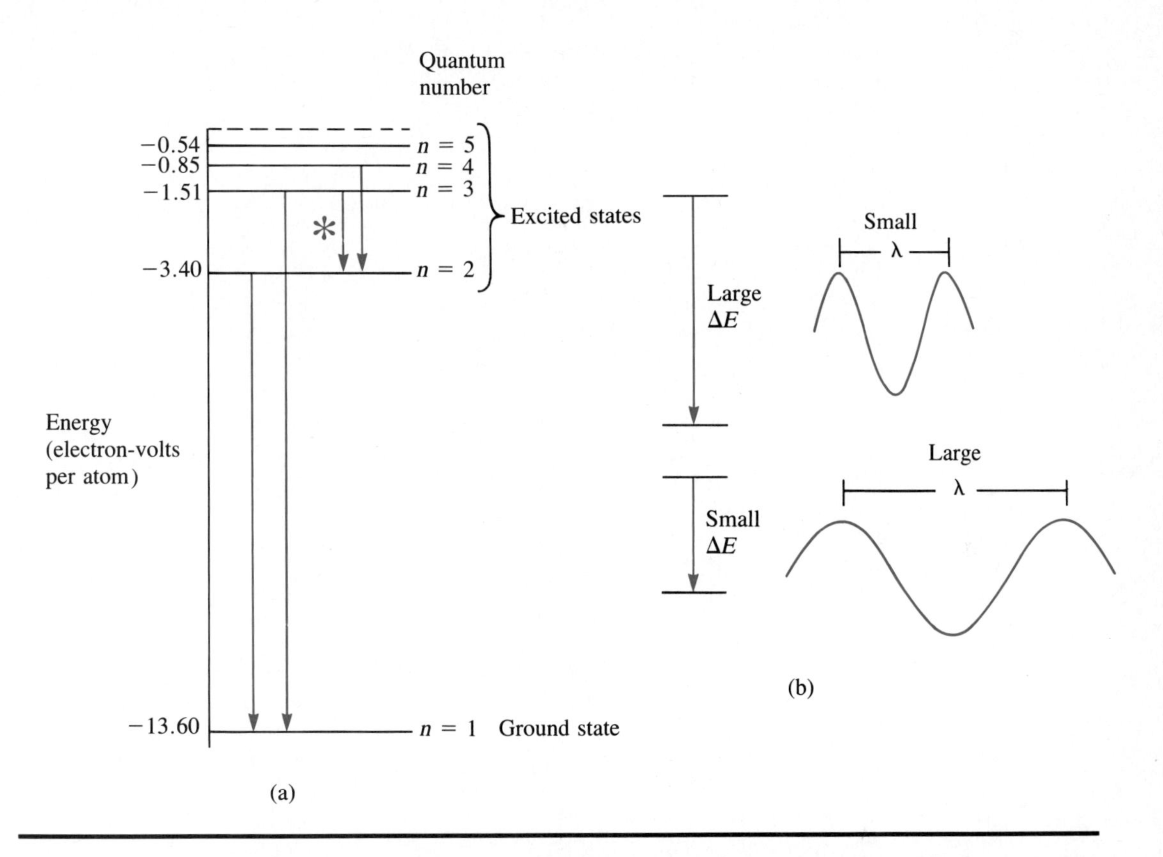

Figure 7.13. (a) Energy of the electronic states of a hydrogen atom and some sample energy changes. The ground state is 13.60 electron-volts more stable than the state in which the electron is completely separated from the nucleus. (b) The inverse relationship between the wavelength of the emitted light and the change in electronic energy. If ΔE is large, λ is small (e.g., in the ultraviolet range).

Example A hydrogen atom has been excited by electrical energy to the $n = 3$ state. It then goes directly to the $n = 1$ state, emitting a photon. Another hydrogen atom is excited to the $n = 2$ state, after which it returns to the ground state, emitting a photon of electromagnetic radiation. Which electromagnetic radiation has the longer wavelength? Explain.

Solution The difference in energy between the $n = 3$ and $n = 1$ states is greater than that between the $n = 2$ and $n = 1$ states. Therefore, the energy of the emitted photon is greater in the first case. A large photon energy implies a large frequency ($E = h\nu$) and a small wavelength ($\lambda = c/\nu$). So the atom in the $n = 2$ state emits radiation of longer wavelength.

Table 7.1 Changes in the Hydrogen Atom's Quantum States and the Corresponding Wavelengths of Light Emission. (The atomic spectrum of hydrogen appears in Figure 7.10.)

ENERGY STATES OF THE ELECTRON		WAVELENGTH OF EMITTED LIGHT (in nanometers)
HIGHER ENERGY, *n*	LOWER ENERGY, *n*	
2	1	121.5
3	1	102.5
4	1	97.2
5	1	95.0
6	1	93.8
3	2	656.3
4	2	486.1
5	2	434.0
6	2	410.2
7	2	397.0
5	3	1282
6	3	1094
7	3	1005

The upper and lower energy states for all the lines shown in Figure 7.10 are given in Table 7.1, which completes the connection between Figures 7.10 and 7.13. Of course, only a fraction of the lines in the complete spectrum of the hydrogen atom are included in Figure 7.10 and Table 7.1.

7.7 QUANTUM MECHANICS

The Bohr theory presented a picture of atoms whose nuclei have electrons circling around them in specific energy levels. The higher the quantum number, the higher the electron's energy and the farther away it is from the nucleus (see Figure 7.14). It was an appealing picture, but soon a serious flaw became apparent. Although the theory worked beautifully in accounting

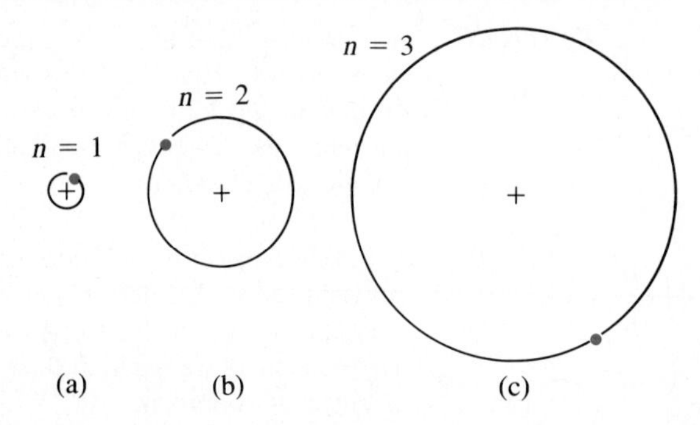

Figure 7.14. The Bohr picture of the hydrogen atom, showing the first three allowed orbits for the electron. (a) The electron is in the n = 1 orbit. (b) The electron is in the n = 2 orbit. (c) The electron is in the n = 3 orbit.

for the atomic spectrum of hydrogen, it was not able to account for the spectra of more complex atoms. Predicted electronic energy levels did not correspond with the known spectral lines. In addition, Bohr's model couldn't account for the bonding of atoms in molecules. A new resolution of these problems was needed, and it came eleven years after Bohr's planetary model was first announced.

Think back for a moment to the models of nineteenth-century physics. Matter had mass and was particulate; light was a wave. Then came quantum theory, and light was seen to have both wave and particle (photon) properties. Louis de Broglie, a young French physicist, completed the cycle in 1924. He asked, If light has both particle and wave properties, couldn't matter also have both wave and particle properties? At first the thought may seem shocking—the floor of your room and the chair in which you sit made of waves? Wouldn't we all get seasick?

With the help of his highly developed intuition and a strong mathematical background, de Broglie proposed an equation relating the wavelength associated with a moving particle to the mass and speed of the particle:

$$\lambda = \frac{h}{mv}$$

This equation implies that any particle of mass m and velocity v will have a wavelike character, the wavelength of which will be equal to h/mv. Take a bullet that weighs 2 grams and is traveling at 300 meters per second. One can calculate that the bullet's wavelength would be 1×10^{-24} nanometer. The electromagnetic spectrum only goes down to a wavelength of 1×10^{-7} nanometer or so. A wavelength of 1×10^{-24} nanometer is below any detectable limit, and the bullet's wave characteristics wouldn't affect any observation that we could make. So, a bullet acts as matter should act.

An electron is much lighter than a bullet; it has a mass of 9×10^{-28} gram. If it were moving at a velocity 0.3% of the speed of light, the electron's wavelength would be 1 nanometer (see Figure 7.15). This is the wavelength

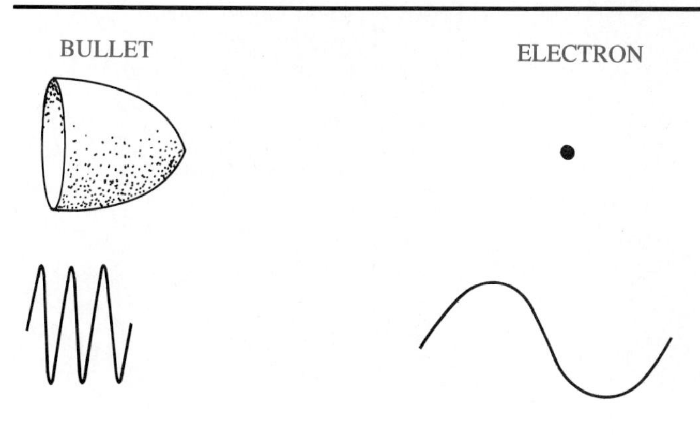

Figure 7.15. Contrast between the wavelength of a bullet and the wavelength of an electron. The difference in wavelengths is actually much greater than that shown here.

THE ELECTRON MICROSCOPE

The electron microscope has become an important research instrument in chemistry and biology during recent decades. It is found in research laboratories, hospitals, and the chemical industry. The key to its operation is the wavelike character of the electrons in the beam that strikes the sample under observation. First, the wavelength of this beam is about one ten-thousandth of the wavelength of visible light, and the smaller the wavelength the greater the ''resolving power,'' or ability to distinguish small objects. Also, the magnification possible with an electron microscope is perhaps 500 times that possible with an

Figure 7.16. Four pictures from an electron microscope. (a) The head of a bluebottle fly. (b) The surface of a bluebottle fly's eye. (c) A DNA molecule. (d) A piece of graphite (carbon).

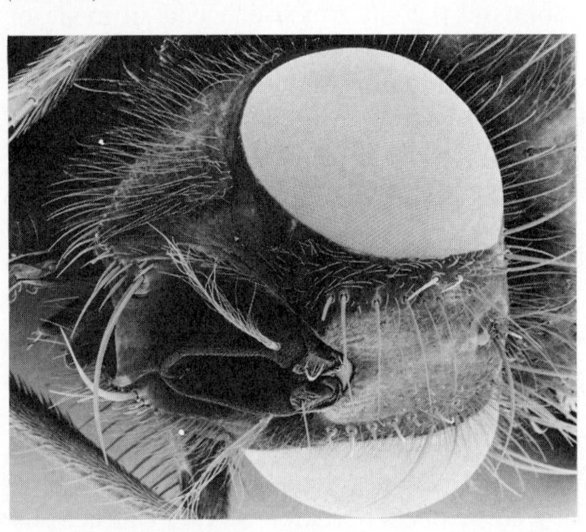

(a)

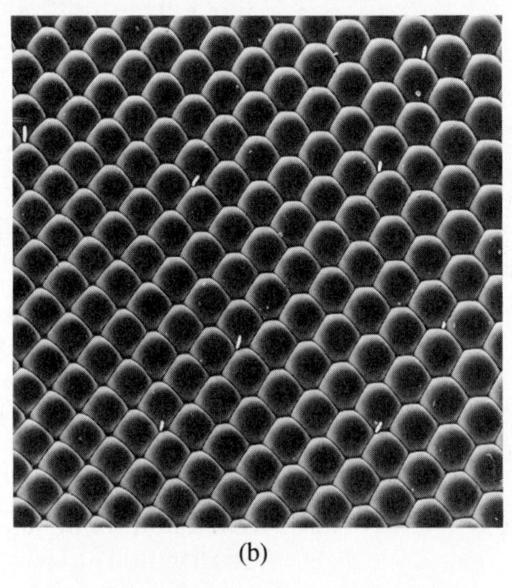

(b)

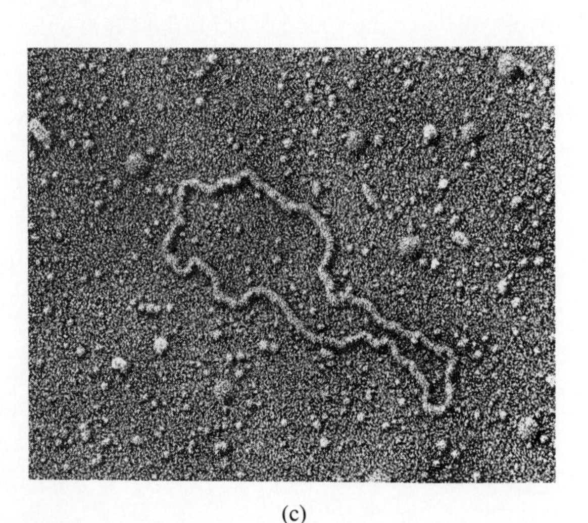

(c)

(d)

optical microscope. These two advantages allow the operator of an electron microscope to observe objects with a diameter as small as 0.3 nanometer, which is the diameter of an individual atom. In routine use, however, the limit of observation is about 100 nanometers.

Because electrons are easily deflected by molecules of air, air must be pumped out of an electron microscope until the pressure is 10^{-7} atmosphere or less. An ordinary microscope has glass lenses for focusing; the electron beam in an electron microscope is focused by magnetic fields. After the electrons hit the sample, they are directed to either a fluorescent screen or photographic film. Scientists have used electron microscopes to examine many kinds of samples, including various surfaces, dislocations in metals, and viruses and bacteria. Usually the sample must be very thin (no more than 250 nanometers thick) so that electrons can go through it. Figure 7.16 gives four examples of pictures from an electron microscope.

---

of a typical x-ray beam. Can an electron beam be reflected like an x-ray beam? The answer is yes. This reflection is the basis of the electron microscope, which has been of such tremendous value to biologists and chemists. A beam of electrons has better resolving power than a beam of visible light does, since the wavelength of an electron is far smaller.

Consider an electron as it floats around the nucleus of an atom. It is a very small, very fast-moving particle that seems to have wave properties. How can we find its position and speed? The natural solution to the problem is to use a beam of light—or more generally, electromagnetic radiation—as a probe (see Figure 7.17). The detail with which we can observe objects depends profoundly on the wavelength of the radiation used. The kind of radio wave employed in radar has a wavelength of a meter or more and provides the crudest means of observation. Any object whose largest dimension is less than a meter will not even show up on a radar screen. Visible light, however, with its wavelengths of several hundred nanometers, gives very fine detail. To paraphrase the late Jacob Bronowski, a scientist and philosopher, "An object will intercept a ray only if it is as long as the wavelength of the radiation; a smaller object simply will not cast a shadow." To get an accurate position of an electron, we need a super-microscope that uses radiation of very short wavelength, such as gamma rays.

Now we encounter a problem. Radiation of extremely short wavelength consists of photons of very great energy. When such a photon strikes an electron on passing through our super-microscope, a considerable amount

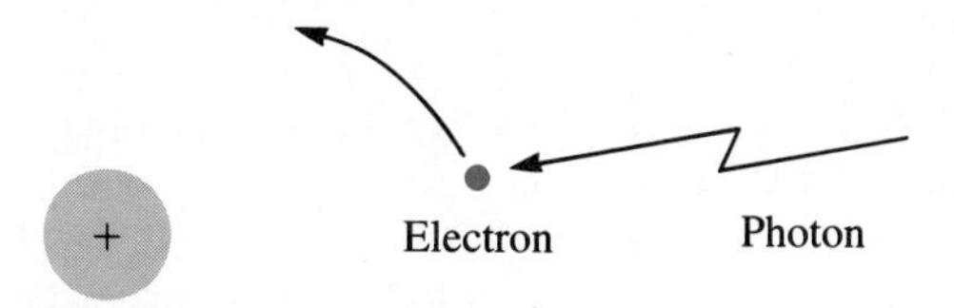

Figure 7.17. A photon of electromagnetic radiation striking an electron in an atom. Since no one has witnessed this process, artistic imagination has been used.

of energy may be transferred to the electron. As a result, the electron may move in a new direction with a very different speed. Any measurement of the speed of this electron must be inexact, because the speed is altered drastically by the process of finding the position. The shorter the wavelength, the more accurately we know the position of the electron but the less accurately we know the speed. At the opposite extreme, for radiation of longer wavelength speed can be accurately determined but position can be located only approximately.

An analogy based on large objects may clarify the problem of the uncertainty introduced by the process of measurement. Imagine you are standing blindfolded next to a river. A small, broken branch is floating downstream near you, and you would like to determine its position and speed. The only method you can think of is to throw stones into the river. Whenever a stone strikes the branch, you hear a sound and can tell about where the branch is. However, the impact of the stone against the branch pushes the branch from you and changes its path. The stone may even slow it down. The next time a stone hits the branch, the branch is not where it would have been if the first stone had not hit it. Therefore, your knowledge of how far the branch would have gone since the last "hit" is only approximate, and you can't calculate an accurate speed (distance traveled divided by the time interval).

At the atomic level, a photon is a projectile that changes the motion of any electron in its way. Any instrument we invent to study the motion of an electron in an atom must use radiation and, therefore, photons. Thus our results must always be uncertain.

Werner Heisenberg stated this dilemma in his famous **Uncertainty Principle** in 1927. In equation form, the principle is

$$\Delta(mv) \cdot \Delta x = h$$

Here the Δ means the uncertainty in measuring the momentum (mv, or mass times velocity) or the position (x) of an electron. This equation states mathematically what we have just discussed regarding electrons in atoms. A small uncertainty in position implies a large uncertainty in momentum and vice versa, because the product of the two is always the same value—Planck's constant. In order to know the exact path of an electron, we need to know its exact position and momentum at several times. The Uncertainty Principle tells us that we cannot even hope to accurately determine electronic pathways.

Example

A person looks through a microscope at a tiny organism moving through a drop of water on a slide. As she does so, visible light strikes the organism and then reflects through the lenses of the microscope into her eye. The organism has a mass a trillion trillion times greater than the mass of an electron. Why isn't the path of the organism changed by the photons of light in the microscope?

Solution When a photon of visible light hits the organism, it is like a ping-pong ball striking a moving bulldozer. The bulldozer continues on its way as though the ping-pong ball never existed, because the energy of the ball is insignificant compared to the energy of the bulldozer. Similarly, since the energy of a photon of visible light is negligible relative to the energy of the moving organism, the photon has no noticeable effect on its path.

One can see the physical basis for the Uncertainty Principle, but it still seems strange that there should be limits in principle on how accurately we can measure the fine details of nature. The implications of the quantum view of matter have fascinated three generations of philosophers and scientists. We may hesitate to believe that our understanding of nature is inevitably uncertain, but our best theories tell us that it is. If there is a more fundamental reality behind the **wave-particle duality** (that is, the exhibition by electrons of both wavelike and particle-like properties), we have not yet discovered it.

7.8 ORBITALS FOR ELECTRONS

Our last remaining task in this chapter on quanta is to see how quantum mechanics can account for the energies of electrons in a way that is consistent with known atomic spectra. We also want our theory to account for the bonding of atoms in molecules. The fundamental answer—one that is the foundation of our current thinking—was proposed by German physicist Erwin Schrödinger in 1926. He said that if an electron is treated as a wave rather than as a particle, its allowed energy states can be described by the mathematics of three-dimensional wave behavior. Schrödinger's famous wave equation can in principle describe the behavior of the electrons in any atom, even the most complicated. In practice, the mathematics is so demanding that exact solutions are impossible; but approximate solutions are possible, and they tell us a great deal about the electronic states of atoms.

The Schrödinger equation is presented here for you to see, since it has such dominance in modern chemistry and physics. It does look impressive!

$$\frac{\delta^2\psi}{\delta x^2} + \frac{\delta^2\psi}{\delta y^2} + \frac{\delta^2\psi}{\delta z^2} + \frac{8\pi^2 m}{h^2}\left(E + \frac{Ze^2}{r}\right)\psi = 0$$

It is probable that over 95% of the chemists in the world cannot solve the Schrödinger equation. Yet the qualitative ideas that come out of it are tremendously helpful. Schrödinger proposed that an electron could be described by a wave function, the symbol for which is the Greek letter ψ (psi). If we can calculate ψ, it is possible to learn the energy of the electron

and its average distance from the nucleus. From the Uncertainty Principle we know that it is impossible to be exact about the path of an electron around the nucleus. However, the wave function, or **orbital,** of an electron describes its energy and where we can, with a specified probability, expect to find it.

Specifying a probability involves making many observations of a phenomenon that can have several outcomes. The **probability** of a particular outcome is the number of times it is observed, divided by the total number of observations. For example, if in 1000 observations an electron is found to be in a certain small region of the atom 28 times, then the probability that the electron will be in this location is $28/1000 = 0.028$. When we describe the orbital of an electron in an atom, we tell where the electron is most likely to be. This is done by giving the region in which the electron races around 95% of the time. In other words, the probability that the electron will be found in this region is 0.95.

Schrödinger's pictures of electron orbitals are very different from the well-defined circular orbits proposed by Bohr. That aspect of Bohr's theory must be discarded. It is no longer reasonable to compare the motions of electrons in atoms with the movement of planets about the sun in well-defined orbits. Both theories, however, emphasize the limits on the energies of the electron in the hydrogen atom. This idea must surely be valid, because from it Bohr was able to predict correctly the many lines in the hydrogen spectrum. The quantum number n proposed by Bohr is retained in the more recent theory but is given a new name—the **principal quantum number.** Just as in the Bohr theory, n determines the energy of the electron. It also determines the number of orbitals that are possible for each allowed energy of the electron. With a principal quantum number of 1 ($n = 1$), only one orbital is possible. This orbital is depicted in Figure 7.18(a). There is a 95% probability of finding the electron within the sphere shown.

When $n = 2$, there are four possible electronic orbitals. One of these orbitals looks like the one shown in Figure 7.18(a), only larger. With the larger quantum number, the electron is farther away from the nucleus on the average. The other three orbitals with $n = 2$ are depicted in Figure 7.18(b). It is possible to have three dumbbell-shaped orbitals, one oriented in each of the three dimensions in space.

A spherical orbital, such as that shown in Figure 7.18(a), is called an ***s* orbital.** When $n = 1$, we have a 1*s* orbital; when $n = 2$, we have a 2*s* orbital; and so on. A dumbbell orbital, as shown in Figure 7.18(b), is called a ***p* orbital.** The two other possible shapes are designated *d* and *f*.

It has been shown that there are n^2 possible orbitals for each value of the principal quantum number. For $n = 3$, there are nine possible orbitals. For $n = 4$, there are sixteen orbitals. The electrons of an atom are found in these orbitals, or energy states. As in the Bohr theory, an electron moves to a new orbital only with the gain or loss of energy. Electricity or a flame can excite an electron to a higher-energy orbital. When an electron moves back to a lower-energy orbital, light is emitted. An additional assumption in this orbital theory is that the maximum number of electrons in any orbital is two. The electrons in an atom must be spread over enough orbitals that each orbital contains no more than two electrons. The importance of this limitation will be discussed in Chapter 8.

Naturally, the most stable electronic arrangements are those with the electrons in the lowest-energy orbitals. A hydrogen atom has one electron,

Figure 7.18. Atomic orbitals. The nucleus is at the center in each instance. (a) An s orbital. (b) A set of three p orbitals.

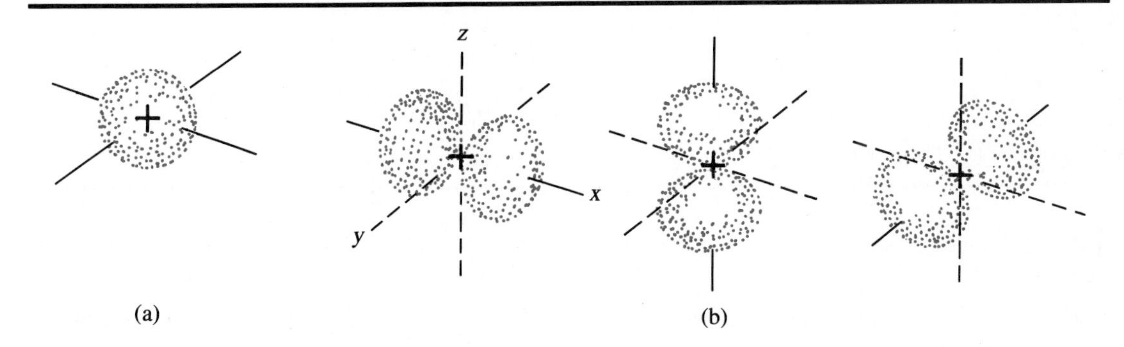

which would be in the 1*s* orbital. The two electrons of a helium atom also both fit into the 1*s* orbital. This arrangement is of lower energy than one in which helium's second electron is placed in the much-higher-energy 2*s* orbital. Lithium, the element with atomic number 3, has three electrons. Two go into the 1*s* orbital, but the third one cannot fit there, so it ends up in the 2*s* orbital.

The method for determining which orbitals will contain electrons is called the *Aufbau, or Building-up, Principle*. In the ground state of an atom, electrons are found in the orbitals of lowest possible energy. By counting the electrons and assigning them to the lowest possible orbitals, one can build a picture of the arrangement of electrons in an atom. However, before we go further, it is necessary to talk about the periodic properties of the chemical elements.

7.9 SUMMARY

Since some difficult concepts have been presented in this chapter, it may be helpful to review the main ideas. To learn about the behavior of electrons in atoms, we make use of electromagnetic radiation, which was once thought to be purely a wave. With the successful interpretation of the photoelectric effect, however, scientists realized that electromagnetic radiation also exhibits a particle-like nature under certain circumstances. These packets of energy were named photons. The need to envision light in terms of a wave-particle duality is a dilemma from which we have not yet escaped.

The presence of a very limited number of wavelengths in the light emitted by excited atoms means that the electrons in the atoms can only have certain energies. Niels Bohr successfully predicted these energies for the hydrogen atom but not for other atoms. Bohr's theory was inadequate because it did not take into account the Uncertainty Principle and the wavelike character of very small particles such as atoms and electrons. This second wave-particle duality was dealt with by Schrödinger, who described the waves associated with an electron in an atom. Recognizing that we can never determine the exact path of an electron in an atom, we discuss only the probabilities of finding the electron in various parts of the atom.

IMPORTANT TERMS

light A form of energy to which the human eye responds.

spectroscope An instrument for separating light or electromagnetic radiation into its component frequencies.

spectrum A colored display of a range of frequencies of visible light. In a more general sense, a spectrum is a range of frequencies of any kind of electromagnetic radiation (see also *electromagnetic radiation*).

wave In this chapter, a regular variation in electric (and magnetic) force in a certain region of space. There are also other kinds of waves, such as sound waves, waves on the surface of a liquid, etc.

wavelength The distance between two repeating, adjacent points in a wave. The symbol for wavelength is λ (lambda).

frequency Number of wavelengths passing a fixed point in one second. The symbol for frequency is ν (nu).

cycle The pattern (one wavelength long) that repeats to make up a complete wave.

electromagnetic radiation A general term for a wave of any frequency or wavelength. Examples are infrared, visible, and ultraviolet radiation.

infrared That region of the spectrum having frequencies just below those of the visible region.

photoelectric effect The release of electrons from a metal surface brought about by shining light on it.

ultraviolet That region of the spectrum having frequencies just above those of the visible region.

photon A packet of energy in a beam of electromagnetic radiation.

quantum When applied to radiation, a photon. More generally, it refers to the theories of electrons in atoms developed by Bohr and Schrödinger.

line spectrum The small set of frequencies of radiation emitted by atoms that have been energized by electricity or heat.

ground state The atomic state in which the electrons have the lowest possible energy.

excited state Any atomic state in which the energy is greater than that of the ground state.

quantum number In the Bohr theory of the hydrogen atom, a number (n) that determines the energy of the electron and the radius of its orbit. It can only be a whole number.

Uncertainty Principle Heisenberg's statement of the limits to the accuracy with which one can determine both the position and the speed of an electron in an atom.

wave-particle duality The exhibition by light and by electrons of both wavelike and particle-like properties.

orbital The wave associated with an electron in an atom. The probability of finding the electron in a particular region of space depends on the amplitude of the wave in that region.

probability The chance that a particular outcome will occur in a series of observations for which several outcomes are possible. Probability is equal to the number of times the given outcome occurs divided by the total number of observations.

principal quantum number A quantum number in the Schrödinger theory of the atom; identical with the quantum number, n, of the Bohr theory.

***s* orbital** An orbital in which the electron is most likely to be located in a region of space shaped like a sphere with the nucleus at the center.

***p* orbital** An orbital in which the electron is most likely to be found in a region of space shaped like a dumbbell with the nucleus at the center.

QUESTIONS

1. a. A radio wave has a frequency of 6×10^7 cycles per second. What is the energy of a photon of this radiation?

b. What is the photon energy of an x-ray having a frequency of 3×10^{17} cycles per second?

c. Which has the longer wavelength, a radio wave or an x-ray?

2. It is the small amount of ultraviolet light in the sun's rays that causes skin to tan. Thinking to fool nature, Lisa decided to use a high-powered infrared heat lamp in order to get a tan. She got a bad burn but not a tan. Why?

3. While welding metal, welders wear thick eye masks. Why do they avoid seeing so much of the beautiful blue light that their flame produces?

4. Would you be aware of radio waves bombarding your skin? Why or why not?

5. In order to expel an electron from an atom of aluminum metal by the photoelectric effect, a photon must have at least 4.2 electron-volts of energy. What is the lowest frequency of light that can be used to expel electrons from the surface of aluminum?

6. The photoelectric threshold for sodium metal is visible light with a frequency of 4.6×10^{14} cycles per second. What minimum energy must a photon have to eject an electron from a sodium atom?

7. Both yellow light and blue light cause electrons to be ejected from sodium metal in the apparatus shown in Figure 7.4(b).

a. For which kind of light is the kinetic energy of the ejected electrons greater? Explain.

b. Does the intensity of the light affect the kinetic energy of the ejected electrons? Explain.

8. Two hydrogen atoms have been energized by electrical energy to the $n = 5$ state. The electron in the first atom then drops to the $n = 4$ state, emitting a photon of electromagnetic radiation. The electron in the second atom goes to the $n = 3$ state, emitting a photon of electromagnetic radiation. With the help of Figure 7.13, explain which electromagnetic radiation has the shorter wavelength.

9. The following statement comes out of the suggestion in quantum mechanics that if light has particle characteristics, then matter could have wave characteristics. From this point of view, explain the rationale of the statement "If a proton and an electron are traveling at the same velocity, the electron has a larger wavelength than the proton."

10. The planets in our solar system are constantly bathed in light from the sun. An astronomer can observe a planet (Jupiter, for example) by looking through a telescope at sunlight that bounces off the planet and comes through the telescope. Explain why the orbit of a planet is exactly the same as it would be if no sunlight interacted with the planet.

11. The de Broglie equation, $\lambda = h/mv$, can be derived mathematically using three equations: $E = h\nu$, $E = mc^2$, and $\nu = c/\lambda$, where c is a velocity term like v. Show the derivation of the de Broglie equation.

12. Occasionally, someone studying quantum mechanics has the mistaken impression that an electron in a p orbital only travels along the surface of the dumbbell shape shown in Figure 7.18(b). Explain the fallacy in this impression.

13. "The chemistry of the Earth and its sun obeys the same laws as does chemistry elsewhere throughout the Universe." On what kinds of observations does this statement depend?

The Periodic Table

*

Since the time of Mendeleev, chemists have come to regard the periodic table of the elements as their most powerful tool for tying chemical facts together. Trends in many physical and chemical properties can be predicted from the positions of elements in a row or column of the periodic table. In Chapter 3 we saw that the elements in a vertical column, more often called a family or **group,** have similar electrical properties and similar formulas in compounds with an element outside the group. For example, the compounds of hydrogen with elements in Group VI (see inside back cover) all have two hydrogen atoms combined with one atom of the Group VI element. The formulas are H_2O, H_2S, H_2Se, and H_2Te (see Table 8.1). Horizontal relationships are also important. The term **period** is used to refer to the elements in a horizontal row. The first period consists of hydrogen and helium, the second period begins with lithium (Li) and ends with neon (Ne), and so on.

Although Mendeleev arranged the elements in order of increasing atomic weight, he was aware that several pairs of neighboring elements—such as K and Ar, and Te and I—were improperly placed. Their properties were not at all like those of the other elements in their groups (see Figure 8.1).

Table 8.1 Formulas of Compounds of Group VI Elements with Two Other Elements

ELEMENT	FORMULA OF HYDROGEN COMPOUND	FORMULA OF BARIUM COMPOUND
oxygen, O	H_2O	BaO
sulfur, S	H_2S	BaS
selenium, Se	H_2Se	BaSe
tellurium, Te	H_2Te	BaTe

After he reversed their order, they fit very well. Mendeleev knew nothing about the internal structure of the atom, but this revised order corresponds exactly to atomic number, or the number of protons in the nucleus of an atom. The atomic number is also equal to the number of electrons in the neutral atom. The number of protons and the number of electrons distinguish atoms of one element from those of another. Moreover, the distribution of electrons in various orbitals is the key to the properties of an element.

We still need a few more terms to describe the periodic table. The elements in columns I through VII (inside back cover) are called **representative elements.** These best exhibit the various trends that we will discuss. At the far right is a single group composed of the **noble gases,** which for the most part have little tendency to react with other elements. They are ''noble'' because they do not mix with the ''common'' elements. Dividing the representative elements into two parts are the **transition elements** in the fourth,

Figure 8.1. Groups VI and VII, with the elements arranged according to increasing atomic weight. Because the criterion is atomic weight, Te and I are placed incorrectly.

VI	VII
O 16.0 8 H_2O	F 19.0 9 HF
S 32.1 16 H_2S	Cl 35.5 17 HCl
Se 79.0 34 H_2Se	Br 79.9 35 HBr
I 126.9 53 HI	Te 127.6 52 H_2Te

KEY

X	Symbol
10.0	Atomic weight
5	Atomic number
HX	Formula of compound with hydrogen

fifth, and sixth periods. Specifically, they are the elements Sc throu
Y through Cd, and La through Hg. They also include, at the tag
the table, atomic numbers 104 through 108, which are the newest ar
elements. Finally, there are the rare earth elements, or **lanthanides,** nun
58 through 71, and the **actinides,** numbers 90 through 103.

8.1 HOW THE PERIODIC TABLE HELPS—METALS AND NONMETALS

Let's look at some physical and chemical properties and how they correlate with the positions of elements in the periodic table. In Chapter 3 we briefly considered one property: the metallic or nonmetallic behavior of an element, as shown by electrical conductivity. The heavy zig-zag line on the periodic chart separates the metals on the left from the nonmetals on the right. Recall that **metals** are generally good conductors, and **nonmetals** are usually poor conductors. Most metals have a shiny appearance and can be drawn into wires or pounded into sheets. A more useful property for defining metallic character is the minimum energy required to remove an electron from an isolated atom of the element. This property is similar to the energy needed to eject an electron from a metal in the photoelectric effect, except that now the atoms must be separate from each other. In other words, atoms of the element must be in the gaseous form. The process for removing an electron is

$$X(g) + \text{energy} \longrightarrow X^+(g) + e^-(g)$$

where X stands for any element in the periodic table and e^- is the symbol for an electron. The added energy is defined as the **ionization energy,** the energy necessary to produce the positive ion of an element.

It seems reasonable to connect a small ionization energy with an element that has a strongly metallic character. Metals conduct electricity because electrons flow easily throughout the metal, and a low ionization energy means that an electron can easily escape from the isolated atom. The general rule is that metals have low ionization energies and nonmetals have high ionization energies.

One important trend in the periodic table is that

Metallic character tends to decrease from left to right in any period.

This trend is illustrated in Figure 8.2(a), in which the ionization energies for the elements in the first and second periods are shown. Despite small irregularities, the ionization energies generally increase from left to right. A second trend is that

Metallic character increases from top to bottom in any group.

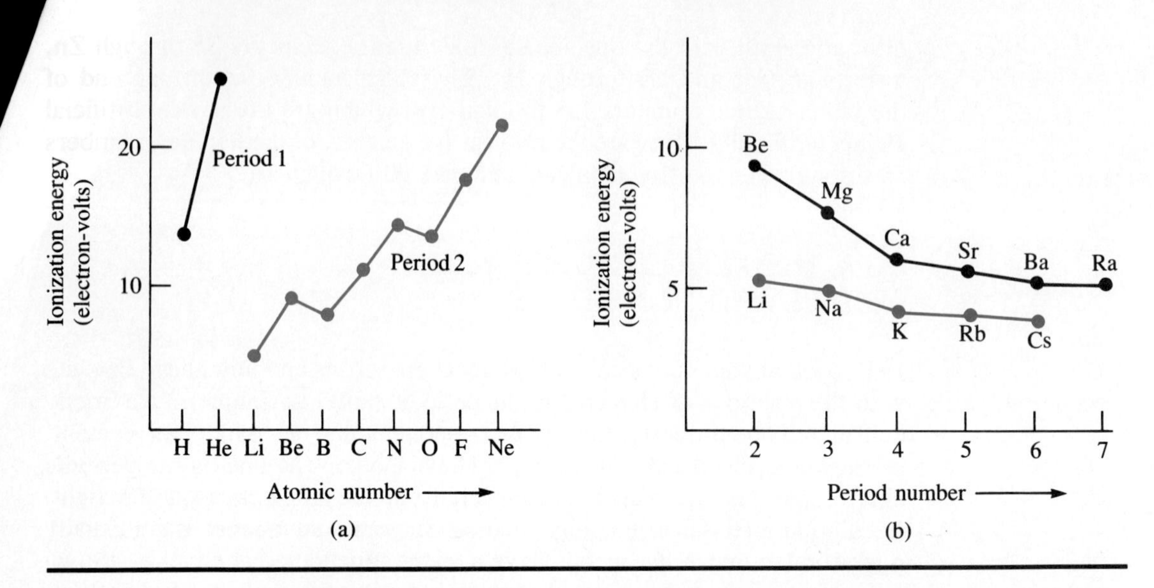

Figure 8.2. (a) Ionization energies of the elements in Periods 1 and 2. (b) Ionization energies of the elements in Groups I and II.

This trend is shown for the metals of Groups I and II in Figure 8.2(b). Combining the two trends, we come to the conclusion that the most metallic element is at the bottom left-hand corner of the periodic table. This element is usually considered to be cesium, Cs, because Fr doesn't occur in nature and can only be made artificially. Similarly, the most nonmetallic element is fluorine, F, close to the upper right-hand corner. Helium, precisely at the upper right-hand corner, forms no chemical compounds and is not included in the trends in metallic character, even though it has the highest ionization energy of any element and is decidedly a nonmetal.

One caution is in order. The trends in metallic character and other trends that we will discuss apply primarily to the representative elements in Groups I–VII. Changes in properties of the transition elements and rare earths are small, and trends are difficult to observe.

Example Which is more metallic, potassium (19) or chlorine (17)? Use the periodic table inside the back cover to find the answer.

Solution Notice that potassium (K) is at the far left of the periodic table in Group I, whereas chlorine (Cl) is at the far right in Group VII. Since metallic character tends to decrease from left to right within a row, you would expect potassium to be more metallic. Also, Cl is in the second period and potassium is in the third. Elements higher up in any group tend to be less metallic. This reinforces the prediction that potassium will be more metallic than chlorine.

SILICON AND SEMICONDUCTORS

Cities and states from coast to coast are seeking high-technology companies. Most states would like to have their own versions of "Silicon Valley," an area south of San Francisco where many modern electronics and computer companies are located.

Why silicon? What's so special about element 14—a grayish, brittle element with a metallic shine? The answer lies in silicon's electrical properties, which make possible the transistor and integrated circuits.

In the periodic table silicon is just to the right of the zig-zag line that separates nonmetals from metals. In the pure state silicon conducts electricity better than most nonmetals but not as well as metals. To make it useful for modern electronic circuits, manufacturers of transistors first grow single crystals of silicon having less than $1 \times 10^{-8}\%$ impurities, and then they add as lit 1×10^{-4} mole % of arsenic or boron to pure silicon. This controlled impurity make the silicon a better electrical conductor, suitable for use in semiconductors for transistors, solar cells, and silicon chips in integrated circuits.

Although silicon is the second most abundant element on the surface of the earth, the silicon used in the manufacture of transistors is fairly expensive because of the high purity required. Commercial use of semiconductor-grade silicon grew rapidly in the 1970s. Improvements in its production kept the cost of silicon constant during an inflationary economic period. Its cost per electrical device decreased rapidly during this time, in part because smaller components required smaller amounts of silicon.

8.2 IONIC CHARGES

We have already seen that different elements in the same group combine in the same ratio with elements in another group. For instance, KCl and NaCl have similar formulas because K and Na are both in Group I. Evidently chlorine, a nonmetal, acts toward these two metals from Group I in exactly the same way: In fact, all the elements in Group I behave identically in this respect. Predictions of formulas arrived at through analogy are especially reliable when one of the elements is a metal from a representative group and the other is a nonmetal.

NaCl and KCl are examples of salts, which are always a combination of a metal and one or more nonmetals. In a salt the metal is in the form of a positive ion and the nonmetal is a negative ion. For example, NaCl consists of Na^+ and Cl^- ions in equal numbers, and KCl contains a 1:1 ratio of K^+ and Cl^- ions. Al_2O_3 has two Al^{3+} ions for every three O^{2-} ions. In each of these formulas the total positive charge balances the total negative charge. Every compound is electrically neutral overall.

There are strong forces holding ions together in salt crystals. These ionic forces are so strong that the formation of salts in the reactions of metals with nonmetals is usually quite exothermic. NaCl and KCl form easily from the elements with just a small amount of heating to get the reactions started.

The equations for the reactions are

$$2\ Na(s) + Cl_2(g) \longrightarrow 2\ NaCl(s)$$

and

$$2\ K(s) + Cl_2(g) \longrightarrow 2\ KCl(s)$$

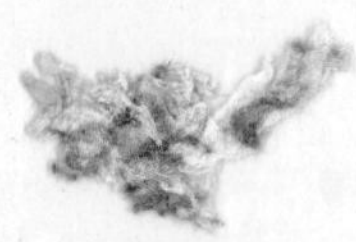

The charges on simple metallic and nonmetallic ions are easily predicted by reference to the periodic table. You may already have noticed that the farther a metal is from the left end of its period the greater the charge is on its positive ion. The farther a nonmetal is from the right end of its period, the greater the charge is on its negative ion. Two general rules can be used to predict the electrical charges of simple ions:

1. The charge on the ion of a representative metal equals the group number of the element.
2. The charge on the ion of a representative nonmetal equals the group number of the element minus 8.

The rule for a metal is easy, but the rule for a nonmetal may seem confusing at first. For fluorine in Group VII, the expected charge on the ion is $7 - 8 = -1$; the formula of the ion is F^-. For O and S, the predicted charge on the ion is $6 - 8 = -2$; the ionic formulas are O^{2-} and S^{2-}, respectively. Table 8.2 gives a few applications of these rules.

Table 8.2 Results of Calculating Ion Formulas for Selected Elements

ELEMENT	GROUP NUMBER	TYPE	CHARGE ON ION	FORMULA OF ION
Br	VII	nonmetal	$7 - 8 = -1$	Br^-
Li	I	metal	$+1$	Li^+
Ca	II	metal	$+2$	Ca^{2+}
Tl	III	metal	$+3$	Tl^{3+}
Se	VI	nonmetal	$6 - 8 = -2$	Se^{2-}
Mg	II	metal	$+2$	Mg^{2+}

Example What are the formulas of the ions of the following elements: Br (35), Li (3), Ca (20), Tl (81), Se (34), and Mg (12)?

Solution First, find the element in the periodic table and determine its group number. If the element is a metal (to the left of the zig-zag line), the charge on the ion is the group number. If the element is a nonmetal (to the right of the dividing line), subtract 8 from the group number to get the charge on the ion.

The case of thallium, Tl, presents a small complication. There are actually two different ions, Tl^+ and Tl^{3+}, found in thallium salts. Two possibilities also exist for a few other metals that are to the right of the transition metal region. The group number gives the maximum positive charge that the ion can have, and in each case the ion of lower charge is two units less than the maximum.

A procedure for arriving at the formulas of simple salts follows readily from the rules for predicting the charges on ions. You need only remember the principle of charge balance: the total positive charge in the formula equals the total negative charge. In Al_2O_3 the positive charge is $(2)(+3) = +6$, and the negative charge is $(3)(-2) = -6$, for an exact cancellation.

Example What is a reasonable formula for the salt lithium sulfide?

Solution To start out, you may have to look at the table of elements inside the front cover to find the symbol for lithium. You may also want to review the discussion of the naming of simple compounds in Section 3.8. Lithium is in Group I, so it forms the ion Li^+. Sulfur, in Group VI, forms S^{2-} ions. In order for the total positive and negative charges to be equal, two Li^+ are needed for one S^{2-}. The formula is therefore Li_2S.

Predicting the oxides of Groups I and II is tricky, because several kinds of oxide are possible, depending on the relative amounts of metal and oxygen that are brought together. Sodium peroxide, Na_2O_2, is one that does not fit the rules. This compound contains the peroxide ion, O_2^{2-}, which consists of two oxygen atoms with an overall charge of -2. Although our generalizations work in the vast majority of cases, nature is always more complicated than any generalization.

8.3 WHY IS THERE A PERIODIC TABLE?

Though we have dealt with only a fraction of the uses of the periodic table, the examples illustrate some of its power. We have seen that metallic character, as defined by ionization energy, follows definite trends in the representative elements. Group numbers for metals and nonmetals determine the charges of the ions of those elements. These in turn allow prediction of formulas of simple salts.

When Mendeleev and others proposed their versions of the periodic table in the 1860s, they knew nothing of the existence of the electron and therefore nothing about how the atom is put together. Without this knowledge, the trends and correlations indicated by the periodic table must have seemed like magic. In later decades, as understanding of the atom increased, questions about why these trends occur and why the periodic table has the shape it does received increasing attention. For example, why are there two elements in the first period, eight in each of the next two periods, and eighteen in each of the next two? Why do the charges on the ions of the representative elements relate to their group numbers? Why do the trends in metallic character exist? No good answers to these questions could be given until the quantum mechanics of the atom was understood. For this reason, we must now take a closer look at orbitals and the order in which they fill with electrons.

8.4 QUANTUM MECHANICS REVISITED

We begin with a review of some terms from Chapter 7. The principal quantum number, n, determines the total energy of the electron and the average distance between the nucleus and the electron in an atom. Values of n can only be nonzero positive integers—1, 2, 3, 4, 5, etc. These limited values reflect the limited electronic energies that are possible. An orbital is the wave associated with an electron in an atom; it tells us the most probable locations of the electron. Two types of orbitals, s and p, have been introduced so far. An s orbital has a spherical shape. When an electron is in an s orbital, it is found somewhere in a spherical region with the nucleus at the center. A p orbital has a dumbbell shape (see Figure 7.18).

To specify any orbital, one must give the value of the principal quantum number n as well as the type of orbital. Thus, possible s orbitals are $1s$, $2s$, $3s$, $4s$, and so on. There is just one s orbital for each value of n. On the other hand, there are three p-type orbitals for each n. Furthermore, the smallest possible n for a p orbital is 2. There is no such thing as a $1p$ orbital, but there are three $2p$ orbitals, three $3p$ orbitals, three $4p$ orbitals, etc.

There are two additional kinds of orbitals, d and f, which are important for analyzing the shape of the periodic table. For a particular value of n there are five d orbitals, and the smallest value of n for which d orbitals exist is 3. There are five $3d$ orbitals, five $4d$ orbitals, five $5d$ orbitals, and so forth. Starting with $n = 4$, there are seven f orbitals for each value of n. Table 8.3 summarizes this information, which seems to have little rhyme or reason to it unless one has learned advanced quantum mechanics.

So far, this description of orbitals applies to all atoms, but in quantum mechanics there is an important distinction between hydrogen and all the other known elements. Recall that Bohr's theory worked well for hydrogen but not for any other element. One reason is that the hydrogen atom has only one electron, whereas the other elements have two or more. The

Table 8.3 Orbitals for the First Six Values of the Principal Quantum Number (The numbers in parentheses are the number of orbitals of each kind.)

PRINCIPAL QUANTUM NUMBER, n	TYPES OF ORBITALS			
1	$1s(1)$			
2	$2s(1)$	$2p(3)$		
3	$3s(1)$	$3p(3)$	$3d(5)$	
4	$4s(1)$	$4p(3)$	$4d(5)$	$4f(7)$
5	$5s(1)$	$5p(3)$	$5d(5)$	$5f(7)$
6	$6s(1)$	$6p(3)$	$6d(5)$	$6f(7)$

moment two or more electrons are present there is a force of electrical repulsion between these negative particles. Although the energy of the electron in a hydrogen atom depends only on the principal quantum number, n, the energy of an electron in any other atom depends on the type of orbital (s, p, d, or f) as well. An approximate energy level diagram for a many-electron atom is given in Figure 8.3. To put together this diagram, scientists analyzed the frequencies of light emitted by many elements, just as Bohr developed his theory to agree with the line spectrum of hydrogen. Some dependence on experimental results was necessary, because it is impossible to solve the Schrödinger equation exactly for a many-electron atom.

It is important to know the relative energies of different orbitals in an

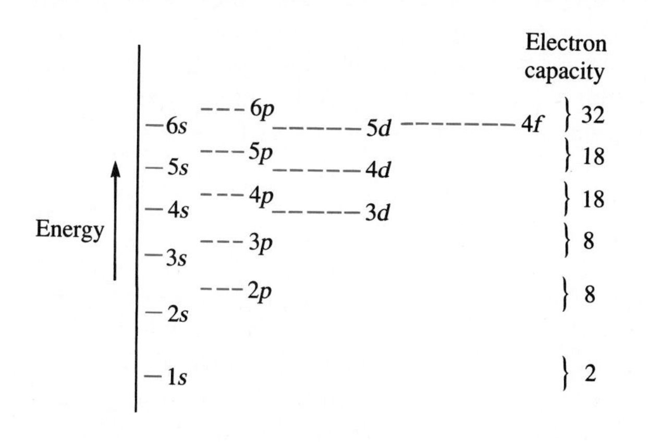

Figure 8.3. An orbital energy-level diagram for many-electron atoms. Each line segment stands for a distinct orbital. The energy differences between orbitals vary from element to element.

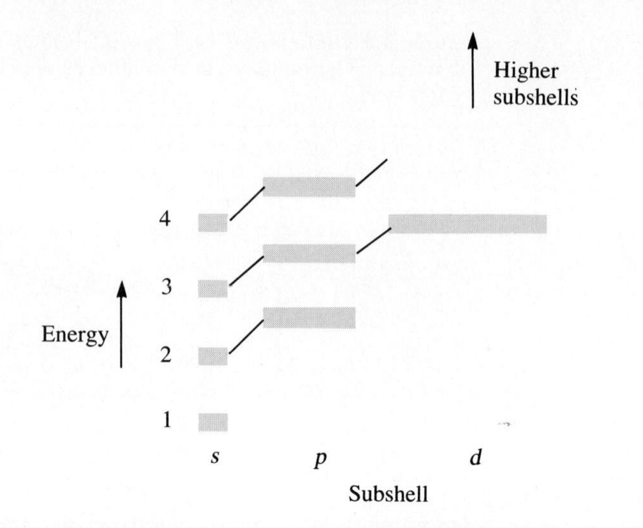

Figure 8.4. A subshell energy-level diagram.

atom because electrons will occupy the orbitals of the lowest energy. Notice that the order of orbitals from lowest to highest energy has a few surprises. In the simplest of worlds we would expect any orbital with $n = 3$ to have a lower energy than any of the $n = 4$ orbitals. According to the diagram, however, the 4*s* orbital has an energy very close to that of the 3*d* orbitals. Depending on the element, sometimes the 4*s* orbital has a lower energy than the 3*d* does, and sometimes it is the other way around. It is always true, however, that for any value of *n* the energies of the orbitals increase in the order $ns < np < nd < nf$. It's just that the energies of some of the shells overlap.

The term **shell** means a complete set of orbitals having the same principal quantum number. For example, the first shell is just the 1*s* orbital, whereas the third shell consists of the 3*s*, 3*p*, and 3*d* orbitals. Another frequently used term is **subshell,** which means a single set of orbitals of the same type and having the same principal quantum number. For instance, the 4*p* subshell consists of the three 4*p* orbitals, and the 4*f* subshell contains the seven 4*f* orbitals. (A subshell energy-level diagram is shown in Figure 8.4.)

8.5 ELECTRONIC CONFIGURATIONS IN ATOMS

Now we are almost ready to determine the orbital for each electron in a many-electron atom. For the lowest energy state—or ground state—of the hydrogen atom, the single electron is in the 1*s* orbital. In which orbital or orbitals do the two electrons in helium reside when the atom is in its ground state? The lowest possible energy results when both electrons are in the

1s orbital; the line spectrum of helium confirms that this is the configuration. Similarly, we might expect the three electrons in lithium, the next element in the periodic table, to be in the 1s orbital for the ground state, but experiments prove this assumption wrong. For example, the energy required to remove one electron from lithium (the ionization energy) is much smaller than the ionization energy of helium. Apparently, at least one electron in lithium is farther away from the nucleus in a higher-energy orbital, where it is less strongly attracted to the nucleus.

It is clear that the electrons in a multi-electron atom do not all pile into the 1s orbital in the ground state. Instead, there is a definite limit to the number of electrons that can be in a single orbital. A very important rule determines the arrangements of electrons in orbitals in all atoms:

The maximum capacity of any orbital is two electrons.

When it comes to occupying orbitals, three's a crowd.

To determine which orbitals of lithium have electrons in them, we imagine that the three electrons are added, one at a time, to a lithium ion that has been completely stripped of electrons. Since lithium has atomic number 3, we start with Li^{3+}. The orbital of lowest energy is filled first, followed by the next-to-lowest orbital, until all three electrons have been used. The idea that the distribution of electrons can be found by adding electrons to orbitals in order of increasing energy has been called the **Aufbau Principle,** after the German word *aufbau,* "to build up." In lithium the first two electrons occupy the 1s orbital, and the third goes into the 2s, which is the next higher orbital on the energy scale (see Figure 8.5). It is this 2s electron that is removed when Li is ionized to Li^+. Because it is a relatively

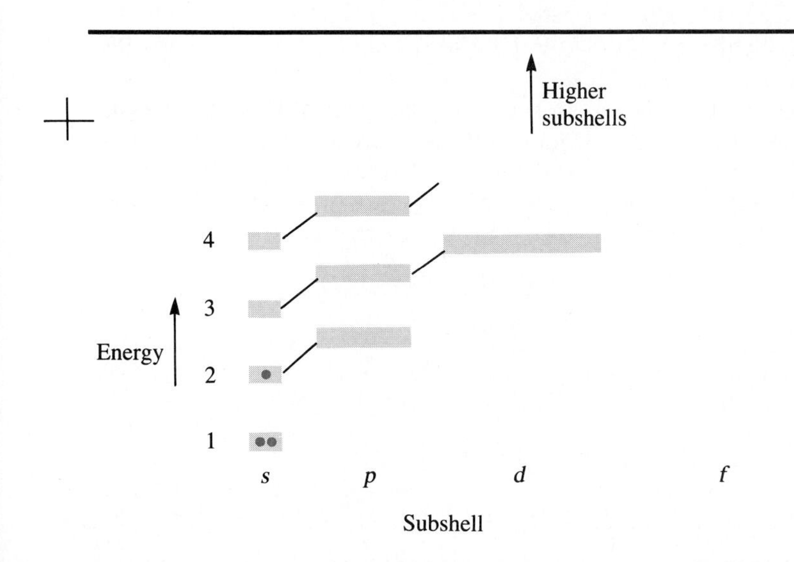

Figure 8.5. Electrons in the subshells of a lithium atom.

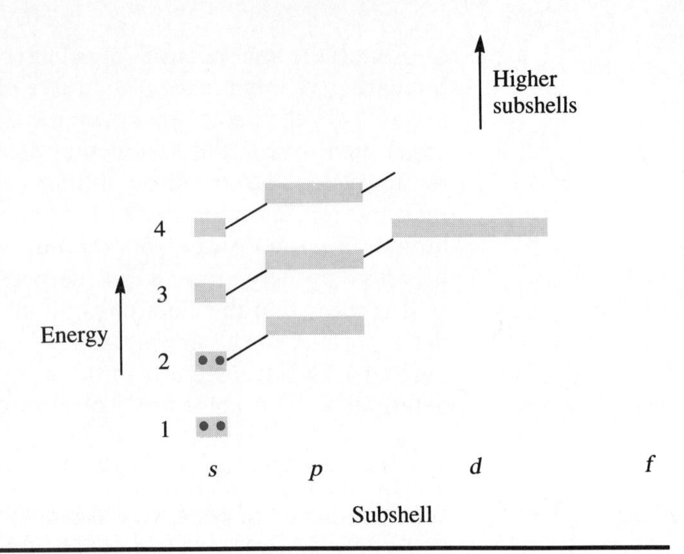

Figure 8.6. Electrons in the subshells of a beryllium atom.

high-energy electron, less energy is needed to rip it from the atom than is required to ionize a $1s$ electron of helium.

In order to understand the periodic table it is useful to know the **electronic configuration** of an atom—the number of electrons in each subshell. The electronic configuration of lithium can be written

$$1s^2\,2s^1$$

In this widely used notation, the symbol "$1s$" stands for a subshell, and the superscript "2" following the symbol gives the number of electrons in the subshell. The symbol "$2s^1$" indicates that a second subshell contains one electron. The subshells are listed from left to right in order of filling, from lowest to highest energy. Note that these symbols stand for subshells rather than for individual orbitals.

Applying the Aufbau Principle to beryllium, the element following lithium, we determine first that there are four electrons, corresponding to the atomic number of 4. The first two are in the $1s$ subshell and the last two fill the $2s$ subshell, to give the configuration $1s^2\,2s^2$ (see Figure 8.6).

Example What is the electronic configuration of a carbon atom?

Solution Carbon has atomic number 6, so the neutral atom has six electrons. Since the $2s$ subshell is filled at beryllium, any additional electrons must enter the $2p$ subshell, which is next in line in Figure 8.4. The electronic configuration is

$$\text{C: } 1s^2\,2s^2\,2p^2$$

Note that the answer accounts for all six electrons; addition of the superscripts gives $2 + 2 + 2 = 6$ electrons.

To go much further with the Aufbau Principle, we must know the numbers of electrons that fill the p, d, and f subshells of orbitals. These subshells consist of 3, 5, and 7 orbitals, respectively. Since each orbital has a maximum capacity of two electrons, the capacities of the subshells are as shown in the following table:

SUBSHELL TYPE	CAPACITY
s	2 electrons
p	6 electrons
d	10 electrons
f	14 electrons

According to these capacities, the electronic configuration of a fluorine atom in its ground state should be $1s^2\ 2s^2\ 2p^5$ (see Figure 8.7). The total number of electrons, 9, checks with the atomic number, and the final five electrons go comfortably into the $2p$ subshell because its capacity is 6.

Example What is the electronic configuration of an aluminum (Al) atom?

Solution The atomic number of aluminum is 13. With the help of the Aufbau Principle and Figure 8.4, we put electrons into subshells, beginning with the subshell of lowest energy—the $1s$ subshell. The electronic configuration for Al is $1s^2\ 2s^2\ 2p^6\ 3s^2\ 3p^1$. Again note that this answer accounts for 13 electrons in all; $2 + 2 + 6 + 2 + 1 = 13$.

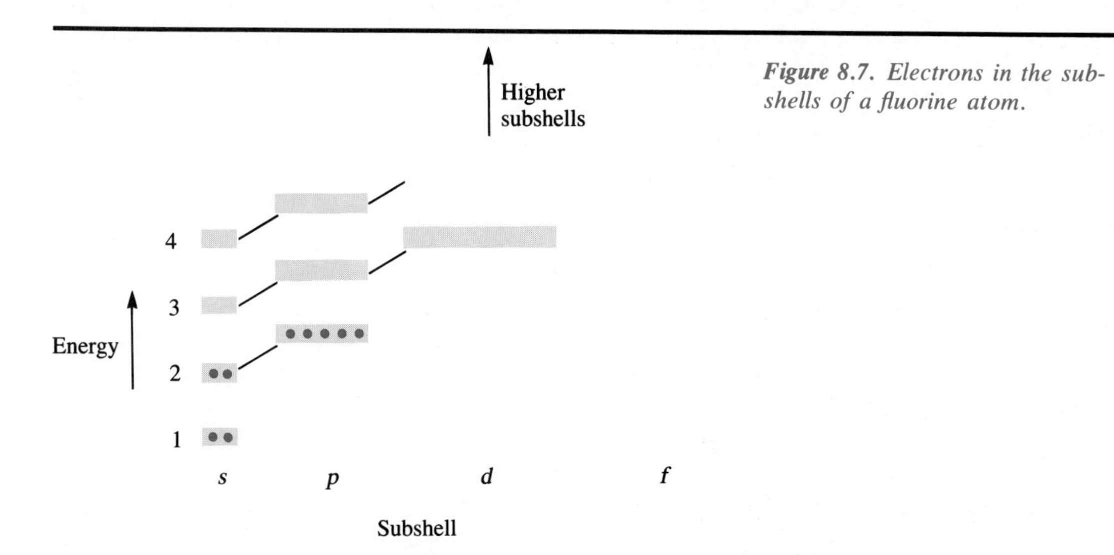

Figure 8.7. Electrons in the subshells of a fluorine atom.

Example Find the error in each of the following electronic configurations, which are supposed to represent the ground states of atoms.

(a) $1s^2\ 2s^2\ 2p^8\ 3s^1$
(b) $1s^2\ 2s^1\ 2p^3$

Solution (a) The $2p$ subshell cannot contain eight electrons, because six is the maximum number. The correct configuration is $1s^2\ 2s^2\ 2p^6\ 3s^2\ 3p^1$.

(b) This configuration might be an excited state, but it cannot be the lowest energy state of an atom because the $2s$ subshell is not filled; yet electrons have been added to the higher-energy $2p$ subshell. The ground-state configuration should be $1s^2\ 2s^2\ 2p^2$.

8.6 ELECTRONIC CONFIGURATIONS AND THE SHAPE OF THE PERIODIC TABLE

There is universal agreement among chemists that the chemical properties of an element are determined primarily by the number of electrons in the **valence shell,** which is the shell of highest principal quantum number. The valence electrons have higher energies than the other electrons in an atom, and they are, on the average, farther from the nucleus. These are the electrons that interact with the valence electrons of other atoms when chemical reactions occur. The inner electrons are so tightly held by the atomic nucleus that they are scarcely affected by chemical changes.

Figure 8.8 gives the electronic configurations of just the valence shells for the elements in the two eight-membered rows of the periodic table: the second and third periods. You can see that elements in the same group have identical valence-shell configurations except for the principal quantum number. For example, oxygen and sulfur each have six valence electrons, two in an s subshell and four in a p subshell. Each row begins with a metal

Figure 8.8. *Periods 2 and 3 of the periodic table, showing the electronic configurations of the valence electrons.*

PERIOD	GROUP I	II	III	IV	V	VI	VII	VIII
2	Li $2s^1$	Be $2s^2$	B $2s^2\ 2p^1$	C $2s^2\ 2p^2$	N $2s^2\ 2p^3$	O $2s^2\ 2p^4$	F $2s^2\ 2p^5$	Ne $2s^2\ 2p^6$
3	Na $3s^1$	Mg $3s^2$	Al $3s^2\ 3p^1$	Si $3s^2\ 3p^2$	P $3s^2\ 3p^3$	S $3s^2\ 3p^4$	Cl $3s^2\ 3p^5$	Ar $3s^2\ 3p^6$

that has a single valence electron and ends with a noble gas that has eight valence electrons. These periods each have eight elements, because that is the number of electrons required to fill the *s* and *p* subshells.

According to the periodic table inside the back cover, Period 4 contains eighteen elements, ten more than each of the preceding rows. Could this increase of ten be related to the fact that the capacity of a *d* subshell is ten electrons? Examination of the subshell energy-level diagram (Figure 8.4) shows that this is precisely the reason. In the aufbau process, the final element in Period 3, Ar, receives a $3p$ electron as the last electron added. The last electron added to potassium at the beginning of Period 4 must be put into the $4s$ subshell, since the $3p$ subshell is filled. The next element, calcium, has two $4s$ electrons. Then the electronic configuration of scandium (Sc), the first transition element, results from the addition of a $3d$ electron, according to the energy-level diagram. Filling the $3d$ subshell accounts for the remaining transition elements. Period 4 is completed by the addition of six electrons to the $4p$ subshell.

By the time the end of the transition elements is reached and the $3d$ subshell has been completed, the $3d$ subshell has become slightly lower in energy than the $4s$ subshell. Therefore, the outermost, or valence, electrons of bromine (Br), for example, are just the two in the $4s$ and the five in the $4p$ subshells. These are the only electrons held loosely enough by the bromine nucleus to get involved in the chemical reactions of bromine. The valence-shell configuration of bromine is closely related to that of the elements above it in Group VII. Each has two *s*-type and five *p*-type valence electrons.

The next period, Period 5, follows a similar pattern. First, two $5s$ electrons are added, then ten $4d$ electrons, and finally six $5p$ electrons. Again, for the elements to the right of the transition elements, the completed *d* subshell sinks to a lower energy and the valence shell of electrons is composed of the $5s$ and $5p$ subshells. In a brief look at the remainder of the periodic table, it can be seen that the extra elements in Periods 6 and 7—the lanthanides and actinides—are the result of filling the $4f$ and $5f$ subshells, respectively, with fourteen electrons each. Every time a brand new kind of subshell appears, the related row of the periodic table becomes longer than any previous row. Table 8.4 gives the electronic configurations of the valence shells for the representative elements, Groups I–VII, plus the noble gases.

Because the chemical properties of representative elements depend primarily on their valence-shell configurations, it may be helpful to underline the valence-shell portion of an electronic configuration. For example, under this scheme, the electronic configurations of nitrogen, sodium, and selenium are

$$\text{N:}\quad 1s^2\,\underline{2s^2\,2p^3}$$

$$\text{Na:}\quad 1s^2\,2s^2\,2p^6\,\underline{3s^1}$$

$$\text{Se:}\quad 1s^2\,2s^2\,2p^6\,3s^2\,3p^6\,3d^{10}\,\underline{4s^2\,4p^4}$$

Table 8.4 Valence-Shell Configurations for the Representative Elements (The value of n is the number of the period in which a given element appears.)

GROUP NUMBER	VALENCE-SHELL CONFIGURATION
I	ns^1
II	ns^2
III	$ns^2\ np^1$
IV	$ns^2\ np^2$
V	$ns^2\ np^3$
VI	$ns^2\ np^4$
VII	$ns^2\ np^5$
noble gases	$ns^2\ np^6$

In the electronic configuration of selenium, note that $3d^{10}$ has been written before $4s^2$ even though these subshells fill in the opposite order. This is done because the $3d$ subshell is filled and has become lower in energy than the $4s$ and $4p$ subshells. For selenium atoms, only the $4s$ and $4p$ electrons should be regarded as valence electrons.

In summary, the valence electrons of Groups I and II are always s-type electrons. The principal quantum number for these electrons equals the number of the period in which the element appears. For Groups III–VII and the noble gases, the valence electrons are in s and p subshells, where again the principal quantum number of the valence shell equals the period number.

8.7 IONIC CHARGE AND GROUP NUMBER

One of the questions raised earlier was why the charge on an ion is closely related to the group number for that element in the periodic table. If the element is a metal, it can form a positive ion by loss of electrons, and the ionic charge equals the group number. A nonmetal can gain one or more electrons to become a negative ion, whose charge equals the group number minus 8.

It turns out that electron shells which are filled with electrons have greater stability than shells which are only partially filled. With atoms having filled shells of electrons there is a maximum attraction between the atomic nucleus and the surrounding electrons. This is why the noble gases are so unreactive—their neutral, separate atoms already have filled valence shells of electrons.

When a metal atom becomes a positive ion, it loses electrons from its valence shell. These electrons are farther from the nucleus than the inner-shell electrons are and therefore have less attraction to the nucleus. Another way of looking at it is that the valence electrons have the highest energy, and thus less energy must be put into the atom to remove these electrons. After the valence electrons have been removed, much more energy is required to strip an inner electron from the atom, because it is closer to the nucleus and is at a lower energy level. Therefore, in chemical reactions only the valence electrons are lost. Because the number of valence electrons equals the group number for a metal, the positive charge on its ion also equals the group number.

Magnesium, for example, is in Group II of the periodic table; its valence electronic configuration is $3s^2$. The two $3s$ electrons in the valence shell can be removed with only a modest input of energy, leaving behind the configuration $1s^2\,2s^2\,2p^6$. In Mg^{2+}, the number of electrons is two less than the number of protons in the nucleus.

The ionization energies of nonmetals are usually too great for them to lose electrons in chemical reactions. Instead, they gain electrons until the orbitals in their valence shells are completely occupied. Any additional electrons would have to enter a subshell of much higher energy, creating a very unstable ion. The limit, therefore, is eight electrons, because that is the capacity of the *s* and *p* subshells making up the valence shell. An atom such as oxygen, in Group VI, has six valence electrons. By adding two electrons, it reaches its upper limit of 8 and a charge of -2. According to the formal rule stated earlier (p. 168), the charge is calculated as $6 - 8 = -2$. The 8 in this formula is, of course, the number of electrons in the completed valence shell. This number is subtracted from the group number, rather than the other way around, so that the result gives both the number of added electrons and the correct sign for the charge on the negative ion.

8.8 TRENDS IN METALLIC CHARACTER

One of the most significant trends in the periodic table is the variation in metallic (or nonmetallic) character of the elements within a group or a period. We have already seen how the ionization energies of gaseous elements show decreasing metallic character from left to right in a period and from bottom to top in a group. The most metallic element is in the lower left corner of the periodic table, and the most nonmetallic element is at the upper right (see Figure 8.9). In general, metals offer the least resistance and nonmetals the most resistance to the loss of electrons. Most often metals lose electrons and nonmetals gain electrons in chemical reactions. The reactions of the elements in Group VII confirm this pattern. Fluorine has the strongest tendency to gain electrons from other substances, chlorine the next strongest tendency, and so on.

What is there about the electronic configurations of metals and nonmetals that causes these differences in behavior? Let's compare lithium, a metal,

INCREASING NONMETALLIC CHARACTER ⟶

← INCREASING METALLIC CHARACTER

INCREASING NONMETALLIC CHARACTER ⟶

Period	Group I	II											III	IV	V	VI	VII	
1	1 H																	2 He
2	3 Li	4 Be											5 B	6 C	7 N	8 O	9 F	10 Ne
3	11 Na	12 Mg					Transition series						13 Al	14 Si	15 P	16 S	17 Cl	18 Ar
4	19 K	20 Ca	21 Sc	22 Ti	23 V	24 Cr	25 Mn	26 Fe	27 Co	28 Ni	29 Cu	30 Zn	31 Ga	32 Ge	33 As	34 Se	35 Br	36 Kr
5	37 Rb	38 Sr	39 Y	40 Zr	41 Nb	42 Mo	43 Tc	44 Ru	45 Rh	46 Pd	47 Ag	48 Cd	49 In	50 Sn	51 Sb	52 Te	53 I	54 Xe
6	55 Cs	56 Ba	57* La	72 Hf	73 Ta	74 W	75 Re	76 Os	77 Ir	78 Pt	79 Au	80 Hg	81 Tl	82 Pb	83 Bi	84 Po	85 At	86 Rn
7	87 Fr	88 Ra	89† Ac															

*Lanthanide series

58 Ce	59 Pr	60 Nd	61 Pm	62 Sm	63 Eu	64 Gd	65 Tb	66 Dy	67 Ho	68 Er	69 Tm	70 Yb	71 Lu

†Actinide series

90 Th	91 Pa	92 U	93 Np	94 Pu	95 Am	96 Cm	97 Bk	98 Cf	99 Es	100 Fm	101 Md	102 No	103 Lw

Figure 8.9. Trends in metallic and nonmetallic character.

with fluorine, a nonmetal. Both are in the second period, and both have a $1s^2$ inner shell. These two inner electrons are much closer to the nucleus, on the average, than the valence electrons are. They act as a kind of shield between the nucleus and the outer valence electrons. The valence electrons do not experience the total pull of the positive nucleus because there are negatively charged electrons of the inner shell in the way. Therefore, the single $2s$ valence electron in lithium does not behave as though it is being attracted to the nucleus by three protons, which is the actual number. Instead, the two inner-shell electrons appear to cancel out the charges of two of the three protons. Roughly speaking, the valence electron in lithium "feels" the attraction of only one proton in the nucleus. This reduced charge is called the **effective nuclear charge.** For the $2s$ electron of a lithium atom, it is $+1$. The effective nuclear charge is always equal to the number of protons in the nucleus minus the number of electrons in inner shells. Although this method of calculation is oversimplified, it serves to illustrate the contrast between metals and nonmetals.

Effective nuclear charge
= atomic number of element − number of inner-shell electrons

This relationship is only approximate, because inner-shell electrons don't completely cancel an equal number of protons in the nucleus.

Looking now at the effective nuclear charge operating on the valence electrons in fluorine, we find that it is considerably larger. Although the inner electrons are still two in number, the number of protons in the nucleus is now nine. The effective nuclear charge for fluorine is $9 - 2 = 7$. Each of the seven valence electrons in fluorine feels an effective nuclear charge of $+7$. The much greater attraction by fluorine for its valence electrons now makes more sense. In fact, as we have seen, fluorine readily gains an extra electron to become the F^- ion, thereby taking on the filled valence shell of a noble gas.

The effective nuclear charge attracting the valence electrons in any other element is calculated the same way. Sodium, directly below lithium in Group I, has an atomic number of 11 and has 10 electrons in its noble-gas inner shells. This results in an effective nuclear charge of $+1$. In fact, all the elements in Group I have an effective nuclear charge of $+1$. Similarly, any representative element has an effective nuclear charge equal to the group number. The larger the group number, the less easily an element will lose electrons. In other words, the larger the group number, the greater the nonmetallic character of the element. This conclusion ties in perfectly with the observation that nonmetals are on the right side and metals are on the left side of the periodic table (Figure 8.9).

The vertical trend in metallic character within a group must still be explained. According to our simple method for calculating effective nuclear charge, this charge is the same for all the elements in a group. Therefore, it cannot be the cause of a vertical trend. Instead, the principal quantum number of the valence electrons appears to be the important factor. Since the principal quantum number of these outermost electrons increases as we go down the group, the average distance between these electrons and the nucleus also increases. The farther away the electrons are, the less strongly they are attracted to the nucleus. For example, the valence-shell configurations of lithium and sodium in Group I are $2s^1$ and $3s^1$, respectively. The weaker attraction between the $3s$ electron of sodium and its nucleus, compared with that between the $2s$ electron of lithium and its nucleus, means that less energy is required to remove the valence electron of sodium. This is exactly what Figure 8.2(b) shows. Sodium is more metallic (loses electrons more easily) than lithium because the valence electron of sodium is farther away from the nucleus.

This trend in metallic character continues right down Group I. Cesium, with the highest principal quantum number for its valence electron, is the most metallic element in the group and, indeed, in the entire periodic table. (Again, we ignore radioactive francium.) On the opposite side of the periodic table, we treat nonmetallic character in a similar fashion. In Group VII, fluorine has the smallest principal quantum number for its valence electrons, which are therefore closest to the nucleus. Of all the elements in Group VII, fluorine attracts an extra electron into its valence shell most strongly and is the least metallic or, in other words, the most nonmetallic. One can analyze other elements between Groups I and VII in the same way.

Example Why is aluminum more metallic than boron, both of which are in Group III? Be as specific as possible by referring to the electronic configurations of these two elements.

Solution Valence electrons in each atom experience an effective nuclear charge of +3, because for each element the number of protons exceeds the number of inner-shell electrons by 3.

$$\text{Effective nuclear charge for boron} = +5 - 2 = +3$$

$$\text{Effective nuclear charge for aluminum} = +13 - 10 = +3$$

The valence-shell configurations of B and Al are $2s^2\,2p^1$ and $3s^2\,3p^1$, respectively. Because aluminum has the higher principal quantum number for its valence electrons, these electrons are farther from the nucleus and are held less tightly, allowing aluminum to lose valence electrons more easily. Therefore, aluminum is more metallic than boron.

Metallic character comes down to the question of how strongly electrons in the valence shell of an element are attracted to the nucleus. Two factors—effective nuclear charge and principal quantum number of the valence electrons—are primarily responsible for the size of this attraction. A large effective nuclear charge makes for a stronger attraction, whereas a large principal quantum number for valence electrons makes for a weaker attraction. The most metallic elements have small effective nuclear charges and large principal quantum numbers for their valence electrons. Highly nonmetallic elements have just the opposite qualities: large effective nuclear charges and small principal quantum numbers for their valence electrons.

IMPORTANT TERMS

group A (vertical) column of the periodic table. This term has the same meaning as a *family of elements*.

period A (horizontal) row of the periodic table.

representative elements The elements in Groups I–VII of the periodic table.

noble gases The elements in the right-most column of the periodic table. The valence shell of electrons for each of these elements is completely filled.

transition elements The elements in the central part of the periodic table, located in Periods 4, 5, and 6.

lanthanides The elements in Period 6 with atomic numbers 58–71.

actinides The elements in Period 7 with atomic numbers 90–103. The actinides and the lanthanides are usually moved out of their proper places and shown at the bottom of the periodic table.

metals The elements on the left side of the periodic table. Metals are generally good conductors of electricity and have low ionization energies.

nonmetals The elements on the right side of the periodic table. Nonmetals have high ionization energies and, with a few exceptions, do not conduct an electric current.

ionization energy The energy required to produce a positive ion by removing an electron from an isolated atom of an element.

shell A complete set of electronic orbitals having the same principal quantum number.

subshell A set of electronic orbitals of only one type (s, p, d, or f) with the same principal quantum number. For example, all three $2p$ orbitals make up the $2p$ subshell; all five $3d$ orbitals comprise the $3d$ subshell of electrons.

Aufbau Principle An approach to finding the distribution of electrons in an atom. Starting with just the nucleus, we add electrons first into the orbital of lowest energy and then into higher-energy orbitals in order. This process of "building up" is finished when the correct number of electrons has been added.

electronic configuration The number of electrons in each subshell of an atom. For example, $3p^4$ means that there are four electrons in the $3p$ subshell of an atom.

valence shell The shell of highest principal quantum number in the electronic configuration of an atom. Electrons in the valence shell have the highest energy and are, on the average, farthest from the nucleus. For the representative elements, valence-shell electrons occupy the s and often the p subshells.

effective nuclear charge The effective pull of an atom's nucleus on its valence-shell electrons. This charge is calculated by taking the number of protons in the nucleus of an atom (atomic number) minus the number of electrons in filled inner shells. For the representative elements, the effective nuclear charge is equal to the group number.

QUESTIONS

1. For each of the following pairs of elements, state which is the more metallic of the two, and briefly explain your reasoning.

a. K or Ca **b.** P or Sb **c.** Be or Ba

2. For each of the following pairs of elements, state which is the more nonmetallic of the two, and briefly explain your reasoning.

a. O or S **b.** B or N **c.** Sn or I

3. What are the expected charges on the ions of these elements?

a. Be **b.** Al **c.** Ra **d.** Te **e.** I

4. What are the expected charges on the ions of these elements?

a. Cl **b.** Rb **c.** Sr

5. Is an element with a relatively large ionization energy metallic or nonmetallic? Explain.

6. Is Se or Cl more nonmetallic? (Since these two elements are in neither the same period nor the same group, your reasoning must be divided into two steps.)

7. What is a likely formula for each of these salts?

a. calcium fluoride **b.** barium fluoride
c. calcium selenide

8. What are likely formulas for compounds of the following pairs of elements?

a. K and S **b.** Ga and Cl
c. Mg and O **d.** Be and F
e. Al and F

9. Give the electronic configuration of each of the following elements:

a. H **b.** He **c.** O **d.** Na **e.** Ar **f.** Se

10. Give the electronic configuration of each of the following elements, and underline the part that constitutes the valence shell.

a. F **b.** K **c.** P **d.** Sr **e.** Se

11. Give the electronic configuration of the expected ion of each element.

a. F **b.** K **c.** Sr **d.** Se

12. Each of the following is an incorrect electronic configuration for a ground state. Find the error and write a correct configuration for the same number of electrons.

a. $1s^1\ 2s^2$
b. $1s^2\ 1p^6$
c. $1s^2\ 2s^2\ 2p^4\ 3s^2\ 3p^1$
d. $1s^2\ 2s^2\ 2p^6\ 3s^2\ 3d^8$

13. Predict whether Na or Mg has the greater tendency to lose one or more electrons and become a positive ion in a chemical reaction. Explain the difference between the elements using the concept of effective nuclear charge.

14. Does S or Cl have the greater tendency to become a negative ion? Explain the difference between the elements in terms of effective nuclear charge.

15. Does N or P have the greater nonmetallic character? Explain the difference between the elements in terms of effective nuclear charge and principal quantum number of valence electrons.

16. The 1*s* electron of a hydrogen atom is considered to be in the valence shell, yet the 1*s* electrons of a sodium atom are not considered to be in the valence shell. Why the difference?

17. Using the electronic configuration of thallium, Tl, explain why this metal can form two different positive ions, Tl^+ and Tl^{3+}.

18. Removing one electron from an atom of carbon takes more than twice as much energy as removing one electron from an atom of sodium, but removing a second electron from carbon (to form C^{2+}) takes only one-half as much energy as removing a second electron from sodium (to form Na^{2+}). Explain this apparent paradox.

9 Covalent Compounds

The idea that an atom consists of a small, dense nucleus along with a cloud of electrons to fill out most of the volume was of great importance in the development of chemistry. Even before this picture became clear, chemists were preoccupied with two key questions: What holds atoms together in compounds, and what accounts for the great variety of different chemicals?

9.1 CHEMICAL COMPOUNDS—TWO CONTRASTING BEHAVIORS

Sodium chloride and gasoline are two familiar substances that have very different properties. Gasoline, however, is not a good choice for a comparison of compounds, because it is a mixture. Instead, we choose octane, C_8H_{18}, an important component of gasoline. At first glance we can see that sodium chloride is a hard, crystalline solid with no odor, and octane is a liquid with a definite odor. Evidently molecules of octane readily escape from the liquid and travel to a person's nose. On the other hand, the atomic-sized particles of sodium chloride either have little tendency to escape or do escape but don't affect our sense of smell. The extremely high melting point (801°C) and boiling point (1413°C) of sodium chloride suggest that the atoms in this compound are strongly bound together. A great deal of thermal energy is required to make the atomic-sized particles of NaCl escape from the sodium chloride structure. By contrast, octane melts at −57°C and boils at 126°C. Less energy is needed to separate octane molecules, even though the molecular weight of octane is about double the formula weight of NaCl.

The electrical conductivities of sodium chloride and octane offer another sharp contrast. When the compounds are solids, neither conducts electricity. In the liquid state, however, sodium chloride is a good conductor, whereas octane is still a very poor conductor. To make this comparison, one must heat the sodium chloride to some temperature above 801°C to change it into a liquid, in which charged particles apparently can move about freely.

Another way conductivities can sometimes be compared is by dissolving each compound in water and then testing the solutions in an apparatus like that shown in Figure 6.2. A water solution of sodium chloride conducts electricity very well, again because of the presence of charged particles. Unfortunately, the comparison fails at this point, since scarcely any octane dissolves in water. Its small tendency to dissolve and the fact that octane is less dense than water have practical consequences; water is useless in extinguishing a gasoline fire. Instead of mixing with gasoline or smothering the fire, water sinks to the bottom and allows the gasoline to continue burning at the surface.

Contrasting properties of sodium chloride and octane are summarized in Table 9.1.

Table 9.1 Comparison of Sodium Chloride and Octane

COMPOUND	MELTING POINT (°C)	BOILING POINT (°C)	ELECTRICAL CONDUCTIVITY IN LIQUID STATE
sodium chloride	801	1413	high
octane	−57	126	very low

Let's examine ethyl alcohol, C_2H_5OH, and see how its properties compare with those of NaCl and octane. Ethyl alcohol appears to be like octane in several respects, since it has relatively low melting and boiling points and as a pure liquid does not conduct electricity. Unlike octane, alcohol does dissolve in water, but the solution does not conduct electricity. This behavior is the opposite of that of a sodium chloride solution. Ethyl alcohol clearly belongs with octane.

What can we say about calcium carbonate, $CaCO_3$? Called *calcite* by geologists, this compound occurs in chalk, limestone, marble, and the stalactites and stalagmites found in some caves. It is a crystalline material, transparent and colorless in large crystals and white when the crystals are very small. When heated, solid calcium carbonate decomposes at about 900°C. Therefore, we cannot readily find the melting point, but it is clear that the compound remains a solid at very high temperatures.

Because it decomposes before it melts, we cannot investigate the electrical conductivity of molten calcium carbonate. The solid has little tendency to dissolve in water, but the small amount that does dissolve makes the water conduct electricity better. On the basis of this limited evidence, calcium carbonate seems to resemble sodium chloride more closely than it does octane.

Example Sugar, or sucrose, is a colorless, crystalline solid. It melts at 185°C and decomposes into a black, tarry mixture at higher temperatures. A solution of sugar in water does not conduct electricity. Does sugar belong in the category with sodium chloride or the one with octane?

Solution Even though it is a solid at ordinary temperatures, sugar melts at a much lower temperature than does sodium chloride. Most significant is the fact that a solution of sugar in water does not conduct electricity. Although certain evidence is not available because of the decomposition, what we know places sugar with octane (and ethyl alcohol).

A compound such as sodium chloride or calcium carbonate, with high melting and boiling points and high electrical conductivity as a liquid or in water solution, is called an **ionic compound.** A compound with low melting and boiling points and low conductivity, such as octane, alcohol, or sugar, is classified as a **covalent compound,** or **molecular compound.** Additional examples of members of this class are water (H_2O), ammonia (NH_3), and carbon dioxide (CO_2). Although it is not perfect, our classification scheme accommodates most compounds.

In view of their ability to conduct electricity when liquid, ionic compounds must contain charged particles that are in fixed positions in the solid but are mobile in the liquid. These are the ions we met earlier. Oppositely charged ions attract each other so strongly that ionic compounds have very high melting and boiling points. On the other hand, covalent compounds require a different model in light of their low melting and boiling points and failure to conduct electricity. These facts tie in nicely with the idea that the molecules of covalent compounds are electrically neutral and attract each other rather weakly. The bonds holding the atoms together within a molecule, however, may be quite strong. These bonds are not altered when a solid melts or a liquid boils, but in chemical reactions they are broken and new bonds are formed.

9.2 SHARED ELECTRONS—THE GLUE IN COVALENT BONDING

Most of the known chemical compounds are covalent compounds. They range from water to the huge cellulose molecule that is found in all plants (see Figure 9.1). In fact, the chemistry of living things is largely the chemistry of covalent compounds. This is not to say that ions are absent or unimportant. If they were, we wouldn't need minerals for a well-balanced diet. Covalent compounds, however, make up the bulk of animal and plant matter.

Examination of their formulas shows that most ionic compounds contain one metallic and one or more nonmetallic elements. On the other hand, the great majority of covalent compounds contain only nonmetals, which are found at the right side of the periodic table.

Chemical bonding involves the most loosely held electrons, or **valence electrons,** of atoms. The energy required to remove one or more valence electrons from an atom of a nonmetal is large, and so even the most loosely held electrons are difficult to remove. As a result, in covalent compounds, electrons are shared between atoms; they are not transferred from one atom to another, as is the case with ionic compounds. A chemical bond formed by the sharing of electrons between atoms is called a **covalent bond.** The simplest example of a covalent bond is in the hydrogen molecule, H_2. If it were possible to obtain a snapshot of a single molecule, using a shutter speed fast enough to stop the motion of the two electrons and two protons, we might see a picture of H_2 like the one in Figure 9.2.

In Figure 9.2 we show the electrons between the two nuclei. Both electrons are attracted to each nucleus at the same time. The result is a pulling

Figure 9.1. Two covalent compounds of greatly differing size: (a) water, and (b) cellulose. The central unit in cellulose is repeated 100 to 3000 times, to give a molecular weight of up to 500,000. Cotton is almost entirely cellulose.

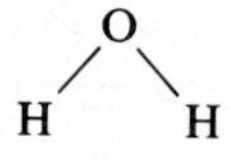

(n = 100 to 3,000)

(a)

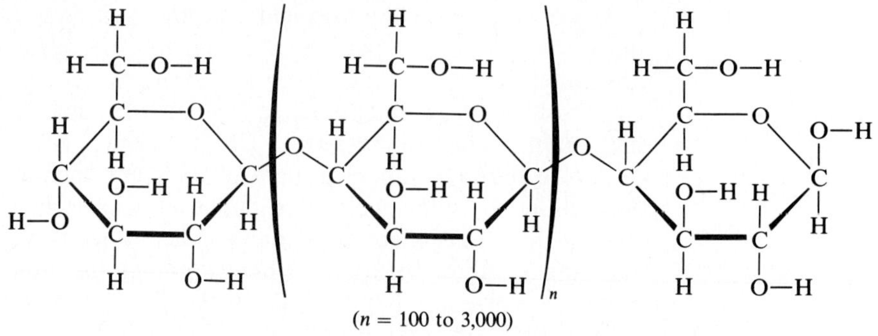

(b)

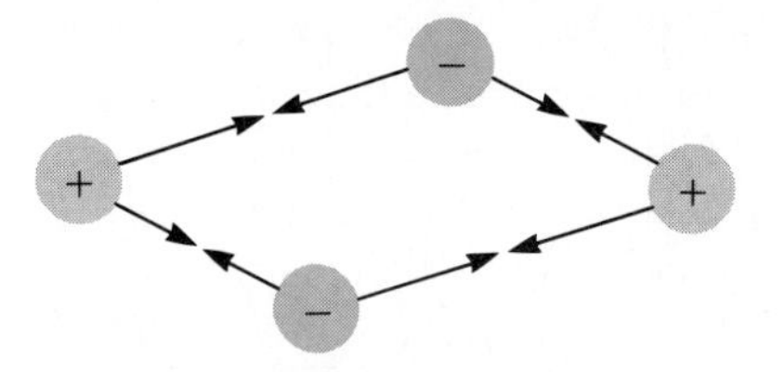

Figure 9.2. A possible arrangement of the electrons and protons in H_2. Attractions pulling the particles together are indicated by arrows.

together of the two hydrogen atoms into a molecule. The strong forces of attraction more than compensate for the smaller repulsions between the two nuclei and between the two electrons. This strong attraction leads to a state of potential energy lower than that of the unbonded, independent atoms.

If, on the other hand, the electrons were on opposite sides of the two nuclei, as shown in Figure 9.3, each electron would be strongly attracted to only one nucleus and would not serve as a glue to hold the nuclei together. It is an experimental fact that hydrogen gas consists almost entirely of H_2 molecules; very few H atoms are present. Evidently arrangements like the one in Figure 9.2 are much more common than the arrangements shown in Figure 9.3. The sharing of pairs of electrons between atoms is a key characteristic of all covalent bonds.

The electrons in a shared pair always come from the outermost **electron shell** (that is, the **valence shell**) of the atoms that bond together. Electrons in inner shells are too strongly held by their own nuclei to be shared with other atoms. At first glance it might seem that all of the valence electrons of an atom would be shared with the neighboring atoms, but this doesn't happen. In the first place, crowding lots of electrons between two atoms would cause strong electrical repulsions among them. As a result, the molecule would have high energy and would not be stable. Experience with many compounds shows that, with a few exceptions, the maximum number of electrons that can be shared between two nonmetal atoms is 6, or three pairs. Even this kind of sharing is rare and occurs in only a small fraction of the millions of known covalent compounds. More often two pairs of electrons are shared, and by far the most common arrangement is the sharing of a single pair.

A second limitation on the sharing of electrons comes from the octet rule. A shared pair of electrons is in the valence shell of each of the two

Figure 9.3. An arrangement of the two electrons that does not give a stable H_2 molecule.

atoms at the same time. If two atoms each contribute an electron to a shared pair, the number of electrons in the outermost shell of each bonded atom has been increased by 1. The **octet rule** says that each atom in a molecule completes its valence shell with a total of eight electrons. This arrangement of electrons is the most stable and has the lowest energy. In general, each atom in a molecule has a noble gas configuration.

∗ (Read this material if you have studied Chapter 8.)

Why do *pairs* of electrons, rather than some other number, tend to reside between atoms in a covalent compound? In atoms, electrons occupy atomic orbitals, each of which has a capacity of 2. The shared electrons in molecules occupy molecular orbitals, which have shapes that are determined by two or more nuclei rather than one. Just like an atomic orbital, a molecular orbital has a capacity of two electrons (Figure 9.4). One shared electron can occupy a molecular orbital, but it will not pull the atoms together strongly. Three electrons would exceed the capacity of the molecular orbital. Two shared electrons fill the orbital and provide maximum stability.

As you saw in Chapter 8, every noble gas beyond helium has eight electrons in its outermost shell. If for some reason an extra electron were added to a noble gas atom, that electron would enter a shell of considerably higher energy. Consider adding an electron to argon, Ar, for example. The resulting negative ion, Ar^-, would have an electronic configuration like that of potassium (K), the next element after argon, but it wouldn't have the extra proton in its nucleus. The added electron would be far from the nucleus, on the average. The fact that Ar^- doesn't exist suggests that Ar has little attraction for an extra electron.

Chlorine, Cl, the element preceding Ar in the periodic table, has one less electron than argon. Chlorine adds only one electron to its valence shell by sharing an electron pair with another atom. Recall that there is another way for Cl to obtain a stable octet of electrons, and that is by gaining complete possession of one more electron, to become the Cl^- ion. Another electron does not add to give Cl^{2-}, however, for the same reason that Ar has no tendency to become Ar^-. Cl^- ions, of course, don't exist by themselves. Instead, they exist in the company of a matching number of positive ions, such as Na^+ ions, in an ionic solid or in a water solution. In both ionic and covalent compounds, then, the octet has a special significance. ∗

Figure 9.4. The average electron density with two electrons in the molecular orbital of H_2. *There is a high probability of finding an electron in the colored area.*

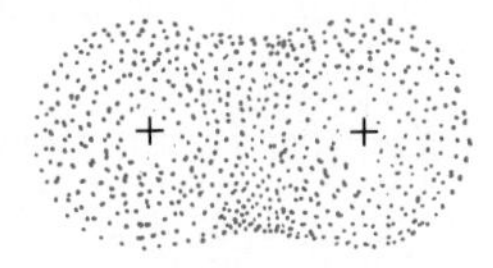

9.3 THE OCTET RULE IN ACTION

The octet rule is such a useful guide to the chemical bonds in covalent compounds that it deserves to be restated.

> *The Octet Rule:* In a covalent compound there are a total of eight shared and unshared electrons in the valence shell of every atom.

There is one major exception to the octet rule: The valence shell of hydrogen can hold only two electrons, rather than eight. Therefore, a hydrogen atom can share only one pair of electrons with another atom. Hydrogen atoms form only one covalent bond. Here the octet rule becomes the duet rule.

Probably the single most important chemical fact about an atom is its number of valence electrons. To show the arrangements of valence electrons in molecules, we will use a convention employed by all chemists—the **electron-dot picture.** For all the elements that we will consider (the representative elements), the number of valence electrons in an electron-dot picture is equal to the group number of the element. The group number refers to the chemical family in the periodic table (see inside back cover).

The symbol for an atom, such as H, C, or S, represents the nucleus plus all the electrons in the filled inner shells. Thus, when C appears in the diagram of a molecule, it stands for a nucleus of six protons and six neutrons and an inner shell of two electrons. S stands for the sulfur nucleus and completed inner shells of two and eight electrons. H stands for the hydrogen nucleus alone. To complete the isolated atom, we add dots to show the valence electrons:

When an atom has up to four valence electrons, we show each one on a different side of the letter symbol. This is an arbitrary arrangement that we use to make it easy to see how many covalent bonds an atom forms. When there are more than four valence electrons, some of them are shown in pairs.

An atom has only a limited number of electronic positions, called **orbitals.** Generally, there are four orbitals for the valence electrons of Groups I–VII in the periodic table. Each orbital can hold two electrons. Electron-dot pictures for atoms of the nonmetals in the first three rows of the periodic table are given in Figure 9.5. Notice that the number of valence electrons is equal to the group number of the element.

Example Draw the electron-dot picture for an atom of the element iodine (atomic number 53).

Figure 9.5. Electron-dot pictures showing the number of valence electrons in some of the nonmetals.

GROUP NUMBER				
III	IV	V	VI	VII
·B·	·C·	·N·	:O·	:F·
	·Si·	·P·	:S·	:Cl·

Solution On the periodic table inside the back cover, iodine, I, is found in Group VII. It has seven valence electrons. The correct picture is

Now let's apply the octet rule to the bonding in a molecule of Cl_2 gas. Chlorine is a good oxidizing agent that is used to keep swimming pools free from dangerous, disease-producing bacteria. Chlorine is quite toxic to bacteria, but unfortunately it is also irritating to many people's eyes.

Each chlorine atom contributes seven valence electrons to a chlorine molecule. In Cl_2 each chlorine atom achieves a completed valence shell of eight electrons by sharing one pair:

:Cl:Cl: or :Cl—Cl:

So, Cl_2 has one shared pair of valence electrons, and each atom also has three unshared pairs of electrons. The most common way of showing bonding is by using a line between two atoms to represent a shared pair of electrons, as in the second drawing above. In the nineteenth and early twentieth centuries, hooks were used to show chemical bonds.

As we discussed earlier, the two electrons between the chlorine atoms are attracted to two nuclei simultaneously. These forces are responsible for holding the atoms together. A rather large amount of energy, 242 kilojoules (kJ), is needed to break the bonds in 1 mole of Cl_2 molecules and to separate all the Cl atoms completely. Breaking other covalent bonds usually requires even more energy.

A line-and-dot drawing like the one shown for Cl_2 is often called a **Lewis structure** in honor of G. N. Lewis (Figure 9.6), who contributed important ideas to the theory of covalent bonding during the 1920s and 1930s. A professor at the University of California at Berkeley, Lewis also made important advances in the field of thermodynamics—the study of energy. He was one of those rare people who have exciting new ideas in more than one field.

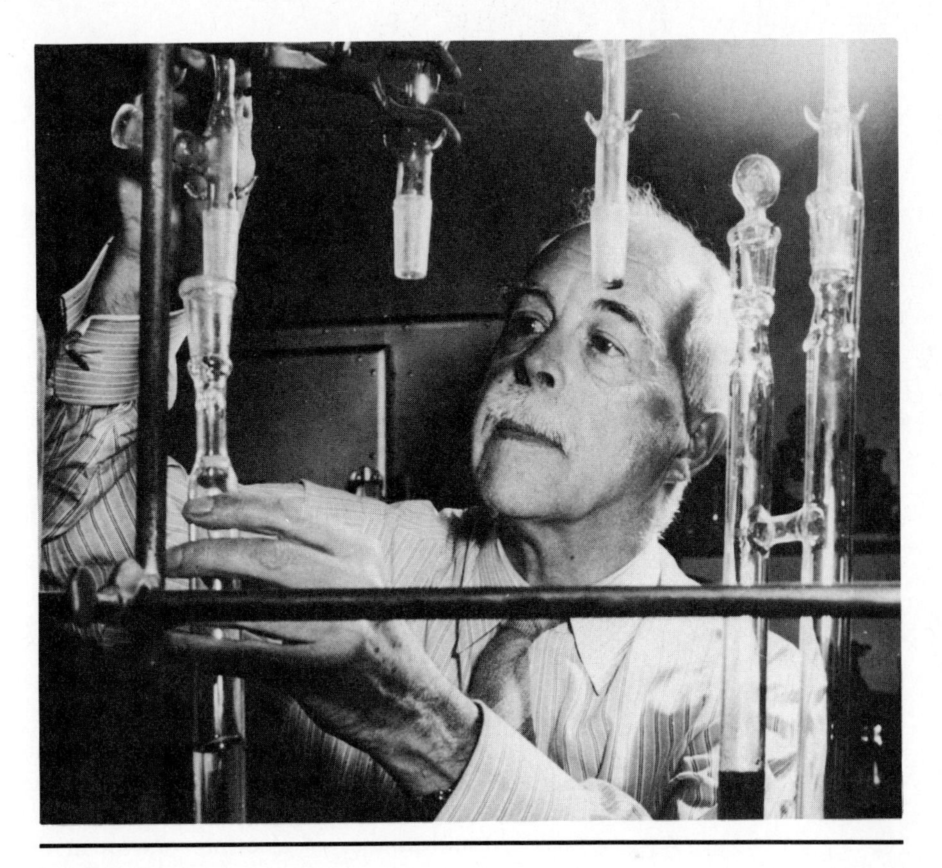

Figure 9.6. Gilbert Newton Lewis.

The Lewis structure for H_2O, a slightly more complicated molecule, can be put together by first counting the number of valence electrons contributed by each atom. The group numbers tell us that an oxygen atom has six electrons in its valence shell, and each hydrogen atom in the H_2O molecule contributes one electron. Writing the dot pictures for the individual atoms suggests how they bond together:

$$H\cdot \quad :\dot{\underset{\cdot}{\ddot{O}}}\cdot \quad H\cdot$$

The formation of one covalent bond between the oxygen and each hydrogen atom will give the oxygen atom a stable octet of electrons and will complete the valence shell of each hydrogen atom with two electrons:

$$\begin{array}{l} :\ddot{O}-H \\ \ \ | \\ \ \ H \end{array}$$

A ball-and-stick model (see Figure 9.7) gives a more realistic picture of the molecule.

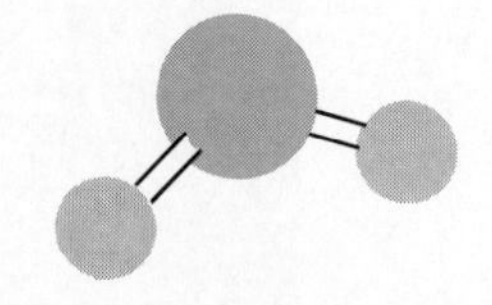

Figure 9.7. A ball-and-stick model of a water molecule.

We can approach the bonding in CCl_2F_2 in a similar fashion. Here are dot pictures for separate C, Cl, and F atoms:

$$\cdot\dot{\underset{\cdot}{\text{C}}}\cdot \qquad :\ddot{\underset{\cdot\cdot}{\text{Cl}}}\cdot \qquad :\ddot{\underset{\cdot\cdot}{\text{F}}}\cdot$$

The carbon atom needs to gain partial possession of four additional electrons to complete the octet. It must form four covalent bonds to do so. Both Cl and F need only one more electron for a filled valence shell. If the carbon atom forms four covalent bonds and Cl and F form one bond each, the sharing of pairs of electrons puts eight electrons in the valence shell of every atom. The Lewis structure is

$$\begin{array}{c} :\ddot{\text{Cl}}: \\ :\ddot{\underset{\cdot\cdot}{\text{F}}}:\ddot{\underset{\cdot\cdot}{\text{C}}}:\ddot{\underset{\cdot\cdot}{\text{Cl}}}: \\ :\underset{\cdot\cdot}{\text{F}}: \end{array} \quad \text{or, preferably,} \quad \begin{array}{c} :\ddot{\text{Cl}}: \\ | \\ :\ddot{\underset{\cdot\cdot}{\text{F}}}-\text{C}-\ddot{\underset{\cdot\cdot}{\text{Cl}}}: \\ | \\ :\underset{\cdot\cdot}{\text{F}}: \end{array}$$

A ball-and-stick model is shown in Figure 9.8.

Although there is no experimental method for distinguishing one electron from another, different symbols are sometimes used to designate the electrons coming from different atoms, simply for convenience in counting electrons. As long as the total number of electrons shown in the Lewis structure equals the sum of the valence electrons provided by the various atoms, all the electrons have been accounted for. In checking the Lewis structure for CCl_2F_2, we have

		TOTAL VALENCE ELECTRONS
1 C atom	4 valence electrons	4
2 Cl atoms	(2 atoms)(7 valence electrons/atom)	14
2 F atoms	(2 atoms)(7 valence electrons/atom)	14
1 CCl_2F_2 molecule		32

Figure 9.8. A ball-and-stick model of a CCl_2F_2 molecule.

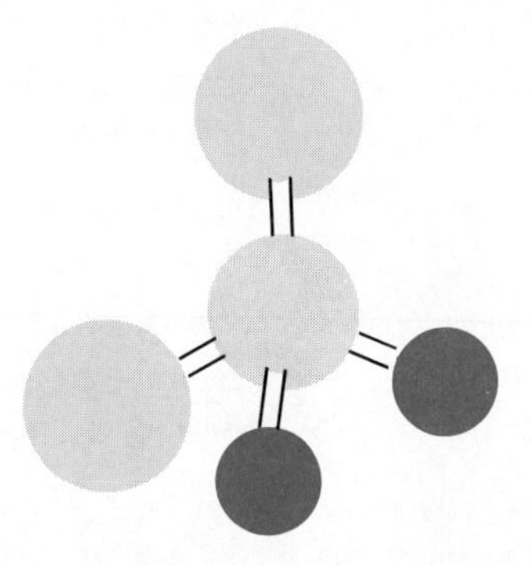

FREONS AND REFRIGERATORS

You have probably been in the same room with the covalent chemical compound CCl_2F_2 many times without even knowing it. Under the name "Freon-12," CCl_2F_2 is the most common refrigerant in use today. Most refrigerators and air conditioners use Freon-12. This chemical is critical to the operation of the compressor that pumps heat from one place to another (see Figure 9.9).

The most important quality of a good refrigerant is the right boiling point, so that it is easy to compress the vapor to a liquid and then change it back to a gas again. Each compression-vaporization cycle can be used to transfer heat from one spot to another, such as from the inside of a house to the outside. CCl_2F_2 boils at $-30°C$, so it vaporizes easily at room temperature unless it is compressed under pressure. CCl_2F_2 is also nontoxic and will not burn. It is remarkably stable, quietly going about the business of cooling our food and our bodies year after year.

Before the discovery of the Freons, the most common refrigerant was sulfur dioxide, which boils at $-10°C$. Unfortunately, SO_2 is a toxic, corrosive gas. If a compressor leaked, people were subjected to a choking, suffocating chemical. It was good riddance when Freon replaced sulfur dioxide refrigerants in the 1930s.

Because the Freons are nontoxic and have no odor or taste, they also became popular as aerosol propellants in spray cans of deodorants and food products. In the 1970s, though, scientists became worried that these stable Freons were slowly making their way up to the stratosphere, where they might interfere with the ozone layer that protects us from the sun's powerful ultraviolet rays. Because of this danger, Freons are seldom used now as aerosol propellants. Inside refrigerator compressors, though, they continue to do good work for us.

Figure 9.9. Diagram of a simple refrigerator. The refrigerant cycles through the system, sometimes as a liquid and sometimes as a low- or high-pressure gas.

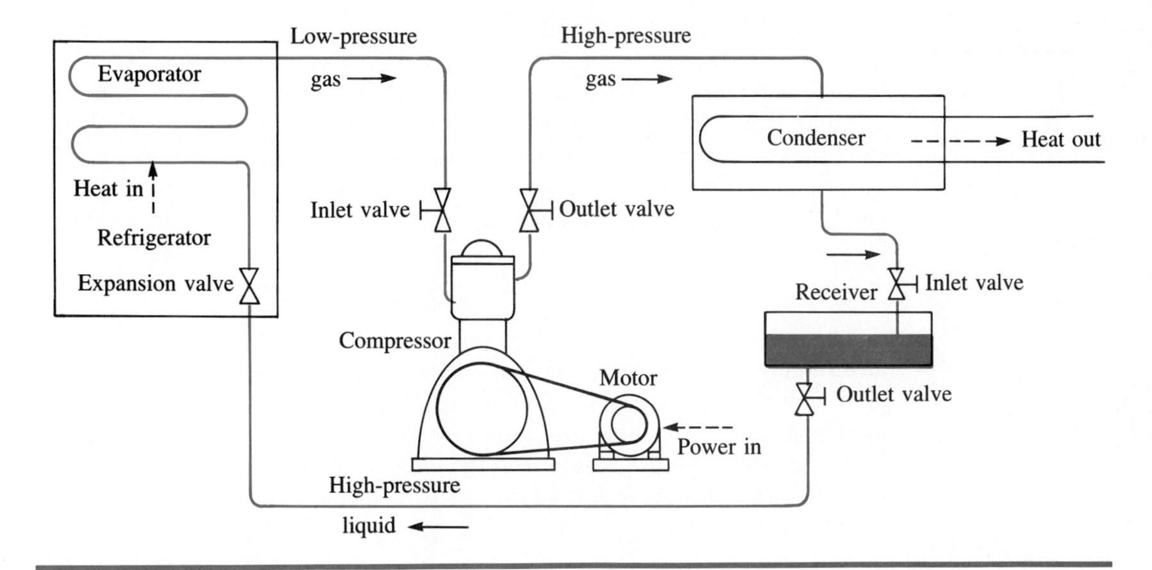

Example Write a reasonable Lewis structure for a molecule of chloroform, $CHCl_3$. This compound became popular in the nineteenth century as an anesthetic used to relieve the pain of childbirth, after it was used by England's Queen Victoria.

Solution We start with dot symbols for each kind of atom in the chloroform molecule:

H· ·C· :Cl·

A hydrogen atom can form only one bond, since two shared electrons fill its valence shell. Chlorine also forms only one bond. The octet of electrons for carbon is reached when four pairs of electrons are shared with other atoms.

The Lewis structure is

```
     :Cl:
      |
  H—C—Cl:
      |
     :Cl:
```

Remember that the number of electrons shown in a Lewis structure must be the same as the total number of valence electrons contributed by the individual atoms. To check the Lewis structure for chloroform, we see that there are one H, one C, and three Cl atoms per molecule.

$$1 \text{ valence electron} + 4 \text{ valence electrons} + (3 \text{ atoms})(7 \text{ valence electrons/atom}) = 1 + 4 + 21 = 26 \text{ total valence electrons}$$

It checks! Remember that each line in the Lewis structure stands for two electrons.

9.4 MULTIPLE COVALENT BONDS

Not all molecules have enough valence electrons to provide each atom with a stable octet when every bond contains just one pair of electrons. A good example is the nitrogen molecule, N_2, in which each nitrogen atom provides five valence electrons. If only one pair of electrons were shared between atoms in a **single bond**, the distribution of electrons in the molecule might look like this:

·N—N·

This structure would give the molecule some stability; however, neither atom obeys the octet rule. If the atoms were to share more electrons to give each a complete valence shell of eight, the atoms would be held together more tightly. At the same time, none of the electrons would be forced into shells having energies greater than the energy of the valence shell. A **triple bond** fills the bill nicely for the nitrogen molecule:

:N≡N:

Now the octet rule is satisfied for each atom, because the six shared electrons reside in the valence shell of each one. A **double bond** is also possible in the right circumstances.

Although double and triple bonds seem to fit the octet rule perfectly, what evidence do we have that they really exist? Measurements of the energies required to break apart different kinds of molecules provide one type of evidence. In order to dissociate 1 mole of N_2 molecules into 2 moles of N atoms, 945 kilojoules of heat energy must be added.

$$N_2 + 945\text{ kJ} \longrightarrow 2\cdot\ddot{\underset{\cdot}{N}}\cdot$$

This energy can be compared with the 160 kilojoules of heat absorbed by 1 mole of hydrazine, H_2NNH_2 (sometimes used as a rocket fuel), when it breaks into 2 moles of NH_2 fragments. The dissociation involves only the breaking of the nitrogen-nitrogen single bond, as the following structures show:

$$\text{H}-\underset{\text{H}}{\underset{|}{\ddot{\text{N}}}}-\underset{\text{H}}{\underset{|}{\ddot{\text{N}}}}-\text{H} + 160\text{ kJ} \longrightarrow 2\cdot\underset{\text{H}}{\underset{|}{\ddot{\text{N}}}}-\text{H}$$

The vastly different energies, 945 kilojoules and 160 kilojoules, demonstrate that the bonds in these two molecules are different. These energies agree nicely with Lewis structures showing triple and single bonds, respectively. The sharing of three rather than one pair of electrons holds the atoms together more strongly, so more energy is necessary to break the triple bond.

A second difference between these molecules of nitrogen and hydrazine is in the distance between nitrogen nuclei in the two molecules: 0.110 nanometer in N_2 and 0.145 nanometer in H_2NNH_2, as measured by x-ray diffraction. The shorter distance in N_2 is consistent with the stronger bonding. Therefore, the distinction between triple and single bonds is more than just a figment of chemists' imaginations.

For carbon dioxide, the octet rule cannot be satisfied if only single bonds are shown, as in

$$:\ddot{\underset{\cdot\cdot}{O}}-C-\ddot{\underset{\cdot\cdot}{O}}:$$

In this structure carbon does not have eight electrons in its valence shell. Yet the problem is not that the total number of electrons is wrong. The carbon contributes four, and each oxygen six, for a total of sixteen valence electrons, which agrees with the number shown. Imagine that one unshared pair on each oxygen atom in this structure shifts to be shared with the carbon atom. Two double bonds are formed, and a reasonable Lewis structure results. Every atom now has an octet of valence electrons.

$$\ddot{\underset{\cdot\cdot}{O}}=C=\ddot{\underset{\cdot\cdot}{O}}$$

Figure 9.10 shows a model of CO_2.

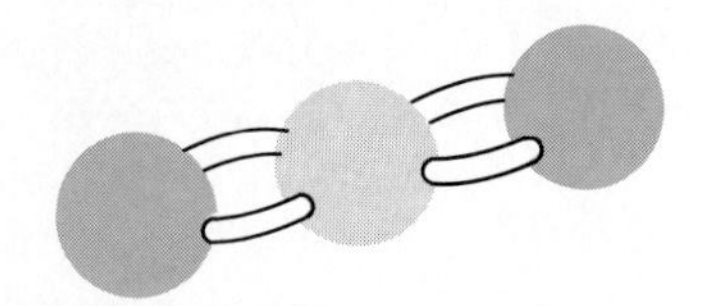

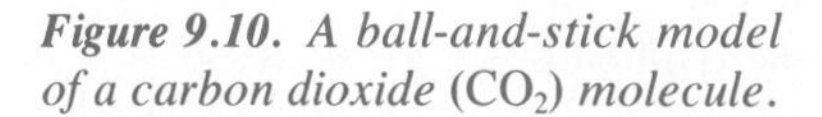

Figure 9.10. A ball-and-stick model of a carbon dioxide (CO_2) molecule.

There are many other molecules that do not have enough electrons to satisfy the octet rule unless double or triple bonds are assumed.

Example Draw a reasonable Lewis structure for acetylene, C_2H_2. Acetylene is the fuel used by metal welders in their oxyacetylene torches. The two carbon atoms participate in a triple bond.

Solution Each carbon atom has four valence electrons, and each hydrogen atom has one, for a total of ten electrons. Knowing that the two carbon atoms are linked together helps a lot.

The correct structure is

$$H—C{\equiv}C—H$$

9.5 RULES FOR WRITING LEWIS STRUCTURES

Most of the basic principles of covalent bonding have now been covered. Before going on to more complicated molecules, let's try to organize the procedure for writing a reasonable Lewis structure for a molecule. Even when the rules given below are used, the method is far from automatic. Often you must try several possibilities before you achieve success.

One more reminder: you must know what atoms are bonded together before you plug electrons into a molecule. Sometimes the arrangement of atoms is obvious, as in H_2O; but in other cases, chemical intuition or experimental evidence is essential. For example, without prior knowledge you might assume that the order of atoms in carbon dioxide was COO rather than the correct OCO.

The following procedure has worked well for many people.

1. Using the standard symbols for the atoms, draw the skeleton for the molecule, and connect with single bonds the atoms that are bonded together.
2. Determine the total number of valence electrons in the molecule by adding together the contributions from each atom.
3. Subtract two valence electrons for each single bond drawn in the first step. Distribute the remaining electrons as unshared pairs so as to give each atom no more than eight electrons.

4. If each atom has an octet (a duet in the case of hydrogen), a reasonable Lewis structure has been written. If not, shift electrons to form double or triple bonds until the octet rule is satisfied for every atom.

Let's apply these rules to formaldehyde, H_2CO, in which both hydrogens and the oxygen are bonded to carbon. Formaldehyde is a very reactive molecule used to make plastics and adhesives. It can also be found in disinfectants and in embalming fluid. Of the products manufactured in the United States, 8% are made in part from formaldehyde.

Step 1. Draw the skeleton.

```
     O
     |
  H—C—H
```

Step 2. Find the total number of valence electrons.

		TOTAL VALENCE ELECTRONS
2 H atoms	(2 atoms)(1 valence electron/atom)	2
1 C atom	4 valence electrons	4
1 O atom	6 valence electrons	6
1 H_2CO molecule		12

Step 3. Subtract two electrons for each single bond shown in Step 1. Distribute 12 − 6 = 6 electrons as unshared pairs.

```
    :O:                  ..
     |                  :O:
  H—C—H      or          |
    ..                H—C—H
```

Step 4. Neither of the structures in Step 3 is satisfactory. In the first, oxygen doesn't have a filled valence shell of electrons; in the second, carbon doesn't. However, by shifting one pair of electrons to form a double bond between C and O, we have a reasonable structure. Notice that the oxygen atom shares two pairs of electrons with carbon and, in addition, there are two unshared pairs of electrons in the valence shell of oxygen.

```
    :O:
     ||
  H—C—H
```

Both the arrangement of atoms and the bonding of formaldehyde are shown in Figure 9.11.

Example Draw a likely Lewis structure for methyl mercaptan, CH_4S, in which three hydrogen atoms and the sulfur atom are bonded to carbon. Methyl mercaptan is a very smelly compound that is added in small amounts to natural gas so that gas leaks can be detected easily and bad explosions prevented.

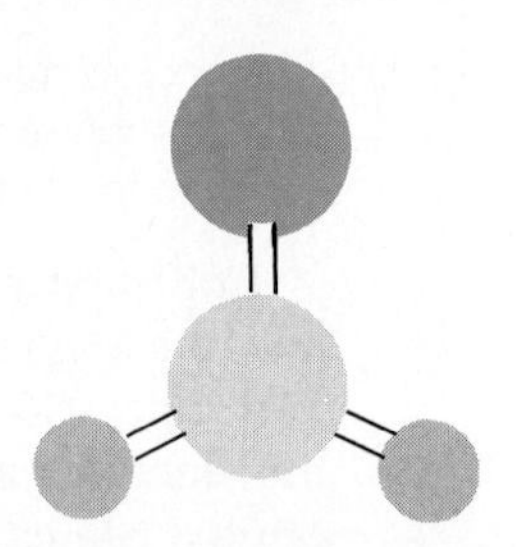

Figure 9.11. A ball-and-stick model of a formaldehyde molecule.

Solution Again, the four-step procedure can be used.

Step 1. Draw the skeleton, using single bonds.

```
     H
     |
  H—C—S—H
     |
     H
```

Step 2. Find the total number of valence electrons.

$$\underset{4\text{ H}}{(4)(1)} + \underset{1\text{ C}}{4} + \underset{1\text{ S}}{6} = 14 \text{ electrons}$$

Step 3. There are five single covalent bonds in the skeleton drawn in Step 1, so we must subtract ten electrons from the total.

$$14 - 10 = 4 \text{ electrons}$$

Since each hydrogen already has a duet and carbon has an octet of electrons, these four electrons must go into the valence shell of sulfur.

```
     H
     |   ..
  H—C—S—H
     |   ..
     H
```

Step 4. Every atom now has an octet (a duet in the case of hydrogen), so our Lewis structure is reasonable.

9.6 EXCEPTIONS TO THE OCTET RULE

One of the chemicals in the polluted air above our cities is nitric oxide, NO, a by-product of the high-temperature combustion in car and truck engines. Nitric oxide presents a special problem to chemists as well as to environmentalists. Because this molecule has an odd number of valence

COMMON SYMBOLS

We see so many symbols from day to day that we tend to take them for granted. Probably everyone who has a driver's license knows the meaning of these two symbols:

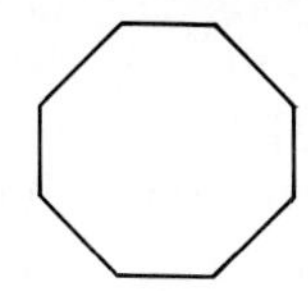

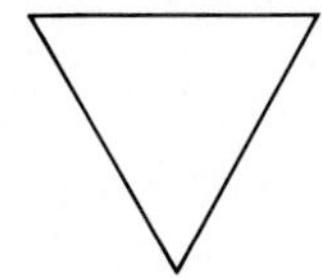

The one on the left means *stop!* On the right is a yield sign, warning of some danger ahead.

Some symbols communicate powerful messages of great meaning:

Others are more specialized but still do a good job of sending the message:

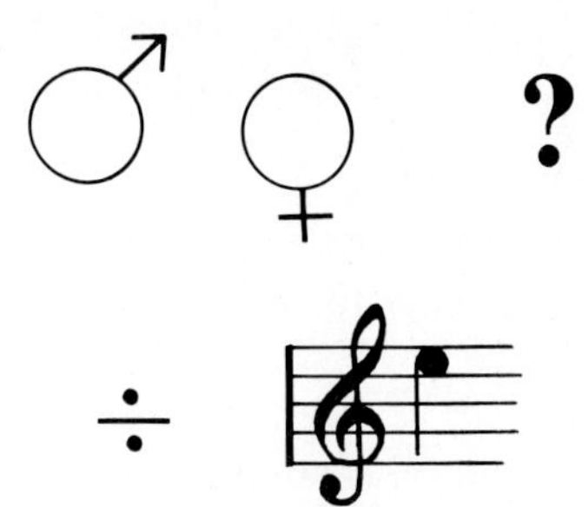

The molecular structures used by chemists can seem very strange until one learns to recognize what the symbols mean, at which point they become rather useful.

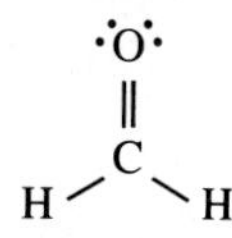

electrons (11 electrons), the octet rule cannot possibly be satisfied for each atom. The best we can do is something like

$$\cdot\ddot{\text{N}}{=}\ddot{\underset{\cdot\cdot}{\text{O}}} \quad \text{or} \quad \underset{\cdot}{\ddot{\text{N}}}{=}\ddot{\underset{\cdot\cdot}{\text{O}}}$$

The unpaired electron makes NO somewhat unstable and chemically reactive toward many other compounds.

There are some other molecules in which more than eight electrons are found in the valence shell of the central atom. For example, in PCl_5, five chlorine atoms are bonded to phosphorus.

:C̈l:
:C̈l̤ \ |
P—C̈l̤:
:C̈l̤ / |
:Cl̤:

Since two electrons are needed for each covalent bond, there are ten electrons in the valence shell of phosphorus. Other elements in the third, fourth, and fifth rows of the periodic table are also capable of having "expanded octets," but we will not analyze them any further.

9.7 MOLECULAR GEOMETRIES

In order for molecules to react together, they must have some way to recognize each other. Chemical recognition depends not only on the nature of the atoms and the kinds of chemical bonds, but also on the shapes, or geometries, of the molecules. These molecular shapes depend on the angles between covalent bonds. The realization that covalent bonds have definite, fixed direction has been of great value in understanding the properties of molecules. For example, we know that the structure of H_2O is bent rather than linear (see Figure 9.12). This simple fact has enormous consequences for the ecology of our lakes and rivers, as we shall see.

Molecular geometries are not obvious in the Lewis structures we have been drawing, because the octet rule contains no information about where shared pairs of electrons are located. Our structures have shown the outside atoms in arbitrary positions about the central atom of a molecule. Yet experiments involving x-ray diffraction show that molecules do have definite shapes.

The term **bond angle** simply means the angle between two covalent bonds from one atom to two neighboring atoms. In H_2O, for example, the oxygen atom is at the intersection of the two covalent bonds. The HOH bond angle is shown below.

O
H H
bond angle

A useful and simple scheme for explaining and predicting bond angles in molecules was developed by four British chemists in the 1940s and 50s. It is called the **VSEPR,** or **valence-shell electron-pair repulsion, theory.** The central assumption of the VSEPR theory is as follows:

> The electron pairs in a valence shell of an atom adopt the arrangement that maximizes the angle between any two pairs.

This generalization seems reasonable—since the negatively charged electrons repel each other, their maximum separation provides the lowest energy. We can think of each pair of electrons as occupying a different electronic orbital, with different orbitals as far as possible from each other. This

Figure 9.12. Model of a water molecule.

principle applies to both shared and unshared pairs of electrons in the valence shell.

Double and triple bonds are treated in a special way. The two pairs of electrons in a double bond and the three in a triple bond are confined to a small region of space between the bonded atoms. They cannot separate and find positions far from each other and yet still remain bonding electrons. To handle this situation, we define a **cluster of electrons** as a single bond, a double bond, a triple bond, or an unshared pair of electrons. For lowest energy, the separation between clusters of electrons is as great as possible. Note that both bonds and unshared pairs are counted as clusters.

Many of the molecules discussed so far, such as H_2O, H_2CO, and CCl_2F_2, consist of a central atom to which all the other atoms are bonded. It is the arrangement of electron clusters in the valence shell of this central atom that determines the bond angles and therefore the geometry of the molecule. If the molecule is more complicated, any atom with two or more bonds is only partially responsible for the overall shape.

What are the stable arrangements of electron clusters? We will consider just three cases: two, three, and four clusters in the valence shell of the central atom. These cases encompass most molecules.

1. Two clusters are on opposite sides of the nucleus in a linear configuration.
2. Three clusters of electrons are at the corners of an equilateral triangle, with the nucleus at the center.
3. Four clusters are at the corners of a tetrahedron, with the nucleus at the center.

The first two arrangements are easy to see because the geometries are in two dimensions, as in Figure 9.13. The third arrangement, a regular

Figure 9.13. Positions of minimum energy for two, three, and four clusters of electrons in the valence shell of the central atom in a molecule. B *stands for an electronic cluster, and* A *stands for the central atom.*

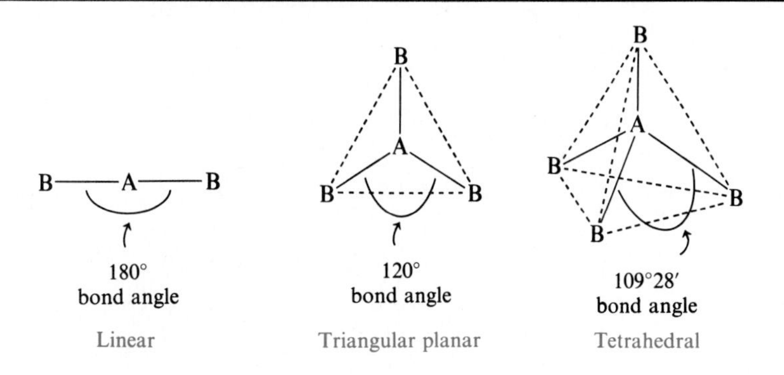

tetrahedron with its four triangular faces, also shown in Figure 9.13, is more difficult to work with because it is based on three dimensions.

Table 9.2 gives the possible combinations of shared electron bonds and unshared electron pairs for these three cases, along with the resulting bond angles. The letter A stands for the central atom in a molecule. One line, two parallel lines, and three parallel lines indicate single, double, and triple bonds, respectively. A pair of dots represents an unshared pair of electrons. Drawing the Lewis structure is an essential first step in predicting the geometry of any molecule, because the Lewis structure reveals the number of electron clusters in the valence shell of the central atom.

The bond angles of 180° and 120° follow from simple geometrical con-

Table 9.2 Combinations of Bonds and Unshared Pairs of Electrons plus Bond Angles for Central Atoms with Two, Three, and Four Clusters of Electrons (The letter A stands for the central atom, and the letter B stands for an outside atom. A pair of dots represents an unshared pair of electrons. Tetrahedral geometry is shown by using a solid wedge for a bond extending in front of the plane of the page and a dashed line for a bond pointing behind the page.)

NUMBER OF ELECTRON CLUSTERS AROUND THE CENTRAL ATOM	ARRANGEMENT OF BONDS AND UNSHARED ELECTRON PAIRS	EXAMPLE	PREDICTED BOND ANGLE
2	B—A≡B	H—C≡N:	180°
2	B=A=B	:Ö=C=Ö:	180°
3	B ‖ A ╱ ╲ B B	:O: ‖ C ╱ ╲ :C̤̈l C̤̈l:	120°
3	B ‖ :A ╲ B	:O: ‖ :N ╲ C̤̈l:	120°
4	B | A B B B	H | C H H H	109°
4	Ä B B B	N̈ H H H	109°
4	:Ä B B	:Ö H H	109°

siderations (Figure 9.13). The angle of 109° cannot be simply drawn in two dimensions, but lines drawn from the center of a tetrahedron to its four corners do in fact make angles of 109° with respect to each other.

Methane, CH_4, is a typical molecule, having four clusters of electrons in the valence shell of the central carbon atom.

```
     H
     |
  H—C—H
     |
     H
```

The fifth entry in Table 9.2 shows the tetrahedral geometry of the methane molecule. Every HCH bond angle should be 109° according to the VSEPR theory. The bond angles measured by electron diffraction are almost exactly 109°.

The Lewis structure of carbon tetrachloride is very similar to that of methane, as the following drawing shows:

```
        ..
       :Cl:
        |
  ..    |    ..
 :Cl—C—Cl:
  ..    |    ..
        |
       :Cl:
        ..
```

CCl_4 is also a tetrahedral molecule. It differs from CH_4 in that each of the outer atoms has four pairs of electrons in its valence shell, but this difference has nothing to do with the shape of the molecule.

What bond angles do we predict for ammonia, NH_3? Its Lewis structure,

```
     ..
  H—N—H
     |
     H
```

shows that the central N atom has four clusters of electrons in its valence shell, but here one of them is an unshared pair of electrons. This pair should occupy one corner of a tetrahedron, as do the three shared pairs. The three-dimensional structure of ammonia is shown in Table 9.2. Each of the three HNH bond angles is predicted to be 109°, in fairly good agreement with the observed 107°.

Closely related to CH_4 and NH_3 is H_2O, which also has four clusters of electrons around the central atom. Two of the clusters are shared, and the other two are unshared pairs of electrons. Remember that a solid wedge indicates that a bond or a pair of electrons extends in front of the plane of the page and a dashed line indicates a bond or pair of electrons points behind the page. Viewed from two possible perspectives, the structure is

```
   ˙O˙            or       ˙ ·.   .·˙
  ◢   ⋱                        O
 H     H                      / \
                             H   H
```

The HOH bond angle in the water molecule is 104.5°, a bit smaller than the predicted 109°. Generally, when the valence shell of the central atom has a mixture of shared and unshared electron pairs, the actual bond angles are somewhat smaller than the predicted values.

In Figure 9.14 the electron-pair arrangements in CH_4, NH_3, and H_2O are shown with the use of balloon models.

Example What value would you predict for the bond angle in formaldehyde, H_2CO?

Solution To use the VSEPR theory, we must first write a satisfactory Lewis structure. This structure was written on page 199.

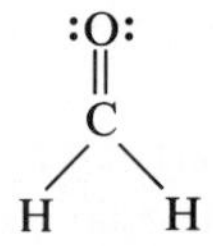

The two electron pairs in the carbon-oxygen double bond count as one electron cluster. This means that there are three electron clusters around carbon, the central atom. The three clusters will be farthest apart when the three bond angles are about 120°.

The shapes of molecules are very important for chemical recognition, especially in biological systems. The complex steroid sex hormones, for example, have very specific geometries that determine their recognition at

Figure 9.14. Electron-pair arrangements with balloon models. The long balloons represent bonding electron pairs, and the round balloons represent the larger space occupied by unshared pairs of electrons. (a) A tetrahedral arrangement of four bonds in a molecule such as CH_4*. The bond angle is 109°. (b) A tetrahedral arrangement of three bonds and one unshared electron pair in a molecule such as* NH_3*. The bond angle is a little less than 109°. (c) A tetrahedral arrangement of two bonds and two unshared electron pairs in a molecule such as* H_2O*. The bond angle is a little less than 109°.*

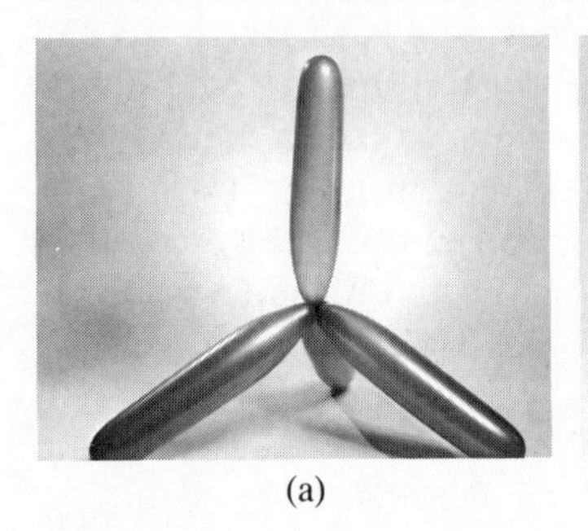

(a)

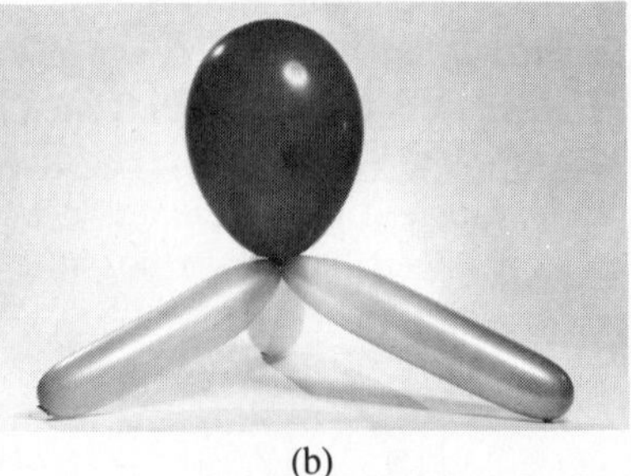

(b)

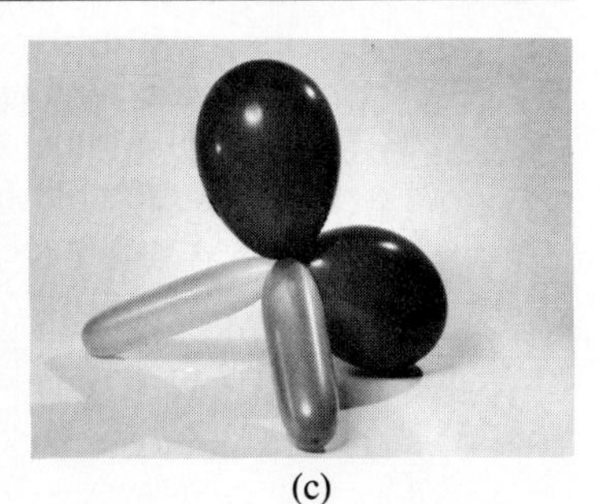

(c)

receptor sites in the body. One of the most popular theories of how we distinguish different odors is based on the idea that our noses have smell receptors of different shapes.

9.8 FORCES BETWEEN MOLECULES

Up to this point in the chapter we have discussed only the chemical bonds within molecules—the bonds that hold the atoms together. Now we will shift to the much weaker forces that help to hold covalent molecules near each other. These forces between covalent molecules are so weak that we won't even call them bonds.

The molecules in a liquid are closely packed but somewhat mobile, whereas the molecules in a gas are much farther apart, moving randomly in all directions. The energy needed to evaporate 1 mole of a typical molecular liquid and thus separate the molecules from each other is just a tenth of the energy required to disrupt 1 mole of chemical bonds and rupture the molecules into atomic fragments. Nevertheless, the relatively weak forces between molecules determine some of the important physical properties of a substance, such as its boiling point.

In Chapter 4 we discussed the equilibrium between the liquid and vapor forms of a substance when it is confined to a closed container. In an open container a liquid boils only when the temperature is raised to the point at which the liquid's vapor pressure equals the pressure of the surrounding air. This temperature is called the *boiling point* of the liquid. Some liquids, such as the Freon-12 used in refrigerators and air conditioners, have such weak forces between molecules that they boil below 0°C. Other liquids, such as water or the ethylene glycol used in antifreeze, have larger intermolecular forces and much higher boiling points. In general, the greater the intermolecular forces, the higher the boiling point.

What causes these weak forces between molecules? Like all of the other important forces in chemistry, they are electrical in nature. In Figure 9.15(a) the average uniform distribution of electrons in a single atom is represented by a symmetrical sphere with the positively charged nucleus at the center. However, the electrons are very light, and this symmetrical electron distribution can easily be disturbed. At any instant the fast-moving electrons might have shifted from their symmetrical distribution for a brief time. In Figure 9.15(b) a few electrons have temporarily moved to the right, creating a very slight excess of negative charge there. The deficiency of electrons on the left means that the protons in the nucleus are more strongly felt in that region, giving it a slight net positive charge. These tiny partial charges, indicated by the Greek letter δ, must be equal in magnitude, because the atom is electrically neutral overall.

Figure 9.15(c) shows what happens when a second atom happens to be very close to the one that has the asymmetrical arrangement of electrons.

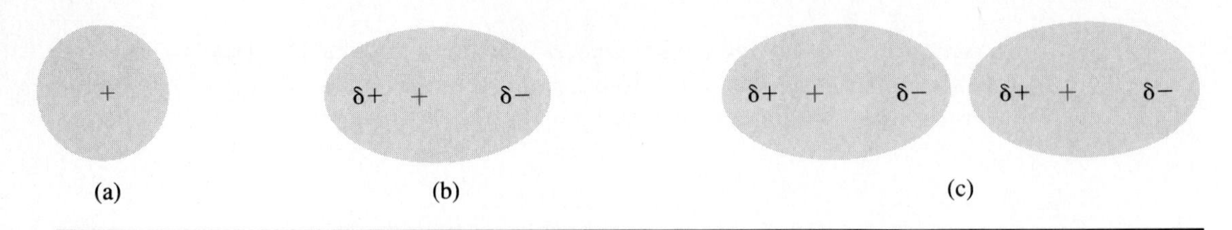

Figure 9.15. Origin of the dispersion force. (a) The average, symmetrical distribution of electrons in an atom. The colored plus sign in the center indicates the nucleus. (b) A short-lived migration of electrons to the right. δ− and δ+ are negative and positive charges, respectively, whose magnitude is a small fraction of the charge on an electron. (c) Formation of a weak electrical attraction between two atoms. The temporary dislocation of charge in the left-hand atom causes a similar situation in the atom on the right.

Electrons in the second atom [shown in Figure 9.15(c) on the right] are repelled by the negative region of the first atom, and they move away to the right. Finally, there is an attraction between the negative side of the first atom and the positive side of the second. Even though the action sounds like the proverbial bootstrap operation, this weak **dispersion force** is believed to operate in every molecular liquid and solid. The explanation just given for atoms also applies to molecules.

There is one remaining question: What determines the size of the dispersion force, which weakly attracts separate molecules to each other? Table 9.3 gives the boiling points of several hydrocarbons, some of which are in gasoline. Simplified structures of two of the molecules—pentane and octane—are shown below.

pentane

octane

It is evident that the boiling point increases as the number of carbon and hydrogen atoms increases. This can only mean that the dispersion force is also increasing. The larger the number of atoms, the larger the surface

Table 9.3 Boiling Points of Some Hydrocarbons

COMPOUND	FORMULA	BOILING POINT, °C
pentane	C_5H_{12}	36
hexane	C_6H_{14}	69
heptane	C_7H_{16}	98
octane	C_8H_{18}	126

area of the molecule is. A large surface area means a large number of electrons and a high probability that some electrons will stray from their average positions. So the dispersion force between molecules, as well as the boiling point of a liquid, becomes larger as the surface area of the molecules increases.

Petroleum refiners make practical use of the different boiling points of the hydrocarbons in gasoline to make our cars run more efficiently. In order for a car to start easily in the winter months, gasoline must be able to vaporize in the carburetor at low temperatures. Therefore, in winter the proportion of hydrocarbons with lower boiling points, such as pentane, is increased so as to increase the vapor pressure of the gasoline. During the summer months, when a higher boiling point is needed, larger molecules are added.

To show the generality of the correlation between boiling point and molecular size or surface area, let's examine the boiling points of the diatomic elements of Group VII in the periodic table.

ELEMENT	°C
F_2	−187
Cl_2	−35
Br_2	59
I_2	184

An iodine molecule has many more electrons and a far larger surface area than a molecule of fluorine does. Again, boiling point increases with molecular size. This relationship works best when the substances are similar. Although special kinds of intermolecular forces exist for certain molecules, the dispersion force is often the most important force between molecules.

Example Predict which has the higher boiling point, CF_4 or CCl_4.

Solution The Lewis structures for these two similar molecules are

```
       ..                      ..
      :F:                    :Cl:
  ..   |   ..            ..   |   ..
 :F — C — F:     and    :Cl — C — Cl:
  ..   |   ..            ..   |   ..
      :F:                    :Cl:
       ..                      ..
```

Which one has the larger surface area and therefore the stronger intermolecular attractions? Chlorine atoms have many more electrons in inner shells than fluorine atoms have. Chlorine atoms are larger than fluorine atoms. The larger the atom, the greater its surface area. We can predict that carbon tetrachloride (CCl_4) has a higher boiling point than carbon tetrafluoride (CF_4). (The data fit our prediction: CCl_4 has a boiling point of 76°C, and CF_4 boils at −129°C.)

9.9 POLAR MOLECULES

Our discussion of dispersion forces in the last section was based on the fact that the distribution of electrons in molecules tends to be symmetrical. Unless the relative electronic positions have become distorted momentarily, we would expect an isolated atom to have an equal electron distribution in all directions, and this is true on the average. However, this is seldom the case for molecules.

Take, for instance, the diatomic molecule ICl, iodine monochloride. It is not obvious that the iodine atom should have the same pull on its electrons as the chlorine atom does. In fact, it doesn't. The nucleus of the chlorine atom has a greater pull on its valence electrons than the iodine nucleus does. Thus a **dipole,** or separation of charge, is set up within the molecule. The shared pair of electrons between the two atoms is pulled closer to the chlorine nucleus, and thus the electron cloud near the chlorine atom has a higher electron density. This is a permanent situation: there is always greater negative charge at the Cl end of ICl than at the I end (see Figure 9.16). Molecules with dipoles are called **polar** molecules; they are said to have **polarity.**

Naturally, the molecules tend to line up in a crystal of ICl so that the opposite electrical charges are closest together. This maximizes the attractive forces and leads to a lower potential energy (see Figure 9.17).

When ICl is heated to 27°C, the dipole forces are no longer able to hold the molecules in a rigid pattern, and so the solid melts. Polar molecules are still quite close together in the liquid state, and dipole forces continue to attract molecules to each other. However, these dipole forces are nowhere near as great as the forces that hold an ionic solid together. In Na^+Cl^- there has been a complete transfer of electrons, leading to ions with full positive and full negative charges. ICl, a covalent compound, melts at 27°C; Na^+Cl^-, an ionic compound, melts at 801°C!

Next let's take the case of carbon tetrachloride, CCl_4. The chlorine atoms attract electrons much more strongly than carbon does. So one might expect the CCl_4 molecule to have a large dipole, but it doesn't. In fact, its dipole is zero! How can this be? The answer has to do with the shape of the CCl_4 molecule.

In the last section we saw that CCl_4 is a tetrahedral molecule.

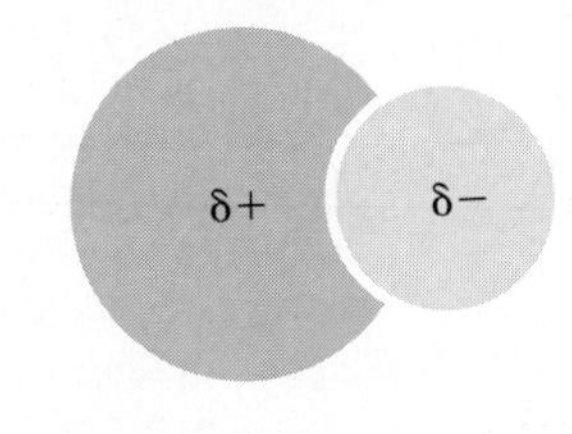

Figure 9.16. Model of an ICl *molecule showing its permanent dipole. The smaller chlorine atom at the right has a stronger pull on the electrons in the molecule. The larger iodine atom is left with a partial positive charge.*

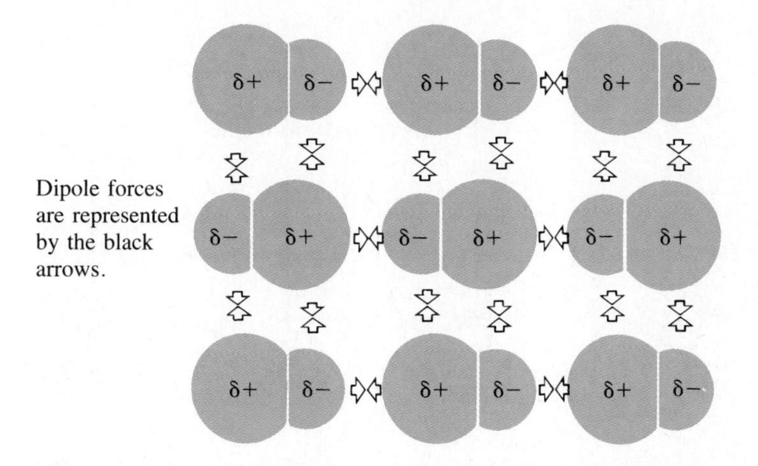

Figure 9.17. The dipole forces in an ICl *crystal.*

$$\mathrm{C^{\delta+}Cl_4^{\delta-}}$$

Each carbon-chlorine covalent bond has an unequal distribution of electrons, but these bond polarities are all pulling in opposite directions. Therefore, they cancel each other out. In mathematical terms, the vectors of the individual dipoles in the bonds add up to zero.

Chemists often use the measured dipole of a covalent compound as evidence of the shape of its molecules. For example, in the case of carbon dioxide, the measured dipole is zero. What does this tell us about the shape of a CO_2 molecule? Oxygen has a much greater pull on the electrons than carbon does. Therefore a partial negative charge sits on each oxygen atom, and a compensating positive charge is on the carbon atom. Only if the molecule has linear geometry—a bond angle of 180°—can these bond polarities pull in opposite directions and exactly cancel each other out. The negative charges on the two oxygen atoms are equivalent to a single negative charge midway between them, which is the exact location of the positive charge. These canceling opposite charges leave the CO_2 molecule with no dipole.

$$\overset{\delta-}{\mathrm{O}}=\overset{\delta+}{\mathrm{C}}=\overset{\delta-}{\mathrm{O}}$$

In contrast, the water molecule has a substantial dipole. The individual bond dipoles of water molecules, with their two O—H bonds, could cancel each other if H_2O molecules were linear. Of course, they are bent, not linear. We know that the bond angle is 104.5°. The VSEPR theory predicts that the four clusters of electrons around the central oxygen atom should produce a bond angle of 109°. One of the experimental foundations of this theory is the measurement of molecular dipoles.

$$\overset{\delta-}{\text{O}} \quad \text{H} \quad \text{H} \quad \delta+$$

The oxygen atom in the water molecule attracts electrons more strongly than hydrogen does. The midpoint of the partial positive charges on the hydrogens is halfway between them, which does not coincide with the partial negative charge on the oxygen atom. This gives the water molecule a dipole with its negative end toward oxygen and its positive end toward the hydrogen atoms.

9.10 HYDROGEN BONDING—WHY WATER IS UNIQUE

In addition to dispersion and dipole forces, there is a third interaction between molecules. Although it affects only those molecules in which a hydrogen atom is attached by a covalent bond to oxygen, nitrogen, or fluorine, this third type of intermolecular interaction has a tremendous effect on our lives. It is called the **hydrogen bond.** Though the hydrogen bond is much weaker than a covalent bond, it is the strongest type of intermolecular force.

When a hydrogen atom shares a pair of electrons with a small nonmetal atom, which has a strong electron pull, the distribution of electrons around the hydrogen nucleus becomes quite meager. It is almost as if the hydrogen atom were a bare proton. Thus the hydrogen nucleus is attracted to the nonmetal atom of a neighboring molecule. An entire chain of molecules can be linked together this way, as shown for hydrogen fluoride in Figure 9.18.

Hydrogen bonds are common in substances containing O—H and N—H covalent bonds, and they have a strong effect on their properties. Hydrogen bonds make molecules much more soluble in water, they stabilize the three-dimensional structures of proteins, and they are a significant force in the

Figure 9.18. The hydrogen bonds (shown by dashed lines) in hydrogen fluoride.

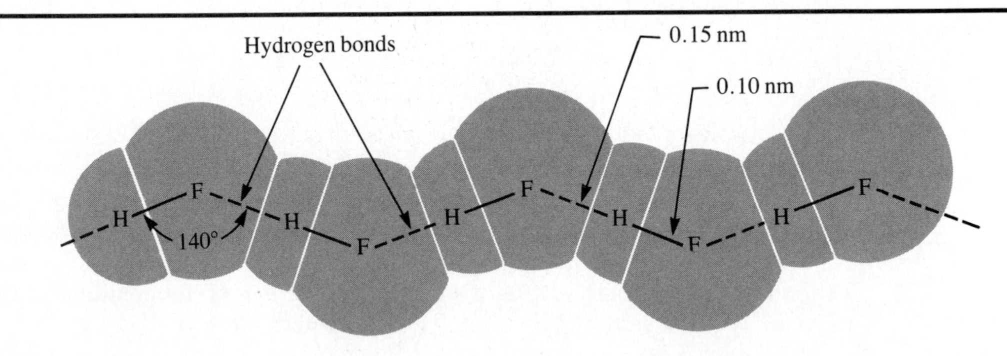

Table 9.4 Boiling Points of Some Covalent Compounds of Similar Size and Mass

COMPOUND	BOILING POINT (°C)
CH_4	−164
NH_3	−33
H_2O	+100
HF	+19

chemistry of the DNA molecules in the chromosomes of our genes.

Perhaps most important of all are the hydrogen bonds that water molecules form with themselves. Because of the moderately strong hydrogen bonds in liquid water, H_2O has a surprisingly high boiling point (see Table 9.4). Water has a boiling point over 250°C higher than that of CH_4.

Imagine how different the earth would be if the boiling point of water were 20°C. Only the coldest parts of our planet would have rivers, lakes, or oceans. There would be no liquid water in our bodies.

We mentioned in Chapter 4 the importance of the density of ice. Ice is less dense than liquid water, so ice floats on water. Lakes freeze on the

Figure 9.19. *(a) The tetrahedral arrangement of four hydrogen atoms around a central oxygen atom in ice. The hydrogen bonds are shown by dashed black lines. (b) The open crystal structure of ice.*

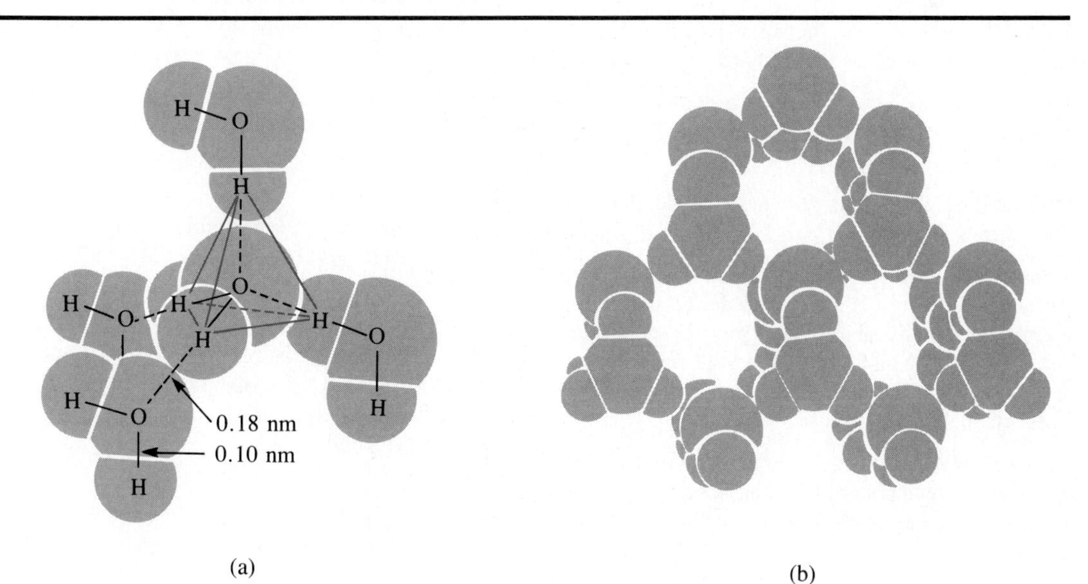

top and insulate the water below. This is very unusual, as most substances are more dense in the solid state than in the liquid state. The open crystal structure of ice results from the tetrahedral geometry of water molecules. Maximum hydrogen bonding occurs when hydrogen atoms from neighboring water molecules bond to the unshared electron pairs on oxygen, as shown in Figure 9.19(a). Melting the ice destroys this open network [see Figure 9.19(b)], and the water molecules move somewhat closer to each other on the average. However, hydrogen bonding is still important in the slightly more dense liquid. What a different world it would be without hydrogen bonds and the tetrahedral geometry of water!

IMPORTANT TERMS

ionic compound A chemical compound containing ions of opposite charge. The strong forces of attraction between these ions give ionic compounds high melting and boiling points. No molecules are present in ionic compounds.

covalent compound, or **molecular compound** A chemical compound in which the atoms are held together in molecules by covalent bonds. Covalent compounds tend to have weak attractions between molecules.

valence electrons The outermost and most loosely held electrons in an atom.

covalent bond A bond between two atoms that consists of a pair of electrons shared by the atoms.

electron shell A group of electrons at about the same average distance from the nucleus of an atom.

valence shell The outermost layer of electrons in an atom.

octet rule A rule stating that each atom in a molecule tends to complete its valence shell with eight electrons by sharing electron pairs with the other atoms to which it is bonded. The most common exception is hydrogen, which completes its valence shell with two electrons.

electron-dot picture A simplified way of drawing an atom or a molecule with emphasis on the valence electrons. The electron-dot picture for an oxygen atom is

:Ö·
  ·

orbitals The specific regions in which electrons are found in atoms. An orbital has a maximum capacity of two electrons. Electrons in higher-energy orbitals are farther from the nucleus on the average.

Lewis structure A description of a molecule in which lines are used for covalent bonds and dots for unshared electrons. The Lewis structure for Cl_2 is

:C̤̈l—C̤̈l:

single bond A single covalent bond having a shared pair of electrons in the valence shell of each bound atom.

triple bond A covalent bond system made up of three single covalent bonds, each having one pair of electrons shared by the two atoms bound together.

double bond A covalent bond system composed of two single covalent bonds, each having one pair of electrons shared by the two bound atoms.

VSEPR (valence-shell electron-pair repulsion) theory The theory that explains bond angles and hence the shapes of molecules on the basis of repulsions of clusters of electrons in the valence shell of the central atom. The average distance between clusters is maximized.

bond angle The angle between two covalent bonds connecting one atom to neighboring atoms. For example,

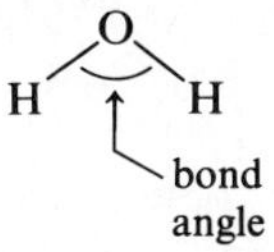

$$\underset{\text{bond angle}}{H-\overset{\displaystyle O}{}-H}$$

cluster of electrons A pair of unshared electrons or a single, double, or triple covalent bond in the valence shell of an atom in a molecule.

dispersion force An intermolecular force caused by temporary unsymmetrical distributions of electrons. Its magnitude depends on the surface area of the molecules.

dipole A partial, permanent separation of electrical charge within a molecule.

polar Possessing a dipole. The adjective is applied to a molecule or a chemical bond.

polarity The condition of having a dipole.

hydrogen bond An important intermolecular attraction between a hydrogen atom in an O—H, N—H, or F—H group within a molecule and an O, N, or F atom in another molecule.

QUESTIONS

1. The table below gives the melting points and electrical conductivities in the liquid state for three compounds. Classify each as ionic or covalent, and give your reasoning.

COMPOUND	MELTING POINT	CONDUCTIVITY OF LIQUID
carbon tetrachloride	−23°C	small
magnesium fluoride	1261°C	large
glycine (an amino acid)	262°C	unknown

2. Identify each covalent bond in each of the following chemicals.

a. CH_4 **b.** HCl **c.** PF_3
d. Na^+Cl^- **e.** $COCl_2$ **f.** H_2NOH

3. A possible instantaneous arrangement of electrons and protons in the H_2 molecule is

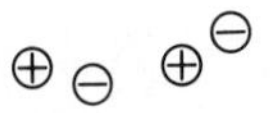

$$\oplus \ominus \quad \oplus \overset{\ominus}{}$$

Predict whether this arrangement would help to hold the two atoms together, and explain your answer.

4. Draw Lewis structures for the following molecules. Note that when H precedes or follows another atom, it is bonded to that atom.

a. NF_3 **b.** SCl_2 **c.** HCN
d. H_2NOH **e.** H_2CCH_2

5. Draw Lewis structures for the following molecules. Note that when H precedes or follows another atom, it is bonded to that atom.

a. CH_2ClF
b. HCOOH (*Hint:* Both oxygen atoms are bound to carbon.)
c. CH_3CCH
d. H_2NNH_2

6. Which bond in each of the following pairs will take more energy to break apart? Briefly explain why you chose your answer.

a. $H-C\equiv C-H \longrightarrow 2H-\dot{\underset{\cdot}{C}}\cdot$ or

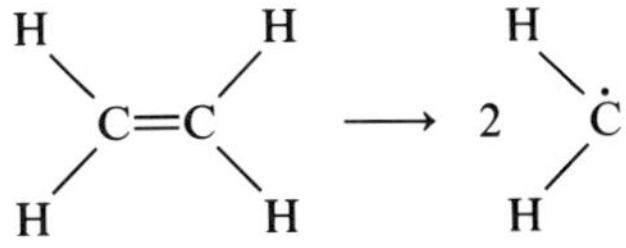

$H_2C=CH_2 \longrightarrow 2\ H_2\dot{C}\cdot$

b. $H_2C=\ddot{O}: \longrightarrow H_2\dot{C}\cdot + \cdot\ddot{O}:$ or

$H_3C-\ddot{O}:H \longrightarrow H_3C\cdot + \cdot\ddot{O}-H$

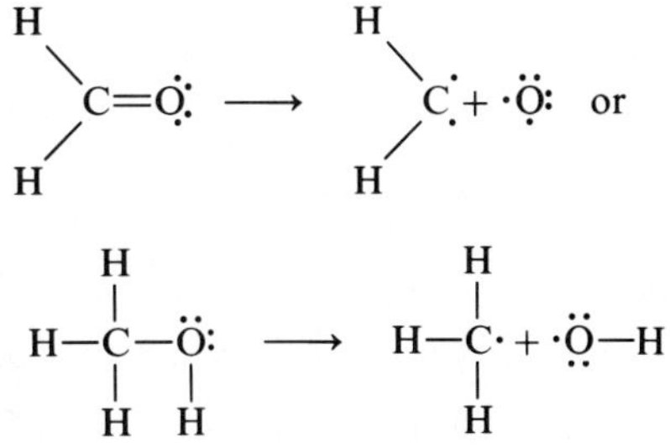

c. 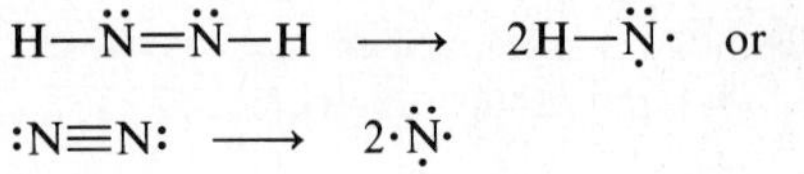 $H-\ddot{N}=\ddot{N}-H \longrightarrow 2H-\ddot{N}\cdot$ or

$:N\equiv N: \longrightarrow 2\cdot\ddot{N}\cdot$

d. 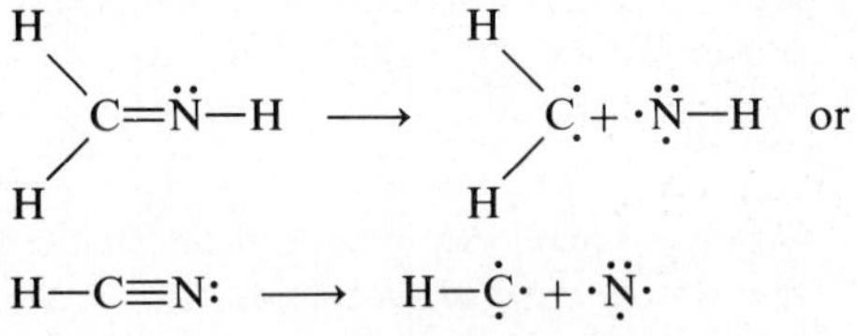 $H_2C=\ddot{N}-H \longrightarrow H_2\dot{C} + \cdot\ddot{N}-H$ or

$H-C\equiv N: \longrightarrow H-\dot{C}\cdot + \cdot\ddot{N}\cdot$

7. Indicate which of the following formulas can be expected to represent a stable compound and which cannot, and give your reason.

a. CS_2 **b.** HF_2 **c.** HCO **d.** H_2I

8. Predict the geometries and bond angles for the following:

a. SCl_2 **b.** PCl_3 **c.** SiF_4 **d.** CH_3CH_3

9. Predict the geometries and bond angles for the following:

a. NF_3 **b.** H_2S **c.** FNO **d.** HCN

10. Which compound in each of the following pairs would be expected to have the higher boiling point? Explain.

a. PCl_3 or PBr_3 **b.** H_2Se or H_2S
c. C_2H_6 or C_4H_{10} **d.** C_2H_6 or H_2O

11. Arrange these compounds in order of increasing boiling point, and explain your rationale: HI, NaCl, HCl.

12. Predict all the bond angles in the following molecules. The skeleton for each molecule is indicated by the order in which the atoms are written.

a. HOOH **b.** H_2CCH_2 **c.** $HC(=O)OH$

13. Ethylene glycol,

$$\underset{\displaystyle OH}{CH_2}-\underset{\displaystyle OH}{CH_2}$$

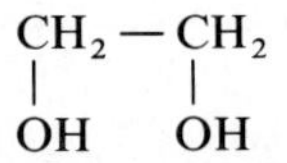

the most common automobile antifreeze, has a boiling point of 198°C. Butane, $CH_3CH_2CH_2CH_3$, with just about the same size and mass, has a boiling point of 0°C. Explain why the boiling of ethylene glycol requires such a high temperature. Use structures to help describe the importance of hydrogen bonding.

10 Salts and Solutions

Salt has been used as a food seasoning and preservative for thousands of years. The word "salt" adds zest to our language as well, in expressions from the Bible's "You are the salt of the earth" to "enjoying salty humor," or "taking it with a grain of salt." Cities have flourished and then declined along with the rise and fall of the salt trade.

10.1 SALTS IN NATURE

Chemists define a **salt** rather broadly as any crystalline compound containing ions. Rocks are made up largely of salts. Although a sample of earth taken at random is so heterogeneous and varied as to resist a simple description, examination with a microscope shows the presence of many kinds of **crystals,** or regularly shaped solid particles. Among the more easily recognized crystals are minerals such as feldspar, rock salt, and ruby. Each of these comes in particles of characteristic geometrical shape. For example, the crystal of ruby shown in Figure 10.1 has six sides of unequal length, like an imperfect hexagon.

Figure 10.1. *A section of ruby found in Africa.*

A ruby is primarily aluminum oxide, Al_2O_3, which is also the chief constituent of bauxite, our main source of aluminum metal. In addition to the Al^{3+} and O^{2-} ions, ruby contains a relatively small number of chromium ions, Cr^{3+}. These occupy locations normally taken by Al^{3+} ions and give ruby its beautiful red color. Sapphire is made of aluminum oxide with a different impurity.

Most of the salts in our environment seem unaffected by the seasons or the weather. We would be surprised to discover a drastic change in the landscape after a rainfall or even a violent thundershower. This implies that these salts have little tendency to dissolve in water. For instance, marble, which is one crystalline form of a salt, calcium carbonate, is unaffected by moisture unless the moisture also contains acids. Chemists have verified in the laboratory that only a tiny amount of calcium carbonate dissolves in a large amount of water. Therefore, we say that calcium carbonate is nearly **insoluble** in water.

On the other hand, some salts found in nature dissolve easily in water; they are said to be fairly **soluble** in water. Sodium chloride, which we use for table salt, is a good example. It is found as a solid only in dry places, such as desert areas and underground formations where water does not flow. In the United States sodium chloride is mined in Michigan and Louisiana and then refined to remove some of the impurities.

Most of the water-soluble salts near the earth's surface were dissolved away long ago and are now in the oceans. The oceans are vast solutions, or homogeneous mixtures (Chapter 2), containing primarily sodium chloride but also other ionic substances such as calcium sulfate and magnesium sulfate. In Chapter 6 you saw data on freezing points which suggested that independent ions are present in solutions of salts. We look now at the characteristics of these ions in the crystalline state.

10.2 IONS IN CRYSTALS

Of the several kinds of forces known to scientists—electrical, magnetic, gravitational, and nuclear—by far the most important in understanding chemistry is the electrical force. Nowhere is this fact better illustrated than in the study of ionic crystals. Their great stability and rigidity result from the strong forces of attraction between oppositely charged neighboring ions. Work must be done to separate these ions, because the potential energy of the ions in the crystal is much lower than that of the completely separated ions.

By looking at the formulas of many ionic compounds, one can see that the great majority consist of one or more metals and one or more nonmetals. The metallic element is present as a positive ion; the nonmetal is a negative ion. If the compound contains several nonmetals, they are bound together by covalent bonds and have one or more extra electrons to make them negative ions.

Often the simplest ionic compounds, containing one metal and one nonmetal, can be prepared in the laboratory by mixing the two elements and adding heat to speed up the reaction. Some examples are as follows:

$$2\,Na(s) + Cl_2(g) \longrightarrow 2\,NaCl(s)$$

$$Mg(s) + Br_2(g) \longrightarrow MgBr_2(s)$$

$$4\,Li(s) + O_2(g) \longrightarrow 2\,Li_2O(s)$$

Remember that (*g*) stands for the gaseous state and (*s*) for the solid state.

Each of these reactions is an oxidation-reduction reaction in which the metal loses electrons (is oxidized) and the nonmetal gains electrons (is reduced)—see Section 6.5. In these reactions electrons always transfer from the metal to the nonmetal, because the amount of energy needed to remove an electron from an isolated atom is much smaller for a metal than for a nonmetal. The transfer of electrons from the nonmetal to the metal is so costly in energy as to be highly unlikely. Similarly, ionic compounds don't form from combinations of several nonmetals, because none of these elements is likely to be present as a positive ion. A positive ion of a nonmetal forms only with a large input of energy.

Example Compounds composed of several nonmetals do not form ionic compounds. What kind of compounds do they form instead?

Solution They form covalent compounds, as we saw in Chapter 9.

The arrangement of electrons in a positive ion formed from a metal on the left side of the periodic table (see inside of back cover) is identical to that of the noble gas at the end of the preceding row. For example, an atom of sodium—the next element beyond neon (at the end of the second row)—loses one electron to form the Na^+ ion, which has the electronic configuration of neon. An atom of magnesium loses two electrons to attain the electronic configuration of neon and become Mg^{2+}. Similarly, the stable ions of potassium and calcium are K^+ and Ca^{2+}, respectively, each with the Ar configuration. The amount of energy needed to remove still more electrons and produce a greater positive charge on the ion is so large that such ions don't form in chemical reactions. The valence electrons of a metal are much easier to remove than any electron from an inner shell.

Many of the common positive ions are listed in Table 10.1, where they are divided into two categories: ions of representative metals and ions of transition metals. A **representative element** is an element in one of the groups labeled with a Roman numeral in the periodic table (inside back cover); a **representative metal** is a metal from Group I, Group II, or the portions of Groups III–VI that are below the heavy zig-zag line. A **transition metal** is a metal in the central region of the periodic table, separating Group II from Group III. As you can see, the ions of Groups I and II do indeed have noble gas configurations, and the charge on the ion equals the group number. Many, but not all, of the ions of the remaining representative metals have a charge equal to the group number. Ions of the transition metals do not have noble gas configurations. Generally speaking, it is difficult to predict the charges on these ions.

The simplest negative ions in crystals are single atoms, usually from Groups VI and VII, which have gained two electrons (Group VI) or one electron (Group VII) to assume noble gas configurations. Examples are O^{2-} and F^-, which have the electron configuration of Ne, and S^{2-} and Cl^-, which have an Ar configuration. Atoms of these elements have considerable attraction for electrons, and many actually release energy when gaining an electron. This release of energy shows the low potential energy and greater stability of the negative ion. Ions with several atoms of nonmetallic elements, such as NO_3^- and SO_4^{2-}, are also very common. The formulas and charges of some negative ions are given in Table 10.2. As discussed

Table 10.1 Common Positive Ions

REPRESENTATIVE METALS	TRANSITION METALS
Li^+, Na^+, K^+,	Cr^{3+}, Mn^{2+},
Mg^{2+}, Ca^{2+}, Sr^{2+}, Ba^{2+},	Fe^{2+} (ferrous), Fe^{3+} (ferric),
Al^{3+},	Co^{2+}, Ni^{2+}, Cu^{2+}, Zn^{2+},
Sn^{2+} (stannous), Pb^{2+}	Ag^+, Cd^{2+}, Hg^{2+} (mercuric)

Table 10.2 Common Negative Ions

IONS HAVING ONE ATOM			IONS HAVING SEVERAL ATOMS				
F^- fluoride	Cl^- chloride	Br^- bromide	OH^- hydroxide	CO_3^{2-} carbonate	NO_3^- nitrate	PO_4^{3-} phosphate	SO_4^{2-} sulfate
I^- iodide	O^{2-} oxide	S^{2-} sulfide	OCl^- hypochlorite	CN^- cyanide	HCO_3^- bicarbonate	CrO_4^{2-} chromate	

in Chapter 3, the name of a negative ion consisting of a single atom comes from the name of the element; the ending *-ide* replaces the last part of the name.

A simple rule makes it possible to predict the formula of any ionic compound consisting of a positive ion, such as any of those in Table 10.1, and a negative ion, such as those in Table 10.2. Because every ionic compound is electrically neutral, the total positive charge in the formula must equal the total negative charge. In other words,

$$\text{(charge on positive ion)} \times \text{(number of positive ions)} = \text{(charge on negative ion)} \times \text{(number of negative ions)}$$

Thus, calcium chloride, composed of Ca^{2+} ions (Table 10.1) and Cl^- ions (Table 10.2), must have the formula $CaCl_2$, so that the $+2$ charge of a calcium ion is balanced by a -2 charge from two chloride ions. Calcium chloride is a common chemical used to melt ice on highways in the northern United States. Similarly, magnesium hydroxide consists of Mg^{2+} ions and OH^- ions. For the charge to be balanced, its formula must have twice as many negative ions as positive ions: $Mg(OH)_2$. The parentheses show that there are two complete OH^- ions for each Mg^{2+} ion. A suspension of magnesium hydroxide in water is called "milk of magnesia" and is used as an antacid and laxative. The formula of sodium phosphate—a compound widely used in industry, pharmaceutical manufacturing, and medicine—must be Na_3PO_4, so as to match the positive charge of three Na^+ ions with the negative charge of one PO_4^{3-} ion. As we have already suggested, these formulas do not indicate molecules of these substances, but rather give the ratio of the two kinds of ions in a crystal.

Example

Give the formulas of the following compounds:

(a) the compound made up of K^+ ions and O^{2-} ions

(b) barium iodide

(c) ferric sulfate

Solution

(a) We need two K^+ ions with one O^{2-} ion to get canceling charges of $+2$ and -2. The formula is K_2O.

(b) The formula is BaI_2, because one Ba^{2+} ion requires two I^- ions to form a neutral compound.

(c) The ions involved are Fe^{3+} and SO_4^{2-}. To balance charges, we pair two Fe^{3+} ions (total charge, +6) with three SO_4^{2-} ions (total charge, −6). The formula is $Fe_2(SO_4)_3$.

As we have seen, a characteristic of some ionic compounds is the presence of negative ions containing several atoms. Table 10.2 has a number of examples. Let's look more closely at the covalent bonding within these ions and their geometries. To draw the Lewis structures of covalent ions, we can use the rules for covalent molecules (Section 9.5) if we make one addition:

> When counting the total number of valence electrons in the ion, be sure to include an extra number of electrons equal to the negative charge on the ion.

These extra electrons are transferred to the covalently bonded atoms during the chemical reaction that produces the ion. For example, when the sulfate ion in sodium sulfate, Na_2SO_4, is originally formed, it receives two electrons in addition to those contributed from the valence shells of four oxygen atoms and one sulfur atom. The electron count is calculated as follows:

		TOTAL VALENCE ELECTRONS
1 S atom	6 valence electrons	6
4 O atoms	(4 atoms)(6 valence electrons/atom)	24
−2 charge	2 extra electrons	2
1 SO_4^{2-} ion		32

Now we can use the usual procedure to write a reasonable Lewis structure and predict the geometry of the sulfate ion. After subtracting eight electrons for the four single bonds between the sulfur and the oxygen atoms, we find that the remaining twenty-four electrons just fill the valence shells of the oxygen atoms as unshared pairs. The result is

```
[       :Ö:       ]2−
         |
   :Ö—S—Ö:
         |
        :Ö:
```

Because of the four clusters of electrons in the valence shell of the central S atom, the sulfate ion has a tetrahedral geometry and bond angles of 109°.

Example Any compound containing a cyanide ion, CN^-, is a deadly poison. What is the Lewis structure of a cyanide ion?

Solution The total number of valence electrons must include the extra one that accounts for the negative charge on the ion. The contributions are

		TOTAL VALENCE ELECTRONS
1 C atom	4 valence electrons	4
1 N atom	5 valence electrons	5
−1 charge	1 extra electron	1
1 CN^- ion		10

The Lewis structure for this ion is closely analogous to that for the nitrogen molecule, N_2. Only a triple bond gives each atom an octet of electrons.

$$:C{\equiv}N:^-$$

The bond angles in four of the most common ions in Table 10.2 are given in Table 10.3.

Table 10.3 Bond Angles in Four Ions Containing Several Atoms

ION	BOND ANGLES
CO_3^{2-}	120°
NO_3^-	120°
PO_4^{3-}	109°
SO_4^{2-}	109°

10.3 A CRYSTAL STRUCTURE

Several characteristics of ionic crystals suggest that the ions within them are arranged in definite patterns. One is the shape of a crystal. Each surface is perfectly flat and has the form of a triangle, square, or other geometric figure. Such order on the surface implies order throughout the crystal. The ions are probably arranged in layers in a regular sequence. Support for this notion comes from the way crystals break apart when struck with a sharp tool. The new crystal faces resulting from the fracture are also flat and have regular geometric shapes.

A second, less commonly known fact is that the different faces of crystals can behave differently in some experiments. For example, the rate of crystal growth may be different in different directions. A common way to grow crystals is to dissolve as much salt as possible in warm water, which is the **solvent,** and then let the crystals deposit from the warm, saturated solution. A **saturated solution** is one that contains all of the **solute,** or dissolved material, that it can hold. If any more solute is added, it simply sinks to the bottom of the solution and remains a solid. As the temperature of the saturated solution drops, the amount of solute the solution can hold usually decreases. Therefore, crystals form and grow. A crystal with faces of different shapes grows more rapidly in some directions than in others.

Figure 10.2(a) is a sketch of a small crystal of alum, $KAl(SO_4)_2 \cdot 12\ H_2O$, which we can imagine to be suspended by a thread in a slowly cooling, saturated solution. Figure 10.2(b) shows the appearance of the crystal quite a bit later. It is much larger, as we would predict, but more important, it has grown more in the directions perpendicular to the square faces than it has in the directions perpendicular to the triangular faces. To account for this difference, we can propose that the layers of ions parallel to the square faces are different from those parallel to the triangular faces. One kind of layer forms more readily and rapidly than the other.

By means of such indirect evidence, chemists long ago suspected that the positive and negative ions in a crystal alternate in some regular fashion. With the advent of x-ray crystallography in 1912, scientists had a much stronger basis for this belief. Analysis of the directions in which x-rays bounce off a crystal reveals the precise arrangement of ions in the crystal. The beauty and fascination of crystals comes from the symmetries produced by their precise geometry.

Crystals come in many shapes, each with a different arrangement of ions. Figure 10.3(a) shows a cluster of sodium chloride crystals found in nature. Many of them are nearly perfect cubes. Well-formed sodium chloride crystals from any source have this cubical shape, as you can see by looking at a few crystals of table salt. The drawing in Figure 10.3(b) shows the **crystal structure,** with the relative sizes and arrangement of ions, in a tiny portion of a sodium chloride crystal. A chloride ion is considerably larger than a sodium ion. If a small piece of the crystal sketched in Figure 10.3(b) were duplicated many times in all directions, a visible crystal would result. It is

Figure 10.2. Growth of an alum crystal. (a) An early stage. (b) A later stage.

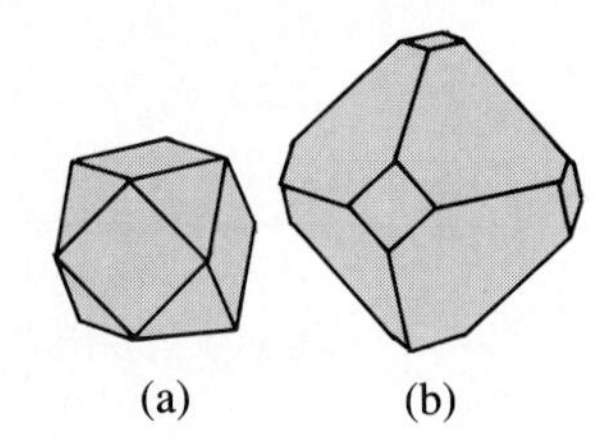

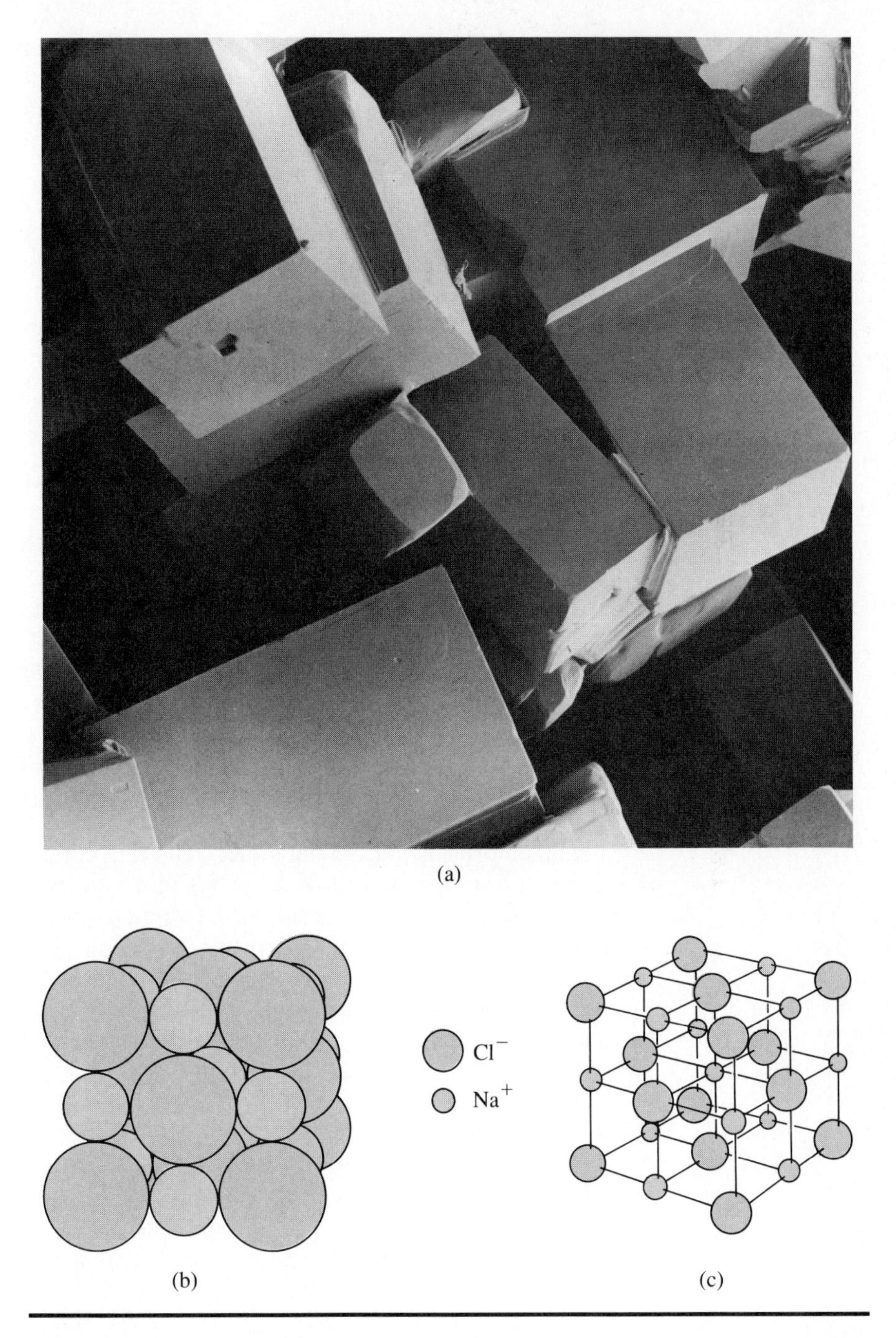

Figure 10.3. *The structure of a sodium chloride crystal. (a) Perfectly shaped crystals found in nature. (b) A model showing the sizes and arrangement of ions. (c) An expanded model of the crystal lattice.*

not hard to see why the observable crystal has the shape of a cube, since the basic crystal unit has the same shape. Figure 10.3(c) gives the same structure but in expanded form, so the locations of the ions can be seen more easily.

The presence of layers of ions is clearly evident in Figure 10.3. Within each layer, Na^+ and Cl^- ions take alternate positions. Adjacent layers are identical, but they are shifted with respect to each other so that ions of the same charge are not in direct contact. As is true of all ionic lattices, any given ion in Na^+Cl^- is closer to ions of opposite charge than it is to ions of the same charge. In the NaCl lattice, each Na^+ ion is surrounded by six Cl^- ions as nearest neighbors, and each Cl^- ion is in contact with six Na^+ ions. Thus, the attraction between nearest neighbors is very large. To be sure, there are repulsions between ions of the same charge, but the distances between these ions are greater. The attractions therefore dominate, and the crystal is extremely stable. Remember that sodium chloride melts at 801°C!

From the drawings in Figure 10.3 it should be clear that a given chloride ion does not "belong" to any one of its six neighboring sodium ions. No unique Na^+Cl^- pairs can be identified in the crystal structure. This is why we say that there is no such thing as an NaCl molecule. The entire crystal might be thought of as a giant molecule, but that point of view doesn't agree very well with our usual concept of a molecule. The formula NaCl only gives the simplest whole-number ratio of the two kinds of ions in the compound. As we noted in Chapter 5, a mole of any ionic compound is 1 gram formula weight, or the sum of the atomic weights expressed in grams in the simplest formula.

Sodium chloride is not the only ionic solid that crystallizes in the pattern described. A number of other compounds in which the relative sizes of the positive and negative ions are about the same as in NaCl adopt the same structure; examples are LiF, MgO, and KBr. There are, however, many other kinds of crystal structures for the great assortment of ionic sizes and charges existing in nature. Although the arrangement of ions may be complicated, each crystal type has a distinctive shape that duplicates the pattern set by a small group of ions within the crystal.

Example

(a) In zinc sulfide, ZnS, each Zn^{2+} ion is surrounded by four S^{2-} ions as close neighbors. How many close neighbors does each S^{2-} have?

(b) Each fluoride ion in CaF_2 is in contact with four calcium ions. How many F^- ions surround each Ca^{2+} ion?

Solution

(a) Because there are equal numbers of Zn^{2+} and S^{2-} ions in any crystal of ZnS, each S^{2-} ion must also have four Zn^{2+} ions as close neighbors. Four sulfide neighbors for each Zn^{2+} and four zinc neighbors for each S^{2-} give a 1:1 ratio of the two kinds of ions. In NaCl, six neighbors for each kind of ion also gives a 1:1 ratio.

(b) In this case there are twice as many F^- as Ca^{2+} ions. If there are four Ca^{2+} ions around each F^-, there must be eight F^- ions around each Ca^{2+}.

In nearly all crystals the ions are arranged in such a way as to maximize attractions between unlike charges and minimize repulsions between like charges. When the crystal contains an ion with several atoms, such as SO_4^{2-} in Na_2SO_4, that ion behaves like a single unit and occupies regular positions in the crystal structure. The bonding holding the crystal together is ionic, just as it is for simpler compounds, such as NaCl.

10.4 WATER SOLUTIONS OF IONIC COMPOUNDS

The largest solution found on earth is, of course, the oceans, where 98% of the earth's water is located. Dissolved substances in the ocean make up 3.5% of its total weight, and they are mostly ionic compounds. The increased density of ocean water (1.025 grams/milliliter at 20°C, compared with 0.998 gram/milliliter for pure water) makes it easier to float and swim in the ocean. The lower freezing point, −1.9°C, means that an ocean is less likely to freeze than is a fresh-water lake.

From the salty taste of seawater we can guess that NaCl is the principal dissolved substance. Na^+ dominates among the positive ions and Cl^- among the negative ions, but many other ions are also present. The three next most numerous ions of each charge are listed in Table 10.4, which gives the number of moles of each ion in 1 liter of seawater. This is a common unit of **concentration,** which in general specifies the amount of solute dissolved in a given amount of solution. The table shows that some of the chloride ions must have come from a source other than NaCl, because the number of moles of Cl^- is greater than the number of moles of Na^+. Though none

Table 10.4 Moles of Ions in 1 Liter of Seawater

POSITIVE ION	MOLES IN 1 LITER	NEGATIVE ION	MOLES IN 1 LITER
Na^+	0.457	Cl^-	0.533
Mg^{2+}	0.052	SO_4^{2-}	0.028
Ca^{2+}	0.010	HCO_3^-	0.002
K^+	0.010	Br^-	0.001

THE COMPOSITION OF SEA WATER

Understanding the oceans has provided many challenges to scientists. One intriguing question is why the composition of seawater has been the same for roughly 200 million years. During all that time, the rivers of the world have been pouring water into the sea at a rate estimated to be 3.3×10^{16} liters of water each year. Dissolved in this river water are primarily Ca^{2+} and HCO_3^- (bicarbonate) ions, along with smaller amounts of Na^+, Cl^-, and some other ions. The amount of HCO_3^-, for example, being swept into the oceans each year is 3×10^{13} moles. What becomes of the water and all the dissolved material? Why doesn't this influx change the composition of seawater?

The reason why the amount of water in the oceans stays constant is not hard to see. The vast surface of ocean water in contact with the air promotes the evaporation of great quantities of water. The water vapor that is formed later condenses as rain and is carried once again to the oceans. The input of water from rivers equals the amount lost through evaporation.

The dissolved ions, on the other hand, cannot evaporate. Other mechanisms must remove them from the ocean water. Otherwise, their concentrations would slowly increase, and the composition of seawater would change.

One way in which Na^+ and Cl^- are removed is by escaping from the surface in droplets of liquid. Small bubbles of air constantly rise to the surface. When they break at the surface, droplets of seawater form and are carried away by breezes, giving the ocean breeze its characteristic feel and smell. The water in the breezes quickly evaporates, leaving tiny particles of NaCl. These can later serve as nuclei for the formation of raindrops.

The concentrations of Ca^{2+} and HCO_3^- are controlled in a very different way. These two ions react in the ocean water:

$$Ca^{2+} + 2\,HCO_3^- \longrightarrow CaCO_3(s) + CO_2(g) + H_2O$$

This equation suggests that the calcium carbonate settles out as a solid and the carbon dioxide escapes as a gas. However, very little of the calcium carbonate in the ocean comes out of solution, or **precipitates,** directly in the way shown by this equation. Instead, tiny organisms, one of which is called *Porites* (a coral), consume the calcium and bicarbonate ions and incorporate calcium carbonate into their skeletons. When the organisms die, the calcium carbonate deposits on the bottom of the ocean. This process occurs especially in shallow, tropical waters, where the sunlight can penetrate and assist the organisms in their important task.

Figure 10.4. Manganese nodules of about 6-centimeter diameter line the floor of the Antarctic Ocean.

Another challenge to scientists has been to devise methods of mining the oceans for useful substances. Even though the concentrations of valuable elements are very small, the tremendous volume of the oceans means that the total amounts of these materials are large. Already the oceans have become a rich source of sodium, magnesium, chlorine, and bromine. Some elements can be found on the ocean floor, where they have precipitated as salts over long periods of time. "Manganese nodules" (see Figure 10.4) contain, by weight, 20% manganese (Mn), 20% iron (Fe), and 1–3% nickel (Ni), copper (Cu), and cobalt (Co) combined. Because these elements are present primarily as the oxides, oxygen accounts for the remainder of a typical nodule. Several companies are engaged in pilot mining operations to test the practicality of dredging these nodules from the sea bottom.

---

of the other positive ions can be paired exactly with any of the negative ions, the total positive charge still equals the total negative charge when all the dissolved ions are considered.

Water has sometimes been called a universal solvent; it is certainly the best solvent for salts. Most salts scarcely dissolve in other solvents. Scientists have extensively investigated the reasons why water has this unique ability to dissolve ionic compounds. Table 10.5 is a balance sheet listing on one side the forces that retard the dissolving of a salt in water and on the other side the force that promotes dissolving. If we can show that the dissolving force is stronger, the reason for the power of water as a solvent will become clear.

Consider first the forces working against the dissolving process. These are the forces holding ions together in a crystal and the force that causes water molecules to attract each other. When a crystal dissolves in water, the crystal structure must be completely broken down, and spaces or holes to accommodate the ions must be created in the water structure. The forces between ions in crystals are very strong and account for their unique properties, such as their high melting points. The force between water molecules comes primarily from hydrogen bonding, some of which must be disrupted in the process of forming holes. Although hydrogen bonding is considerably stronger than other forces between covalent molecules, it is weaker than

Table 10.5 Balance Sheet for Dissolving an Ionic Compound in Water

FACTORS OPPOSING SOLUTION	FACTOR FAVORING SOLUTION
1. Ionic crystal forces—strong. 2. Hydrogen bonding between water molecules—moderate.	1. Hydration of ions—strong.

the forces between ions. The entries in the left-hand column of Table 10.5 are the forces working against the dissolving process.

The two factors on the left are certainly significant, but on the other side of the balance sheet is an important force strongly favoring dissolving. The ions in solution are hydrated, or tightly bound by water molecules. In other words, each **hydrated ion** has a tight sheath of about a half-dozen water molecules around it (see Figure 10.5). A positive hydrated ion owes its stability to the electrical attraction between the positive charge of the ion and the negative end of the water dipole. In Figure 10.5 each oxygen atom, which has a partial negative charge, is oriented toward the K^+ ion. On the other hand, the negative ion in the same solution (Cl^-, for example) attracts the positive end of several water molecules to form its hydration layer. Each ion with its attached water molecules moves as a unit in the solution. Considerable energy is released when the process of hydration occurs. This energy makes up for the energy that must be put into the crystal to break it apart.

The force of hydration is listed in the right-hand column of Table 10.5. It is competitive with the forces on the left side of the balance sheet: ionic crystal forces and hydrogen bonding between water molecules.

Which side of the balance sheet predominates? It is difficult to make absolute predictions, but one fact is clear: no other common liquid approaches water in its ability to hydrate ions. Liquids other than water dissolve only

Figure 10.5. A hydrated potassium ion. The negative side of each water molecule points toward the positive ion.

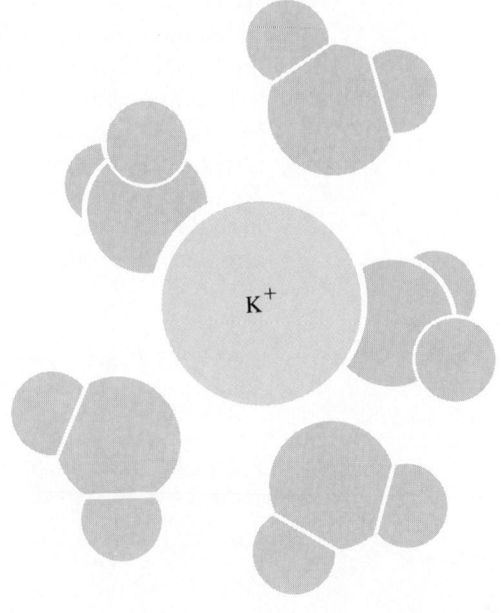

small amounts of salts. Although the factors opposing the dissolving process are nearly as strong for other liquids as for water, the factor favoring dissolving is much weaker. Other liquids, which have much smaller polarities than water does, do not form layers of solvent molecules around ions. The factor that promotes the dissolving of a salt is stronger for water than for any other liquid. Water is unique, and life as we know it could not exist without it.

Example The octane molecule, C_8H_{18}, has a much smaller polarity than the ammonia molecule, NH_3, does. Will liquid octane or liquid ammonia dissolve more table salt? Why?

Solution With its bigger dipole, ammonia should form layers around ions more effectively than octane can. This difference tends to make ammonia the better liquid for dissolving salts. Even though ammonia molecules are held together by hydrogen bonds and resist the formation of holes to accommodate ions, the tendency of ammonia molecules to attach themselves to ions is much more significant.

10.5 OTHER KINDS OF SOLUTIONS

When chemists are deciding what liquid might be good for dissolving a particular substance, they often go back to an old saying, "Like dissolves like." This simply means that a solute is likely to be fairly soluble in a solvent only if the two have similar polarities. A solute with polar molecules is likely to be nearly insoluble in a solvent whose molecules have a low polarity, and vice versa. If both solute and solvent have high polarity, or if both have low polarity, the solute should be reasonably soluble. This rule applies mainly to solutes and solvents that are molecular compounds and allows us to go beyond solutions of salts in water.

One can make sugar solutions so concentrated (more than 80% sugar) that they are thick syrups. Why is water capable of dissolving so much sugar? The sugar, or sucrose, molecule has a rather complicated structure:

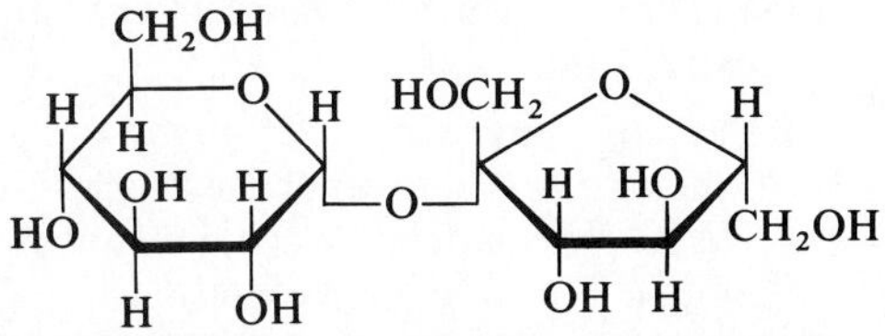

(This drawing uses a shorthand notation; carbon atoms are understood to be located at the intersections of two or more lines.) What is important to note here is that the molecule contains many OH groups, which make it quite polar. Furthermore, there is hydrogen bonding between sugar molecules, just as in water. When sugar dissolves in water, the hydrogen bonds between sugar molecules and some of the hydrogen bonds between water molecules must be broken apart. At the same time, however, new hydrogen bonds form between sugar and water molecules. Although energy is absorbed in breaking the original hydrogen bonds, a roughly equal amount of energy is released as the new hydrogen bonds form. This is a reasonable tradeoff, and a considerable amount of sugar dissolves.

If you try to dissolve sugar in liquid octane, the result is very different. The C—H bonds in octane are only slightly polar, and the whole molecule has little polarity. Since sugar and octane are different in their polarities, they are a poor match to form a solution. Very little sugar dissolves in octane, because there are no strong attractions between sugar and octane molecules to make up for the breaking of hydrogen bonds between sugar molecules.

A solute like iodine, on the other hand, is quite soluble in octane. I_2 has no dipole and is similar to octane. The I_2 molecules in solid iodine are held together by the dispersion force, as are the molecules in liquid octane. The strength of this force in the two pure substances is about the same as it is between iodine and octane molecules in the solution. Whenever the forces between like and unlike molecules are approximately equal, the two substances are likely to form a solution with ease. As would be expected, iodine is not very soluble in water.

Example

Predict whether each of the following solutes will be fairly soluble in the given solvent, and give a brief explanation:

(a) hydrogen fluoride, HF, in octane, C_8H_{18}

(b) barium bromide, $BaBr_2$, in chloroform, $CHCl_3$

(c) ethyl alcohol, C_2H_5OH, in water

Solution

(a) Hydrogen fluoride is highly polar and has hydrogen bonding between molecules; octane is nonpolar. Because these molecules are distinctly unlike, very little hydrogen fluoride dissolves in octane.

(b) Barium bromide, an ionic compound, is completely different from chloroform, which has limited polarity. Very little barium bromide should dissolve in chloroform.

(c) Because there is one OH group in each molecule of ethyl alcohol, hydrogen bonding exists between the molecules. In this respect the two liquids are similar. We predict that ethyl alcohol dissolves easily in water. In fact, the two liquids form solutions in any proportion.

The solubility of a compound in several liquids may reveal useful and even unexpected information about the bonding in the compound. Beryllium chloride, $BeCl_2$, is very soluble in water. This is not surprising. The compound consists of a metal and a nonmetal and thus should be ionic, and many ionic compounds dissolve in water. What is surprising, though, is that $BeCl_2$ is also soluble in a number of liquids that have only limited polarity, such as carbon disulfide, CS_2. CS_2 is a linear molecule (like CO_2) and has no dipole. Also, the melting and boiling points of $BeCl_2$—405° and 488°C, respectively—are rather low compared with those of many ionic compounds. All this evidence suggests that beryllium chloride has properties of both covalent and ionic compounds. Chemists regard $BeCl_2$ as intermediate between the two types. Apparently, when the compound forms the two valence electrons in a beryllium atom are not completely transferred to two chlorine atoms to give Be^{2+} and 2 Cl^-. There is some sharing of electrons in covalent bonds, although they are very polar covalent bonds.

A key fact about beryllium is that it is very close to the dividing line between metals and nonmetals in the periodic table. It is barely on the metal side of the line. Other metals in similar positions—such as Al, Ge, Sn, and Sb—behave similarly. For example, $SnCl_4$ melts at −33°C, boils at 114°C, and dissolves in both water and slightly polar solvents. Because of its very low melting and boiling points, it seems more covalent than ionic. Whenever a compound consists of a metal and a nonmetal, one or both of which are near the metal/nonmetal borderline, the bonding is probably not completely ionic or covalent.

10.6 CONDUCTION, FREEZING, AND COLOR

You have already seen how Arrhenius used data on the freezing points of solutions to argue that ions exist in salt solutions (Chapter 6). Arrhenius's own experiments involved measurements of the electrical conductivities of solutions. Typically, when the dissolved salt was NaCl, KBr, KNO_3, or $BaCl_2$, the conductivity was large and suggested to Arrhenius that nearly all the solute was in the form of ions. This interpretation agreed well with the conclusion drawn from freezing-point data. Despite the fact that two independent sets of evidence pointed to the same conclusion, many other chemists were not convinced until much later; even scientists often hate to give up the notions that they have accepted as truth.

A few of the solutions studied by Arrhenius (for example, CdI_2 in water) had abnormally low conductivities. Arrhenius proposed that separation of these salts into ions is only partial, and he was able to calculate the number of ions formed. The result agreed with his calculation from freezing-point data on the same solutions. It appears that CdI_2 and the other compounds that dissociate incompletely are in the intermediate class—partially ionic and partially covalent. In addition to Cd^{2+} and I^- ions, the solution contains CdI_2 molecules.

Several experiences from everyday life relate to the conductivities and freezing points of solutions of salts in water. The danger of standing in wet shoes or in a puddle of water while working with an electrical fixture is widely appreciated. It can result in electrocution if you accidentally touch a "hot" wire. The electricity can find a conducting path through your body (which is largely a dilute salt solution in water) and the moisture around your feet. If the water were absolutely pure, the danger would be less, because pure water is a poor electrical conductor. However, most water contains ionic impurities. The water supplied to a household faucet contains low concentrations of several salts and is a moderately good conductor.

The practice of spreading salt on roads in winter to melt ice and improve driving conditions is fairly common in northern states. Behind this practice is the principle of lowering the freezing point with a dissolved substance. As long as the air temperature is not far below the normal freezing point of water, 0°C, a concentrated solution of salt in water will not freeze. For example, suppose we have a salt solution that freezes at -10°C. If the solution, at a temperature of -5°C, is poured onto ice that is at the same temperature, the ice melts, because the temperature is above the freezing point of the solution. It is like mixing ice and pure water at 5°C, which causes the ice to melt quickly.

When salt crystals are thrown onto an icy road, no solution is present initially. Very soon, however, the salt picks up moisture and forms a tiny amount of concentrated solution, and the melting process begins. Salt, NaCl, is clearly more effective than an equal number of moles of, say, sugar would be. NaCl separates into ions and gives twice the freezing-point reduction provided by sugar. (We are, of course, ignoring the impracticality and expense of using sugar on our roads.)

A salt such as $CaCl_2$ is even better than NaCl. Each mole of $CaCl_2$ yields 3 moles of ions, $Ca^{2+} + 2Cl^-$, when dissolved in water. Thus, a concentrated solution of $CaCl_2$ has a very low freezing point. In fact, the highway departments of most states with severe winters use a mixture of calcium chloride and sand. The increase in automobile corrosion produced by a solution of calcium chloride or sodium chloride is certainly a disadvantage of this practice; another is the pollution of rivers.

The colors of a few ionic compounds in water solution lead to an interesting question. Does each ion have a distinctive color, or does the color of the solution depend on the particular combination of positive and negative ions? The question is easily answered by looking at the colors of four solutions: K_2CrO_4, KCl, Na_2CrO_4, and NaCl, each in water. The first and third solutions are yellow; the second and fourth are colorless. The simplest hypothesis is that K^+, Na^+, and Cl^- are colorless and CrO_4^{2-} is yellow. This would account for the yellow color whenever CrO_4^{2-} is present and the absence of color when K^+ and Cl^- or Na^+ and Cl^- are the solutes.

Such a simple hypothesis, however, will not explain the tastes of ionic compounds. Suppose we were to taste carefully a pinch of some of the various ionic compounds. First of all, some are very toxic, and tasting them would lead to dramatic and unpleasant results. Others are much less toxic. Sodium chloride (table salt) has a universally recognized taste. Do our taste buds respond primarily to the sodium ion or the chloride ion? Potassium chloride, which for health reasons is often added to the sodium chloride we buy in a grocery store, has a different taste. Since sodium chloride and potassium chloride both contain the chloride ion, perhaps the positive ion is responsible for the different tastes. Yet sodium sulfate, Na_2SO_4, has a taste different from that of sodium chloride. It is not possible to say that some ions have no taste and others have distinctive tastes. Perhaps every ion contributes to our sensation of taste.

IMPORTANT TERMS

salt A crystalline solid containing ions. To most chemists, *salt* and *ionic compound* mean the same thing.

crystal A solid particle having a regular geometric shape.

insoluble Refers to a substance that does not dissolve in a given liquid to form a solution.

soluble The quality of dissolving in a particular liquid to yield a solution.

representative element An element in one of the groups labeled with a Roman numeral (I–VII) in the periodic table.

representative metal A metal, from any of Groups I through VII, that is below the zig-zag line in the periodic table.

transition metal A metal from the central region of the periodic table, between Group II and Group III.

solvent The liquid in which a substance dissolves to make a solution.

saturated solution A solution that contains all the dissolved material it can hold.

solute The substance that dissolves in a solvent.

crystal structure A particular arrangement of ions in a crystal. The sodium chloride structure is one example.

concentration The amount of solute present in a stated amount of solution. Concentration is often expressed as the number of moles of solute dissolved in 1 liter of solution.

precipitate To come out of solution as solid particles. When used as a noun, *precipitate* refers to the particles.

hydrated ion An ion and its closely attached water molecules in a water solution.

QUESTIONS

1. By referring to a periodic table, identify the noble gas having the same electronic configuration as the common ion for each of the following elements.

a. Na **b.** Li **c.** I **d.** Al **e.** Sr

2. Give the formulas of the ionic compounds made from the following pairs of ions.

a. Pb^{2+} and Br^- **b.** Fe^{2+} and PO_4^{3-}
c. Sn^{2+} and CO_3^{2-} **d.** Li^+ and S^{2-}

3. Give the formulas of the following ionic compounds. (You may want to review the rules for naming compounds, given in Chapter 3. Tables 10.1 and 10.2 may also help.)

a. potassium bromide

b. aluminum trichloride
c. calcium oxide
d. ferrous nitrate
e. cobalt carbonate
f. chromium hydroxide

4. For each of the following ions, write the Lewis structure and predict the bond angles where applicable.

a. OH^- **b.** OCl^- **c.** PO_4^{3-}
d. HCO_3^- (H is attached to O.)

5. Sodium sulfate, Na_2SO_4, consists of Na^+ ions and SO_4^{2-} ions; the S and O atoms in SO_4^{2-} are held together by covalent bonds. Would you expect sodium sulfate to have a high or a low melting point? Why? Will it conduct electricity in the molten state? Explain.

6. a. In the cesium chloride crystal structure, each chloride ion is surrounded by eight cesium ions. How many nearest neighbors does each cesium ion have?

b. In the crystal structure of titanium oxide, each titanium ion is in contact with six oxide ions and each oxide ion is in contact with three titanium ions. What is the formula of this ionic compound? (The symbol for titanium is Ti.)

7. Which is a better conductor of electricity, ocean water or water from a fresh-water lake? Why?

8. Predict whether each of the following solutes will be soluble or insoluble in the stated solvent, and explain your answers.

a. glucose, $C_6H_{12}O_6$ (contains five OH groups), in ethyl alcohol, C_2H_5OH

b. sulfur, which forms nonpolar S_8 molecules, in octane

c. sodium hydroxide in carbon disulfide

9. Antimony trichloride, $SbCl_3$, melts at a temperature of 73°C and is soluble in both water and chloroform, $CHCl_3$. Based on this evidence and the positions of the two elements in the periodic table, discuss whether this compound is best classified as ionic, covalent, or intermediate between the two.

10. Ethyl alcohol, C_2H_5OH, has weaker forces between molecules than water does. Does this difference by itself make alcohol less able or more able to dissolve salts than water is? Explain. Is this one fact sufficient to allow you to draw a conclusion about the relative solubilities of salts in alcohol and water? Discuss.

11. Hypothetically, if 1 gram of NaCl cost the same as 1 gram of KCl, which would be more economical for use on icy roads in the winter? Explain.

12. When nickel sulfate, $NiSO_4$, is dissolved in water, the solution has an intense green color. Describe other observations you would make about solutions of salts of your choice in order to determine whether the Ni^{2+} ion or the SO_4^{2-} ion is responsible for the green color.

Organic Chemicals

When chemists think of covalent compounds, they instinctively turn to the world of organic chemistry. Over two million organic chemicals have already been identified, and the synthesis and discovery of new ones is proceeding at a tremendous rate. Originally, organic compounds were defined as those found in plants and animals. As time went on, chemists discovered that all organic compounds contained atoms of the element carbon. They were also able to synthesize new carbon compounds, which were very much like the ones occurring naturally.

Today, **organic chemistry** is defined as the study of compounds containing carbon; inorganic chemistry is the study of compounds not containing carbon. Although chemists still isolate many organic chemicals from plant and animal sources, most are synthesized from coal and petroleum.

In this chapter you will see examples of many types of organic molecules. Usually it is not necessary to memorize the formulas or structures of these molecules. Many are presented only as illustrations of a general idea. The basic goal is simply to make sure that each example makes sense, without worrying about the details. Your instructor can provide guidance as to which formulas and structures you should learn.

11.1 CARBON—THE BASIS OF ORGANIC CHEMISTRY

Since the element carbon is part of every organic chemical, a good place to begin is with the natural forms of carbon itself. There are two forms of carbon in nature—diamonds and graphite (see Figure 11.1). The diamond is virtually pure carbon. Geologists tell us that graphite is formed from coal if the coal is left in the ground long enough. Graphite is one of the softest known materials, whereas diamond is one of the hardest.

As unusual as the difference between its two forms may be, carbon is unique in another way. Carbon atoms can link with each other in chains to form large molecules. Chemical bonds holding carbon atoms together are strong. For example, to break carbon-carbon single bonds, about 347 kilojoules per mole of bonds is required. By contrast, breaking silicon-silicon bonds requires only 222 kilojoules per mole, even though silicon is just below carbon in the same family of the periodic table. The great

Figure 11.1. *Two different forms of carbon: graphite and diamonds.*

hardness of diamond comes from the strength of the carbon-carbon bonds, which must be fractured when even a tiny part of a diamond is worn away by grinding. The vast number and variety of organic compounds are a result of the endless number of ways in which carbon and other atoms can be linked together.

By putting together the proper kinds and numbers of atoms in the proper way one can produce organic chemicals with the sourness of vinegar or the sweetness of sugar, the smell of peppermint or the fragrance of a rose, the pain-killing power of aspirin or the sting of a bee. Choose the right combinations of atoms and you will have compounds of every color imaginable. They can be as transparent as cellophane or as opaque as rubber.

The majority of organic chemicals are made of molecules containing atoms of only three elements—carbon (C), hydrogen (H), and oxygen (O). Examples are the hydrocarbons, including petroleum and natural gas, and the carbohydrates, an important group of foods.

```
                                         H
                                         |
                                     H—C—O—H
                                         |
                                         C———O
                                    H  / |     \   O—H
                                    | /  H      \  |
    H  H  H  H  H  H                C    |       \ C
    |  |  |  |  |  |                |\  O—H  H   / |
 H—C—C—C—C—C—C—H             H—O \  |    |  /  H
    |  |  |  |  |  |                   C————C
    H  H  H  H  H  H                   |    |
                                       H    O—H
```

hexane, a hydrocarbon

glucose, a carbohydrate

Three other key elements found in the organic compounds of plants and animals are nitrogen (N), phosphorus (P), and sulfur (S). Figure 11.2 shows a fanciful version of the periodic table as it might appear to organic chemists and biochemists, who tend to view the world as though it were composed primarily of these six elements. In this periodic table, emphasis is given to the elements prominent at the surface of the earth. Therefore, elements present in plants and animals assume major importance, along with elements in the water and minerals found at the earth's surface.

Example Classify each of the following compounds as organic or inorganic:

(a) H_2SO_4

(b) $C_2H_4O_2$

Solution (a) There are no carbon atoms in sulfuric acid; it is an inorganic compound.

(b) This carbon-containing compound is clearly organic. It has the formula of acetic acid, the sour substance in vinegar.

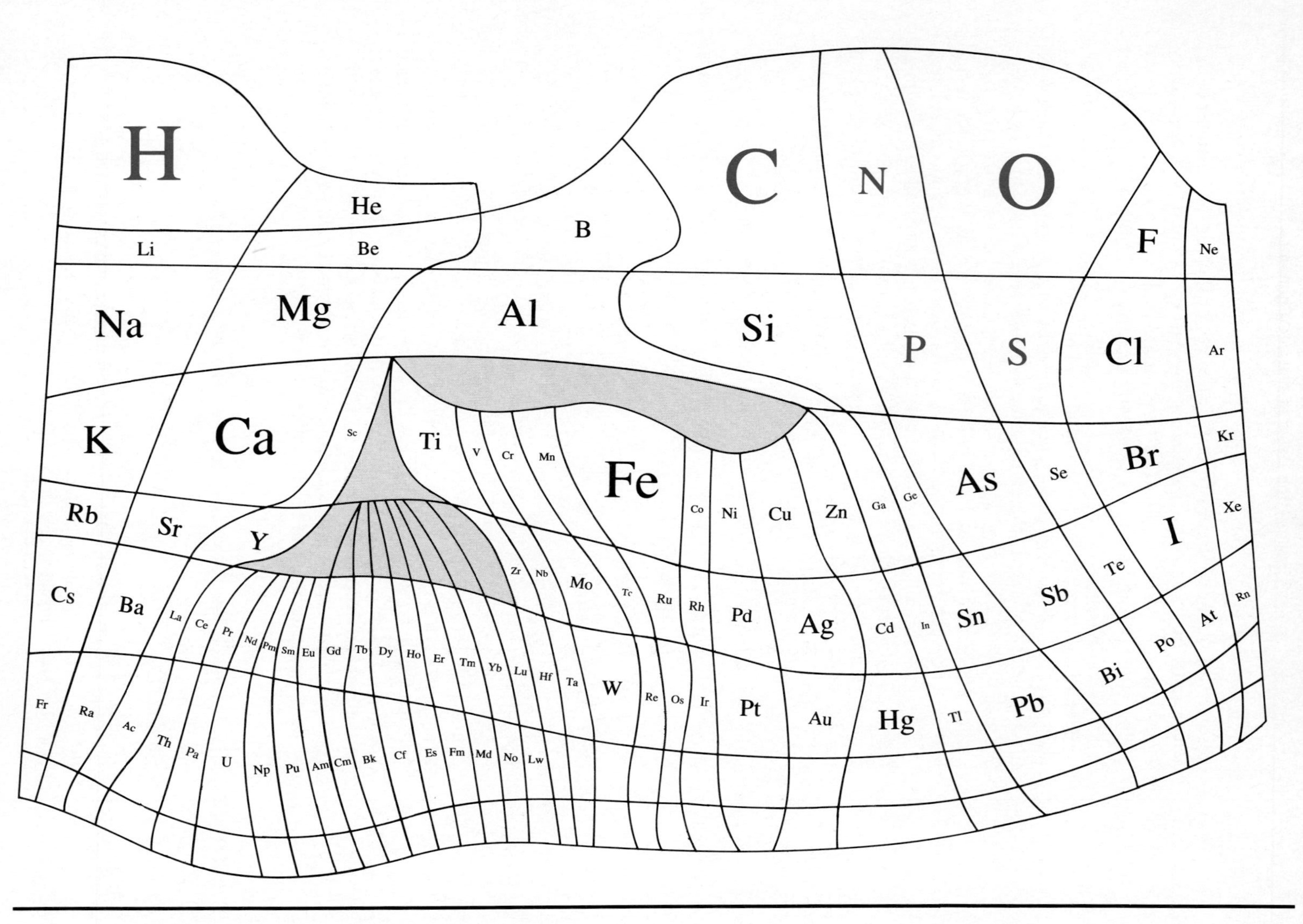

Figure 11.2. *A periodic table that emphasizes the abundance of elements at the earth's surface.*

11.2 THE STRUCTURES OF ORGANIC CHEMICALS

Most organic chemicals are covalent compounds, whose atoms are held together by the sharing of electrons in covalent bonds. Once you know the number of electrons needed to complete the valence shells of just four elements—carbon, hydrogen, oxygen, and nitrogen—it is not difficult to draw structures for most organic chemicals. You have only to remember the number of electron pairs that each of these elements shares with other atoms (see Table 11.1). The remaining electrons in the valence shell are unshared pairs. For example, chemists speak of **tetravalent carbon,** since carbon shares four pairs of electrons with other atoms and has no unshared pairs in its valence shell. Nitrogen, on the other hand, generally has one unshared pair and shares three pairs of electrons with adjacent atoms. Oxygen has two unshared and two shared pairs, and, as always, hydrogen is unique in filling its valence shell with one shared pair.

Table 11.1 Number of Shared Pairs of Electrons in the Valence Shells of Elements Commonly Found in Organic Chemicals

ELEMENT	NUMBER OF SHARED ELECTRON PAIRS
carbon	4
nitrogen	3
oxygen	2
hydrogen	1

Figure 11.3. *The structures of some important organic chemicals. Remember, it is* not *the intention that every structure be learned.*

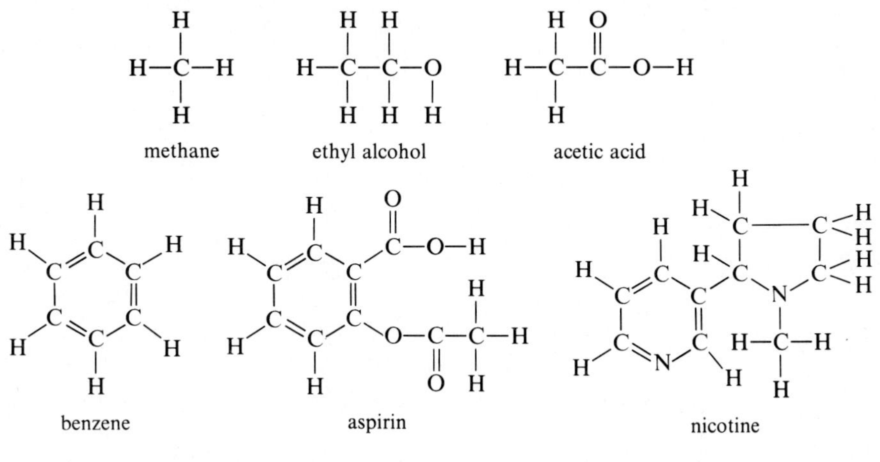

methane

ethyl alcohol

acetic acid

benzene

aspirin

nicotine

THE STRUCTURES OF DIAMOND AND GRAPHITE

Diamond cutting is an exacting job, with thousands of dollars resting on the right tap of the hammer. If the cutter picks the wrong place to cleave a stone, it can shatter. The job calls for nerves of steel. Graphite, on the other hand, has such good cleavage planes (planes that slide past one another with minimum resistance) that it flows easily. Graphite is an important lubricant and, when mixed with clay, is used as the "lead" in pencils. Yet, both diamond and graphite are carbon, and in both materials every carbon atom has four shared pairs of electrons.

Why don't diamonds have good cleavage planes? Each carbon atom connects to other carbons through single bonds that go in all three dimensions, as shown in Figure 11.4(a). To cleave the crystal apart, we have to break strong covalent bonds.

Graphite is quite a different beast—it has planes of interconnected carbon rings, as depicted in Figure 11.4(b). The bonds within each plane are strong, but adjacent planes are held together only by relatively weak dispersion forces. There are no covalent bonds between them, and the planes slide easily past each other. It's unlikely that graphite jewelry will ever catch on.

Figure 11.4. Carbon atoms in (a) diamond and (b) graphite.

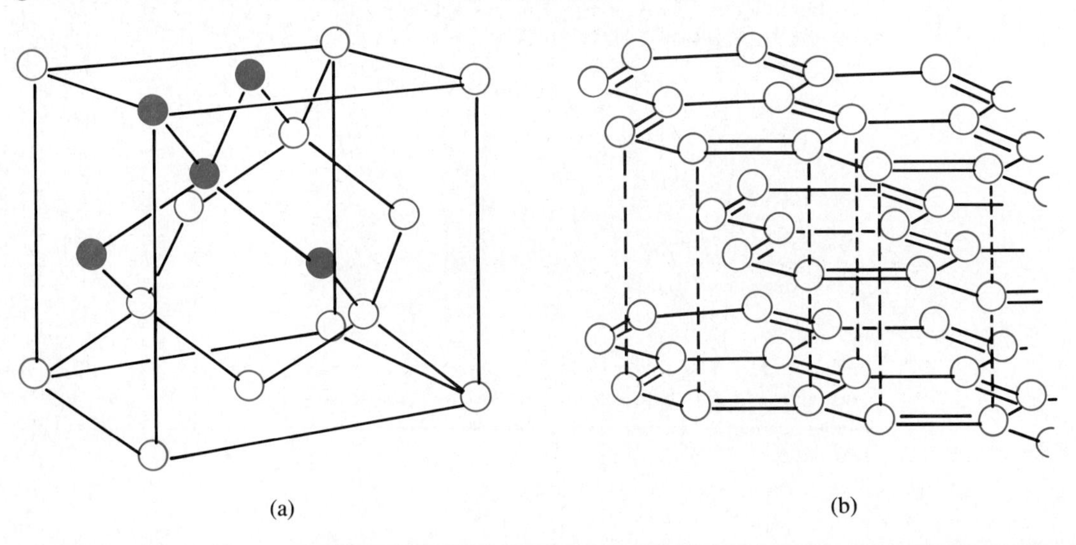

The number of shared pairs of electrons can be made up of any combination of single, double, or triple bonds (Chapter 9). For example, a carbon atom can have four single bonds, a double bond and two single bonds, two double bonds, or some other combination that adds to four shared pairs of electrons. Unless the element holds a positive or negative electrical charge, these simple rules apply amazingly well. The pattern makes understanding the structures of organic chemicals fairly straightforward. The structures in Figure 11.3 vary from simple to complex, but in every case each carbon atom has four lines extending from it, each hydrogen has one, each oxygen has two, and each nitrogen has three. (Unshared pairs of electrons are not shown, following a scheme that will be explained shortly.)

The properties of a compound correlate with its structure. Seeing a structural formula allows a chemist to make predictions about the chemical and physical properties of the compound itself. Books on organic chemistry often have as many structures as they have sentences. These pictures convey much useful information to the practiced eye of a chemist. To someone who doesn't know the rules, though, they may look like graffiti. Since we will be talking later about many organic compounds important to your health and the economy, it is worthwhile now to discuss the ground rules for writing their structures.

11.3 TOWARD SIMPLER STRUCTURAL FORMULAS

A complete molecular structure shows the electrons in the valence shells of all atoms, as well as showing how the atoms themselves are arranged. When there are lots of atoms, writing complete structures can be pretty boring and can take a great amount of space. Therefore, chemists have simplified the task. One example of this simplification has already been discussed—using a line between two atoms to show a shared pair of electrons. This practice is standard in all fields of chemistry. In addition, organic chemists commonly use three other kinds of abbreviations in writing the structures of molecules.

1. Chemists omit the valence electrons that aren't directly involved in chemical bonds (that is, the unshared pairs). We know that the electrons are there—the octet rule still applies—but we don't bother to show them. Chemists sometimes call such a simplified structure a **Kekulé structure,** after the German chemist who first used the notation. August Kekulé, who lived in the nineteenth century, had never heard of electrons and didn't know what held atoms together. Yet we still find it useful to follow his way of drawing structures.

```
    H  :O:               H  O
    |  ||                |  ||
 H—C—C—Ö—H           H—C—C—O—H
    |      ··            |
    H                    H
```

Lewis structure for acetic acid

Kekulé structure for acetic acid

2. Often when carbon atoms are connected by bonds, chemists draw only the lines representing the bonds, omitting the symbol C between neighboring

To interpret the right-hand structures, imagine a carbon atom at each intersection of lines. Because only three pairs of electrons are explicitly indicated for each carbon atom, you must also imagine a single bond extending from each carbon atom to a hydrogen atom, which is also not shown.

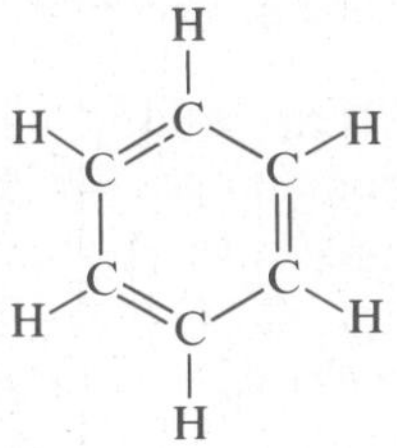

can be written

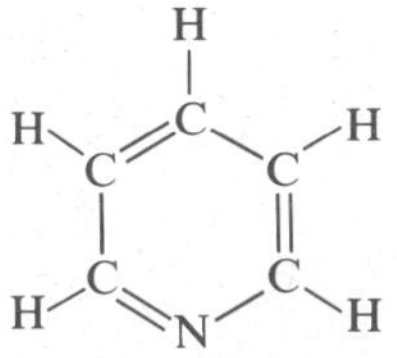

can be written

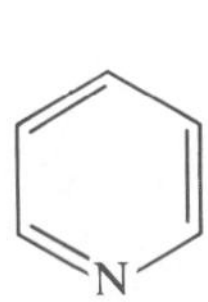

bonds. They assume that a carbon atom is at the end of each line (for a single bond) or pair of lines (for a double bond), unless another atom is shown. Hydrogen atoms attached to carbon can also be omitted for simplicity's sake. You have to remember that carbon always shares four pairs of electrons and hydrogen shares one pair.

3. Even drawing out all the lines can become a chore. Sometimes it is easier to condense the structural drawing by leaving out the lines altogether. In this case, though, you can still make clear which atoms are attached to which other ones by writing the hydrogen atoms directly following the carbon or oxygen or nitrogen to which they are attached.

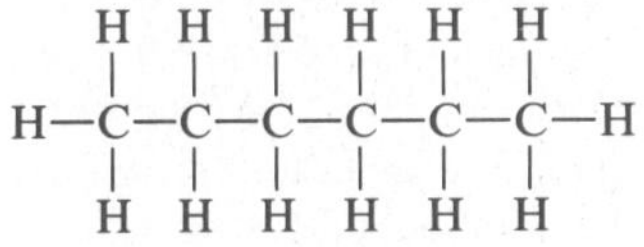

```
   H  H  H  H  H  H
   |  |  |  |  |  |
H—C—C—C—C—C—C—H
   |  |  |  |  |  |
   H  H  H  H  H  H
```

becomes $CH_3CH_2CH_2CH_2CH_2CH_3$

Kekulé structure for hexane

formula for hexane with bonds omitted

Clearly, the second and third simplifications can't be applied at the same time to a group of atoms. Either some atoms or some lines can be omitted, but not both. It is quite common, however, to use the second simplification for one part of a molecule and the third for another part of the same molecule. A summary of these three simplifications in the structural formulas of organic compounds follows.

1. Unshared pairs of electrons are omitted. For example,

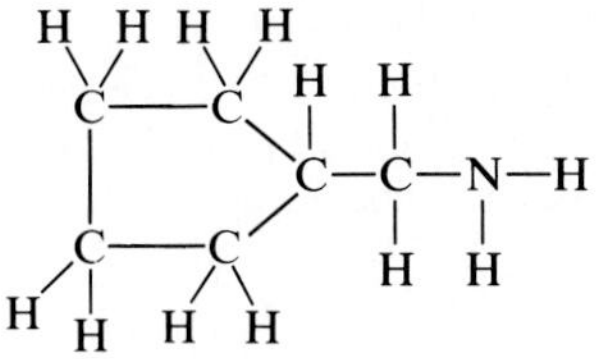

(structure) is used instead of (structure)

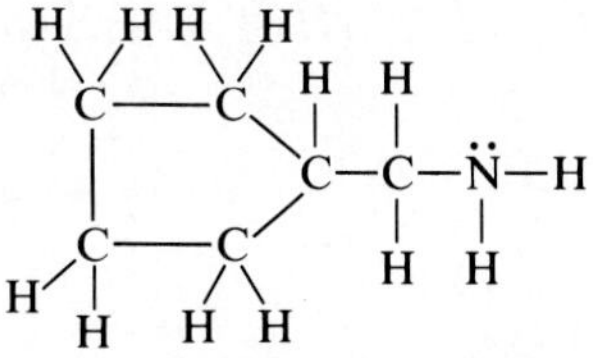

2. The symbols for carbon and hydrogen atoms are not written unless it's convenient to show them. It is assumed that carbon is at the end of each covalent bond unless another element is written. When an implied carbon atom has less than four shared pairs of electrons, bonds to hydrogen atoms make up the difference.

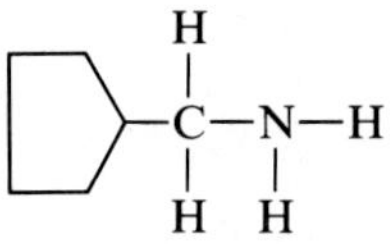

(structure) is used instead of (structure)

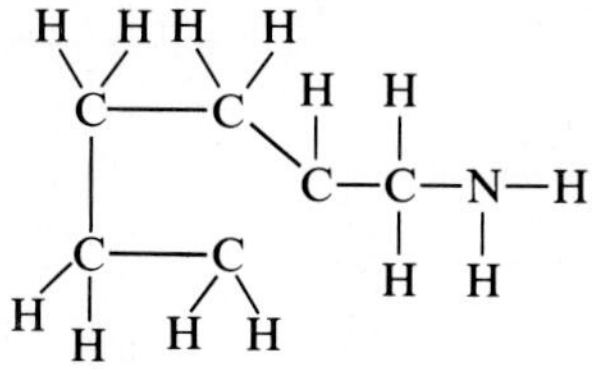

3. Lines between atoms are omitted when no confusion over the structure will result. Hydrogen atoms directly follow the atom to which they are bonded.

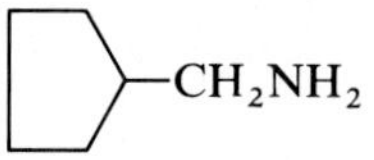

(cyclopentyl)–CH_2NH_2 is used instead of (structure)

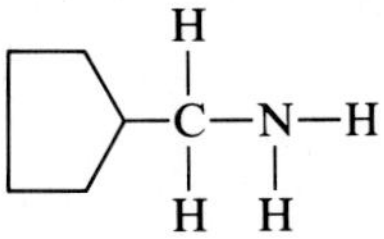

Using these simpler ways of writing structures, a chemist would normally write two of the structures shown in Figure 11.3 as follows:

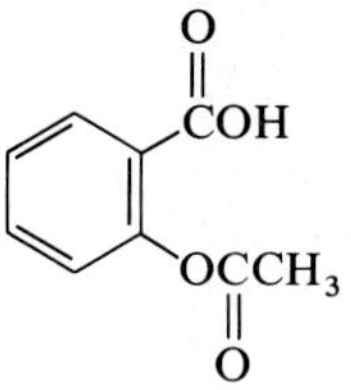

aspirin

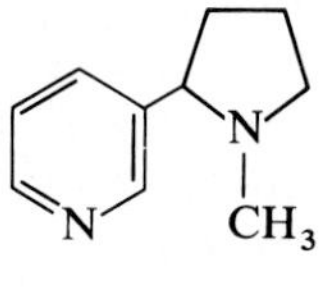

nicotine

Example Draw a complete Kekulé structure for acetaminophen, $C_8H_9NO_2$, which is an active ingredient of the popular pain-killers Excedrin and Tylenol. The simplified structure for acetaminophen is

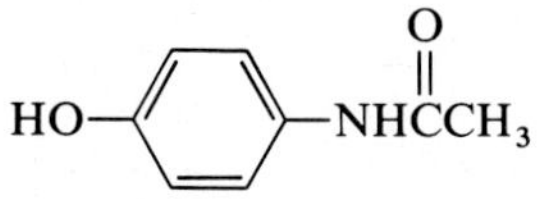

HO–(ring)–NHCCH$_3$ (with C=O)

Solution In this structure all three of the simplifications are used. Only the electrons in covalent bonds have been drawn. The carbon atoms in the ring and the hydrogen atoms attached to them have been left out. Some of the right-hand part of the structure has been written without bonds. The complete Kekulé structure is

```
          H       H
           \     /
            C—C          O  H
           //    \\      ||  |
H—O—C            C—N—C—C—H
           \     /     |       |
            C=C        H      H
           /     \
          H       H
```

It was no accident that August Kekulé, the father of structural formulas, first studied architecture. The orderly lines of Kekulé structures grew out of combining chemical intuition with an architect's line drawings.

11.4 HYDROCARBONS FROM OIL—ALKANES

Other than green plants, the most important source of organic compounds is petroleum. It formed long ago from lush plant growth that became buried in sediments; in a way, petroleum can be thought of as buried sunshine. Many of the compounds in petroleum are **hydrocarbons,** compounds composed only of the elements carbon and hydrogen. These range from the simplest hydrocarbon, methane (CH_4), to complex compounds containing over 40 carbon atoms. Gasoline alone contains more than 100 different hydrocarbons. Crude oil is a sticky, smelly, complex liquid. It is distilled and refined on a gigantic scale to make our hydrocarbon fuels, lubricants, solvents, petrochemicals, and asphalt (see Figure 11.5).

Methane is the simplest member of a series of related hydrocarbons called **alkanes,** which have only single bonds between carbon atoms. The next member is ethane, C_2H_6, followed by propane, C_3H_8, and butane, C_4H_{10}. Propane and butane are familiar fuels, usually supplied in tanks under pressure as liquefied petroleum gas (LP gas). Methane is the chief component of natural gas. Each of these compounds differs from the next by one carbon atom and two hydrogens. The compounds in this series have properties that vary in a regular and predictable manner; the series is homologous—a family of **homologs.**

The nomenclature used for simple inorganic compounds (Chapter 3) won't work with hydrocarbons, because each of these compounds contains only carbon and hydrogen. Each alkane homolog has a different name, but each name ends with the same three letters, *-ane* (see Table 11.2).

The names of the first four alkanes listed in Table 11.2 were invented

(a)

(b)

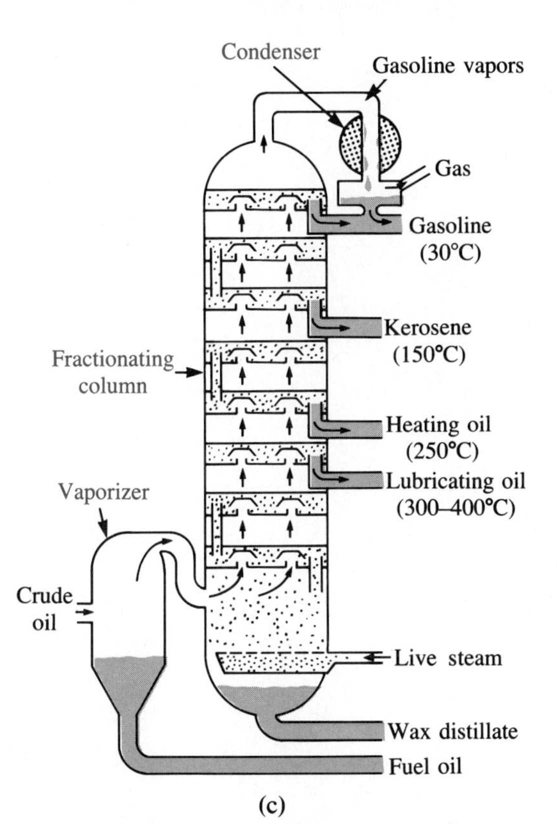

(c)

Figure 11.5. (a) Pumping unit of an oil well. (b) A modern oil refinery. (c) Diagram of a distilling column, showing the various levels at which petroleum fractions are removed.

Table 11.2 The First Eight Normal Alkanes

NAME	MOLECULAR FORMULA	KEKULÉ STRUCTURE	CONDENSED FORMULA	BOILING POINT (°C)	DENSITY (in grams/milliliter), 20°C
methane	CH_4	H H—C—H H	CH_4	−162	
ethane	C_2H_6	H H H—C—C—H H H	CH_3CH_3	−89	
propane	C_3H_8	H H H H—C—C—C—H H H H	$CH_3CH_2CH_3$	−42	
butane	C_4H_{10}	H H H H H—C—C—C—C—H H H H H	$CH_3CH_2CH_2CH_3$	0	
pentane	C_5H_{12}	H H H H H H—C—C—C—C—C—H H H H H H	$CH_3CH_2CH_2CH_2CH_3$	36	0.626
hexane	C_6H_{14}	H H H H H H H—C—C—C—C—C—C—H H H H H H H	$CH_3CH_2CH_2CH_2CH_2CH_3$	69	0.659
heptane	C_7H_{16}	H H H H H H H H—C—C—C—C—C—C—C—H H H H H H H H	$CH_3CH_2CH_2CH_2CH_2CH_2CH_3$	98	0.684
octane	C_8H_{18}	H H H H H H H H H—C—C—C—C—C—C—C—C—H H H H H H H H H	$CH_3(CH_2)_6CH_3$	126	0.703

Table 11.3 Greek Prefixes

PREFIX	NUMBER
pent-	5
hex-	6
hept-	7
oct-	8

by the chemists who discovered them. In the names of the rest of the compounds listed, the prefix comes from the Greek numbering system (see Table 11.3). Just as a pentagon has five sides, a molecule of pentane has five carbon atoms.

Actually, there are two butanes, each with the formula C_4H_{10}. The one in Table 11.2 is called *normal* butane, or *n*-butane. **Normal** means that there is no branching in the carbon chain. No carbon atom is bonded to more than two other carbon atoms. In the other butane, called isobutane, the chain of carbon atoms is not continuous; it has a branch.

```
   H  H  H  H              H     H     H
   |  |  |  |              |     |     |
H—C—C—C—C—H          H—C——C——C—H
   |  |  |  |              |     |     |
   H  H  H  H              H     |     H
                              H—C—H
                                 |
                                 H
```

n-butane
boiling point 0°C

isobutane
boiling point −12°C

Notice that both *n*-butane and isobutane have the same formula, but they have different structures. We call these two compounds **isomers.** Isomers are common and very important in organic chemistry. Some isomers have similar properties, as do *n*-butane and isobutane, with their comparable boiling points. In other cases, isomers may differ greatly from each other.

11.5 UNSATURATED HYDROCARBONS—ALKENES

There is another group of hydrocarbons, the **alkenes,** which can be converted into alkanes by reaction with hydrogen in the presence of finely divided metals.

$$\text{alkene} + H_2 \xrightarrow{\text{Ni}} \text{alkane}$$

The alkenes are not saturated with hydrogen as the alkanes are, so alkenes are called **unsaturated** hydrocarbons. They take on H_2 because of the presence of a double bond between carbon atoms, which is the key feature of an alkene. Take, for example, the simplest alkenes—ethylene and propylene.

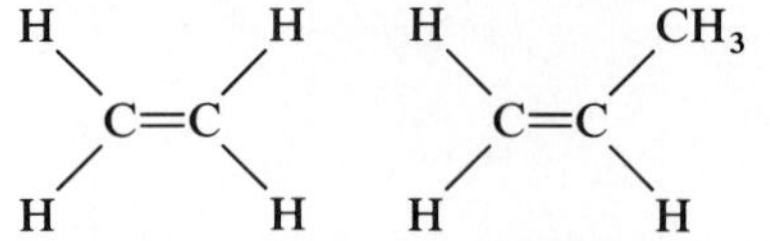

$$\underset{\text{ethylene}}{\mathrm{H_2C{=}CH_2}} \qquad \underset{\text{propylene}}{\mathrm{H_2C{=}CH(CH_3)}}$$

As in all organic compounds, each carbon atom of ethylene and propylene has four shared pairs of electrons, and each hydrogen has one shared pair. There are three clusters of electrons in the valence shell of a doubly bonded carbon atom, and the bond angles are very close to 120° (see Table 9.2, page 204). Ethylene is a flat molecule, with each atom in the same plane.

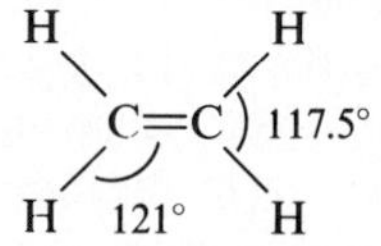

$$\mathrm{H_2C{=}CH_2} \quad (\angle \mathrm{H{-}C{-}H} = 117.5^\circ;\ \angle \mathrm{H{-}C{=}C} = 121^\circ)$$

Because of the carbon-carbon double bond, alkenes have chemical reactions unknown to alkanes. The major reactions of alkenes involve addition of chemicals across the double bond (see Figure 11.6). That is, the chemical adding to an alkene splits into two parts; one part bonds to one of the doubly bonded carbon atoms, and the other part bonds to the other doubly bonded carbon. At the same time, the double bond becomes a single bond.

The addition of hydrogen across the carbon-carbon double bond of alkenes is an important part of food chemistry. Over the years Americans have become used to cooking with and eating butter and other fats. We prefer not to dip our bread into equally nutritious, cheaper oils made from corn

Figure 11.6. Addition reactions of alkenes.

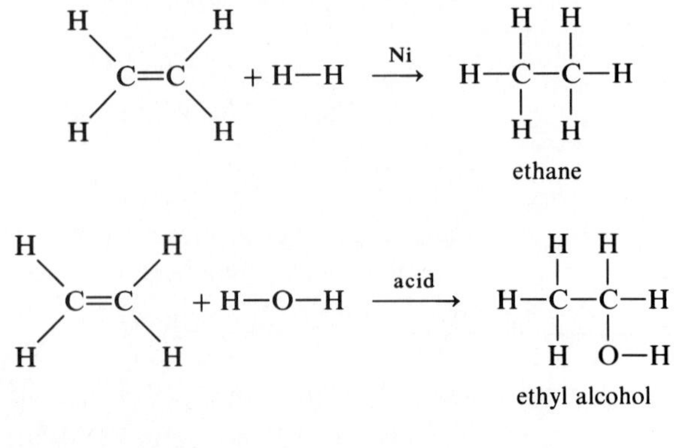

$$\mathrm{H_2C{=}CH_2} + \mathrm{H{-}H} \xrightarrow{\text{Ni}} \underset{\text{ethane}}{\mathrm{H{-}CH_2{-}CH_2{-}H}}$$

$$\mathrm{H_2C{=}CH_2} + \mathrm{H{-}O{-}H} \xrightarrow{\text{acid}} \underset{\text{ethyl alcohol}}{\mathrm{H{-}CH_2{-}CH_2{-}O{-}H}}$$

or cottonseeds. The flavor is different because of different trace constituents. Yet the major constituents of these vegetable oils have the same structures as those of butter fat, except that the oils are unsaturated; they have more C-C double bonds in their carbon chains. It turns out that unsaturation tends to lower the melting point. Unsaturated oils are liquid at room temperature, whereas saturated fats are solid.

Because of our prejudice against eating oils, the hardening of oils is important in producing oleomargarine as well as cooking fats such as Crisco and Spry. Chemical engineers react the oils with hydrogen in the presence of finely divided nickel metal to produce the hardening. Our prejudice works against a good diet, since our bodies can convert saturated fats into cholesterol, a factor in diseases of the heart and blood vessels. We metabolize unsaturated oils by other pathways.

Of all the organic chemicals produced in this country, ethylene is the leader. The U.S. chemical industry produces about 30 billion pounds of it each year. That's about 50,000 liters of ethylene gas for every U.S. citizen. Big business indeed! The ethylene is converted into over 100 billion pounds of chemicals and polymers, such as polyethylene (see Figure 11.7). About

Figure 11.7. Some commonly used polyethylene containers.

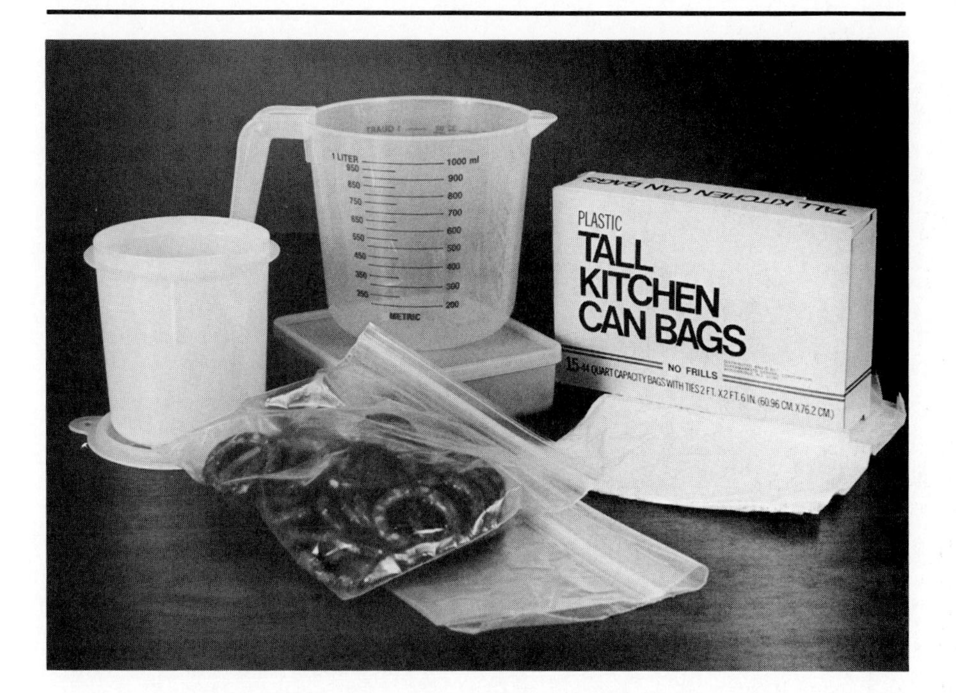

45% of the total tonnage of organic chemicals produced in the United States is made from ethylene. At 14 billion pounds per year, propylene is big business too. It stands in third place among organic chemicals.

Ethylene and propylene are produced from petroleum, mostly by heating ethane and propane, thereby cracking them apart.

$$CH_3CH_3 \xrightarrow{\text{heat}} CH_2{=}CH_2 + H_2$$

$$CH_3CH_2CH_3 \xrightarrow{\text{heat}} CH_3CH{=}CH_2 + H_2$$

These reactions are just the reverse of the addition reactions.

U.S. industry uses about 60% of its ethylene to make polymers: plastics, fibers, elastomers, adhesives, and coatings. The other 40% is used to make other organic chemicals, the most important being ethylene oxide.

```
            CH2=CH2
               |
               | O2
               ↓
               O
              / \
          CH2—CH2        ethylene oxide
         /     |     \
        ↙      ↓      ↘
detergents  antifreeze  polyester fibers
```

11.6 FUNCTIONAL GROUPS

The way organic chemicals are put together determines their properties. In other words, the exact structure of a compound relates to its chemical qualities.

Consider the two compounds with the molecular formula

```
    H  H                      H       H
    |  |                      |       |
H—C—C—O                  H—C—O—C—H
    |  |  |                   |       |
    H  H  H                   H       H
```

ethyl alcohol	dimethyl ether
boiling point: 78°C	boiling point: −23°C
melting point: −117 C	melting point: −138°C
burns smoothly	explodes easily

Both follow the bonding rules that we discussed earlier. But they are very different compounds! Ethyl alcohol can be stored easily in bottles; it boils at 78°C. Dimethyl ether, however, must be stored in tanks because it is a gas at room temperature; it boils at −23°C. These two compounds are examples of isomers—compounds that have the same molecular formula but different structures.

Notice that ethyl alcohol has an oxygen-hydrogen bond, whereas dimethyl ether has all of its hydrogen atoms linked to carbon. You may recall the requirement for hydrogen bonding between molecules (Chapter 9): in order for hydrogen bonding to occur, a hydrogen atom must be bonded to a nitrogen, oxygen, or fluorine atom. This requirement is satisfied for ethyl alcohol but not for dimethyl ether. Therefore, ethyl alcohol has a much higher boiling point, because a lot of energy is required to disrupt the hydrogen bonds between the alcohol molecules.

Every covalent compound that has an —O—H group shares some specific chemistry. For example, all of them react with metals in Group I of the periodic table, liberating hydrogen gas.

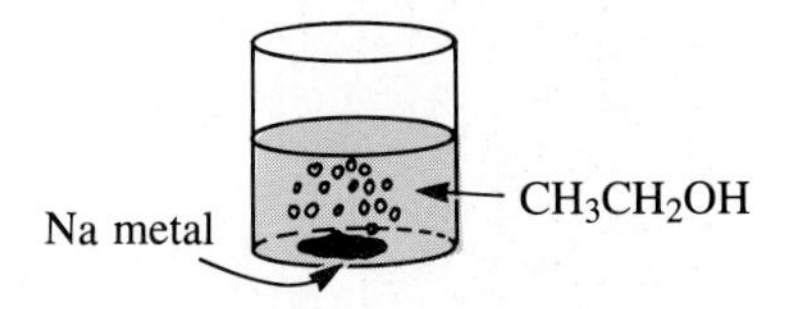

$$2\ CH_3CH_2OH + 2\ Na \longrightarrow 2\ CH_3CH_2ONa + H_2$$

This is the same kind of reaction that occurs when sodium metal reacts with water, which also has —O—H bonds. Evolution of hydrogen gas, which is explosive, can be dramatic if the reaction mixture gets hot enough. As in other explosions, heat, light, and sound result.

Dimethyl ether, on the other hand, does not react with sodium metal at usual temperatures. No H_2 gas forms.

$$CH_3{-}O{-}CH_3 + Na \longrightarrow \text{no reaction}$$

Any organic compound that contains the —O—H group is called an **alcohol.** A compound that has an oxygen atom attached to two carbon atoms is called an ether.

These specific arrangements of atoms relate to the chemical properties or functions of the compounds containing them. Chemists define a **functional group** as a specific group of atoms within a molecule that correlates with specific chemical properties. In the section on alkenes we saw another functional group:

$$\diagdown\!\!\!\!\diagup C{=}C \diagup\!\!\!\!\diagdown$$

The double-bonded pair of carbon atoms is responsible for the special properties of this class of compounds. Textbooks on organic chemistry owe their organization to the functional-group approach. Usually each functional group has a different chapter devoted to it. Table 11.4 lists a few of the most significant functional groups that are found in organic chemicals.

Table 11.4 Common Functional Groups

FUNCTIONAL GROUP[a]	CLASS OF COMPOUND
—O—H	alcohol
—O—	ether
$-\overset{\vert}{\underset{\vert}{C}}-H$, $-\overset{\vert}{\underset{\vert}{C}}-\overset{\vert}{\underset{\vert}{C}}-$, etc.	alkanes (and most other organic compounds)
$>C=C<$	alkene[b]
$-\overset{O}{\overset{\Vert}{C}}-H$	aldehyde
$-\overset{O}{\overset{\Vert}{C}}-$	ketone
$-\overset{O}{\overset{\Vert}{C}}-O-H$	carboxylic acid
$-\overset{O}{\overset{\Vert}{C}}-O-$	ester
$-N<$	amine[b]

[a]Assume that a carbon atom is at the end of a dangling line.
[b]Alkenes and amines can have either a hydrogen or a carbon atom attached to a dangling line.

Example With the help of Table 11.4, write (1) the complete structure (including all atoms and valence electrons) and (2) a simplified structure for

(a) an alkene with six carbon atoms

(b) an aldehyde with three carbon atoms

Solution (a) The functional group that must be included in the alkene molecule is

$$>C=C<$$

and four other carbon atoms must be present. The easiest way to satisfy these conditions is to place six carbon atoms in a chain and then put a double bond between any pair of them. Hydrogen atoms are then added to complete the valence shells of all carbon atoms. Among many possible answers is the following:

```
    H  H  H  H  H  H
    |  |  |  |  |  |
H—C—C=C—C—C—C—H
    |        |  |  |
    H        H  H  H
```

complete structure

$CH_3CH{=}CHCH_2CH_2CH_3$

simplified structure

(b) For an aldehyde the functional group is

```
   O
   ‖
—C—H
```

Two additional carbon atoms must be in the molecule.

```
    H  H :O:
    |  |  ‖
H—C—C—C—H
    |  |
    H  H
```

complete structure

```
          O
          ‖
CH3CH2CH
```

simplified structure

11.7 THE ALCOHOLS

Ethyl alcohol is probably the best known alcohol. It is also known as *grain alcohol* or simply *alcohol;* its preferred chemical name is *ethanol.* One of the most important industrial organic chemicals, ethanol is used as a solvent, as a raw material for the production of other chemicals, and as a component of gasohol. Gasohol is a mixture of gasoline and ethanol. A small amount of ethanol increases the octane rating of gasoline considerably. It has been common to see posted on a lead-free gasoline pump the statement "Contains Ethanol."

Of course, ethanol is an essential ingredient of all alcoholic beverages. Because of the demand for and high taxes on alcoholic beverages, the government closely supervises the production of ethanol made by fermentation. Battles between "moonshiners" and "revenoors" have become part of American folklore. The nontaxed industrial ethanol is denatured by the addition of obnoxious, toxic compounds that discourage its use in beverages. Ethyl alcohol itself is also a toxic compound. Wine can only reach about 12% alcohol content by natural fermentation, since yeasts die at higher alcohol concentrations. Stronger liquor must be made by adding distilled spirits.

CH_3CH_2OH	CH_3OH
ethanol ethyl alcohol alcohol	methanol methyl alcohol wood alcohol

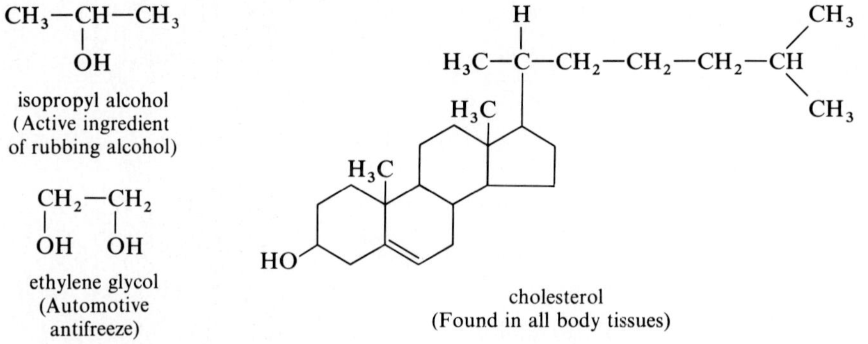

$CH_3-CH-CH_3$
$\quad\quad\;\; |$
$\quad\quad\; OH$

isopropyl alcohol
(Active ingredient
of rubbing alcohol)

CH_2-CH_2
$|\quad\quad\;\; |$
$OH\quad OH$

ethylene glycol
(Automotive
antifreeze)

H
$H_3C-C-CH_2-CH_2-CH_2-CH$ (CH_3)(CH_3)
H_3C
H_3C
HO

cholesterol
(Found in all body tissues)

Figure 11.8. Some interesting and important alcohols.

distillation of wood. Ingestion of methanol is very dangerous; small amounts can cause blindness and death. Many accidents occur each year when people mistake this extremely toxic compound for its less harmful relative.

Some scientists believe that methanol will be an important fuel in the future. We already know how to make it from coal. Methanol is an "ultra-clean" liquid fuel that would have no difficulty meeting environmental regulations. Besides being nonpolluting, methanol burns in an automobile engine better than 100-octane gasoline. Also, it would not be very difficult to modify car engines to accommodate methanol. You may wonder why methanol is such a good fuel when it is so poisonous, but consider how toxic and flammable gasoline is. Figure 11.8 shows some other important alcohols. Bear in mind that for these and later examples it is not necessary to learn or memorize all of the structures.

11.8 ORGANIC ACIDS AND ESTERS

Organic acids, like inorganic acids, react with bases and are sour to the taste. Almost everyone has tasted them. Vinegar is a dilute solution (around 5%) of acetic acid in water. The sour taste of a lemon or a grapefruit comes from citric acid.

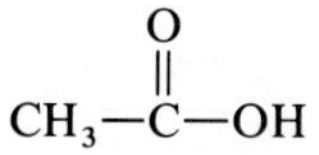

$$CH_3-\overset{\overset{\displaystyle O}{\|}}{C}-OH$$

acetic acid

$$\begin{array}{l} \quad\quad\quad\quad\quad\quad\;\; O \\ \quad\quad\quad\quad\quad\quad\;\; \| \\ \quad\quad\;\, O\;\; CH_2-C-OH \\ \quad\quad\;\, \|\quad\;\; | \\ HO-C-C-OH \\ \quad\quad\quad\quad\; | \\ \quad\quad\quad\;\; CH_2-C-OH \\ \quad\quad\quad\quad\quad\quad\;\; \| \\ \quad\quad\quad\quad\quad\quad\;\; O \end{array}$$

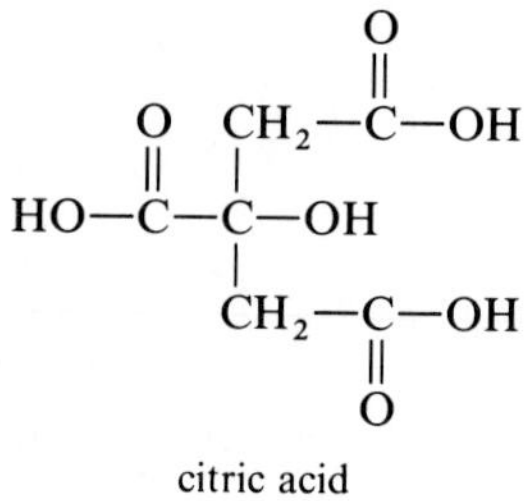

citric acid

The functional group of organic acids is the

$$-\overset{\overset{\displaystyle O}{\|}}{C}-OH \quad \text{or} \quad -CO_2H$$

group. This group of four atoms is known as a carboxyl group, and we call a compound containing this group a **carboxylic acid.** A carboxylic acid is a much "weaker" acid than hydrochloric acid (HCl) or nitric acid (HNO_3). This means that in water solution only a small fraction of carboxylic acid molecules dissociate into a hydrogen ion (H^+) and a negative ion.

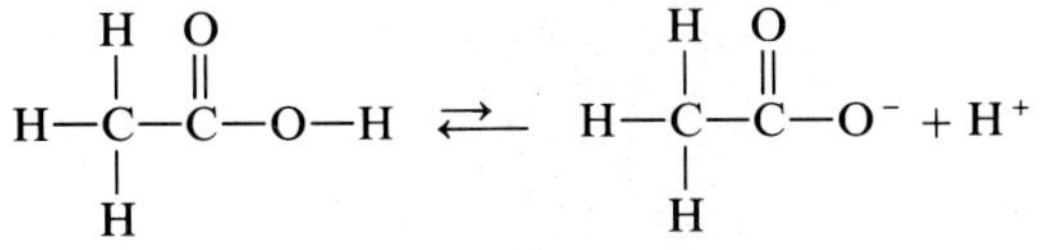

$$H-\underset{\underset{\displaystyle H}{|}}{\overset{\overset{\displaystyle H}{|}}{C}}-\overset{\overset{\displaystyle O}{\|}}{C}-O-H \rightleftharpoons H-\underset{\underset{\displaystyle H}{|}}{\overset{\overset{\displaystyle H}{|}}{C}}-\overset{\overset{\displaystyle O}{\|}}{C}-O^- + H^+$$

acetic acid — acetate ion

Hydrochloric and nitric acids, on the other hand, dissociate completely when they dissolve in water.

Another carboxylic acid that almost everyone has ingested is aspirin. Only recently have scientists begun to understand why aspirin is an effective pain killer, although it has been used for over ninety years. With the careful testing and licensing procedures for new drugs nowadays, aspirin would probably be available only by doctor's prescription if it were discovered in the 1980s. The structures of aspirin and some other common carboxylic acids are given in Figure 11.9.

When a carboxylic acid reacts with an alcohol in the presence of a strong inorganic acid, an **ester** is formed.

$$CH_3\overset{\overset{\displaystyle O}{\|}}{C}OH + HOCH_2CH_3 \overset{H^+}{\rightleftharpoons} CH_3\overset{\overset{\displaystyle O}{\|}}{C}OCH_2CH_3 + H_2O$$

ethyl acetate, an ester

The simplest member of the alcohol family is *methyl alcohol*, or *methanol*. Some people still call it *wood alcohol* because it was once made by the

Figure 11.9. *Some common carboxylic acids.*

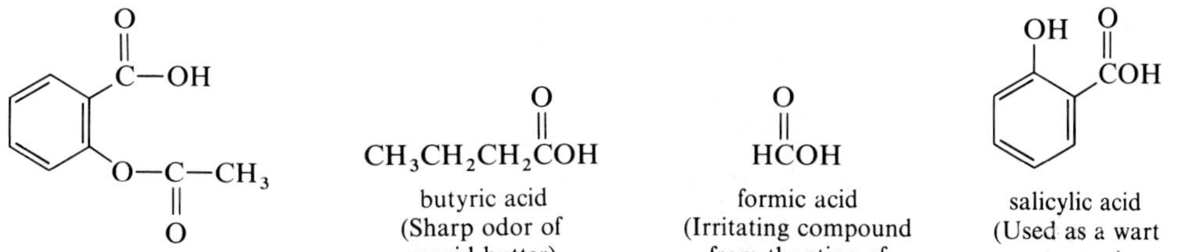

aspirin (acetylsalicylic acid): benzene ring bearing $-\overset{\overset{\displaystyle O}{\|}}{C}-OH$ and $-O-\underset{\underset{\displaystyle O}{\|}}{C}-CH_3$

$$CH_3CH_2CH_2\overset{\overset{\displaystyle O}{\|}}{C}OH$$

butyric acid (Sharp odor of rancid butter)

$$H\overset{\overset{\displaystyle O}{\|}}{C}OH$$

formic acid (Irritating compound from the sting of ants and nettles)

salicylic acid (Used as a wart remover): benzene ring bearing OH and $\overset{\overset{\displaystyle O}{\|}}{C}OH$

All esters have the functional group

$$\begin{array}{c} O \\ \| \\ -C-O- \end{array}$$

Ethyl acetate, one of the simplest esters, has a pleasant, fruity odor. In fact, it is the major volatile constituent of pineapples. Ethyl acetate is used as a solvent for many cosmetic products; among them is fingernail polish. See Figure 11.10 for some other common esters.

To understand the naming of esters, we must look at the names of some alkyl groups, such as ethyl and methyl. The names of alkyl groups are derived from those of the parent alkanes.

$$\begin{array}{cccc}
\begin{array}{ccc} & H & \\ & | & \\ H- & C & -H \\ & | & \\ & H & \end{array} &
\begin{array}{ccc} & H & \\ & | & \\ H- & C & - \\ & | & \\ & H & \end{array} &
\begin{array}{cccc} & H & H & \\ & | & | & \\ H- & C- & C & -H \\ & | & | & \\ & H & H & \end{array} &
\begin{array}{cccc} & H & H & \\ & | & | & \\ H- & C- & C & - \\ & | & | & \\ & H & H & \end{array} \\
\text{methane} & \text{methyl group} & \text{ethane} & \text{ethyl group}
\end{array}$$

Notice that an **alkyl group** cannot exist alone. Alkyl groups are not stable molecules, because one of the carbon atoms within an alkyl group is bonded to only three atoms. Carbon needs four shared pairs of electrons to be stable. But alkyl groups are found as component parts of organic structures. The "dangling" bond of an alkyl group will be linked to some other atom—oxygen, nitrogen, chlorine, sulfur, or another carbon atom.

There are two different kinds of propyl groups, and they have different names. The *n*-propyl group attaches at an end carbon atom; the isopropyl group attaches at the middle carbon.

$$\begin{array}{ccc}
\begin{array}{ccccc} & H & H & H & \\ & | & | & | & \\ H- & C- & C- & C & -H \\ & | & | & | & \\ & H & H & H & \end{array} &
\begin{array}{ccccc} & H & H & H & \\ & | & | & | & \\ H- & C- & C- & C & - \\ & | & | & | & \\ & H & H & H & \end{array} &
\begin{array}{ccccc} & H & H & H & \\ & | & | & | & \\ H- & C- & C- & C & -H \\ & | & | & | & \\ & H & & H & \end{array} \\
\text{propane} & n\text{-propyl group} & \text{isopropyl group}
\end{array}$$

Example

The structure of isopentyl acetate appears in Figure 11.10. Isopentyl acetate is the common artificial banana flavoring in inexpensive candies. Which alcohol and which carboxylic acid would you have to react together to synthesize this ester?

Figure 11.10. Three esters that have fruity odors.

$$\begin{array}{ccc}
\begin{array}{l} O \\ \| \\ CH_3COCH_2CH_2CH_3 \end{array} &
\begin{array}{l} O \\ \| \\ CH_3CH_2CH_2COCH_3 \end{array} &
\begin{array}{l} O \\ \| \\ CH_3COCH_2CH_2CHCH_3 \\ | \\ CH_3 \end{array} \\
\text{propyl acetate} & \text{methyl butyrate} & \text{isopentyl acetate} \\
\text{(Pear odor)} & \text{(Apple odor)} & \text{(Banana odor)}
\end{array}$$

Solution Look first at the structure of isopentyl acetate.

$$\text{from } CH_3\overset{\displaystyle O}{\overset{\|}{C}}OH \nearrow \quad \boxed{CH_3\overset{\displaystyle O}{\overset{\|}{C}}}\,OCH_2CH_2\underset{\displaystyle CH_3}{\underset{|}{C}}HCH_3 \quad \nwarrow \text{ from } HOCH_2CH_2\underset{\displaystyle CH_3}{\underset{|}{C}}HCH_3$$

The boxed part at the left side of the structure comes from the carboxylic acid. In this case it is acetic acid. The group enclosed in the balloon at the right comes from the alcohol, which here is isopentyl alcohol. Notice that the name of the alkyl group does not change.

$$\underset{\text{acetic acid}}{CH_3\overset{\displaystyle O}{\overset{\|}{C}}OH} + \underset{\text{isopentyl alcohol}}{HOCH_2CH_2\underset{\displaystyle CH_3}{\underset{|}{C}}HCH_3} \overset{H^+}{\rightleftharpoons} \underset{\text{isopentyl acetate}}{CH_3\overset{\displaystyle O}{\overset{\|}{C}}OCH_2CH_2\underset{\displaystyle CH_3}{\underset{|}{C}}HCH_3} + H_2O$$

11.9 CYCLIC HYDROCARBONS

We have already seen two of the sources of the tremendous diversity in organic compounds. One is the variety of functional groups that can be present in a molecule. The second is the many ways in which chains of carbon atoms can be organized in either continuous or branched form. Organic molecules can also be composed of atoms joined together to form closed chains, or rings; these are the **cyclic** organic chemicals.

Both alkanes and alkenes can form cyclic compounds. Cyclohexane, C_6H_{12}, is a common cycloalkane; most of it is converted into nylon.

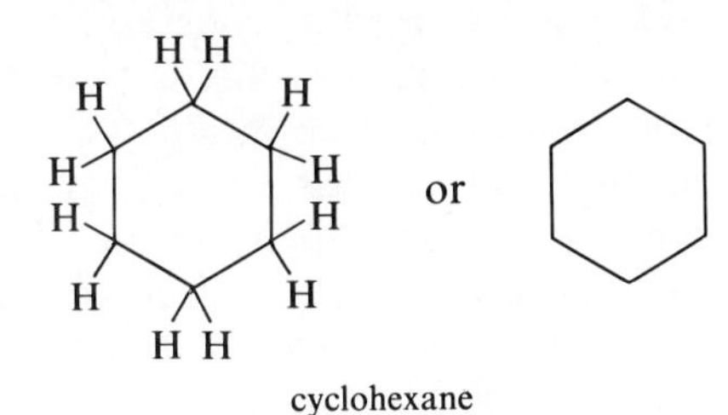

cyclohexane

Example Cyclopropane, C_3H_6, has been a useful general anesthetic. Draw a simplified structure for cyclopropane.

Solution With three carbon atoms there is only one ring possible—a ring with three carbon atoms. Therefore, the simplified structure of cyclopropane is

Since each carbon atom must have four bonds, there are two hydrogen atoms attached to each carbon, as the formula demands.

Cyclopentene is a representative cycloalkene. The double bond has the same kinds of chemical reactions whether it is in a ring or part of an open chain of carbon atoms.

cyclopentene

The most important cyclic hydrocarbons are the **aromatic** compounds (see Figure 11.11). The first ones discovered had distinct, sometimes pleasant odors, thus the name *aromatic*. Later, chemists learned that these aromatic compounds contain a six-membered ring with three carbon-carbon double bonds, the so-called **benzene ring.** The simplest chemical that contains a benzene ring is called *benzene*.

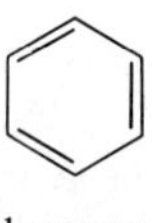

benzene

With three double bonds, you might expect benzene to be highly reactive, but it isn't. The arrangement of bonds in benzene gives the compound great stability. By 1850, compounds with benzene units were recognized to be important, yet no one could figure out benzene's structure. Finally, in 1865, August Kekulé solved that puzzle. It was the first time that a ring of carbon atoms had been proposed, and the discovery made Kekulé famous in the scientific world.

Although Kekulé's achievement was great, his structure for benzene is no longer regarded as satisfactory. Modern chemists, using quantum mechanics, have shown that the second pair of electrons in each double bond in benzene cannot be located between specific carbon atoms. Instead, the electrons are shared equally by all six carbon atoms. Since it can be confusing to draw benzene this way, many chemists often still use the Kekulé picture, although it now conveys a different message than it did in

Figure 11.11. *Two important aromatic hydrocarbons.*

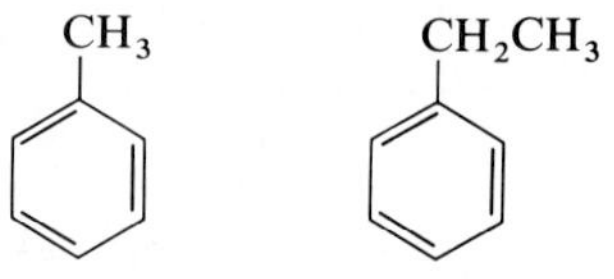

CH_3 CH_2CH_3

methylbenzene (toluene) ethylbenzene

ORIGIN OF THE BENZENE RING

In 1890, on the twenty-fifth anniversary of his discovery of the benzene structure, Kekulé (Figure 11.12) gave a famous speech in Berlin. His account of the way the idea came to him follows.

"Certain ideas at certain times are in the air; if one man does not enunciate them, another will do so soon afterwards.

"It was said that the benzene theory appeared like a meteor in the sky, absolutely new and unheralded. Gentlemen! The human intellect does not operate in that way. Something absolutely new has never been thought, certainly not in chemistry. . . .

"Perhaps it will interest you, if I let you know, through highly indiscreet disclosures from my inner life, how I arrived at some of my ideas. . . .

"I was sitting writing at my textbook but the work did not progress; my thoughts were elsewhere. I turned my chair to the fire and dozed. Again the atoms were gamboling before my eyes. This time the smaller groups kept modestly in the background. My mental eye, rendered more acute by repeated visions of the kind, could now distinguish larger structures of manifold conformation: long rows, sometimes more closely fitted together all twining and twisting in snake-like motion. But look! What was that? One of the snakes had seized hold of its own tail, and the form whirled mockingly before my eyes. As if by a flash of lightning I awoke; and this time also I spent the rest of the night in working out the consequences of the hypothesis.

"Let us learn to dream, gentlemen, then perhaps we shall find the truth.

"But let us beware of publishing our dreams till they have been tested by the waking understanding."

Figure 11.12. August Kekulé.

Kekulé's time. Other chemists write benzene with a circle to show the six shared electrons.

benzene

11.10 COMBUSTION OF ORGANIC CHEMICALS

It is a rare carbon compound that does not burn. The taming of fire from burning wood, coal, oil, and other organic materials could be considered the most significant technological advance ever made by mankind. Much of the energy that fuels our world comes from the **combustion,** or burning, of hydrocarbons, as you saw in Chapter 5. No matter which hydrocarbon burns, the end products of complete combustion are carbon dioxide and water. (See Table 11.5.)

When too little O_2 is present, carbon monoxide forms instead of carbon dioxide; incomplete combustion generates less heat (see Table 11.6). The advantage of burning the fuel completely is obvious; more heat is gained.

Table 11.5 Examples of Complete Combustion

REACTANTS		PRODUCTS	HEAT PRODUCED PER GRAM OF HYDROCARBON (in kilojoules)
$CH_4 + 2\ O_2$	$\longrightarrow$	$CO_2 + 2\ H_2O$	53.5
$2\ CH_2{=}CHCH_3 + 9\ O_2$	$\longrightarrow$	$6\ CO_2 + 6\ H_2O$	44.7
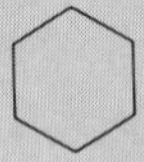 + $9\ O_2$	$\longrightarrow$	$6\ CO_2 + 6\ H_2O$	43.9
$2\ C_6H_6 + 15\ O_2$	$\longrightarrow$	$12\ CO_2 + 6\ H_2O$	40.5

Table 11.6 Examples of Incomplete Combustion

REACTANTS		PRODUCTS	HEAT PRODUCED PER GRAM OF HYDROCARBON (in kilojoules)
$CH_4 + O_2$	$\longrightarrow$	$CO + 2\ H_2O$	32.2
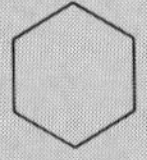 + $6\ O_2$	$\longrightarrow$	$6\ CO + 6\ H_2O$	23.8
$2\ C_6H_6 + 9\ O_2$	$\longrightarrow$	$12\ CO + 6\ H_2O$	18.8

The combustion of gasoline in an automobile engine is never complete, so some carbon monoxide comes out of the exhaust pipe. Besides being poisonous, carbon monoxide has no odor. Being in a closed space with a car engine running is dangerous because of the possibility of carbon monoxide poisoning.

Example Write a balanced chemical reaction for the complete combustion of propane, a constituent of LP gas.

Solution In complete combustion, the products will be carbon dioxide and water. Propane has the molecular formula C_3H_8 (see Table 11.2). Balancing the atoms of each element in the chemical reaction by inspection, we get

$$C_3H_8 + 5\,O_2 \longrightarrow 3\,CO_2 + 4\,H_2O$$

Five molecules of oxygen are necessary to completely burn a molecule of propane. (Refer to page 50 in Chapter 3 if you want a review in balancing equations.)

* 11.11 THE REACTIONS OF ORGANIC COMPOUNDS (*optional*)

More than two million different organic chemicals exist. It would be impossible even to name them all if we didn't have some systematic nomenclature. In the same way, the concepts of molecular structure and the functional group have made it possible to catalog and make sense of organic chemistry. We have classes of compounds organized by functional group, and we also have a few key classes of chemical reactions. Simple reactions can be placed in one of the six classes shown in Table 11.7. These categories are assets to chemists trying to predict the chemical changes that compounds undergo. We have already seen examples of all but one of them, even though we didn't identify the classes at the time. The one class we have not considered is the *acid-base reaction*, which will be the subject of Chapter 12. Here in this chapter we have spent a good deal of space discussing the combustion of hydrocarbons—an example of a chemical *oxidation reaction*.

Addition and elimination reactions are common with unsaturated molecules. The hardening of oils by an *addition reaction* with hydrogen gas is an example.

$$\text{alkene} + H_2 \xrightarrow{\text{Ni}} \text{alkane}$$

Elimination reactions are the opposite of addition reactions. Consider the industrial synthesis of ethylene from petroleum:

$$CH_3{-}CH_3 \xrightarrow{\text{heat}} CH_2{=}CH_2 + H_2$$

Table 11.7 Classes of Chemical Reactions

I.	Acid-base
II.	Oxidation-reduction
III.	Addition
IV.	Elimination
V.	Rearrangement
VI.	Substitution

The reactant, ethane, loses two hydrogen atoms in the reaction. These two hydrogen atoms are eliminated from the organic molecule. Once outside, they are free to go their own way as a hydrogen molecule.

A *rearrangement reaction* produces an isomer of the reactant. The identical atoms are present in the product, but they are assembled differently. During the production of gasoline, rearrangement reactions produce a higher-quality gasoline.

$$\underset{n\text{-hexane}}{CH_3CH_2CH_2CH_2CH_2CH_3} \longrightarrow \underset{\text{isohexane}}{CH_3CH_2CH_2\underset{\displaystyle CH_3}{\underset{|}{C}}HCH_3}$$

Finally, for an example of a *substitution reaction*, consider the synthesis of an ester from a carboxylic acid and an alcohol.

$$CH_3\overset{\overset{\displaystyle O}{\|}}{C}OH + HOCH_2CH_3 \overset{H^+}{\rightleftharpoons} CH_3\overset{\overset{\displaystyle O}{\|}}{C}OCH_2CH_3 + HOH$$

The —OH group of acetic acid is no longer present in the product, ethyl acetate. Instead, an $—OCH_2CH_3$ group is in its place. The $—OCH_2CH_3$ group has replaced, or substituted for, the —OH group. *

IMPORTANT TERMS

organic chemistry The study of carbon compounds.

tetravalent carbon Another name for carbon, which describes the fact that carbon atoms always share four pairs of electrons.

Kekulé structure Drawing of a molecular structure in which all atoms are shown and covalent bonds are drawn as lines between atoms. Only those valence electrons which are part of the chemical bonds are shown; unshared electron pairs are not shown.

hydrocarbon A covalent compound containing only the elements carbon and hydrogen.

alkanes A family of hydrocarbons in which all the atoms are linked by single bonds.

homologs The members of a series of compounds whose structures and properties vary in a regular and predictable way.

normal Refers to a carbon chain that doesn't branch.

isomers Compounds that have the same molecular formula but different molecular structures.

alkenes A group of hydrocarbons whose members contain one or more carbon-carbon double bonds.

unsaturated Refers to compounds containing double or triple bonds.

alcohol An organic chemical that contains the —OH functional group.

functional group A specific group of atoms within a molecule that correlates with specific chemical properties.

carboxylic acid An acidic organic compound that contains the functional group

$$—\overset{\overset{\displaystyle O}{\|}}{C}OH$$

ester A derivative of a carboxylic acid containing the functional group

$$—\overset{\overset{\displaystyle O}{\|}}{C}O—$$

alkyl group A saturated hydrocarbon group that can be part of a molecule.

cyclic Containing one or more rings of atoms.

aromatic Containing one or more benzene rings.

benzene ring A cyclic structure containing six carbon atoms and three double bonds. A hydrogen atom or a functional group is bonded to each carbon atom.

combustion Burning.

QUESTIONS

1. A supermarket shelf contains bottles of safflower oil that say in bold print, "Highest in Unsaturates." Why should a consumer want to buy unsaturates?

2. The simplest aldehyde, formaldehyde, has the molecular formula CH_2O. Draw a Lewis structure and a Kekulé structure for formaldehyde.

3. Classify each of the following compounds as organic or inorganic:

a. NO_2
b. $HC_2H_3O_2$
c. $C_{12}H_{22}O_{11}$ (sucrose)
d. CrO_3
e. $C_9H_{11}NO_2$

4. Draw a complete Kekulé structure for each of the following:

a. CH_3CHCH_2OH (OH on the middle carbon)

propylene glycol, ingredient in many cosmetics

b. $HSCH_2CHCOH$ (C=O on the last carbon; NH_2 on the CH)

cysteine, an amino acid

c. H_2N–(benzene ring)–$COCH_2CH_3$ (C=O)

benzocaine, a local anesthetic

d. $(CH_3CH_2)_2NCH_2CH_2OC$–(benzene ring)–NH_2 (C=O)

procaine, a local anesthetic

e. (benzene ring with CH_3)–$CN(CH_2CH_3)_2$ (C=O)

a common mosquito repellent

f. $CH_3CH_2CH_2CHCHCH_2OH$ (CH_2CH_3 and OH substituents)

another mosquito repellent

g. (barbiturate ring: N–H, C=O, N–H, C=O, C=O) with $CH_2CH{=}CH_2$ and $CHCH_2CH_2CH_3$ (CH_3)

Seconal, a short-acting barbituate

h. (H_3C, CH_3, CH_3, CH_3, CH_3, CH with C=O)

retinal, Vitamin A

i. (CH_3, $CHCH_2CH_2CH_2CH(CH_3)_2$, CH_3, CH_3, HO)

cholesterol

5. Write a simplified structure for each of the following:

a. a carboxylic acid with five carbon atoms
b. an alcohol with five carbon atoms
c. an amine with four carbon atoms
d. an ester with three carbon atoms
e. an ether with five carbon atoms
f. a ketone with four carbon atoms

6. Write a simplified structure for

a. an aromatic compound containing six carbon atoms
b. a cyclic compound having a four-membered ring, a carboxyl group, and six carbon atoms in all

7. Identify the professional areas of three chemists teaching in the chemistry department at your university or college. Use these categories: analytical chemistry, biochemistry, inorganic chemistry, organic chemistry, physical chemistry.

8. Certain hair conditioners contain di-*n*-butyl ether. Draw the structure of this compound.

9. In each case draw complete Lewis structures and Kekulé structures for two isomers with the given molecular formula.

a. C_5H_{12} **c.** C_2H_7N **e.** $C_8H_8O_2$
b. C_5H_{10} **d.** C_3H_8O

10. Write a balanced chemical reaction for the complete combustion of the following:

a. $CH_3CH_2CH_2CH_2CH_3$

b. $CH_3C(CH_3)_2CH_2CH(CH_3)CH_3$ (drawn as: $CH_3CCH_2CHCH_3$ with CH_3, CH_3 above and CH_3 below)

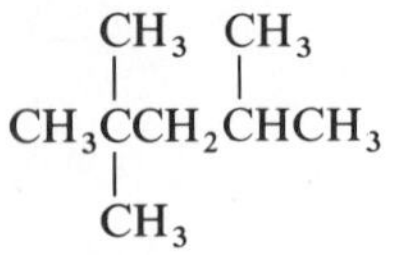

c. CO

d. (benzene ring with CH_3)

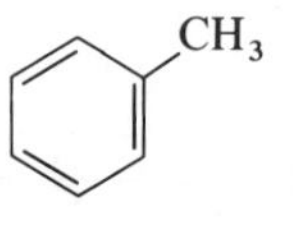

e. (cyclopentene ring)

f. △

g. graphite

11. Methanol may be an important fuel in the future. Write a balanced chemical reaction for the complete combustion of methanol.

12. Locate the functional groups in each molecule, and classify the molecule in each case.

a. CH_3CHCH_2OH with OH on the second carbon

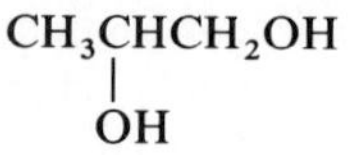

b. HO–(benzene ring)–CH=O, with CH_3O on the ring

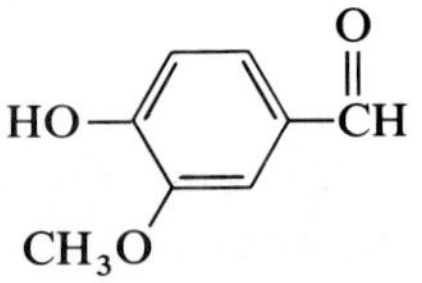

vanillin, the active ingredient of artificial vanilla flavoring

c. (benzene ring)–C(=O)N$(CH_2CH_3)_2$, with CH_3 on the ring

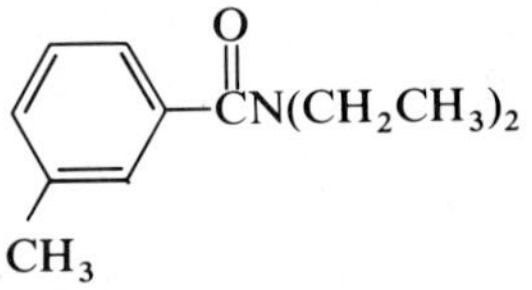

a common mosquito repellent

d. (cyclohexene ring with H_3C, CH_3, CH_3; conjugated chain with CH_3, CH_3, ending in –CH=O)

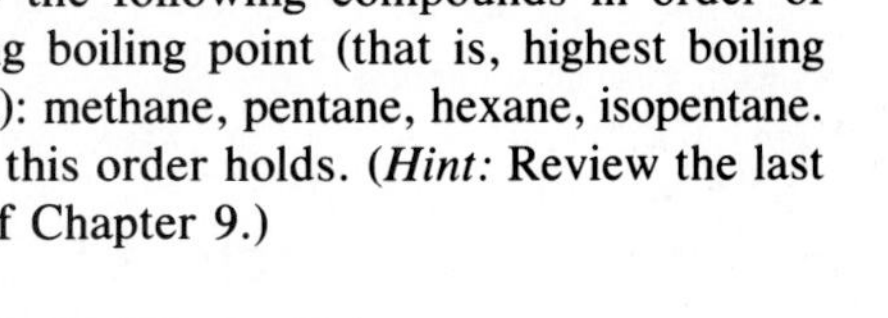

retinal, Vitamin A

13. Place the following compounds in order of decreasing boiling point (that is, highest boiling point first): methane, pentane, hexane, isopentane. Tell why this order holds. (*Hint:* Review the last section of Chapter 9.)

14. a. Oil of wintergreen is

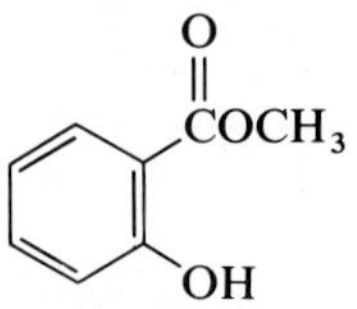

(benzene ring with $C(=O)OCH_3$ and OH)

It is synthesized from salicylic acid,

O
‖
COH

OH

Which alcohol would you use in the synthesis?

b. Oil of jasmine is

O
‖
CH_2OCCH_3

Show the structure of the carboxylic acid and the alcohol that one would use in its synthesis.

15. Isobutane and propane are used as aerosol propellants in shaving-cream containers. What advantage do these compounds have over hexane? Over methane?

16. The formula of compound X is C_3H_6O. In the presence of finely divided metals, compound X reacts with H_2 to form a compound whose formula is C_3H_8O. Compound X also reacts with sodium metal to form hydrogen gas. Write a complete Lewis structure for compound X that is consistent with this evidence.

Optional Section

17. Write a balanced chemical reaction for each of the following cases. Show the structures of the products.

a. addition of H_2 to

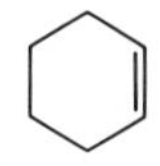

b. addition of water to

c. elimination of HBr from $CH_3CHBrCH_3$

The Chemistry of Acids and Bases

Grapefruit, lemons, and pickles taste sour, whereas soap tastes bitter and has a slippery feeling. The chemicals that give fruit and pickles their sour taste are acids. Bases produce the bitter taste of soap. A hundred years ago chemists routinely tasted the chemicals that they isolated from nature. Now we know that many chemicals, including acids and bases, can damage the body. Tasting them is a dangerous way to identify them; far better ways are available. We can safely identify acids and bases by observing their chemical reactions.

An **acid** changes the color of certain dyes through chemical reactions (litmus from blue to red, for example). Acids react with many metals, dissolving the metal and producing hydrogen gas, as well as with solid calcium carbonate (limestone or marble), dissolving the solid and forming bubbles of carbon dioxide gas. Citric acid, which was discussed in Chapter 11, is nutritious in small amounts. We ingest it whenever we eat oranges, lemons, or other citrus fruit. It has a familiar sour or biting taste. Vinegar, which contains acetic acid, has the same kind of sour taste. Uncured cucumber pickles are prepared by heating cucumbers with vinegar, salt, sugar, and spices. Sauerkraut and some pickles are made by fermentation in a salt solution. Lactic acid bacteria grow in this mixture and convert the food's sugars to lactic acid and acetic acid.

$$\underset{\text{acetic acid}}{CH_3{-}\overset{\overset{\displaystyle O}{\|}}{C}{-}OH} \qquad \underset{\text{lactic acid}}{CH_3{-}\overset{\overset{\displaystyle OH}{|}}{\underset{\underset{\displaystyle H}{|}}{C}}{-}\underset{\underset{\displaystyle O}{\|}}{C}{-}OH}$$

A **base** is also able to change the color of some dyes (litmus from red to blue, for example). Bases react with acids to cause the properties of acids to disappear. They initially were defined as those substances which react with acids to form salts; they were the "base" of the salt. When dissolved in water, many bases feel slippery to the touch, as they slowly dissolve layers of skin. Some familiar bases are baking soda, milk of magnesia, ammonia water, and lye. Some bases are good household cleaners, but they can also be dangerous, especially to small children who don't know how to handle them properly. Figure 12.1 shows some common household bases.

Figure 12.1. Bases useful in the home.

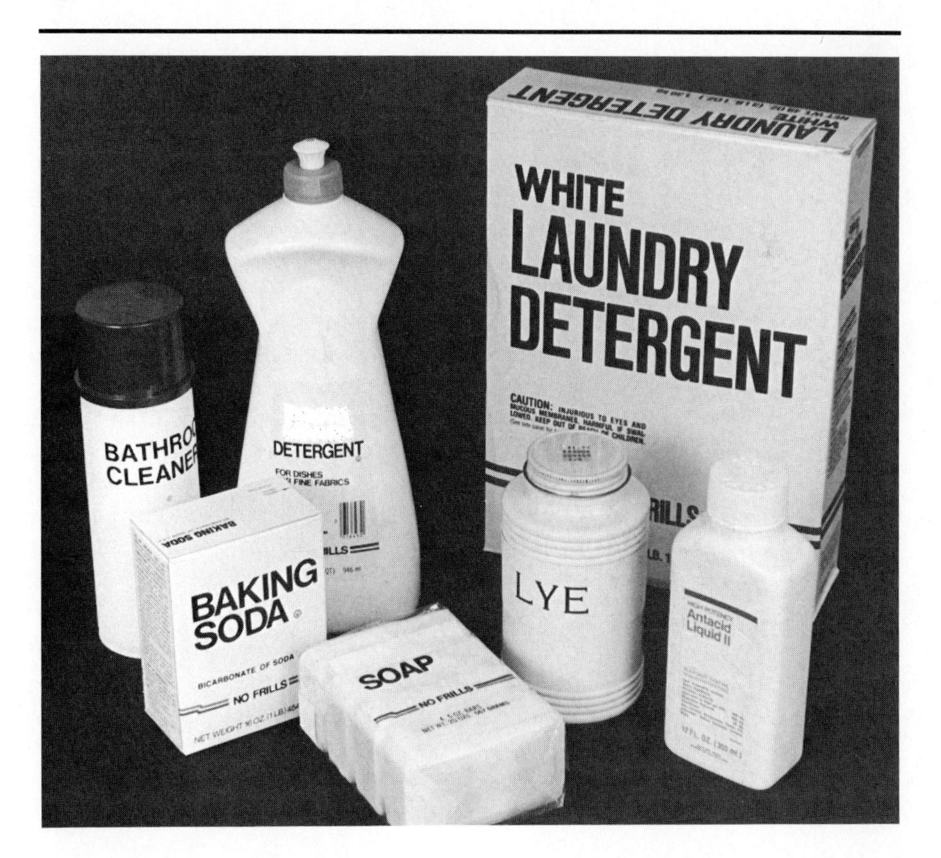

Acids and bases destroy, or neutralize, each other. Antacids, for example, contain bases such as aluminum hydroxide that react with excess stomach acid. Neutralization of the excess hydrochloric acid (H^+Cl^-) in the stomach lessens the discomfort of an upset stomach. You can destroy spilled acid by neutralizing it with a base such as baking soda, $NaHCO_3$, or lime, $Ca(OH)_2$.

Acids and bases are big business in the United States. Our annual production of sulfuric acid (40–45 million tons per year) is greater than that of any other chemical. Most of the H_2SO_4 is used to prepare phosphoric acid (H_3PO_4) by reacting it with rocks that contain large amounts of phosphate salts. The phosphoric acid is in turn transformed into phosphate fertilizers such as ammonium phosphate. All growing plants require phosphate as a nutrient. Thus, the fortunes of the sulfuric acid industry depend largely on the fortunes of U.S. farmers, who adjust their purchases of fertilizer up or down depending on anticipated market conditions for their products. Food is inexpensive and plentiful in the United States partly because the use of phosphate fertilizers helps to produce high crop yields.

12.1 ACID-BASE THEORY

Reactions between acids and bases were recognized early in the history of chemistry. They form an important class of chemical reactions, like the oxidation-reduction reactions discussed in Chapter 6. After considerable disagreement among chemists over what element is common to all acids, an English scientist, Humphrey Davy (see Figure 12.2), showed in 1810 that all acids contain hydrogen.

According to one modern definition, an acid is any compound that dissociates in water solution to give a hydrogen ion, H^+, and, as a by-product, a negative ion that is different for each acid. The hydrogen ion, which actually is closely bound to one or more water molecules in solution, is responsible for the peculiar characteristics of an acid. In the pure state most acids—such as hydrogen chloride, HCl, and sulfuric acid, H_2SO_4—are covalent compounds consisting of neutral molecules. In water solution, however, the bond between hydrogen and its neighboring atom breaks, leaving both electrons with the neighboring atom. The hydrogen, now having no electron (compared with one electron in the neutral atom), has a positive charge, and the remaining fragment has a negative charge. HCl and H_2SO_4 dissociate to form ionic solutions when they are dissolved in water:

$$\text{H—Cl}(g) \longrightarrow \text{H}^+(aq) + \text{Cl}^-(aq)$$

or

$$\text{H—Cl}(g) + \text{H}_2\text{O} \longrightarrow \text{H}_3\text{O}^+(aq) + \text{Cl}^-(aq)$$

$$\text{H—O—}\underset{\text{O}}{\overset{\text{O}}{\overset{|}{\underset{|}{\text{S}}}}}\text{—O—H}(l) \longrightarrow \text{H}^+(aq) + \text{H—O—}\underset{\text{O}}{\overset{\text{O}}{\overset{|}{\underset{|}{\text{S}}}}}\text{—O}^-(aq)$$

Figure 12.2. Humphrey Davy.

$$\text{or}\quad \mathrm{H{-}O{-}\underset{\underset{\displaystyle O}{|}}{\overset{\overset{\displaystyle O}{|}}{S}}{-}O{-}H}(l) + \mathrm{H_2O} \longrightarrow \mathrm{H_3O^+}(aq) + \mathrm{H{-}O{-}\underset{\underset{\displaystyle O}{|}}{\overset{\overset{\displaystyle O}{|}}{S}}{-}O^-}(aq)$$

The symbol (*aq*) stands for **aqueous,** which refers to water solution. As usual, (*g*) stands for a gas and (*l*) for a liquid. Chapter 10 discussed the forces of hydration of ions in water solution. The ion H_3O^+, formed by the combination of a hydrogen ion and a water molecule, is called the **hydronium ion.** It has an oxygen atom connected to three hydrogen nuclei by covalent bonds. For the sake of simplicity, we will use H^+ as the acidic component in water solutions, even though in reality all hydrogen ions are hydrated. Notice that in H_2SO_4 the hydrogen atoms are bonded to oxygen atoms, not to the sulfur, contrary to the implication of the formula as normally written.

Experimental evidence for the dissociation into ions of a covalent HCl molecule is much the same as the evidence for ionic solutions of salts. As a gas, HCl does not conduct electricity very well. However, hydrogen chloride dissolved in water, or hydrochloric acid, is a good electrical conductor.

The simplest definition of a base is a substance that produces hydroxide ion, OH^-, in water solution. Common examples are NaOH, KOH, and $Ca(OH)_2$, all of which are ionic compounds and solids. As they dissolve in water, these compounds dissociate into a metallic ion (Na^+, K^+, or Ca^{2+}, respectively) and OH^-. In this respect they behave like most ionic compounds. Their ability to change the color of red litmus to blue and to cause a slippery feeling on skin is given to them by the OH^- ion. Basic substances are often called **alkaline,** from the old word *alkali*, used for centuries to signify a base.

When solutions of an acid and a base are mixed in just the right proportions, the resulting solution does not have the slippery feeling of the base, the sour taste of the acid, or the ability to alter the colors of dyes such as litmus. In addition, heat is produced, as indicated by a temperature rise of several degrees; reactions of acids with bases are exothermic. The acid and base have destroyed each other, or, as chemists like to say, have neutralized each other. The reaction in every case is the same: the H^+ of the acid combines with the OH^- of the base to give H_2O. For example, as shown in Figure 12.3, aqueous hydrochloric acid and aqueous potassium hydroxide react according to the equation

$$\underbrace{H^+(aq) + Cl^-(aq)}_{\text{hydrochloric acid solution}} + \underbrace{K^+(aq) + OH^-(aq)}_{\text{potassium hydroxide solution}} \longrightarrow \underbrace{H_2O}_{\text{water}} + \underbrace{K^+(aq) + Cl^-(aq)}_{\text{potassium chloride solution}}$$

An acid that is completely ionized in water solution, such as hydrochloric acid, is called a **strong acid.** Any hydroxide that is soluble in water is a **strong base,** because it is also completely ionized in solution. Some examples of strong acids and strong bases are given in Table 12.1.

During the development of the theory of acids and bases, chemists were fond of saying that an acid plus a base gives a salt plus water. The equation showing the reaction between hydrochloric acid and potassium hydroxide solution includes the ions of a salt, $K^+(aq)$ and $Cl^-(aq)$, as products. These ions, however, were also present in the original solutions of acid and base. They can be regarded as a by-product of the reaction. Water, on the other hand, forms as the result of a new covalent bond between H^+ and OH^-. Water is the essential product of the acid-base reaction.

Figure 12.3. *Reaction between an acid and a base.*

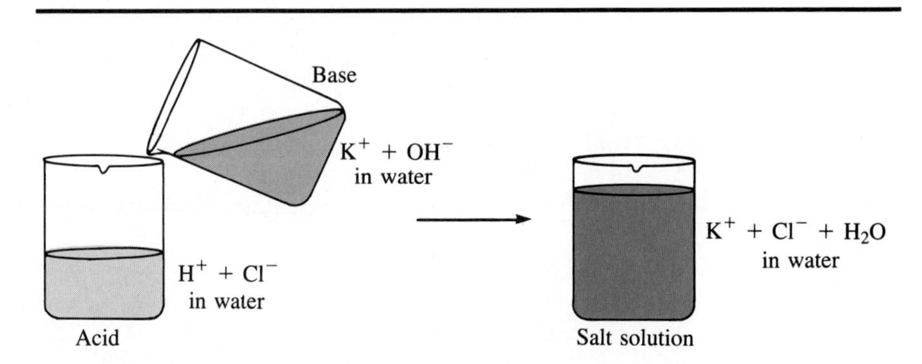

Table 12.1 Examples of Strong Acids and Strong Bases

NAME	FORMULA	IONS IN SOLUTION
Strong Acids		
hydrochloric acid	HCl	$H^+ + Cl^-$
hydrobromic acid	HBr	$H^+ + Br^-$
hydriodic acid	HI	$H^+ + I^-$
sulfuric acid	H_2SO_4	$H^+ + HSO_4^-$
nitric acid	HNO_3	$H^+ + NO_3^-$
Strong Bases		
sodium hydroxide	$NaOH$	$Na^+ + OH^-$
potassium hydroxide	KOH	$K^+ + OH^-$
calcium hydroxide	$Ca(OH)_2$	$Ca^{2+} + 2\ OH^-$
barium hydroxide	$Ba(OH)_2$	$Ba^{2+} + 2\ OH^-$

The equation for a strong acid–strong base reaction can be written by means of a simple procedure. First, with the help of Table 12.1, write on the left side the ions present in the solutions of acid and base. Second, write H_2O plus the ions of the by-product salt on the right side. Finally, balance the equation by placing numbers in front of certain ions or molecules as necessary.

Example Write the chemical equation for the neutralization of hydrobromic acid, $HBr(aq)$, by sodium hydroxide solution, $NaOH(aq)$.

Solution According to Table 12.1, H^+ and Br^- are provided by the acid, and Na^+ and OH^- are present in the base solution. As always, H_2O is formed, and the by-product salt ions, Na^+ and Br^- in this case, remain in the final solution. The reaction is written:

$$H^+(aq) + Br^-(aq) + Na^+(aq) + OH^-(aq) \longrightarrow H_2O + Na^+(aq) + Br^-(aq)$$

From now on we will omit all the (aq) symbols for brevity:

$$H^+ + Br^- + Na^+ + OH^- \longrightarrow H_2O + Na^+ + Br^-$$

The equation is already balanced.

Example Write the equation for the reaction between nitric acid, $HNO_3(aq)$, and calcium hydroxide solution, $Ca(OH)_2(aq)$.

Solution Again, the ions are found in Table 12.1. If there is enough nitric acid to react with both of the available hydroxide ions from $Ca(OH)_2$, the unbalanced equation is

$$H^+ + NO_3^- + Ca^{2+} + 2\,OH^- \longrightarrow H_2O + Ca^{2+} + NO_3^- \qquad \text{Incorrect}$$

The balanced equation is

$$2\,H^+ + 2\,NO_3^- + Ca^{2+} + 2\,OH^- \longrightarrow 2\,H_2O + Ca^{2+} + 2\,NO_3^- \qquad \text{Correct}$$

To balance the equation, we multiplied both H^+ and NO_3^- by 2 because both were in the same acid solution (nitric acid). In the balanced reaction, the overall charge on each side of the equation is zero. All solutions must be electrically neutral, so all equations must have an overall charge of zero on each side.

12.2 CHEMICAL EQUILIBRIUM

Not all acids and bases are totally ionized in aqueous solution. For example, carbonic acid, H_2CO_3, which is present in all carbonated beverages and also in our bloodstream, is only partially ionized. Many of the carbonic acid molecules are not ionized in water solution; only a small fraction dissociate into H^+ and HCO_3^- ions. Organic carboxylic acids also dissociate to only a small extent in water solution. Carbonic acid and the many carboxylic acids are "weaker" acids than hydrochloric acid or sulfuric acid. In order to make sense of the differences between solutions of strong acids and those of weaker acids, we must develop a fuller understanding of the concept of chemical equilibrium. Then we will return to the discussion of the difference between strong and weak acids.

To most people, **equilibrium** implies a balanced or stable state of affairs. When a rowboat or canoe is moving along in equilibrium, it is best if no one disturbs the equilibrium by standing up and rocking the boat. An equilibrium can also be reached by achieving a balance between opposing forces. Two sides battling against each other in a game of tug of war (see Figure 12.4) are in a state of equilibrium as long as there is a perfect standoff, with no shift in either direction. If either side pulls the other into the lake, equilibrium no longer exists, because the pull on the rope is greater in one direction.

To chemists, equilibrium means all this and more. The notion of there being unchanging amounts of liquid and vapor when the two are at equilibrium in a closed container was discussed earlier (Chapter 4). At the molecular level, however, we saw that there is lots of activity. A certain number of molecules escape from the surface of the liquid and become vapor. At the same time an equal number of vapor molecules strike the surface and stay there to become part of the liquid. Because the number of molecules evaporating equals the number condensing, the amounts of liquid and vapor remain constant. In effect, evaporation and condensation cancel each other. However, the action is dramatic at a molecular level (see Figure 12.5).

Figure 12.4. A tug of war that has reached a state of equilibrium.

In a **chemical equilibrium** the reactants and products of a chemical reaction are at a point where there is no overall change. Let's look at the decomposition of HI into its elements, hydrogen and iodine, as given in this balanced chemical equation:

$$2\,HI(g) \rightleftharpoons H_2(g) + I_2(g)$$

Hydrogen iodide, HI, is a colorless gas that is stable at room temperature but breaks down at high temperatures. It is easy to tell that HI decomposes. A flask that initially contained HI at about 400°C soon becomes violet, as I_2 (which has an attractive violet color) forms. As shown in Figure 12.6, the flask turns darker and darker until finally the color remains constant. From the fact that the change in color has stopped we can tell that the amount of iodine in the flask has become constant. According to other evidence, plenty of HI remains, and its concentration also remains constant.

At chemical equilibrium, the reaction has reached a balanced, stable state that does not change over time. In this example, even in the early stages of the chemical reaction, some molecules of H_2 and I_2 combine to form HI. However, HI molecules are breaking up more rapidly than they are forming, so the direction of the overall reaction is from left to right. Finally, at equilibrium, the speeds of reaction in the two directions are equal and the amounts of all three substances stay the same. The rapid reactions in both directions are shown by arrows in both directions.

A chemical reaction can be useful only if a reasonable amount of product is formed. In other words, conditions must be such that the reaction reaches

Figure 12.5. Molecular action.

Figure 12.6. The changing appearance of a flask containing HI, H_2, *and* I_2 *as the reaction* $2\ HI(g) \rightleftharpoons H_2(g) + I_2(g)$ *approaches equilibrium.*

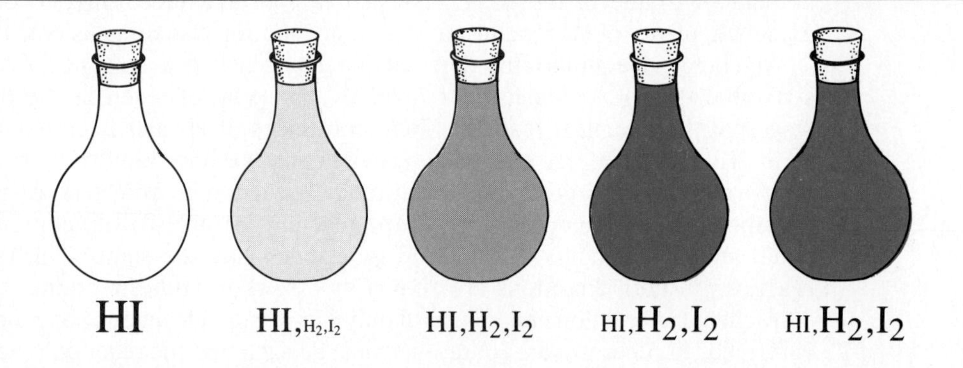

OXYGEN LEVELS IN RIVERS AND LAKES

Equilibria are found everywhere in the world around us. For example, every body of water contains dissolved oxygen, which comes from the air in contact with the water. To a good approximation, the oxygen in the air and the oxygen in a flowing river are in equilibrium with each other. The process can be written

$$O_2(g) \rightleftharpoons O_2(aq)$$

Suppose that after equilibrium has been established the temperature of the water increases slightly. This change disturbs the equilibrium and causes a decrease in the concentration of dissolved oxygen. In other words, the equilibrium position is shifted toward oxygen gas, which is less soluble in water at the higher temperature.

This fact has important implications for the kinds of fish that can live in rivers and lakes. Desirable game fish such as trout and bass require water at a temperature low enough so that the concentration of dissolved oxygen is relatively large. For example, brown trout flourish (see Figure 12.7) when the water is between 55° and 65°F but die when it reaches about 78°F. At the higher temperature their activity increases at the same time that there is less oxygen to support them. Fish such as carp that do not require as much oxygen can survive at higher temperatures.

Figure 12.7. Cold, nonpolluted waters favor game fish.

equilibrium when the amount of products is greater than the amount of reactants. No reaction can proceed past the point of equilibrium. A statement about the **position of equilibrium** indicates whether an equilibrium lies to the left or to the right. At temperatures around 25°C, scarcely any HI decomposes into H_2 and I_2, no matter how long we wait. The position of equilibrium in this circumstance lies far to the left. For another reaction, the combining of carbon and oxygen to give carbon dioxide, the opposite is true. The equilibrium position is far to the right; the amount of reactants remaining is not even measurable.

The position of equilibrium, on the side of reactants or on the side of products, depends on the driving force for the reaction. In other words, it is a matter of energy and molecular disorder, as we saw in Chapter 4. There are two kinds of driving forces for chemical changes.

1. Changes tend to occur in the direction of lower potential energy with the evolution of heat energy. In other words, exothermic processes are favored.
2. Changes tend to occur in the direction of greater molecular disorder.

Whether or not the position of a chemical equilibrium favors formation of the products depends for the most part on the relative energies of the reactants and the products. This, in turn, depends on the reactants' and products' fundamental chemical structures or, in other words, on the stabilities of their chemical bonds. Many reactions proceed almost completely to products simply because the products are so much more stable than the reactants. Other reactions proceed because the molecular disorder is greater in the products than in the reactants. In either case the driving force, and therefore the position of equilibrium, depends on experimental conditions, such as concentration and temperature.

12.3 DISTURBING A CHEMICAL EQUILIBRIUM

Many chemical reactions for which we find a large concentration of products at equilibrium are exothermic; they give off heat energy. Thus the position of equilibrium depends on the temperature at which the reactions occur. For instance, at room temperature the formation of water from hydrogen and oxygen gases is virtually complete.

$$2\ H_2(g) + O_2(g) \longrightarrow 2\ H_2O(g)$$

However, very little water forms at 3700°C. In fact, water vapor at this temperature would decompose nearly completely into hydrogen and oxygen. When one is talking about the position of equilibrium for a reaction, it is essential to specify the temperature.

The position of a chemical equilibrium can also be changed by adding or removing a reactant or product. If extra reactant is added to a mixture of reactants and products at equilibrium, some of the added reactant combines with the other reactants to give more products. The reaction shifts to the right until a new condition of equilibrium is established. If there is some way to remove product, thereby decreasing its concentration, the reaction also shifts to the right. Finally, removing reactant causes the reaction to go to the left. All these effects can be summarized in a single statement:

> When a reaction is at equilibrium, adding or removing a reactant or product causes the reaction to go in the direction that will partly undo the change. The process stops when a new state of equilibrium is reached.

If a substance involved in the equilibrium is added, some of it is consumed by subsequent reaction, and the increase is partly undone. If a substance is removed, more of it is formed, and the decrease is partly undone. This generalization applies to all chemical reactions.

Example One of the several reactions that occurs in exhaust systems of automobiles is

$$2\,NO(g) + O_2(g) \rightleftharpoons 2\,NO_2(g)$$

If this reaction is at equilibrium, what is the effect of each of the following?

(a) adding O_2

(b) removing NO

(c) adding NO_2

Solution (a) When O_2 is added, the reaction is pushed in the right-hand direction, so some of the added O_2 is consumed.

(b) Removing NO causes the reaction to shift to the left. Additional NO is produced, so the original change is partly undone.

(c) Adding NO_2 (a product) pushes the reaction leftward and part of the added NO_2 is used up.

12.4 WEAK ACIDS

Now that we have considered the fundamental ideas of chemical equilibrium, we can return to acids and bases. We saw earlier that some acids, such as hydrochloric acid and sulfuric acid, are completely dissociated in water solution. Other weaker acids, such as carbonic acid, dissociate to only a small extent in water. The dissociation of carbonic acid is given by the following equation:

$$H_2CO_3 \rightleftharpoons H^+ + HCO_3^-$$

In the equilibrium between the dissociation of H_2CO_3 and the recombination of H^+ and HCO_3^-, recombination of the two ions is more likely. Thus, the equilibrium position lies on the side of neutral H_2CO_3 molecules. The bicarbonate anion has a strong attraction for a proton. Even though rapid dissociation-association reactions occur constantly, the equilibrium position is on the reactant side, far to the left. An acid such as carbonic acid is classified as a **weak acid.** The equilibrium position for the ionization of weak acids contrasts with that for strong acids. In fact, the concentration of neutral molecules in $HCl(aq)$, $HNO_3(aq)$, and other strong acids is too small to be measured. The position of these equilibria is almost completely on the side of dissociated ions.

Several weak acids are encountered in nature and in commerce. Carbonic acid, a component of carbonated drinks and blood, is also present in lake, river, and ocean water. Acetic acid is a constituent of vinegar, and phosphoric acid is used in the manufacture of phosphate fertilizers. Some important weak acids are given in Table 12.2.

The last entry in this table illustrates that some acids with more than

Table 12.2 Some Important Weak Acids

NAME	IONIZATION EQUILIBRIUM
carbonic acid	$H_2CO_3 \rightleftharpoons H^+ + HCO_3^-$
acetic acid	$CH_3COOH \rightleftharpoons H^+ + CH_3COO^-$
sulfurous acid	$H_2SO_3 \rightleftharpoons H^+ + HSO_3^-$
phosphoric acid	$H_3PO_4 \rightleftharpoons H^+ + H_2PO_4^-$
dihydrogen phosphate ion	$H_2PO_4^- \rightleftharpoons H^+ + HPO_4^{2-}$

one hydrogen atom can ionize in successive stages. Phosphoric acid, H_3PO_4, loses one H^+ to become $H_2PO_4^-$, which can then undergo another ionization to form HPO_4^{2-} and yet another ionization to form PO_4^{3-}. Each successive ionization has less tendency to occur than the preceding one. In fact, the dissociation of H_3PO_4 is over 100,000 times more favorable than that of $H_2PO_4^-$. Table 12.2 includes only one example of multiple ionization, because multiple ionizations are not significant in the specific applications we will discuss.

In summary, we can say

1. A strong acid has more ions than molecules in its water solution.
2. A weak acid has more molecules than ions in its water solution.

Example Ammonia gas (NH_3) dissolves in water to become the weak base ammonium hydroxide.

$$NH_3 + H_2O \rightleftharpoons NH_4^+ + OH^-$$

Is the position of this equilibrium largely on the right or on the left?

Solution Like weak acids, weak bases have more molecules than ions in water solution. A solution of ammonia in water produces few hydroxide ions. The equilibrium position is largely on the left.

When a weak acid reacts with a strong, completely ionized base such as aqueous sodium hydroxide, most of the hydrogen ions needed for the reaction are not present in the solution. They are covalently bound within molecules of the weak acid. The chemical equation must be written with the nonionized form of the weak acid on the left side. Part of the reaction involves a pulling of the H^+ from the weak acid. The reaction between acetic acid and aqueous calcium hydroxide, for example, is written

$$2CH_3\overset{\overset{\displaystyle O}{\|}}{C}OH + Ca^{2+} + 2OH^- \longrightarrow 2H_2O + Ca^{2+} + 2CH_3\overset{\overset{\displaystyle O}{\|}}{C}O^-$$

Acetic acid is represented as

$$\overset{\displaystyle O}{\overset{\|}{CH_3COH}} \quad \text{rather than } H^+ + \overset{\displaystyle O}{\overset{\|}{CH_3CO^-}}$$

because a very large fraction of the acid is in molecular form. Structural formulas of chemicals show the way they actually exist. When the chemicals are ionized, they are written as ions. When they exist primarily as covalent molecules, they must be written as molecules (see Figure 12.8).

The Kekulé structure of acetic acid is

$$\begin{array}{ccccccc} & & H & & O & & \\ & & | & & \| & & \\ H & - & C & - & C & - & O-H \\ & & | & & & & \\ & & H & & & & \end{array}$$

You may have noticed already that the only proton taken away from the acetic acid molecule in the acid-base reaction is the proton bound to an oxygen atom. The protons attached to carbon are much less acidic; they are never involved in acid-base reactions that take place in water solution. The acidic protons of weak acids are almost always attached to an oxygen or a nitrogen atom. Every weak acid found in Table 12.2 has the acidic hydrogen bound to an oxygen atom.

Example Write the equation for the reaction that occurs when solutions of phosphoric acid and sodium hydroxide are mixed. Assume that only one hydrogen ion from each molecule of phosphoric acid reacts.

Solution

$$\underset{\text{phosphoric acid}}{H_3PO_4} + \underset{\text{sodium hydroxide}}{Na^+ + OH^-} \longrightarrow \underset{\text{water}}{H_2O} + \underset{\text{sodium dihydrogen phosphate}}{Na^+ + H_2PO_4^-}$$

Figure 12.8. Reaction between a weak acid and a strong base.

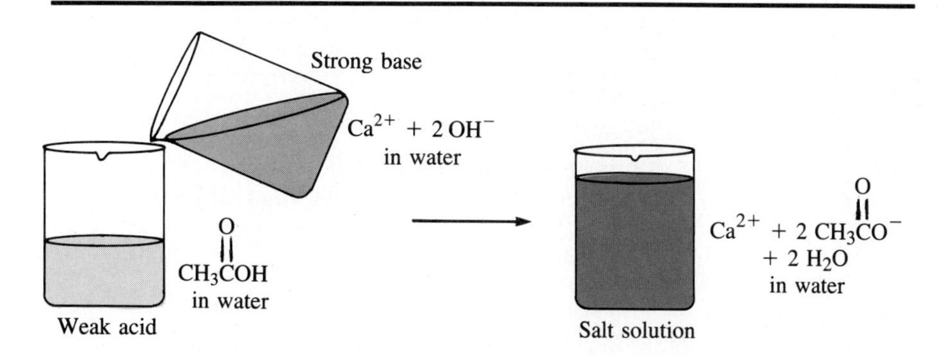

Because it is a weak acid (see Table 12.2), phosphoric acid is shown on the left side of the equation as a covalent molecule that is not ionized. As usual, aqueous sodium hydroxide is ionized, $Na^+ + OH^-$. Water is a product, along with a solution of an ionized salt as a by-product.

12.5 ACIDIC AND BASIC SOLUTIONS—HOW ACID IS ACID?

Even in the purest water there are very low concentrations of H^+ and OH^-, because H_2O itself ionizes ever so slightly:

$$H_2O \rightleftharpoons H^+ + OH^-$$

This equation tells us that H^+ ions and OH^- ions are formed in equal numbers. Therefore, their concentrations are equal. Careful measurements with pure water have shown that the concentrations of H^+ and OH^- are each equal to 1×10^{-7} mole in 1 liter of water. This is a tiny, tiny number. The equilibrium does indeed lie very far to the left.

Any water solution containing H^+ and OH^- in concentrations of 1×10^{-7} mole per liter is said to be a **neutral solution.** We have seen that an acid ionizes in water to give H^+ and a negative ion. Therefore, when an acid is dissolved in water, the concentration of hydrogen ion increases to some value greater than 1×10^{-7} mole per liter. This leads to the definition of an **acidic solution** as one in which the concentration of hydrogen ion is greater than 1×10^{-7} mole per liter. Similarly, a **basic,** or **alkaline, solution** is defined as one in which the concentration of hydroxide ion is greater than 1×10^{-7} mole per liter.

Example The H^+ concentration of a particular water solution is 1×10^{-3} mole per liter (this is the H^+ concentration in some soft drinks). Is this solution acidic, or is it basic?

Solution 1×10^{-3} is larger than 1×10^{-7}. Remember that a small magnitude for a negative exponent means a relatively large value for the entire number. Therefore, the solution has a greater concentration of H^+ than does a neutral solution. It is acidic.

Let's look at the concentration of OH^- in an acidic solution. The extra H^+ in this acidic solution causes the equilibrium

$$H_2O \rightleftharpoons H^+ + OH^-$$

to shift to the left. The reaction of some of the extra H^+ with OH^- results in a decrease in the concentration of OH^- to a value below 1×10^{-7} mole per liter. In brief, an acidic solution has a hydrogen ion concentration greater than 1×10^{-7} mole per liter and a hydroxide ion concentration less than 1×10^{-7} mole per liter. Similarly, in a basic solution some of the excess hydroxide ion reacts with a portion of the tiny amount of hydrogen ion present to make the hydrogen ion concentration less than 1×10^{-7}. The concentrations of H^+ and OH^- in acidic and basic solutions are summarized in Figure 12.9.

Chemists often specify the acidity or basicity of a solution by giving a number on the pH scale. The **pH** of a solution is just the exponent, with the sign changed, in the number representing the hydrogen ion concentration. For instance, the pH of a solution with a hydrogen ion concentration of 10^{-2} mole per liter is 2. A solution with an H^+ concentration of 10^{-11} has a pH of 11. A solution with an H^+ concentration of $10^{-5.2}$ has a pH of 5.2. From what we have said, it follows that

> An acidic solution has a pH of less than 7, and a basic solution has a pH of more than 7.

Figure 12.10 gives the pH scale and also shows the pHs of some common solutions.

The pH scale is convenient for expressing the acidity of water solutions because it uses simple numbers to give a quantitative measure of acidity. Just like the Richter scale for describing the magnitude of an earthquake, the pH scale has a logarithmic basis. This means that each change of one unit on the scale signifies a tenfold change in the acidity of a solution. So a soft drink with a pH of 3.5 has ten times the acid concentration of tomato juice, which has a pH of 4.5. Lemon juice with a pH of 2.4 is 10^5, or 100,000, times more acidic than human blood, which has a pH of about 7.4.

The fairly acidic pH of lemon juice indicated in Figure 12.10 is caused by citric acid. Vinegar and beer contain acetic and carbonic acids, respectively.

Figure 12.9. The concentrations of H^+ *and* OH^- *in acidic and basic solutions.*

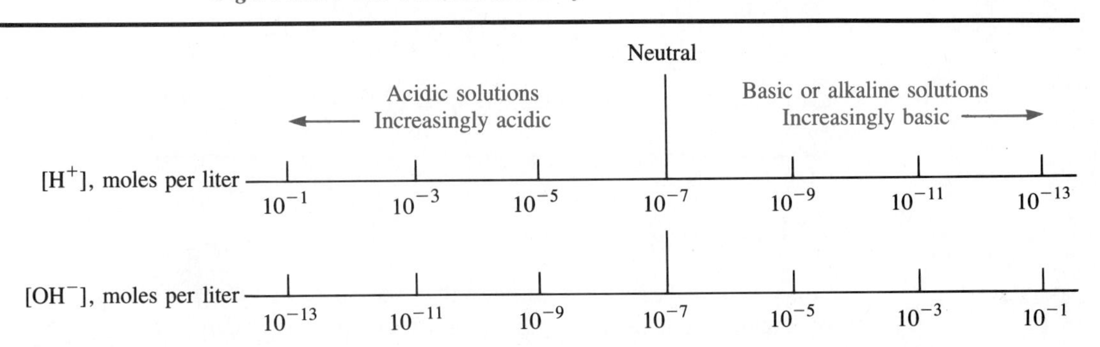

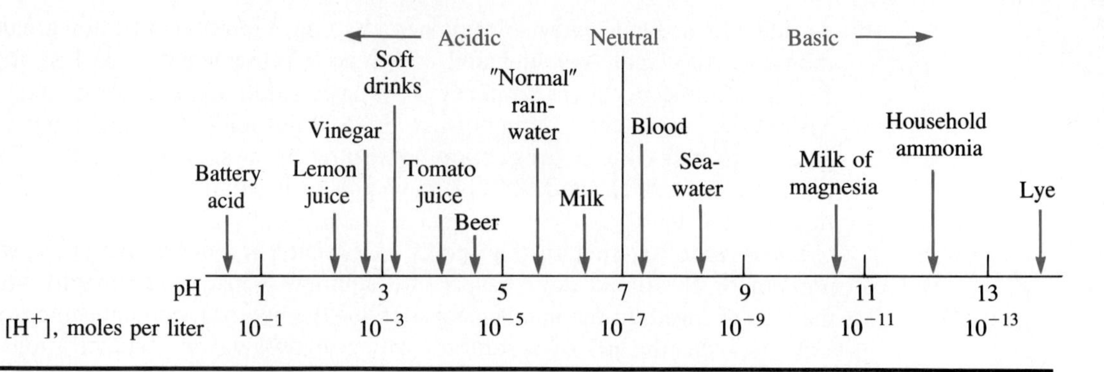

Figure 12.10. The pH scale and the pHs of some familiar solutions.

Seawater owes its slightly basic character to bicarbonate ion, HCO_3^-, which reacts to some extent with water to give OH^- ion:

$$Na^+ + HCO_3^- + H_2O \rightleftharpoons Na^+ + H_2CO_3 + OH^-$$

Example

(a) The H^+ concentration in Lake Ontario is about 1×10^{-8} mole per liter. What is the pH of Lake Ontario?

(b) The pH of tomato juice is about 4. What is the H^+ concentration of tomato juice?

(c) Which is more acidic—Lake Ontario water or tomato juice? By how much?

Solution

(a) An H^+ concentration of 1×10^{-8} mole per liter corresponds to a pH of 8. This means that Lake Ontario is slightly alkaline.

(b) A pH of 4 means that tomato juice has an H^+ concentration of 1×10^{-4} mole per liter. It is mildly acidic.

(c) Tomato juice is more acidic. In order to see how much more acidic, we have to see by what factor 1×10^{-8} mole per liter must be multiplied in order to get the greater acid concentration of 1×10^{-4} mole per liter.

$$(1 \times 10^{-8}) \cdot (\text{factor}) = 1 \times 10^{-4}$$

$$\text{factor} = \frac{1 \times 10^{-4}}{1 \times 10^{-8}} = 1 \times 10^4$$

Tomato juice is more acidic than the water of Lake Ontario by a factor of 10^4, or 10,000.

* 12.6 ACID RAIN (*optional*)

Because of the increasing number of lakes in upper New York State and New England that no longer have any fish, the American public is hearing a good deal about the problem of acid rain. The problem is even more severe in eastern Canada and is a source of international friction, because most investigators believe that polluted air from the United States crosses the border, adding significantly to the acid rain falling on Canada. Scientific understanding of the process responsible for acid rain is still limited. This is perhaps not surprising when one considers the very complicated natural system involved.

Curiously, the term **acid rain** does not simply mean rain with a pH of less than 7. As shown in Figure 12.10, normal rainwater is somewhat acidic. It has a pH of about 5.6, which is well within the acid range. The cause of this acidity is dissolved carbon dioxide, which comes from the small amount of carbon dioxide naturally present in air. Carbon dioxide in water forms carbonic acid, which, as we mentioned earlier, is a weak acid. The sequence is as follows:

$$CO_2(g) + H_2O(l) \rightleftharpoons H_2CO_3(aq)$$

$$H_2CO_3(aq) \rightleftharpoons H^+(aq) + HCO_3^-(aq)$$

The acidity of normal rainwater is relatively small, though, because H_2CO_3 is present in low concentration and it is a weak acid. Fish have adapted to this level of acidity and flourish in lakes with a pH of 5.6 or slightly greater.

Data collected over the past twenty-five years show that in certain areas of North America the rain and snow is, on the average, considerably more acidic than pH 5.6. The lowest pHs are found in the eastern Great Lakes region, but there are also problem areas in the west and southwest (see Figure 12.11). A correlation between high acidity in lakes and reduced fish populations has been well established, both in Scandinavia and in North America. Game fish such as bass, trout, and perch are severely affected by a pH of 5 and are wiped out by a pH of 4.5.

There is little doubt that the increased acidity of rainfall is associated with human activity. Although historical data on the pH of rainwater are scanty, it appears that the pH has been decreasing dramatically since about 1930, when it was 5.6 or greater. Ice cores taken from considerable depths in the Arctic region are revealing, because they contain one layer for each year of precipitation. They show that acidity began to increase as early as 1800, near the start of the industrial revolution. *

* 12.7 SOURCES OF ACID RAIN (*optional*)

Two processes, both associated with combustion, seem to be responsible for acid rain. The first is the burning of sulfur along with carbon in sulfur-containing coal. The product is sulfur dioxide, SO_2. The second process consists of several reactions between N_2 and O_2, both present in the air

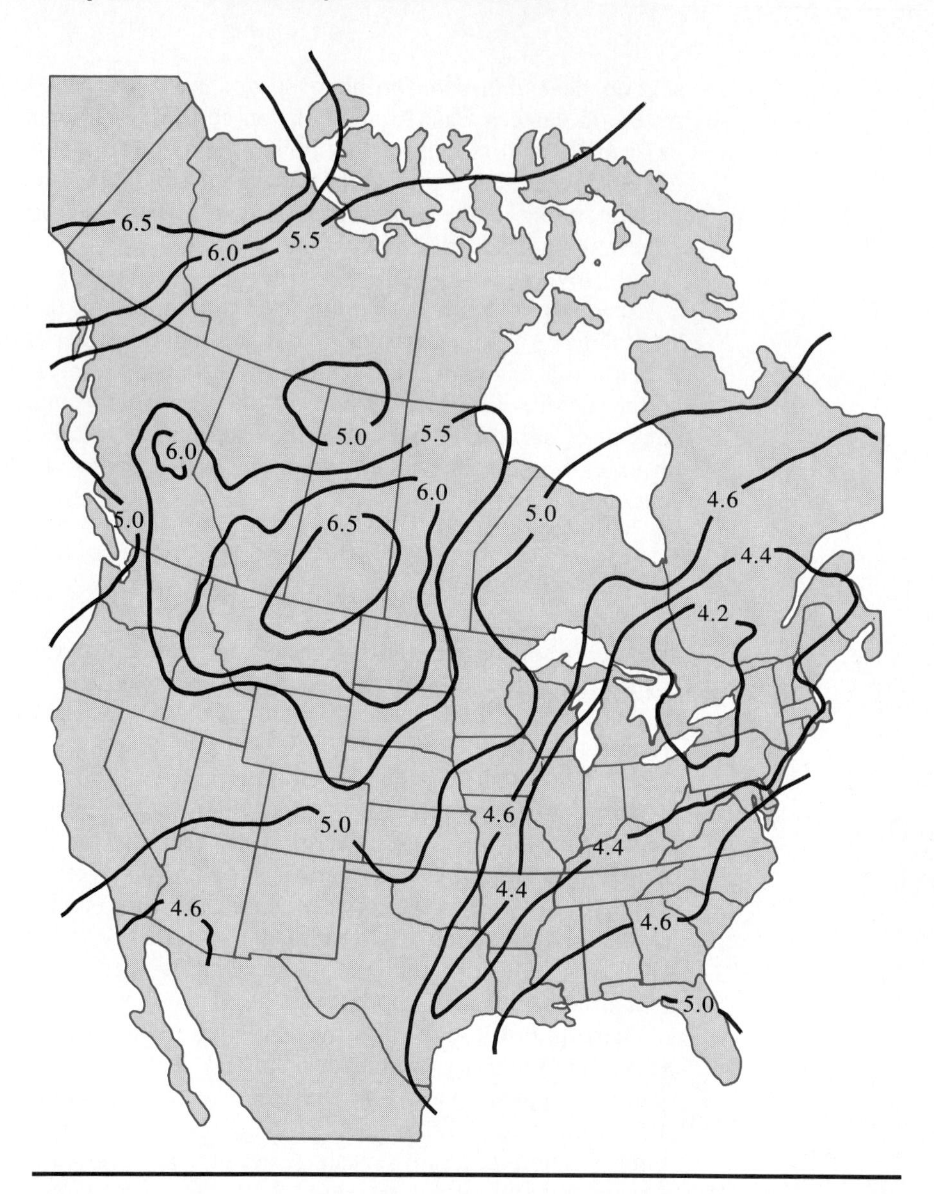

Figure 12.11. The average pH of annual precipitation in the United States (1979) and how it varies with location. (Source: National Atmospheric Deposition Program)

used for combustion. Oxides of nitrogen, mainly NO and NO_2, are formed at high combustion temperatures. Both the sulfur and the nitrogen oxide are converted into acids by subsequent reactions.

Let's look more closely at the SO_2 situation. Not all of the SO_2 in the air can be blamed on humans—there are some natural sources too, such as volcanoes. But most of the SO_2 does come from the burning of fossil fuels. In the United States over half of the SO_2 can be traced to electric utilities, and most of the remainder comes from other industrial sources. By far the greatest emission occurs in the Ohio River Valley—Illinois, Indiana, Ohio, Pennsylvania, West Virginia, and Kentucky—where high-sulfur coal is burned in power plants. Gas from the smokestacks is carried by the prevailing winds to the east, which is the target of the heaviest concentration of acid in the rain.

Some of the SO_2 reacts with droplets of water or water vapor in the air to give sulfurous acid, H_2SO_3, which is a weak acid.

$$SO_2(g) + H_2O(g) \longrightarrow H_2SO_3(aq)$$

Some of this H_2SO_3 ionizes until an equilibrium is established:

$$H_2SO_3(aq) \rightleftharpoons H^+(aq) + HSO_3^-(aq)$$

Because H_2SO_3 is a weak acid, only a low concentration of hydrogen ion is produced by this reaction. Nevertheless, any increase in SO_2 concentration increases the concentration of H_2SO_3 and therefore the H^+ concentration. Much of this acid is carried to the ground by rain and snow.

An even more important source of acidity, however, is the strong acid sulfuric acid, which is the result of oxidation of SO_2 while it is airborne. The oxidation of SO_2 to SO_3 by O_2 in air is shown by the reaction

$$2\,SO_2(g) + O_2(g) \longrightarrow 2\,SO_3(g)$$

The reaction of SO_3 with moisture in the air is given by

$$SO_3(g) + H_2O(g) \longrightarrow H_2SO_4(aq)$$

Figure 12.12. *A tall smokestack is used to avoid high pollution levels locally.*

Even though SO_2 emissions in the United States have not increased significantly in recent decades, the severity of the acid-rain problem has increased. This puzzling fact is probably related to the installation of extremely tall smokestacks at many power plants (see Figure 12.12). The object was to alleviate air pollution in the local area, and in this respect the tall stacks have been very successful. The final effect, however, was simply to send the problem elsewhere. The offending oxides of sulfur and nitrogen and the resulting acids do eventually return to earth after having been carried by the wind for many miles. In the meantime they and their oxidation products are aloft for days and are completely converted into acids. Before the introduction of tall stacks the oxides returned to the surface much sooner and often reacted with compounds in the soil to become harmless substances. In short, more acid has been formed in the air from the same amount of oxide because of the tall smokestacks.

Whereas the rain in the eastern part of the country contains more H_2SO_4 than HNO_3, HNO_3 dominates in the precipitation falling in the Rocky Mountains. This nitrogen-containing acid comes from oxides of nitrogen, which in turn form from N_2 and O_2 at the high temperature of coal-fired boilers and internal-combustion engines. About 45% of the nitrogen oxides is emitted by motor vehicles, 28% by electric utilities, and 27% by other combustion processes. Parts of the western United States are probably also receiving acid deposition in a dry form. *

* 12.8 THE FATE OF ACID RAIN (*optional*)

Fortunately, most of the acid in rain and snow never reaches rivers and lakes. Perhaps 90% of it is neutralized by compounds in the soil. But in the spring, the rapid melting of snow sends much of the water directly to rivers and lakes before it can percolate through the soil.

Wherever there is limestone (calcium carbonate, $CaCO_3$) in the soil, it can react with the acid:

$$H^+ + HSO_4^- + CaCO_3 \longrightarrow Ca^{2+} + HCO_3^- + HSO_4^-$$

Since limestone and marble are popular building materials, the acid rain also damages buildings and sculptures (see Figure 12.13). It is estimated that acid rain causes $2 billion worth of damage to buildings in the United States every year.

Other compounds in soil can also neutralize the acids. One effect is to release Ca^{2+}, Na^+, K^+, or Mg^{2+} ions, which find their way into lake water. These are generally not harmful, but another ion, Al^{3+}, that is also released is toxic to fish and some plants. In simplified form, the reaction producing this aluminum ion is

$$3\,H^+ + 3\,HSO_4^- + Al(OH)_3(\textit{in a complex form}) \longrightarrow 3\,H_2O + Al^{3+} + 3\,HSO_4^-$$

(a)

(b)

Figure 12.13. An effect of acid rain. (a) Taken in 1910, this photo shows the effect of 400 years of weathering on a gargoyle on Lincoln cathedral in England. (b) In 1984, only 74 years later, acid rain and other atmospheric pollution have worn the figure to a barely recognizable remnant.

The ability of soil to neutralize acid varies greatly from place to place. Some soils, such as those in the Adirondack Mountains in New York State, are very thin and cannot react with much of the acid to which they are exposed. Therefore, the lakes in that region are particularly susceptible to high acidity. It is unfortunate that many other scenic lakes, such as those in southern Ontario and Nova Scotia and on the Minnesota-Ontario border, are in equally fragile surroundings. Not only are the fish in these lakes at risk, but the spruce, pine, aspen, and birch trees seem also to be sensitive to acid rain. Acid rain could have a profound effect on the tourist industry in these areas. In Canada, the billion-dollar sport fishing industry is threatened, and so is the forestry industry, which employs one out of every ten Canadians.

A problem of such severity naturally brings about demands for public action. Not unexpectedly, opposing views emerge. One group argues that the causes of acid rain have been amply demonstrated and that emissions of sulfur and nitrogen oxides must be reduced. The other side maintains that our understanding of the process is very limited and that more research must be done before drastic action is taken. Annual reports of the congressionally mandated National Acid Precipitation Assessment Program began to appear in 1983. About $100 million is being spent annually in the United States on research relevant to the acid-rain problem. *

IMPORTANT TERMS

acid A substance that ionizes to give hydrogen ion (H^+) in water solution.

base A substance that ionizes to give hydroxide ion (OH^-) in water solution.

aqueous Refers to a substance dissolved in water.

hydronium ion The H_3O^+ cation, which is formed by reaction of a proton and a water molecule.

alkaline Refers to any substance or solution that is basic in character.

strong acid An acid that is completely ionized in water solution.

strong base A base that is completely ionized in water solution. All metal hydroxides are strong bases, although some of them are not very soluble in water.

equilibrium A state in which opposing forces balance exactly.

chemical equilibrium A condition of no overall change in a chemical reaction. At the molecular level the reaction still proceeds in both directions, but the opposing rates are equal.

position of equilibrium The relative amounts of reactant and product when a reaction is at equilibrium. For example, when the ratio of products to reactants is very large, the position of equilibrium is far to the right, or product side, of the reaction.

weak acid An acid for which only a small fraction of the molecules are ionized at equilibrium in water solution.

neutral solution A solution in which the concentrations of hydrogen ion and hydroxide ion are both equal to 1×10^{-7} mole per liter.

acidic solution A solution in which the hydrogen ion concentration is greater than 1×10^{-7} mole per liter (and the hydroxide concentration is less than 1×10^{-7} mole per liter).

basic solution, or **alkaline solution** A solution in which the hydroxide concentration is greater than 1×10^{-7} mole per liter (and the hydrogen ion concentration is less than 1×10^{-7} mole per liter).

pH The level of acidity of a water solution, expressed by the exponent in the hydrogen-ion concentration with the sign changed. A basic solution has a pH of greater than 7, and an acidic solution has a pH of less than 7.

acid rain Rain and snow with high acidity (low pH), which threatens wildlife in the lakes and forests of Canada, the eastern part of the United States, and parts of northern Europe.

QUESTIONS

1. Write the equation for the dissociation of each of the following strong acids when it is dissolved in water.

a. $HNO_3(l)$
b. $HClO_4(l)$
c. $HIO_3(s)$

2. Write the equation for the dissociation of each of the following strong bases when it is dissolved in water.

a. $Ba(OH)_2(s)$
b. $LiOH(s)$
c. $Sr(OH)_2(s)$

3. Write chemical equations for the reactions between the following strong acids and bases. Show substances in ionic form where appropriate.

a. nitric acid and potassium hydroxide in water solution
b. $HI(aq)$ and $Ba(OH)_2(aq)$
c. $H_2SO_4(aq)$ and $NaOH(aq)$ (Assume that both Hs from H_2SO_4 react.)
d. $Ca(OH)_2(aq)$ and $HBr(aq)$

4. Place each of the compounds listed below in one of these categories: base, strong acid, weak acid.

a. H_2CO_3 **b.** $Ca(OH)_2$ **c.** HCl
d. H_3PO_4 **e.** $NaOH$ **f.** $HCOOH$

5. The reaction

$$2\,SO_2(g) + O_2(g) \longrightarrow 2\,SO_3(g)$$

is at equilibrium in a closed container. What is the effect of each of the following?

a. removing O_2
b. adding SO_3
c. adding SO_2

6. The Haber process, which is very important in the production of chemical fertilizers, uses this reaction:

$$N_2(g) + 3\ H_2(g) \longrightarrow 2\ NH_3(g)$$

At 400°C the equilibrium position for this reaction is such that approximately equal amounts of reactants and products are present.

a. If H_2 is added to a vessel in which this reaction is at equilibrium, will the reaction shift to the left or to the right? Explain.

b. Would removal of NH_3 from the equilibrium system promote formation of additional NH_3? Explain.

7. The reaction

$$CaO(s) + CO_2(g) \longrightarrow CaCO_3(s)$$

is proceeding from left to right within a closed container.

a. Is the pressure in the container increasing or decreasing? Explain.

b. Is the speed at which CaO and CO_2 are combining greater than or less than the speed at which $CaCo_3$ is decomposing? Explain.

c. How will these rates change with respect to each other as chemical equilibrium is approached?

8. Write equations for the reactions between the following weak acids and strong bases. Use appropriate formulas (ions or molecules).

a. aqueous acetic acid and sodium hydroxide

b. $H_2SO_3(aq)$ and $KOH(aq)$ (Assume that only one H^+ from H_2SO_3 reacts.)

c. $H_3PO_4(aq)$ and $NH_3(aq)$ (Assume that two H^+ ions from H_3PO_4 react.)

9. Classify each of the solutions described below as acidic or basic.

a. $[H^+] = 1$ mole/liter
b. pH = 10
c. $[OH^-] = 0.003$ mole/liter
d. $[H^+] = 1 \times 10^{-5}$ mole/liter
e. pH = 6

10. Classify each of the solutions described below as acidic, basic, or neutral.

a. $[OH^-] = 1 \times 10^{-6}$ mole/liter
b. $[H^+] = 1 \times 10^{-9}$ mole/liter
c. pH = 7
d. pH = 1.38
e. $[OH^-] = 1 \times 10^{-11}$ mole/liter

11. What is the pH of each solution described below?

a. $[H^+] = 1 \times 10^{-6}$ mole per liter
b. $[H^+] = 1 \times 10^{-2}$ mole per liter
c. $[H^+] = 1 \times 10^{-10.5}$ mole per liter

12. Explain why the addition of $NaHCO_3$ to a lake with a pH of 4 would decrease the acidity of the lake water. (When $NaHCO_3$ dissolves in water, Na^+ and HCO_3^- ions are present.)

13. Carbon dioxide provides the fizz (the bubbles) in soft drinks. When a bottle or can of soft drink is opened to the air, the higher pressure in the container is released.

a. Write the equation for the equilibrium of CO_2 between the solution and the gas phase before the container is opened.

b. Explain what causes CO_2 to be liberated when the pressure is released.

14. If the concentration of CO_2 in the air were to decrease markedly, would the acidity of rainwater increase or decrease? Explain.

Optional Sections

15. Some of the SO_2 that is put into the air dissolves in lakes and causes an increase in acidity. Another part of the SO_2 is first oxidized to SO_3, which then dissolves and increases the acid content of the lakes. Would the dissolving of a certain number of moles of SO_2 or of an equal number of moles of SO_3 cause a greater decrease in the pH of the lake water?

16. There is considerable pressure from environmentalists for legislation requiring coal-burning power plants to install devices that decrease greatly the amount of SO_2 emitted from their stacks. The benefits of such action would be the elimination of many of the after-effects of acid rain. On the other side of the issue, the expense of SO_2 scrubbers would result in a much higher cost to consumers for electricity, especially in the Ohio River Valley. Another effect would be reduced profitability of electric utilities, translating into a loss of jobs in the industry and lower dividends for stockholders. Discuss your stand on this issue.

17. The cost of research to determine the sources and long-term effects of acid rain is very large and is borne primarily by the government. This research is one item among many that increases the troublesome federal deficit. In your view, is this cost justified?

Mechanisms of Chemical Reactions *

Some chemical reactions are slow; others are very fast. A piece of scrap iron left outside can take many years to turn completely into rust. At the other extreme, under the right conditions gunpowder or a stick of dynamite takes only a fraction of a second to change into products after it starts to explode. The chemical reactions in our bodies cover an equally wide range of speeds. Think of a baseball player standing confidently at home plate as the pitcher sends a blazing fastball traveling at 80 or 90 miles per hour. Once the batter decides to swing, a sequence of incredibly fast physiological processes begins. In about 0.04 second, the batter's nerve impulses are transmitted into muscle action. Then a host of chemical reactions that drive muscles bring the bat across home plate before the ball can reach the catcher's mitt. By contrast, the reactions causing a person's body to age occur over decades. Very, very slowly do changes such as wrinkles in the skin appear.

Figure 13.1 gives other examples of greatly differing speeds of reaction. Here we compare speeds by looking at the times needed for one-half of the original reactants to change into products. A short time (at the bottom

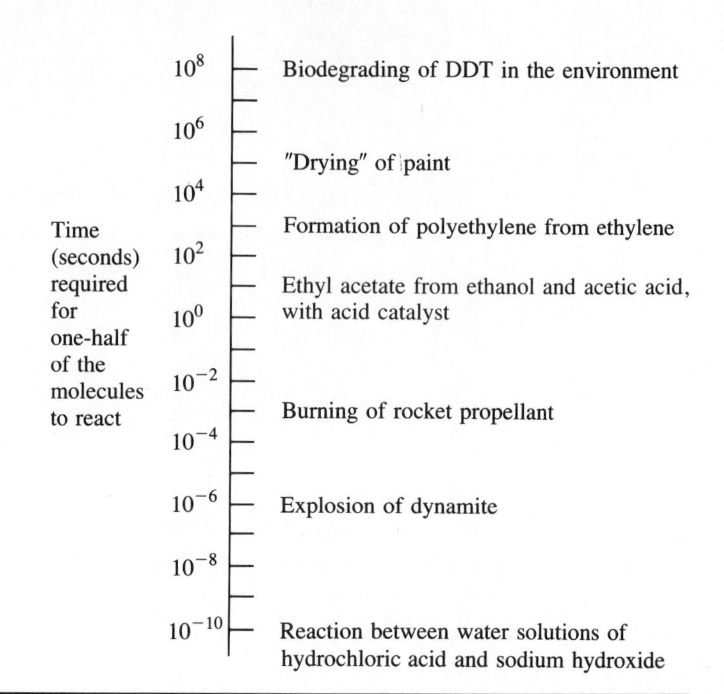

Figure 13.1. A comparison of times needed for one-half of the reactants to change into products. Each step up on the time scale represents an increase by a factor of 10 in the time required for the reaction.

of Figure 13.1) means a very fast reaction. Reactions between acids and bases are among the fastest known. Although the drying of new paint may appear to be simply evaporation of a solvent, chemical reactions are also involved.

Many chemists are actively engaged in trying to explain this great variety of reaction rates. An understanding of these rates gives insight into the specific steps that occur in a reaction. Moreover, in order to design industrial equipment to produce a useful chemical, an engineer must know the speeds of the reactions involved. The term **reaction rate** means the change in concentration of a reactant or product in a unit time interval, such as 1 minute or 1 second. It is tempting to think that a reaction for which the equilibrium lies far to the right would take place rapidly. Unfortunately, the chemical world is not that simple. Look around you at all the things that are combustible—objects made of wood, cardboard, or paper are probably in sight. All of these are capable of reacting with oxygen at temperatures of 20–25°C to form carbon dioxide and water. The driving force for these reactions is large, and yet the reactions themselves are imperceptibly slow. Only with the help of a spark or a match do combustible materials burn with considerable speed. Principles very different from those governing equilibria apply to the speeds of chemical reactions.

13.1 PATHWAYS OF CHEMICAL REACTIONS

The **pathway of a reaction,** or its exact sequence of steps, is the key to its speed. In life, the path we follow to accomplish certain tasks may or may not be important. Does it really matter which side of the bed we get out of in the morning (unless one side is against the wall)? In other situations the path can make a lot of difference. Driving from one town to another by a gravel road is usually much slower than taking a paved highway, even though the distance may be shorter. If a chemical reaction is to occur in a reasonable amount of time, it must follow a favorable pathway.

There are several pathways by which sugar can react with oxygen. One pathway occurs when sugar and oxygen are heated to a high temperature under high pressure in a closed vessel. Carbon dioxide and water form in a rapid combustion reaction. A more enjoyable way of reacting sugar with oxygen is to eat a lump of sugar or a bar of candy, which is mostly sugar. The initial reactants are sugar and inhaled oxygen. The final products are again carbon dioxide, which we exhale, and water, which we either exhale as vapor or excrete in our urine. Along the way, however, almost thirty intermediates form and then are consumed in later steps. The intermediates in this biochemical pathway are completely different from the intermediates in the direct combustion of sugar. Different pathways always involve different intermediates and different reaction speeds.

The important characteristic of an **intermediate** is that it is neither one of the initial reactants nor one of the final products. It is produced in one step of the reaction pathway and is used up in a later step. In a way the process is like taking a flight from Dallas to Boston with a stop in Atlanta (see Figure 13.2). The stop in Atlanta is an intermediate point. You won't get to Boston without touching down in Atlanta first, but Atlanta is not your final destination.

Another term for the pathway of a reaction is **reaction mechanism**—the sequence of steps leading from the original reactants to the final products. Intermediates are formed in between. In each step the reacting molecules collide and rearrange into new combinations. These new molecules emerge from the collision as the products. There is no way that the reactants in any step can form products without coming together first.

One of the most difficult problems facing a chemist is to discover the mechanism for a reaction, because the intermediates usually have very small concentrations, which are close to or below the limits of detection. It may be necessary to guess what the intermediates are, without positive identification. However, available evidence indicates that most reactions consist of several steps, even though the overall reaction may look very simple.

As an example let's examine a reaction that plays a major role in the formation of smog over many of our larger cities. Its mechanism is still

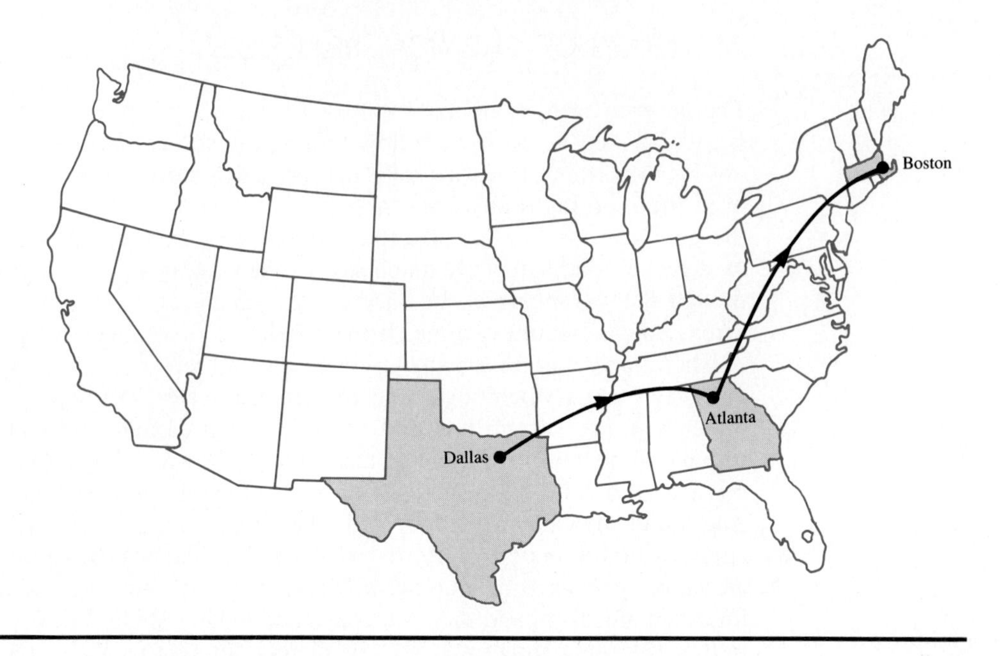

Figure 13.2. A flight from Dallas to Boston with an intermediate stop in Atlanta.

somewhat uncertain despite decades of investigation. It is the reaction between nitric oxide (NO), formed in automobile and truck engines, and the oxygen it encounters in the air after it escapes from an exhaust pipe.

$$2\,NO + O_2 \longrightarrow 2\,NO_2$$

The NO_2 produced gives smog its yellowish-brown color.

From the balanced chemical equation it looks as though two molecules of NO collide with one O_2 molecule to give two molecules of NO_2. You might imagine that during the collision the bond between the oxygen atoms in O_2 breaks and two new bonds between N and O form. The mechanism would be the same as the overall balanced equation. It's a reasonable hypothesis, but chemists who have studied the reaction don't accept it. For one thing, the probability that three molecules of gas will come together in one place at exactly the same instant is very small. Although this possibility cannot be dismissed completely, it is over a thousand times more likely that two molecules will collide or that a molecule with very high energy will break apart. We can devise several mechanisms in which just two molecules collide in each step.

For the first mechanism we assume that N_2O_2 is an intermediate. It is produced in the first step and consumed in the second.

$$2\,NO + O_2 \longrightarrow 2\,NO_2 : \text{Possible Pathway \#1}$$

	$NO + NO \longrightarrow N_2O_2$	Step 1
	$N_2O_2 + O_2 \longrightarrow NO_2 + NO_2$	Step 2
	$NO + NO + \cancel{N_2O_2} + O_2 \longrightarrow \cancel{N_2O_2} + NO_2 + NO_2$	Overall reaction
or	$2\,NO + O_2 \longrightarrow 2\,NO_2$	**Net reaction**

Each of these steps involves the collision of two molecules and is called an *elementary reaction*. An **elementary reaction** shows exactly which molecules collide to produce products in a single chemical event. In step 1 above, an N_2O_2 molecule forms directly from a collision of two NO molecules. No other intermediates form along the way (see Figure 13.3).

Any reaction mechanism consists of a series of elementary reactions. The elementary reactions must add to give the overall reaction, as illustrated in the steps above. Each molecule of N_2O_2 formed in the first step is used up in the second step. None is left over at the end, and we cancel the equal numbers of molecules of N_2O_2 on each side of the next-to-last equation to see the net reaction.

A second possible mechanism or pathway for the reaction of NO and oxygen involves a different intermediate and a different set of elementary reactions.

$$2\,NO + O_2 \longrightarrow 2\,NO_2 : \text{Possible Pathway \#2}$$

	$NO + O_2 \longrightarrow NO_3$	Step 1
	$NO_3 + NO \longrightarrow NO_2 + NO_2$	Step 2
	$NO + O_2 + \cancel{NO_3} + NO \longrightarrow \cancel{NO_3} + NO_2 + NO_2$	Overall reaction
or	$2\,NO + O_2 \longrightarrow 2\,NO_2$	**Net reaction**

Once again the intermediate, NO_3 in this case, is used up in the second step after forming in the first step. Therefore it does not appear in the equation for the net reaction.

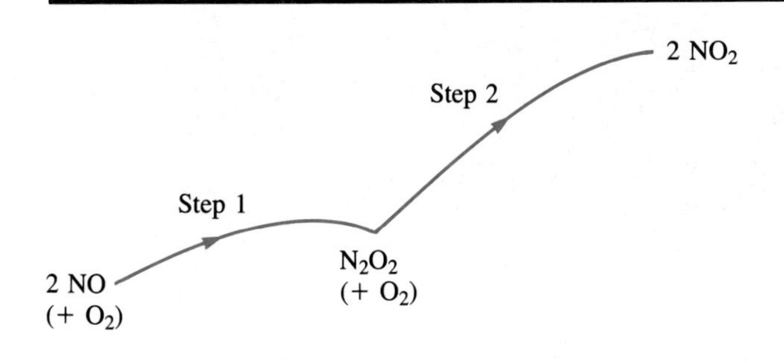

Figure 13.3. *A route from* 2 NO + O_2 *to* 2 NO_2. *An intermediate,* N_2O_2, *is formed along the way.*

A third possible mechanism has an oxygen atom as the intermediate.

$$2\ NO + O_2 \longrightarrow 2\ NO_2 : \text{Possible Pathway \#3}$$

	$NO + O_2 \longrightarrow NO_2 + O$	Step 1
	$O + NO \longrightarrow NO_2$	Step 2
	$NO + O_2 + O + NO \longrightarrow NO_2 + O + NO_2$	Overall reaction
or	$2\ NO + O_2 \longrightarrow 2\ NO_2$	Net reaction

During the many years of investigation of this reaction, some scientists felt that the first pathway was most likely, while others favored the direct reaction between 2 NO and O_2, despite the low probability of a collision involving three molecules. In polluted air, which contains many chemicals, more complicated mechanisms are important. In this environment the conversion of NO to NO_2 involves intermediates from water vapor and incompletely burned hydrocarbons.

The intermediates (N_2O_2, NO_3, and O) proposed in the above three mechanisms have one thing in common: they are very unstable and therefore can react rapidly with a variety of chemicals. Very soon after a molecule of an intermediate forms in the first step, it is gobbled up in the second step. Molecules of the intermediate simply don't accumulate, and the concentration of these molecules remains very small. This is why it is difficult to detect intermediates, even with the sensitive instruments of modern chemistry (see Figure 13.4). So, it can rarely be proved that a particular mechanism is correct. Usually, all a chemist can say is that a particular

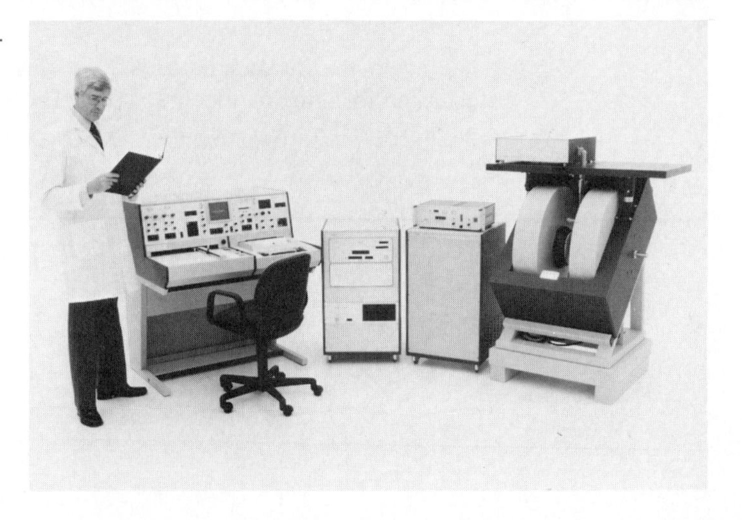

Figure 13.4. *An electron-spin resonance spectrometer. This instrument can detect an intermediate having an unpaired valence electron.*

mechanism is consistent with all the evidence. He or she must admit that some other mechanism might also be consistent with the facts. However, knowing reaction pathways helps a chemist to predict rates of reaction and the effect of environment on a reaction.

Example A possible mechanism for a reaction consists of four steps:

$$ICl \longrightarrow I + Cl$$
$$Cl + H_2 \longrightarrow HCl + H$$
$$H + ICl \longrightarrow HCl + I$$
$$I + I \longrightarrow I_2$$

What is the net reaction corresponding to this mechanism?

Solution To find the overall reaction, add the elementary reactions together. Then cancel intermediates that appear on both sides of the resulting equation.

$$ICl \longrightarrow I + Cl$$
$$Cl + H_2 \longrightarrow HCl + H$$
$$H + ICl \longrightarrow HCl + I$$
$$I + I \longrightarrow I_2$$

$$ICl + \cancel{Cl} + H_2 + \cancel{H} + ICl + \cancel{I} + \cancel{I} \longrightarrow \cancel{I} + \cancel{Cl} + HCl + \cancel{H} + HCl + \cancel{I} + I_2$$

or $$2\,ICl + H_2 \longrightarrow 2\,HCl + I_2$$

Example Ozone, O_3, which can be recognized by its pungent odor, sometimes forms in the air during a thunderstorm. It soon decomposes into regular oxygen molecules according to the reaction $2\,O_3 \rightarrow 3\,O_2$. The mechanism for this reaction involves an oxygen *atom* as the only intermediate. Write a possible mechanism for the reaction.

Solution In any reaction mechanism, the reactants in the first step can only be substances that are reactants in the overall reaction, because these are the only substances present at the beginning. Since only ozone is present initially, it must start the first step. Assume that O_3 dissociates to give the intermediate (O):

$$O_3 \longrightarrow O_2 + O$$

Next, O can react with a second molecule of O_3 to give more of the product, O_2. The complete mechanism becomes

$$O_3 \longrightarrow O_2 + O$$
$$O + O_3 \longrightarrow O_2 + O_2$$

$$2\,O_3 \longrightarrow 3\,O_2 \qquad \text{Net reaction}$$

FIRST THINGS FIRST

Chemists cannot write a mechanism for a reaction unless they know the reactants and the products. In other words, they must first know the overall balanced chemical equation. There are stories of chemists who were so confident of their chemical knowledge that they studied the mechanisms of reactions without ever bothering to isolate the products. They wrote fancy sequences of elementary reactions describing chemistry that had no basis in the real world, because they had guessed wrong as to the reaction products. When their mistakes were exposed, a few well-known chemists were rather embarrassed.

Writing a mechanism for a chemical reaction without first knowing the products is like buying a ticket to an unknown destination. Can you imagine this conversation?

Customer: I want to buy a ticket.

Ticket agent: Where to?

Customer: Oh, I don't know. Just sell me a ticket.

Ticket agent: Huh?

Customer: On second thought, I would like to go with an intermediate stop in San Francisco.

Ticket agent: But you don't know where you want to end up?

Customer: Does it matter?

Ticket agent: Why don't you come back tomorrow after you've thought it over.

The intermediate cancels from each side to yield the given net reaction. Although we can't prove that this mechanism is correct, it agrees with the conditions stated in the problem.

Each of the elementary reactions shown in the three possible pathways for the reaction of NO with O_2 has two molecules of reactants colliding together. Such a reaction is called a **bimolecular** (two-molecule) **reaction.** Bimolecular elementary reactions are by far the most likely in reaction mechanisms. Also possible is a **unimolecular** (one-molecule) **reaction,** in which only one molecule reacts. An example is the first step in the ozone reaction described above. Occasionally chemists propose a **termolecular reaction** (one with three molecules on the left side), but this type of reaction is improbable, especially in the gaseous phase, as we have already suggested.

Example Which of the following cannot possibly be an elementary reaction?

(a) $N_2(g) + 3\ H_2(g) \rightarrow 2\ NH_3(g)$

(b) $H_2(g) + I_2(g) \rightarrow 2\ HI(g)$

(c) $2\ C_8H_{18}(g) + 25\ O_2(g) \rightarrow 16\ CO_2(g) + 18\ H_2O(g)$

Solution Parts (a) and (c) cannot possibly be elementary reactions, because the likelihood that either 4 or 27 molecules will come together at exactly the same time in a single place is extremely small. The reaction in part (b) could be an elementary reaction.

13.2 FAST AND SLOW ELEMENTARY REACTIONS

We have seen that different reactions can have vastly different speeds (Figure 13.1) and that most reactions we observe in the laboratory or in nature have mechanisms consisting of a series of elementary reactions. According to all the evidence we have, the elementary reactions that make up a mechanism can also have different speeds. For example, the proposed mechanism for the decomposition of N_2O, or laughing gas, at high temperature consists of one slow step and one much faster step. The equation for the decomposition of N_2O is

$$2\,N_2O \longrightarrow 2\,N_2 + O_2$$

The mechanism proposed for this reaction consists of two steps:

$$N_2O \longrightarrow N_2 + O \qquad \text{Slow}$$

$$O + N_2O \longrightarrow N_2 + O_2 \qquad \text{Fast}$$

In this mechanism there is just one intermediate, an oxygen atom. When we add the two steps and cancel out the oxygen atom on each side, we get the net reaction. From data on how the speed of the reaction changes when the concentration of N_2O is altered, chemists have concluded that the first step is much slower than the second step. This first step acts like a bottleneck. The final products, N_2 and O_2, cannot form any faster than N_2 and O form in the first step. The bottleneck step is usually called the **rate-determining step.**

A useful analogy can be drawn between the effect of a rate-determining step on a mechanism and that of a pipe of narrow diameter on a flow of water. In Figure 13.5 an arrangement of water tanks and pipes is used to represent the mechanism of the decomposition of N_2O. The flow is started when water is poured into the uppermost tank, which is like adding reactants to the first step of the reaction mechanism. A small-diameter pipe connects this first tank to the next one. The third tank is joined to the intermediate tank by a pipe of large diameter. Because of the difference in the diameter of the drain pipes, the first tank drains very slowly, and the intermediate tank empties rapidly. Therefore, very little water accumulates in the intermediate tank. The last tank (holding the final products of the drainage) fills at just about the same rate as the first tank empties. The speed of the overall process is the same as the speed of the first step, so the first step is rate-determining.

Let's return to the mechanism for the decomposition of N_2O. The intermediate O atoms are produced relatively slowly. A split second after an O atom forms, it reacts with an N_2O molecule in the second step to give an O_2 molecule and an N_2 molecule, the final products. Because of the negligible time delay, the final products form at the same speed as the rate-determining step occurs.

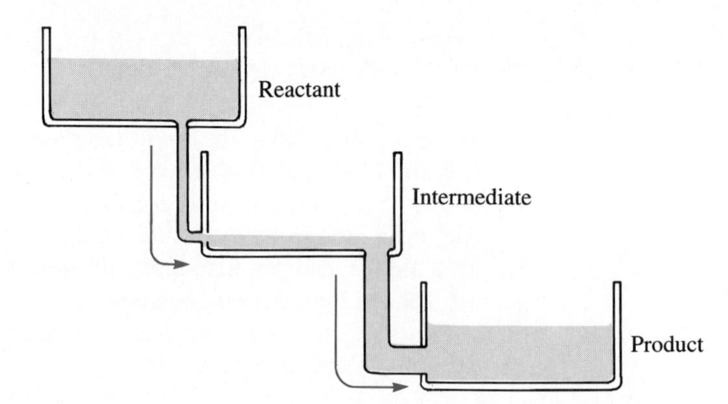

Figure 13.5. *Flow of water through interconnected tanks representing the reactant, intermediate, and products for the decomposition of N_2O. Because the first pipe has the smaller diameter, flow through that pipe is rate-determining.*

13.3 REACTION SPEED AND MOLECULAR COLLISIONS

The speed of a slow reaction can almost always be increased by increasing the concentrations of the reactants. In Chapter 6 we mentioned a reaction for making hydrogen gas:

$$\underset{\text{iron}}{Fe} + \underset{\text{sulfuric acid}}{H_2SO_4} \longrightarrow \underset{\text{hydrogen}}{H_2} + \underset{\text{ferrous sulfate}}{FeSO_4}$$

Early chemists discovered that one way to make this reaction go faster is to use sulfuric acid with a higher concentration—that is, with more moles of acid per liter of solution. Acid of higher concentration causes hydrogen to bubble out of the solution more rapidly. In the same way, the reaction between NO and O_2 goes faster when the concentration of either reactant is increased. If the freeways around Los Angeles are unusually congested, more cars than usual are sending NO into the air, and the smog forms more rapidly.

Some experimental numbers will show how concentration and reaction rate are connected. Table 13.1 gives some data from six experiments on another reaction occurring in polluted air:

$$CO(g) + NO_2(g) \longrightarrow CO_2(g) + NO(g)$$

In this table, "initial concentration" means the concentration of a reactant just after the reactants are mixed together to start the reaction.

In the first three experiments the concentration of NO_2 was held constant at 0.1 mole per liter. Thus the variation in rate must be caused by the

Table 13.1 Rates for the Reaction Between Carbon Monoxide and Nitrogen Dioxide at 430°C

INITIAL CONCENTRATION (in moles/liter)		REACTION RATE (in moles/liter of CO_2 formed per second)
CO	NO_2	
First Set		
0.1	0.1	0.06
0.2	0.1	0.12
0.3	0.1	0.18
Second Set		
0.2	0.1	0.12
0.2	0.2	0.24
0.2	0.3	0.36

change in concentration of CO. We see that when the concentration of CO doubles or triples, the reaction rate doubles or triples. Similarly, the second series of experiments shows that the rate is directly proportional to the concentration of NO_2. For the reaction of CO and NO_2 there is a simple relationship between the concentration of each reactant and the rate of reaction.

What makes the reaction rate double when the concentration of either CO or NO_2 doubles? This reaction takes place in the gaseous state, and our picture of a gas is a space thinly populated with molecules that are in rapid motion. Many collisions between molecules occur each second. In this reaction two kinds of molecules, CO and NO_2, must collide in order for the reaction to occur. During a collision between reactant molecules, which lasts the incredibly short time of about 10^{-13} second, bonds reshuffle to produce new molecules, which then fly apart. Figure 13.6 shows a simplified version of this process. When molecules smack together at great speeds,

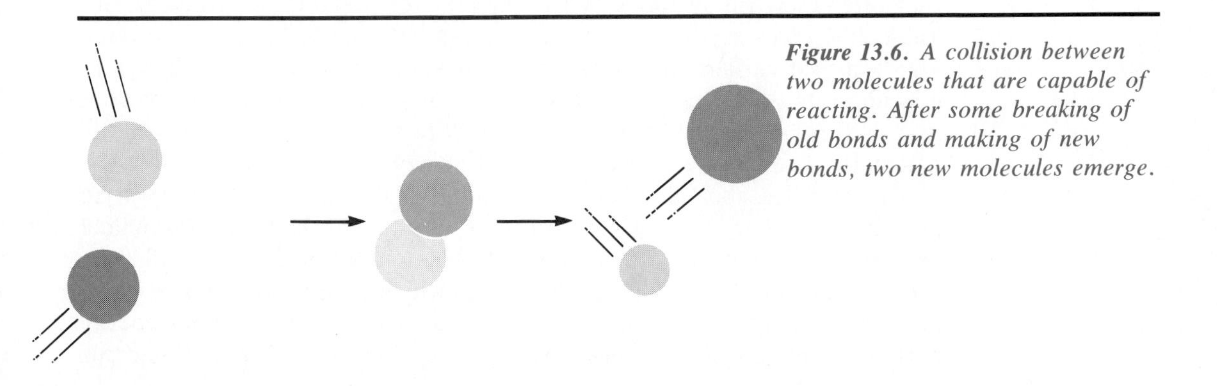

Figure 13.6. A collision between two molecules that are capable of reacting. After some breaking of old bonds and making of new bonds, two new molecules emerge.

changes can occur; chemistry can happen. The more collisions there are between reactant molecules every second, the greater the rate of reaction is. And the number of collisions depends on the number of molecules of each reactant packed into the container. If an experimenter doubles the concentration of one reactant, the number of collisions per second doubles, and the reaction rate doubles. The same should be true if the concentration of a different reactant is doubled. Therefore, we can explain the data in Table 13.1 by recognizing the connections between reaction rate, collision rate, and concentration.

Although our discussion so far has applied to reactions among gases, similar considerations apply to the rates of reactions taking place in liquid solutions. Because molecules are more crowded in a liquid, the molecules of reactants will bump into solvent molecules many times before they collide with each other. Nevertheless, the rate at which they collide depends on their concentrations.

Higher concentration of reactants $\Longrightarrow$ More collisions $\Longrightarrow$ Faster reaction rate

13.4 REACTION RATE AND TEMPERATURE

Increasing the temperature increases the speed of nearly all chemical reactions. According to the kinetic theory of gases (Chapter 4), the number of collisions per second between molecules of a gas increases as the temperature is raised. The frequency of collisions in a liquid also increases with temperature. A simple explanation, therefore, is that reaction rates increase with temperature because the frequency of collisions between reactant molecules increases. Calculations based on the kinetic-molecular theory, however, show that this cannot be the whole explanation.

With a 10° rise in temperature, the average speed of the molecules in a gas increases by 1–2%. Since the frequency of collisions between molecules depends on molecular speed, the frequency also should increase by about 1–2%. However, measurements of the rates of various reactions show that they often increase by 200 or 300% when the temperature increases by 10°. Table 13.2 gives some data for the reaction between CO and NO_2, which we discussed earlier. This table makes it amply clear that the change in collision rate is not enough to explain the much larger change in reaction rate. Nevertheless, chemists were reluctant to throw out the entire kinetic-molecular theory, since it had explained many other observations so well.

The Swedish chemist Svante Arrhenius, who contributed so much to our understanding of ions, was the first scientist to make an acceptable analysis of the dependence of reaction rates on temperature. In 1889 he showed that there is an energy "hill" over which each elementary reaction must be "pushed" to completion. Arrhenius called this potential energy barrier the **energy of activation,** usually given the symbol, E_a. When this

Table 13.2 Effect of Temperature on the Speed of the Reaction $CO + NO_2 \rightarrow CO_2 + NO$ and on the Collision Rate in a Gas

TEMPERATURE	RELATIVE REACTION RATE	RELATIVE COLLISION RATE
600 K	1.0	1.00
650 K	7.9	1.04
700 K	46.4	1.08
750 K	214	1.12
800 K	821	1.15

energy is large, reactions have a difficult time and their rates tend to be slow. On the other hand, a low value of E_a usually means a fast reaction rate. Hikers who have to cross high mountains must expend much energy to get to their destinations. The trip will take a long time. If only a small hill is in the way, the trip will be faster, even though the horizontal distance is the same. Climbing upward is always much slower than traveling on level ground.

As rapidly moving reactant molecules collide, some of their kinetic energy changes into potential energy, and they move along the curve in Figure 13.7 to the top of the energy barrier. This transition stage, the top of the energy hill between reactants and products in an elementary reaction, is called the **activated complex.** It contains some partially broken bonds that

Figure 13.7. The energy of activation for two elementary reactions. (a) The products have lower energy; the reaction is exothermic. (b) The products have higher energy; the reaction is endothermic.

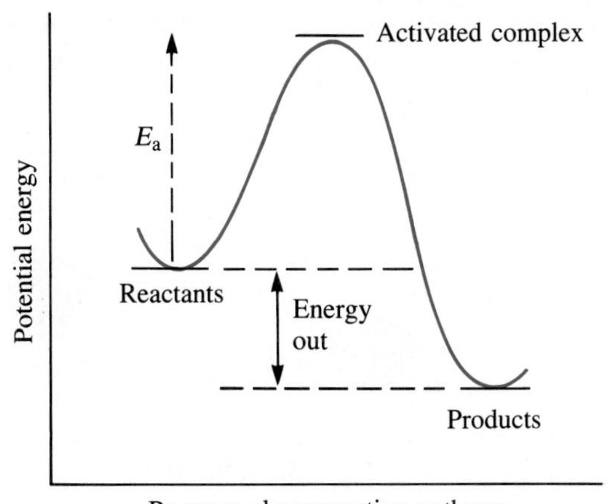

(a)

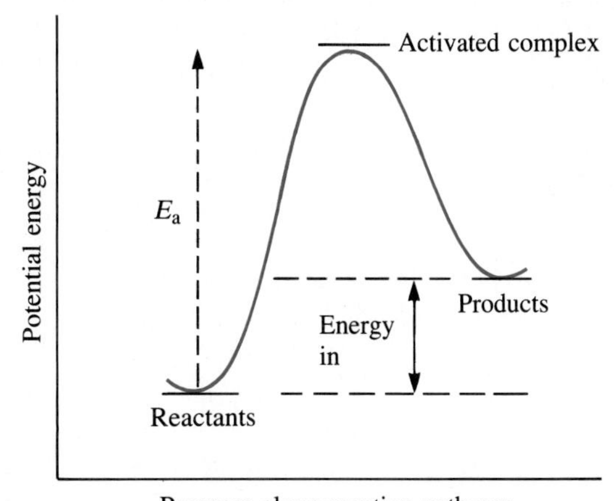

(b)

were present in the reactants and some partially formed bonds that will be present in the products. After passing through this transition stage, the atoms separate into the new combinations which are the products. Figure 13.7(a) shows an exothermic reaction, for which the energy of the products is lower than that of the reactants, and Figure 13.7(b) corresponds to an endothermic reaction.

When a reaction pathway consists of a series of elementary reactions, each elementary reaction has an energy barrier to cross, at the top of which is an activated complex. Let's return to our earlier travel analogy (Figure 13.2). When a person takes a plane from Dallas to Boston with a stopover in Atlanta, the Atlanta airport is an intermediate landing point. There would be two transition stages (or activated complexes) where the plane would be at its highest altitude. One would come when the airplane was at its highest altitude between Dallas and Atlanta—say, about 30,000 feet above central Mississippi. The other transition stage would be somewhere in the air between Atlanta and Boston. Notice that it would be impossible for the airplane to stop at either transition stage.

In order for a pair of colliding reactant molecules to change into products, they must have a kinetic energy at least as large as E_a. If their kinetic energy is smaller than E_a, when they collide they merely bounce apart without reacting, as shown in Figure 13.8.

For many reactions, the energy of activation is larger than the average kinetic energy of the reactant molecules. Therefore, more often than not the collisions between them are ineffective and do not lead to reaction. Typical values for the energy of activation in a chemical reaction are from 40 to 170 kilojoules per mole of reactant. For example, E_a for the reaction between CO and NO_2 is 132 kilojoules per mole. By contrast, the average kinetic energy of the reactant molecules is 2.5 kilojoules per mole at 300 K and 5 kilojoules per mole at 600 K.

It is not difficult to see why a reaction might have a large activation energy. The electrons in the outer part of a molecule give the surface a negative charge, which repels the surface of any other molecule. If a reaction

__Figure 13.8.__ A collision in which the two reactant molecules do not have enough energy to react. The kinetic energy is less than E_a. The molecules separate unchanged.

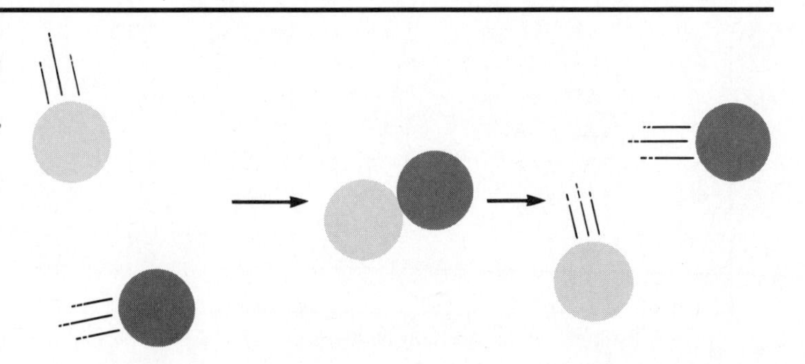

is to take place, this defensive barrier must be overcome, and that process requires energy. Chemical bonds are also being reshuffled in the activated complex. There will be unstable combinations of atoms along the way.

If the average kinetic energy of reactant molecules is typically much less than the activation energy, how can any molecules have the energy necessary for reaction? The answer is that there is a tremendous spread of molecular energies in any gas. At any instant some molecules have five times the average energy, and a few even have ten times the average energy. Figure 13.9 shows the approximate distribution of energies for the molecules of any gas at two temperatures, 300 K and 800 K. To make our example specific, we'll assume the gas is nitric oxide, NO.

Suppose that in order to react during a collision a nitric oxide molecule must have an energy greater than E_a in Figure 13.9. At 300 K only a small fraction of the molecules have such energies. Therefore, only a small fraction of them can react when they collide. At the higher temperature, however, almost half of the NO molecules have at least the critical energy, and the reaction rate will be far greater.

The data of Table 13.2 now begin to make some sense. Reaction rates are slow at low temperatures because most of the collisions between reactants don't result in products. Only a small fraction of the molecules have the large kinetic energy necessary for reaction. When the temperature is increased by 10 or 50°C, the fraction of the molecules having energies of E_a or more increases severalfold.

Figure 13.9. *Distribution of kinetic energy for* NO *at two temperatures. Gray and colored areas show the molecules that have an energy greater than a critical value,* E_a.

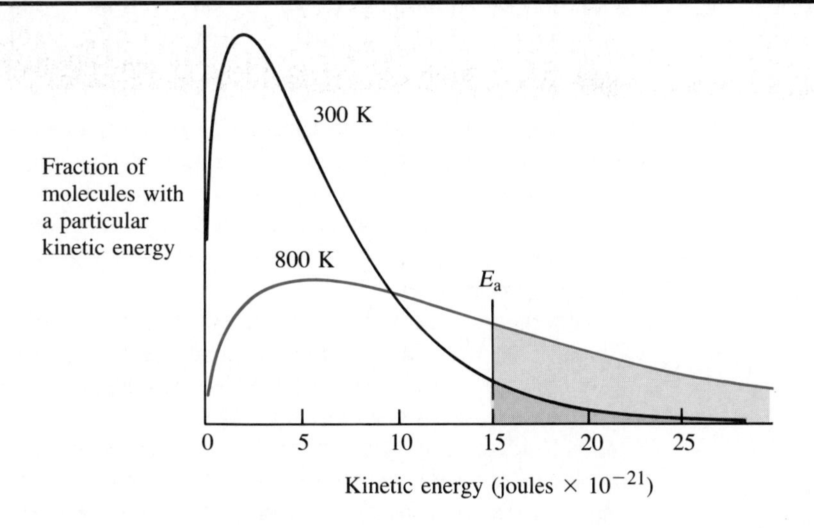

Example The activation energies for two elementary reactions are as follows:

$$H + HBr \longrightarrow H_2 + Br \qquad E_a = 15 \text{ kJ/mole}$$

$$CO + O_2 \longrightarrow CO_2 + O \qquad E_a = 213 \text{ kJ/mole}$$

If the collision rates for the two sets of reactants are about the same, which reaction is likely to be faster at any given temperature?

Solution The fraction of molecules with enough energy to reach the activated complex will always be larger for the reaction with the lower activation energy (E_a). The first reaction should be faster at all temperatures.

13.5 COMPOSITION OF THE ATMOSPHERE

Throughout this chapter many of the examples of chemicals and their reactions have involved common air pollutants, such as NO, NO_2, and SO_2. As we have indicated, the combustion of fuels by industrialized societies has been a major source of these pollutants, but what many people fail to realize is that there are also natural sources of many of them. Before we look at more of the chemistry of air pollution, we will consider the components of unpolluted air.

As far as scientists can tell, the composition of the earth's atmosphere has been nearly constant for millions of years, although early in the history of the earth the makeup may have been quite different. Only "clean dry air near sea level" has a well-defined composition. "Clean" refers to air that has not received industrial emissions. "Dry" means without moisture, which is not included because the water content varies so much—from 0 to 4% as the humidity changes. "Near sea level" is important because the composition begins to change above an altitude of about 40 miles. Table 13.3 gives the percentages of eleven gases in the atmosphere.

The mole % of nitrogen is given as 78.09. This means that in any sample of air, 78.09% of the moles are nitrogen. The number 780,900 ppm (parts per million) signifies that out of every 1,000,000 moles of air 780,900 are nitrogen. Because 1 mole of any substance contains the same number of molecules, the values in the table also give the relative numbers of molecules of the various gases.

Together, nitrogen and oxygen make up 99% of the earth's atmosphere. Because of the stable triple bond in the nitrogen molecule, nitrogen is unreactive, and most of the N_2 in the air does not participate in chemical reactions. A small fraction, however, is acted upon by microorganisms in the soil, becoming nitrate ion, NO_3^-. Another process slowly converts this inorganic nitrogen back into N_2. Even though O_2 is much more reactive, again only a small fraction of the atmosphere's oxygen leaves or enters the

THE PHLOGISTON THEORY

In the eighteenth century and earlier, scientists had very little idea as to what substances were in the air. One of their more bizarre ideas, from our current perspective, was that air contains a substance called "phlogiston" as a result of the burning of wood, candles, or any other combustible material. According to the phlogiston theory, a candle that is burning in a closed jar releases phlogiston into the surrounding air. Eventually the candle goes out when the air becomes saturated with phlogiston.

Along with many others, Joseph Priestley, the discoverer of oxygen gas, believed in the phlogiston theory. Because combustible substances burn much more brightly in pure oxygen than in air, Priestley referred to his newly discovered gas as "dephlogisticated air." Even Priestley recognized that he was better at making accurate observations in the laboratory than at explaining them. It remained for a French chemist, Antoine Lavoisier, to explain the true function of oxygen in the process of combustion.

air each year. Oxygen is removed through chemical and physiological combustion and is replaced through the photosynthesis of green plants.

Argon, the next component in order of decreasing concentration, is so unreactive that there is scarcely any cycling of it between the atmosphere and other parts of the environment. It and the other noble gases are left over from prehistoric times.

Carbon dioxide is continuously cycled through the atmosphere by the actions of photosynthesis, respiration, and fuel combustion. The oceans also dissolve large quantities of CO_2, some of which eventually ends up in sediments. Methane, CH_4, is a stable trace component at the parts-per-

Table 13.3 Composition of the Atmosphere (clean dry air near sea level)

SUBSTANCE	CONCENTRATION	
	mole %	ppm
nitrogen	78.09	780,900
oxygen	20.94	209,400
argon	0.93	9,300
carbon dioxide	0.0318	318
other noble gases	0.00243	24.3
methane	0.00015	1.5
hydrogen	0.00005	0.5
nitrogen oxides	0.000025	0.25
carbon monoxide	0.00001	0.1
ozone	0.000002	0.02
sulfur dioxide	0.00000002	0.0002

million level. It is formed in oxygen-deficient environments, such as swamps, from decaying organic matter. Just how it leaves the atmosphere is not known for certain. It may migrate to high altitudes and be destroyed there.

Some of the bad actors in polluted air come not only from smokestacks and exhaust pipes but also from natural sources, although the amounts are very small. What are the natural sources of pollutants? For nitrogen oxides they are the soil, where nitrogen and oxygen are combined by microorganisms, and the atmosphere, in which the two elements combine when lightning occurs. Carbon monoxide comes in part from volcanoes. Hydrogen sulfide, H_2S, with its familiar rotten-egg smell, is the major source of the sulfur in the atmosphere. It comes primarily from swamps. Hydrogen sulfide is rapidly oxidized to sulfur dioxide by excess oxygen. There is so little hydrogen sulfide in the air that it doesn't even appear in Table 13.3. Ozone, O_3, forms in the upper atmosphere and slowly makes its way to the surface of the earth.

13.6 OXIDATION OF SO_2—UNCATALYZED AND CATALYZED

In Chapter 12 we said that a key reaction in the formation of acid rain is the oxidation of SO_2 to SO_3:

$$2\,SO_2 + O_2 \longrightarrow 2\,SO_3$$

Following this reaction, the SO_3 reacts with water to give sulfuric acid.

$$SO_3 + H_2O \longrightarrow H_2SO_4$$

The sulfuric acid dissolved in rainwater brings acid rain. Now we are in a position to appreciate the mechanism for the first part of this process, the formation of SO_3 from SO_2.

On a sunny day sulfur dioxide can react directly with O_2 in an elementary reaction to give SO_3 and an oxygen atom as the products.

$$SO_2 + O_2 \longrightarrow SO_3 + O \qquad \textbf{(Equation 13.1)}$$

However, the activation energy for this reaction is moderately large, and the reaction is quite slow in the atmosphere. This step is followed by one or more fast reactions which use up the oxygen atoms. The reaction between an oxygen molecule and a sulfur dioxide molecule is rate-determining and controls the overall rate of acid-rain formation.

We can predict that increasing the concentration of either SO_2 or O_2 will increase the rate of the reaction. A higher concentration means more frequent collisions between the reactant molecules. Furthermore, like most reactions, this one will go faster when the temperature is raised. At a high temperature a large fraction of the SO_2 and O_2 molecules collide with a kinetic energy greater than E_a, the activation energy. Both effects are illustrated in Figure 13.10.

Chemists have observed that when air is doubly polluted with SO_2 and oxides of nitrogen the oxidation of SO_2 is faster than when the oxides of nitrogen are absent. It appears that oxides of nitrogen catalyze the reaction. A **catalyst** increases the rate of a reaction by making an alternative pathway possible. Often the catalyst is a reactant in an early step of the mechanism and a product of a later step. Therefore, there is no change in the amount of catalyst present, and the overall reaction is the same as it would be without the catalyst. The alternative pathway has a much lower activation energy than does the pathway for the uncatalyzed reaction, as shown in parts (a) and (b) of Figure 13.11. As always, the lower the activation energy, the faster the reaction.

Let's return now to the catalyzed oxidation of SO_2 in the atmosphere. One possible pathway for the catalysis by NO_2 is

$$NO_2 + SO_2 \longrightarrow SO_3 + NO$$

$$NO + O_2 \longrightarrow NO_2 + O$$

The overall reaction is

$$\cancel{NO_2} + SO_2 + \cancel{NO} + O_2 \longrightarrow SO_3 + \cancel{NO} + \cancel{NO_2} + O$$

or

$$SO_2 + O_2 \longrightarrow SO_3 + O$$

Figure 13.10. *The effects on the rate of a chemical reaction of raising the temperature and increasing the concentrations of reactants.*

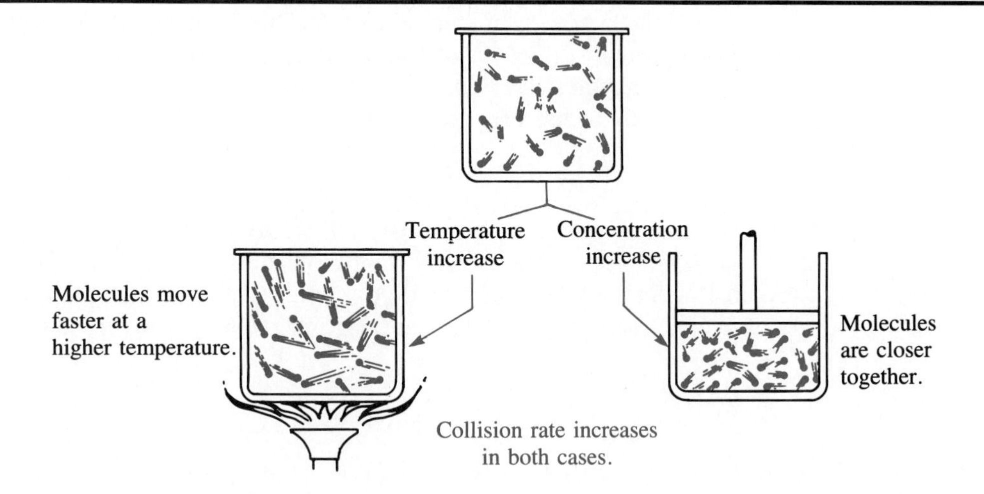

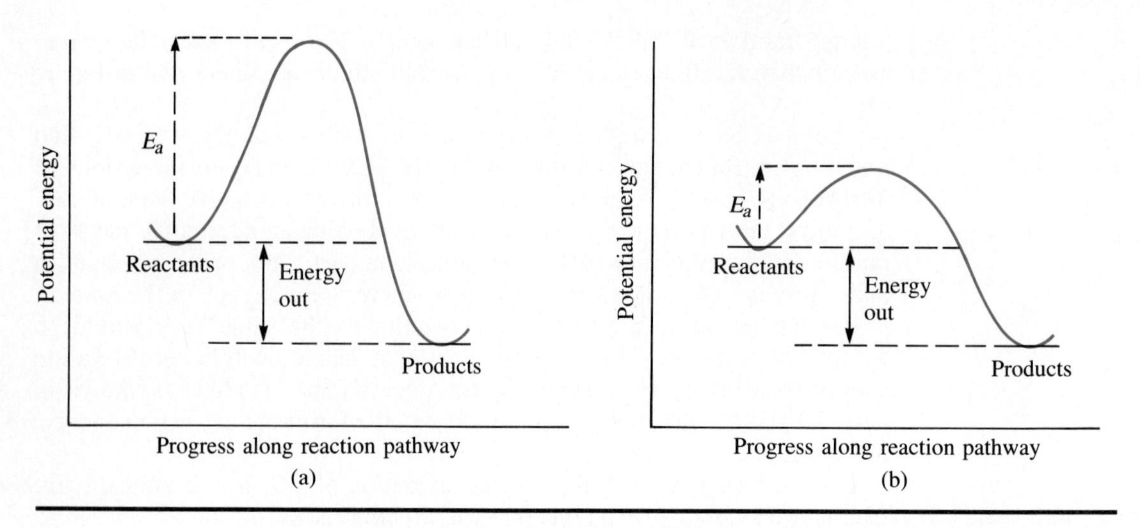

Figure 13.11. (a) The energy of activation for an uncatalyzed chemical reaction. (b) The smaller energy of activation for the same overall reaction when a catalyst participates and allows a different pathway.

The oxidation reaction is the same as before; NO and NO_2 appear on both sides of the equation and cancel out. NO_2 serves as a catalyst in this pathway. The reason the overall process is now faster is that every step is faster than the single step shown in Equation 13.1. Each step in the catalyzed reaction pathway has a lower activation energy than the single step does.

Much industrial research is aimed at finding new catalysts. Catalysts are responsible for the success of many reactions vital to our nation's industries. The refining and "cracking" of petroleum into gasoline, the making of plastics, and the synthesis of ammonia for fertilizers all would be impossibly slow without catalysts. Most of the chemical reactions in our bodies are catalyzed by enzymes—we couldn't live without them.

A careful look at the first step in the catalyzed reaction

$$NO_2 + SO_2 \longrightarrow SO_3 + NO \qquad \textbf{(Equation 13.2)}$$

shows that an oxygen atom is transferred from NO_2 to SO_2. In order for this transfer to occur, the two reactant molecules must collide so that an oxygen atom in the NO_2 molecule is close to the S atom of the SO_2 molecule. A reasonable orientation for the reacting molecules is shown here:

O=S(–O) + O—N=O ⟶ [O=S(–O)···O···N=O] ⟶ O=S(–O)—O + N=O

The structure within brackets is the activated complex for Equation 13.2. The structure has a dashed line on the left to suggest partial formation of the new S—O bond and has a dashed line on the right to represent the partially broken N—O bond.

As we have seen, one requirement for a successful collision between reactant molecules in an elementary reaction is that their kinetic energy be at least as large as the activation energy for that step. A second requirement is that the reactant molecules have a **favorable orientation,** or relative positioning of the atoms in the activated complex. Usually one or more new bonds will form in the products. If so, the atoms that will form a new bond must be next to each other when the reactant molecules collide. If the orientation is not favorable, the reactants will bounce apart without reacting even when they have the necessary activation energy.

Example One of dozens of reactions that may occur in a smog-ridden area is the linked oxidation of carbon monoxide and nitric oxide:

$$CO + NO + O_2 \longrightarrow CO_2 + NO_2$$

One suggested mechanism for this reaction involves the unstable molecular fragment, HO, as a catalyst.

$$CO + HO \longrightarrow CO_2 + H$$
$$H + O_2 \longrightarrow HO_2$$
$$NO + HO_2 \longrightarrow HO + NO_2$$

(a) Verify that these steps add to give the correct net reaction.
(b) Identify the intermediates in the mechanism.
(c) Tell why HO is considered to be a catalyst in this reaction pathway.

Solution (a) The sum of the three steps in the mechanism is

$$CO + \cancel{HO} + \cancel{H} + O_2 + NO + \cancel{HO_2} \longrightarrow CO_2 + \cancel{H} + \cancel{HO_2} + \cancel{HO} + NO_2$$

HO, H, and HO_2 cancel on both sides to leave the desired net reaction.

(b) The intermediates are H, which is produced in the first step, and HO_2, which is formed in the second step. The other substances that are formed are the final products, CO_2 and NO_2, and the regenerated catalyst, HO, which is present in much greater amount than either intermediate.

(c) Even though HO is involved in the mechanism, its concentration doesn't change, because it is both used up and formed. Presumably, the pathway made possible by the catalyst has a lower activation energy and a greater rate than the uncatalyzed reaction.

IMPORTANT TERMS

reaction rate The change in the concentration of a reactant or product per unit of time (usually 1 minute or 1 second).

pathway of a reaction or **reaction mechanism** The sequence of simple reactions that convert the reactants into products. Most reactions consist of several of these simple steps, which add to give the overall reaction.

intermediate A compound, often unstable, that is formed in one step of a pathway and then is used up in a later step. It does not appear in the equation for the net reaction.

elementary reaction A single, simple step in a reaction mechanism. The reactants in an elementary reaction transform directly (through an activated complex) into the products.

bimolecular reaction An elementary reaction in which there are two reactant molecules. Bimolecular reactions are much more common than unimolecular or termolecular reactions.

unimolecular reaction An elementary reaction having a single reactant molecule.

termolecular reaction An elementary reaction in which there are a total of three reactant molecules.

rate-determining step The slowest step in a reaction mechanism. The rate of the overall reaction equals the rate of the rate-determining step.

energy of activation The potential energy barrier in any elementary reaction. The kinetic energy of the reactants must equal or exceed the energy of activation in order for the products to form.

activated complex The short-lived complex, containing partially broken and partially formed bonds, at the top of the energy barrier.

catalyst A substance that increases the rate of a reaction but is not a reactant or a product in the equation for the net reaction. The catalyst is consumed in one step of the mechanism and is reformed in a later step, so there is no net change in its concentration.

favorable orientation The relative position of reactant molecules in the activated complex that allows a close approach of those atoms which will be bonded together in the products.

QUESTIONS

1. One of several possible mechanisms for a reaction consists of the following steps:

$$I_2 \longrightarrow 2\,I$$
$$H_2 + I \longrightarrow HI + H$$
$$H + I \longrightarrow HI$$

Identify the intermediates, and write the equation for the net reaction corresponding to this mechanism.

2. Clearly explain the difference between *activated complex* and *intermediate*.

3. Which of the following could be elementary reactions? Explain your answers.

a. $N_2O \rightarrow N_2 + O$

b. $Fe + 5\,CO \rightarrow Fe(CO)_5$

c. $2\,CO + O_2 \rightarrow 2\,CO_2$

d. $O_3 + NO \rightarrow NO_2 + O_2$

4. At temperatures well below 430°C (Table 13.1), the mechanism for the reaction

$$CO(g) + NO_2(g) \longrightarrow CO_2(g) + NO(g)$$

may involve the formation of the intermediate, $NO_3(g)$. Write a pathway consisting of two bimolecular elementary reactions that add up to give the correct net reaction. Show NO_3 as an intermediate.

5. The elementary reaction

$$Cl(g) + Cl(g) \longrightarrow Cl_2(g)$$

has an activation energy very close to zero. At room temperature, is the fraction of the collisions resulting in reaction of the Cl atoms closer to 0 or to 1? Explain.

6. The reaction $H_2 + I_2 \rightarrow 2\ HI$ may have this mechanism:

$$I_2 \longrightarrow 2\ I \qquad \text{Step 1}$$

$$I + H_2 \longrightarrow H_2I \qquad \text{Step 2}$$

$$H_2I + I \longrightarrow 2\ HI \qquad \text{Step 3}$$

The third step is rate-determining. Which step is likely to have the largest activation energy? Explain.

7. A possible mechanism for the oxidation of SO_2 to SO_3 in the presence of NO is

$$NO + O_2 \longrightarrow NO_3$$

$$NO_3 + SO_2 \longrightarrow SO_3 + NO_2$$

Explain why it would be wrong to say that NO is simply serving as a catalyst in this sequence.

8. For the reaction

$$2\ ICl + H_2 \longrightarrow 2\ HCl + I_2$$

other mechanisms are possible besides the one given earlier in this chapter. Write a sequence of two bimolecular elementary reactions involving just a single intermediate, HI, and verify that they add to give the correct net reaction.

9. From 10 to 20 miles above us is an atmospheric layer rich in ozone, which forms under the action of ultraviolet light from the sun. One of the processes by which this ozone decomposes is

$$O_3 + O \longrightarrow 2\ O_2$$

There is evidence to show that this reaction is catalyzed by CF_2Cl_2, which formerly was used in aerosol cans and then released into the atmosphere. The first step in the catalyzed pathway is probably

$$CF_2Cl_2 \xrightarrow{\text{ultraviolet light}} CF_2Cl + Cl$$

The second and third steps of the mechanism add up to

$$O_3 + O \longrightarrow 2\ O_2$$

These steps involve Cl as a catalyst; the intermediate is assumed to be ClO.

a. Write a second step and a third step that satisfy the above conditions.

b. Explain why Cl can be regarded as a catalyst.

c. For the sake of discussion, assume that step 2 of this mechanism has a greater activation energy than step 3. In each of these elementary reactions the number of reactant molecules (or atoms) that collide in 1 second is very large, but only a certain fraction of these collisions result in the formation of products. For which reaction (step 2 or step 3) is this fraction greater? Explain.

d. In either of these two steps, is there a need for a favorable orientation of reactant molecules during a collision in order for a reaction to occur? Discuss.

10. Explain why the structure

$$\begin{matrix} O & & & & O \\ & \backslash\backslash & & / & \\ & & S\text{---}N & & \\ & / & & \backslash\backslash & \\ O & & & & O \end{matrix}$$

rather than the one given on page 312, would *not* be a reasonable activated complex for the reaction

$$NO_2 + SO_2 \longrightarrow SO_3 + NO$$

PART TWO

Applications

SECTION ONE

Chemistry and Your Life

Chemistry and the Consumer— Plastics, Colors, and Keeping Clean

Chemicals and chemical processes have been important to our material well-being throughout human history. Learning to use fire was a major step toward civilization. Primitive man found deposits of native copper and learned how to fashion it into tools and cooking vessels. Later, people found that heating certain rocks with charcoal produced iron. Without knowing it, they were converting ore into metal by a chemical reaction.

Recently, we have learned to synthesize plastics for a great variety of uses, from kitchen utensils to automobile parts, at a much lower cost than that of using traditional materials. Synthetic detergents allow us to wash dishes and clothes in hard water at lukewarm or cool temperatures, with a saving of energy.

Most of the chemicals that consumers use are organic compounds, and about 90% of the synthetic organic compounds that our society produces come from petroleum and natural gas. The many hydrocarbons in these resources are valuable not only for their energy content but also as starting materials for compounds ranging from methanol, CH_3OH (the simplest alcohol), to polyethylene, a plastic made from ethylene, $CH_2{=}CH_2$. Although other fuels may replace gasoline, heating oil, and natural gas in the future, new raw materials for our chemical industry may be more difficult to find.

Consumer chemicals tend to be molecules of many atoms. Simple molecules, such as water and carbon dioxide, play key roles in nature, but in the world of technology larger molecules are more often important. The functional part of a soap is an organic ion having as many as 18 carbon atoms; a plastic or adhesive has hundreds of atoms per molecule.

14.1 KEEPING CLEAN WITH SOAP

The procedure for making **soap** remained much the same from classical times to the U.S. colonial period. Pioneer families made their own soap from wood ashes saved from their fires and from animal **fat** left from cooking and butchering (see Figure 14.1). About once a year they would devote several days to soap making. Someone would put the ashes in a barrel with a spigot at the bottom and run water through them to dissolve the lye. This solution of lye was boiled in a kettle, fat was added, and the heating continued for a good many hours.

The essential ingredients in the kettle were triglycerides (esters of glycerol and long-chain carboxylic acids) from the fat and a base, mainly potassium carbonate, from the ashes. Upon heating in the kettle, the fat and the base reacted to give a soap and glycerol as a by-product (see Figure 14.2). This process is an example of a saponification (soap-making) reaction. Later, sodium hydroxide—a stronger base than potassium carbonate—was used in the commercial manufacture of soap.

It is important to realize that soap is a sodium salt of a long-chain carboxylic acid, often called a **fatty acid** since it comes from fat. In other words, it is an ionic compound, containing a positive sodium ion and a

Figure 14.1. Soap-making as it was done in pioneer times. (a) Adding lye to fat. (b) After the mixture is boiled and then cooled, soap forms and is cut into cakes.

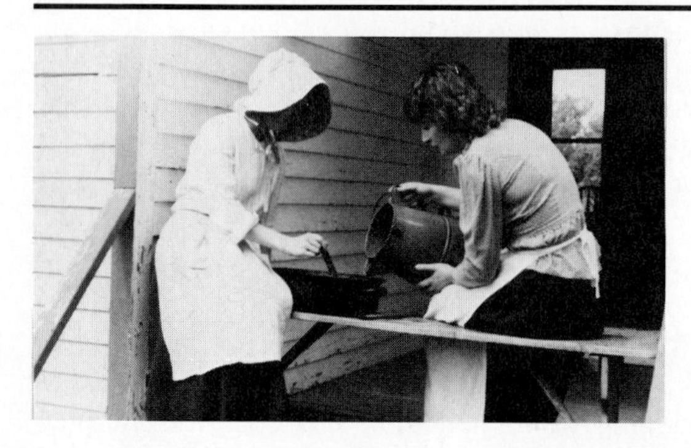

(a)

(b)

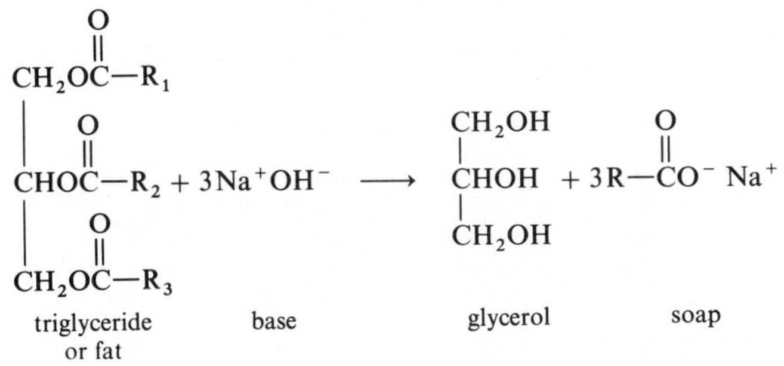

$$\begin{array}{l} \quad\;\; \text{O} \\ \quad\;\; \| \\ \text{CH}_2\text{OC—R}_1 \\ |\quad \text{O} \\ |\quad \| \\ \text{CHOC—R}_2 + 3\text{Na}^+\text{OH}^- \longrightarrow \begin{array}{l}\text{CH}_2\text{OH} \\ | \\ \text{CHOH} \\ | \\ \text{CH}_2\text{OH}\end{array} + 3\text{R—}\overset{\text{O}}{\overset{\|}{\text{C}}}\text{O}^-\ \text{Na}^+ \\ |\quad \text{O} \\ |\quad \| \\ \text{CH}_2\text{OC—R}_3 \end{array}$$

triglyceride or fat — base — glycerol — soap

Figure 14.2. A reaction that produces soap, the sodium salt of a fatty acid.

negative carboxylate ion. The negative charge on this ion resides on the carboxylate group

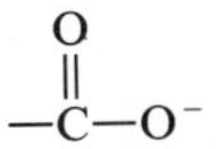

$$\begin{array}{c} \text{O} \\ \| \\ \text{—C—O}^- \end{array}$$

The remainder of the ion is a normal alkyl chain containing 11 or more carbon atoms. Like all alkanes, this hydrocarbon portion of the ion is not attracted to water molecules. Alkanes do not dissolve in water, because they are nearly nonpolar and water is very polar. The general rule is that "like dissolves like." (Alkanes dissolve easily in a variety of other organic liquids that are also nonpolar.) The end of the soap ion with the negative charge, on the other hand, is strongly attracted to the positive end of the water molecule. So one end of the soap ion "likes" water, while the other does not. This contrast gives soap its special properties, as we shall shortly see.

In Figure 14.2 only one example of the R group or alkyl chain is given for the portion of the ester that comes from a carboxylic acid. Fats and vegetable oils, however, contain a half-dozen different alkyl chains. In each triglyceride R_1, R_2, and R_3 are often different, and each is identified with a particular carboxylic acid. The principal fatty-acid constituents of the triglycerides in beef tallow, for example, are oleic, palmitic, and stearic acids. Table 14.1 provides information about these and some of the other fatty-acid constituents of fats. A few of the acids contain one or more double bonds along the alkyl chain. Such an acid is called an **unsaturated fatty acid.** The acids having no double bonds are called **saturated fatty acids,** as they are saturated with hydrogen.

The qualities of a bar of soap can be controlled by the choice of fats from which it is made, because the salts of different fatty acids have different properties. As the length of the alkyl chain increases, the soap feels harder to the touch. Sodium laurate (from lauric acid) is soft and can be deformed easily. Sodium stearate is much harder. As you saw in Chapter 9, the larger the surface area of a hydrocarbon molecule in the liquid state, the greater

Table 14.1 Important Fatty-Acid Constituents in Soaps

NAME	NUMBER OF CARBON ATOMS	FORMULA	SOURCE
lauric acid	12	$CH_3(CH_2)_{10}COOH$	coconut oil
palmitic acid	16	$CH_3(CH_2)_{14}COOH$	beef tallow, cottonseed oil, lard, palm oil
stearic acid	18	$CH_3(CH_2)_{16}COOH$	beef tallow, lard
oleic acid	18	$CH_3(CH_2)_7CH{=}CH(CH_2)_7COOH$	beef tallow, cottonseed oil, lard, palm oil
linoleic acid	18	$CH_3(CH_2)_4CH{=}CHCH_2CH{=}CH(CH_2)_7COOH$	cottonseed oil, lard

the dispersion forces between molecules. Likewise, the stronger the forces between alkyl chains of soap molecules, the harder the soap is.

Sodium oleate is softer than sodium stearate, even though both have 18 carbon atoms and similar molecular sizes. Similarly, a triglyceride containing three stearic acid molecules is a solid fat at room temperature, whereas one made from oleic acid is a liquid oil. The presence of a double bond in the middle of the alkyl chain reduces the force between molecules. The central part of an oleic acid molecule found in nature has this geometry:

```
     H       H
      \     /
       C=C
      /     \
  \ /         \ /
  CH2         CH2
```

As a result, every oleic acid molecule has a permanent kink in it. Neighboring oleic-acid fragments in oils or soaps do not pack together very closely, and the forces between molecules are weakened. The hydrocarbon chains in stearic-acid fragments, on the other hand, are more flexible and can pack together more tightly.

A second property of soaps also depends on the fatty acid. The longer the alkyl chain, the less soluble the soap is in water. Sodium stearate has the firmness needed in a soap, but it is fairly insoluble in lukewarm water. The salts of the 12- and 14-carbon acids are considerably more soluble. Soap manufacturers have discovered that using a mixture of beef tallow and some coconut oil gives a soap bar with the right hardness and a reasonable solubility in water.

14.2 HOW SOAP WORKS

Because of its dual polar and nonpolar nature, soap interacts with water in a remarkable way and produces its familiar cleansing action. When soap dissolves in water, the anions clump together to form **micelles** (see Figure 14.3). In the interior of each micelle, the hydrocarbon ends of a group of anions come together. Here, because of the attractions between hydrocarbon chains, they are in a "friendlier" environment than when they are individually surrounded by water molecules. This arrangement leaves the carboxylate groups, which are water-seeking, at the surface of the micelle, in contact with the water. An increase in stability results, and more soap can dissolve in the water than if all the anions were separated from each other. To balance the negative charges of the carboxylate groups, an equal number of Na^+ ions are distributed at the surface of each micelle. For simplicity these are not shown in Figure 14.3.

Suppose that you pour soapy water onto a surface to be cleaned—a tabletop, a greasy pan, some soiled clothes, or your hand. Micelles make it possible to bring droplets of grease into a soap solution. The mechanical action of a washing machine or other scrubbing motion breaks the grease layer into small bits, around which micelles form (as shown in Figure 14.4). By this process, grease, which is quite insoluble in water, is made soluble in a soap solution. The grease droplet is now in a hydrocarbon region with which it is compatible. This unusual kind of mixture is often called an *emulsion*. Rinsing away the soap solution also washes away the grease, leaving clean clothes, dishes, or hands. Micelles are just one example of super-molecular organization.

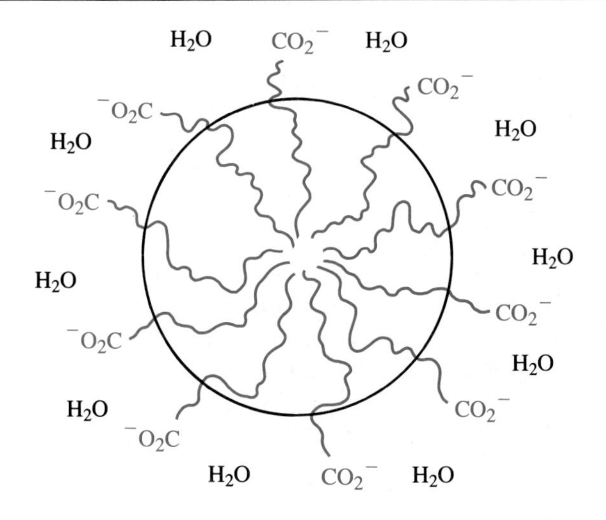

Figure 14.3. A simplified diagram of a micelle.

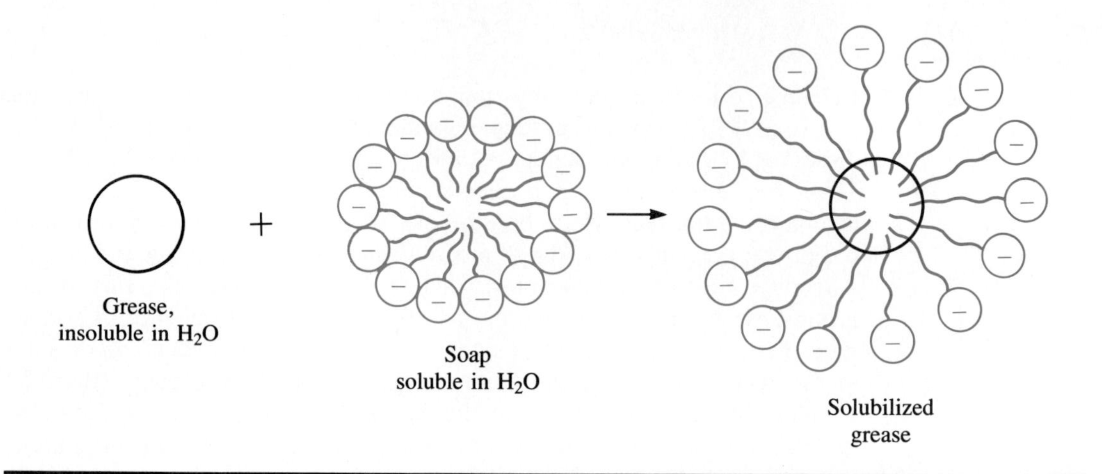

Figure 14.4. A diagram showing how a droplet of grease is made soluble in a soap solution. Close to the carboxylate groups at the exterior of each micelle is a layer of sodium ions to neutralize the negative charges. For simplicity these are not shown.

Grease is naturally soluble in organic solvents (''like dissolves like''). Because some natural fabrics, such as wool, tend to shrink when washed in a water solution, the practice of dry cleaning with an organic solvent was begun many years ago. It has not replaced washing in water because of the expense. Originally, dry cleaning was done by dipping clothing into gasoline or another highly flammable petroleum solvent, which dissolved away the grease and dirt. Later, a less flammable solvent also distilled from petroleum was substituted to reduce the risk of fire. Currently many dry-cleaning establishments use a synthetic solvent called tetrachloroethylene, $CCl_2{=}CCl_2$, which is nonflammable. After treatment with the solvent, the clothes are spun to remove most of the liquid and then are placed in a drying cabinet to complete the process.

14.3 DETERGENTS AND BLEACHES

Until the 1940s most families used soap for all their washing. However, any household with hard water experienced a messy problem: whenever soap was added to water in a bathtub, washing machine, or dishpan, an unsightly gray solid precipitated. Often some of this material settled on the clothes being washed. It was also responsible for the ring left in a bathtub or wash basin after the water had drained. Hard water contains Ca^{2+} and/or

Mg^{2+} ions, which react with soap anions to give a precipitate. For example, with Ca^{2+} the reaction is

$$2\,CH_3(CH_2)_{14}COO^- + Ca^{2+} \longrightarrow \underset{\text{insoluble in water}}{(CH_3(CH_2)_{14}COO)_2Ca}$$

One way to avoid the difficulty is by using soft water, which has Na^+ ions in place of Ca^{2+} and Mg^{2+}, but four or five decades ago water softeners were not as common as they are now. The other way is to use a detergent instead of soap.

Another reason for replacing soap is that the fat from which it is made may be in short supply in some countries. As we will discuss in Chapter 16, fat is one of three major ingredients of food. Wherever there is malnutrition, it does not make sense to use fat in the less-critical area of soap production.

In the 1930s chemists discovered how to synthesize detergents from petroleum. Detergents do not form precipitates in hard water. Like soap, a **detergent** contains a large ion or molecule that has an oil-seeking part and a water-seeking part. One class of detergent is very similar to soap. In place of a carboxylate group, $—COO^-$, it has a sulfate group, $—OSO_3^-$. An example is

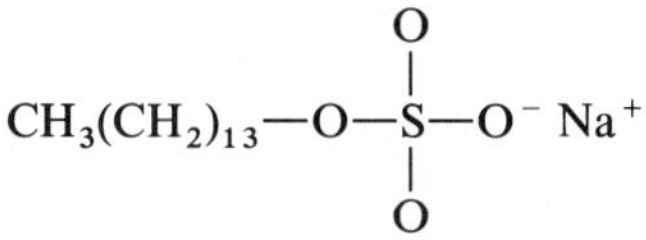

$$CH_3(CH_2)_{13}—O—\overset{\overset{\displaystyle O}{|}}{\underset{\underset{\displaystyle O}{|}}{S}}—O^-\ Na^+$$

Fortunately, the calcium and magnesium salts of this anion are fairly soluble in water. Therefore, when the detergent dissolves in hard water, no precipitate forms. Because its structure is so similar to that of soap, a detergent cleans in the same way. By the late 1940s detergents had become very popular for washing laundry and dishes. A detergent popular with many manufacturers today is an alkylbenzenesulfonate, an example of which is

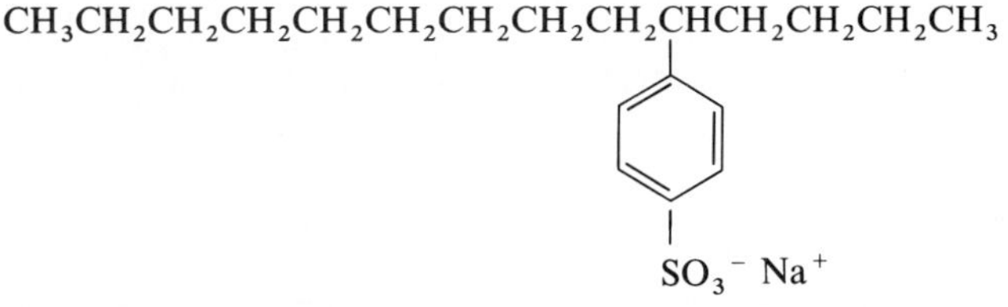

$$CH_3CH_2CH_2CH_2CH_2CH_2CH_2CH_2CH_2\underset{\underset{\displaystyle \underset{|}{C_6H_4}}{|}}{C}HCH_2CH_2CH_2CH_3$$
$$SO_3^-\ Na^+$$

Most products contain a mixture of alkylbenzenesulfonates with different numbers of carbon atoms in the alkyl group.

Although the active portion of the most widely used detergents is an anion, there are some detergents in which the oil-seeking alkyl group is part of a cation. A popular kind of cationic detergent is an alkyltrimethyl-ammonium bromide,

$$\mathrm{R{-}\overset{+}{N}(CH_3)_3}\quad \mathrm{Br^-}$$

As usual, the R group has an affinity for grease, dirt, and oil, but here the positively charged nitrogen atom provides the attraction for water. The cations form micelles around tiny particles of grease in much the same way as the anions of soap and the alkylbenzenesulfonates do.

A few detergents are nonionic, such as a dimethylalkylphosphine oxide,

$$\mathrm{R{-}\overset{\overset{\displaystyle O}{\|}}{P}(CH_3)_2}$$

The alkyl group (R) has the typical 12–20 carbon atoms and constitutes the oil-seeking part of the detergent. Even though the end of the molecule with the P—O group has no charge, it attracts water molecules by means of hydrogen bonding. A hydrogen atom of a water molecule closely approaches the oxygen atom of the detergent, so micelles can again be formed. Nonionic detergents are especially effective and gentle in cleaning fine fabrics and dishes.

The soap suds that we see so frequently are just large collections of bubbles. They occur whenever the surface of a soap or detergent solution is agitated, giving us proof that the cleaning agent is there and doing its work. Whether or not the suds assist significantly in cleaning is open to doubt. Certainly, too many suds can be a problem. If a clothes washer becomes filled with suds, the clothes don't come in contact with the bulk of the detergent solution as much as they should.

The efficacy of cleaning is improved by adding another substance, called a *builder,* to the detergent. The builder does such things as keep the pH in a range in which micelles readily form, promote the breakup of dirt into small particles, and prevent dirt from lodging again on clothes or dishes once it has been removed. A favorite builder for many years has been sodium tripolyphosphate,

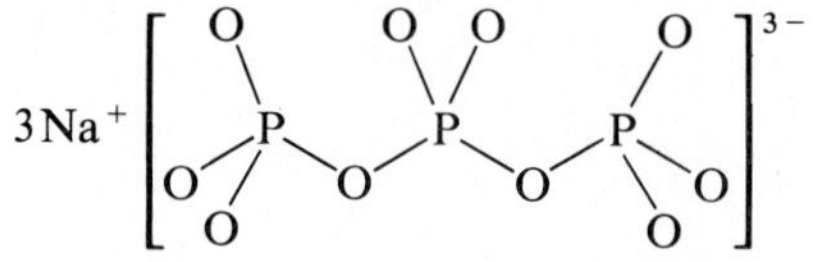

$$3\,\mathrm{Na^+}\left[\mathrm{O_3P{-}O{-}PO_2{-}O{-}PO_3}\right]^{3-}$$

In water solution tripolyphosphate ions slowly react with water to give hydrogen phosphate and dihydrogen phosphate ions, HPO_4^{2-} and $H_2PO_4^-$, respectively. When these ions are carried into lakes, they serve as powerful nutrients for algae, which subsequently grow and multiply until the water looks like a greenish-brown soup. Eventually there are not enough nutrients for all the algae, and many of them die and decay. Because decay is an oxidation reaction, the dissolved oxygen in the water is depleted to the point where fish no longer can survive. The process is called *eutrophication*.

Because of strong public concern over eutrophication about three decades ago, detergent producers began searching for other builders to substitute for phosphate. This search has been made difficult by the fact that the ways in which a builder accomplishes its tasks are not well understood. Although the phosphate content of commercial detergents has been greatly reduced, the search for a satisfactory replacement continues.

Other additives in detergents improve performance in a variety of ways, some of which have more to do with marketing than with keeping clean. For example, an **optical brightener** makes white laundry appear even whiter (and, by implication, cleaner). These compounds are fluorescent, which means that they absorb light of certain wavelengths and then emit light of somewhat longer wavelengths. Newly washed clothing that appears dingy or off-white is reflecting most of the visible light striking it but absorbing a very small amount in a narrow range of colors. A properly chosen optical brightener fluoresces (radiates) just those wavelengths or colors that are absorbed by the fabric. This compensation makes the item appear truly white, although it may be no cleaner.

To make clothes whiter or to remove stains, people occasionally add a **bleaching agent** to the washer along with some detergent. Frequently the color of the stains is caused by a number of double bonds in the molecules of the staining substance. Most bleaches are oxidizing agents, which change colored compounds into colorless substances having fewer double bonds. The main hazard with bleaches is that they may also oxidize fibers and weaken the fabric, especially if used repeatedly.

A bleach that has been used for many years contains sodium hypochlorite, NaOCl, or calcium hypochlorite, $Ca(OCl)_2$, as the active ingredient in water solution. These compounds ionize in water to yield Na^+ or Ca^{2+} ion and hypochlorite ion, ClO^-, which has a strong tendency to oxidize colored compounds by removing electrons from them. Some other compounds containing chlorine also have this ability to destroy colored substances by oxidation.

Other bleaches depend on active oxygen rather than active chlorine. An example is potassium monopersulfate, K_2SO_5, in which the monopersulfate ion has the following arrangement of atoms:

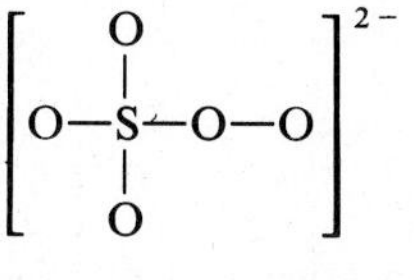

$$\left[\begin{array}{ccccc} & & O & & \\ & & | & & \\ O & - & S & \leftarrow O - & O \\ & & | & & \\ & & O & & \end{array}\right]^{2-}$$

The single bond between two oxygen atoms is called a *peroxide linkage* and is weaker than most single bonds. It breaks readily, releasing oxygen, which can then oxidize colored materials. Potassium monopersulfate is less harmful to fabrics than are bleaches containing hypochlorite. At the same time, the monopersulfate bleach is less effective in removing color. Another popular agent in nonchlorine bleaches is sodium perborate, often written

$NaBO_2 \cdot H_2O_2 \cdot 3H_2O$. Although the structure of this compound is not completely understood, it too contains a peroxide linkage and provides oxygen for oxidizing stains.

14.4 COLORS AND DYES

Color not only enhances the beauty of the world but also conveys important information; a color-blind person suffers from a severe handicap. Color is important to the well-being of many animals. Bright colors guide hummingbirds and bees to flowers, where they find nectar and pollen, and chameleons protect themselves from predators by changing their colors to blend with the background.

Sunlight is often called white light, because it appears neutral and is not endowed with that quality we call color. Actually, it is a mixture of all the colors of the rainbow, as can be demonstrated by passing sunlight through a prism. In Figure 7.1 a simple spectroscope for displaying the colors present in any source of light was diagrammed; it is repeated here as Figure 14.5. At one end of the spectrum of colors lies red, followed by yellow, green, blue, and then violet at the other end.

The basis for color in any object is selective absorption of light. Some substance in the object absorbs one or more colors from the white light coming from the sun or from an electric light. The color we see is the combination of colors that are not absorbed but instead are reflected by or transmitted through the object.

Suppose we observe a piece of blue glass as illustrated in Figure 14.6. White light coming from the left enters the glass, where various molecules or ions absorb red, yellow, and green light. In this process, energy from the light is transformed into potential and kinetic energy of the molecules or ions within the glass. Because most of the red, yellow, and green have

Figure 14.5. A diagram of a simple spectroscope. With the aid of lenses to focus the rays, light from a source is split by a prism into beams of different colors, which strike a photographic plate at different positions.

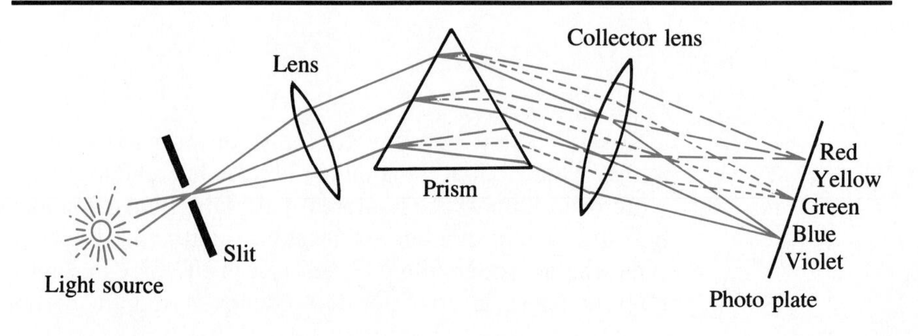

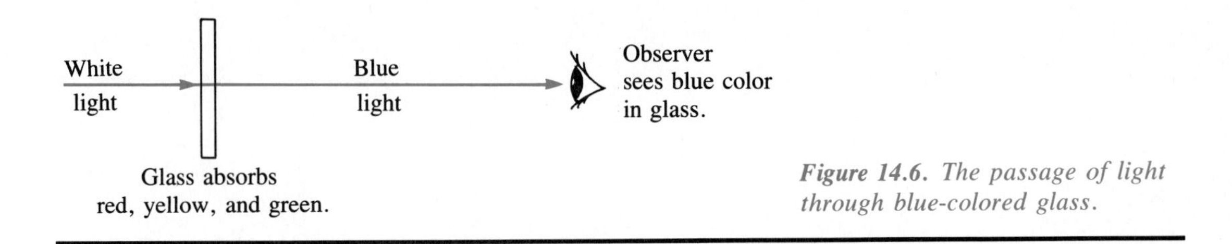

Figure 14.6. The passage of light through blue-colored glass.

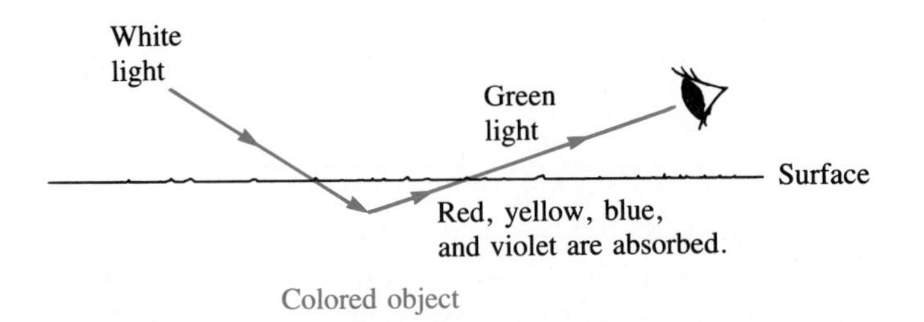

Figure 14.7. The path followed by a light ray directed at an opaque object with a green color.

been removed by the time the light rays emerge from the right side of the glass, the only colors remaining are blue and violet. Standing on the right, we see just these colors and report that the glass has a blue-violet color. Similarly, glass that absorbs yellow and green light transmits red, blue, and violet. It has a purplish-red, or magenta, color.

Most colored objects that we view are opaque. The explanation of their colors, however, is similar to the one presented above. White light hits the surface of the object—a green shirt, for example—and continues a short distance into the interior (see Figure 14.7). Along the way one or more colors are absorbed. In this opaque object, however, most of the light is reflected within a short distance from the surface. When light rays are redirected, many head for the surface and continue to lose the colors that were partially absorbed on the way in. When these rays emerge from the surface, they contain only the complementary colors, which we see when the light rays enter our eyes.

When just one or two colors are absorbed by an object, we see the remaining colors not separately but in combination. For example, a mixture of red and green appears yellow. So when blue is absorbed, all the other

Table 14.2 Complementary Colors

PRIMARY COLOR	COMPLEMENT
red	cyan (blue-green)
green	magenta (blue-red)
blue (indigo)	yellow

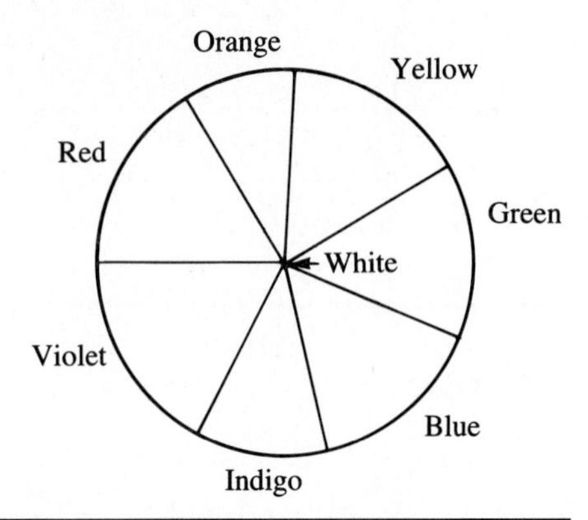

Figure 14.8. A color wheel.

colors are passed along to our eyes, but we see yellow. Because they mix to give neutral or "white" light, blue and yellow are said to be **complementary colors.** Similarly, removal of any color from white light leaves that color's complement. Table 14.2 lists some pairs of complementary colors.

A color wheel (see Figure 14.8) provides another way to display complementary colors. White is at the center of the wheel, and any two colors joined by a straight line through the center are complementary. Artists and designers sometimes use the color wheel to predict the effect of combining various colors. A combination of red and yellow, for instance, appears to the human eye as orange, the color between the two on the color wheel.

Figure 14.9. Dyeing cloth in the nineteenth century. The fabric hangs over the rods laying across the tank and passes through the solution of dye as workers move the rods.

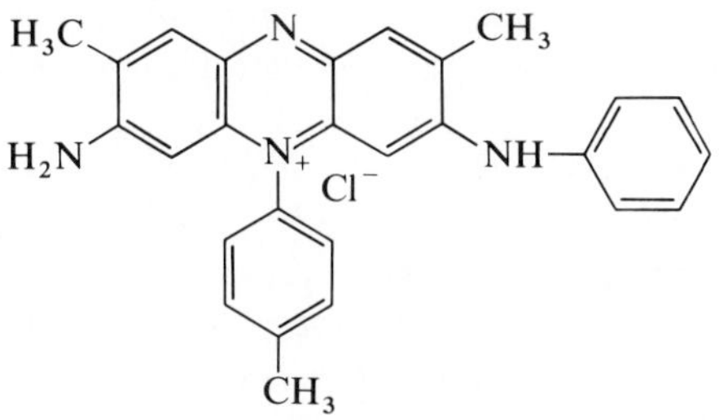

Figure 14.10. Molecular structure of mauve, or mauveine, the first synthetic dye.

A consumer encounters color in fabrics, colored foods, and some cosmetics, to mention only a few examples. Organic dyes are responsible for nearly all of these colors. The word *dye* implies a colored substance that adheres to cloth, and most of our examples will involve dyes for fabrics. However, anything that imparts a color to another substance can be called a dye.

Before the middle of the nineteenth century all dyes came from natural sources—usually plants, but sometimes insects. Through difficult, often expensive processes, dyes such as indigo, saffron, and henna were extracted from large quantities of plants. Some of these dyes adhered well to fabrics and retained their colors; others washed out rather easily or faded quickly in sunlight. The results of a dyeing operation were uncertain, because the purity of different batches of the same dye was difficult to control. Figure 14.9 shows one step in the dyeing of fabric in the nineteenth century.

Figure 14.11. William Henry Perkin.

In 1856 the first synthetic dye, mauve (see Figure 14.10), was prepared by the English chemist William Henry Perkin (see Figure 14.11). For the next few decades, the history of organic chemistry was largely the history of dye synthesis. For the first time it became possible for everybody, not only the wealthy, to own brightly colored clothes, and people responded with enthusiasm. Many important organic reactions were discovered as chemists sought new dyes. With increased understanding of the structures of molecules, organic chemists were able to produce dyes having a great variety of hues and intensities. Nowadays nearly all dyes are synthesized from benzene and other aromatic compounds along with inorganic compounds that provide nitrogen, sulfur, or halogen atoms.

14.5 COLORED MOLECULES

What is there about the structure of a dye such as mauve that makes it colored? This is a crucial question for dye chemists, because a thorough understanding allows the synthesis of a dye tailormade to any specification. When Perkin made mauve, the concepts of structure and bonding had yet to be invented. In 1865, however, a major insight into the structure of benzene set the stage for new understanding. In 1876 in Germany, where the dye industry was to flourish, Otto Witt pointed out some of the structural features of organic molecules that cause them to be colored. These serve as a useful guide even today, despite the more sophisticated theories of bonding now available.

All colored organic compounds contain a **chromophore,** a color-producing group of atoms having at least one double bond. Groups such as —N=O and —N=N— are simple chromophores; Figure 14.12 shows some more complex examples as well. The chromophore enables the molecule to absorb light in the visible range of the spectrum. Many dyes also contain an **auxochrome,** a simple group such as —NH_2 or —OH, which has some influence on the color. Finally, the molecule usually contains a network of carbon atoms connected by alternating single and double bonds. Benzene rings, among other structures, serve this function.

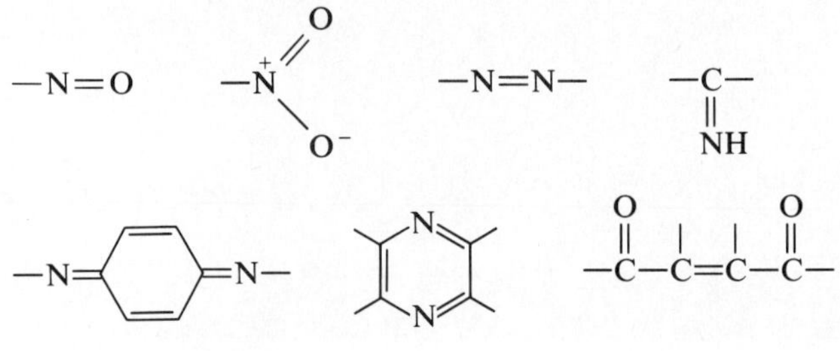

Figure 14.12. *Examples of chromophores, groups of atoms that are primarily responsible for color in organic molecules.*

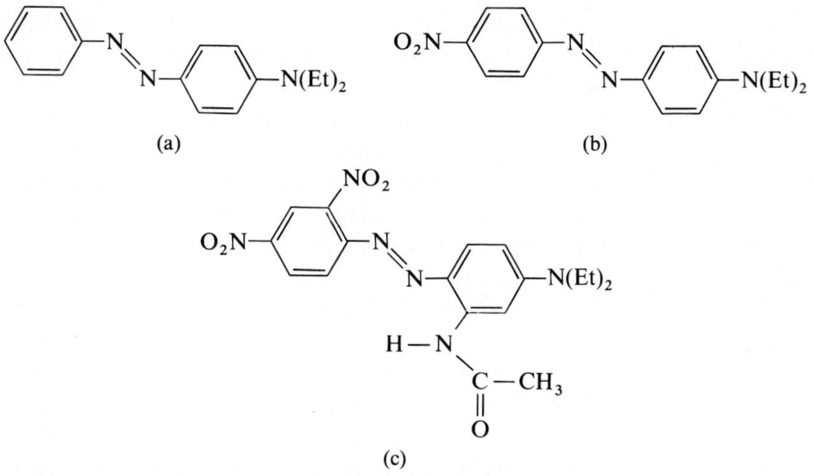

(a) (b) (c)

Figure 14.13. Three azo dyes with the same basic structure but very different colors. (a) A dye with an intense yellow color. (b) With the addition of a nitro group, the yellow turns to red. (c) With more additions, the red changes to blue.

We will use a widely used class of dyes, the azo dyes, to illustrate how chemists can vary the color by changing certain groups. The chromophore in all azo dyes is —N═N—, and an auxochrome common to our examples is

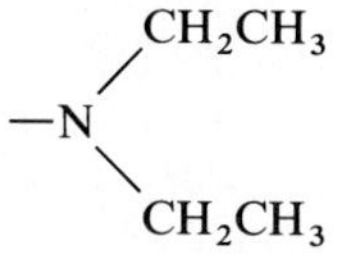

$$-N\begin{matrix} CH_2CH_3 \\ CH_2CH_3 \end{matrix}$$

abbreviated $-N(Et)_2$. Part (a) of Figure 14.13 shows the structure of an azo dye that has an intense yellow color. This molecule absorbs violet and blue light, so the light reaching an observer's eye is green, yellow, and red, a combination that appears mainly yellow. With the addition of a nitro group, $-NO_2$, as in Figure 14.13(b), the color changes to red. This molecule absorbs all the other colors (violet, blue, green, and yellow). Placing still more groups in just the right places, as in Figure 14.13(c), causes the color to change to blue. In this case all colors except violet and blue are being absorbed. Considering all the chromophores and different organic functional groups that can be placed into dye molecules, it is not surprising that about 3500 dyes are used commercially.

14.6 DYEING CLOTH

Even the most attractive dye is of little use if it cannot be made to stick to a fabric. The *fastness* of a dye—its tendency to remain on a fabric—is as important as its color. Because fabrics differ so much in their molecular structures, each dye must be designed for a particular fabric. Cotton, for example, can only hold onto dye molecules by means of dispersion forces and hydrogen bonds, which are very weak compared with ionic and covalent bonds. The basic structural unit in cotton and in paper, both of which are almost pure cellulose, is the simple sugar glucose. Many glucose units are linked together to make a very large molecule. Some of the dyes for cotton and paper have molecular sizes similar to that of glucose, and they fit snugly into the empty spaces found in cellulose. Because of the weak attractions, however, these dyes can be washed out over a period of time. For paper this is not too important, but for cotton it is a drawback.

A better dye for cotton is one that forms insoluble particles within the fibers. Since this kind of dye is insoluble in water, it is first converted into an ionic form to make it soluble and then dissolved in water. As shown in Figure 14.14, after the cotton fabric has been impregnated with dye and then removed from the solution, oxygen in the air changes the dye molecules back into the insoluble form. In this form they have more attraction for themselves than for the cellulose chains, and so they band together in aggregates that are too large to escape from the pores of the cloth. Therefore, very little dye is lost when the fabric is washed.

Dyes attach to nylon, wool, or silk in a completely different way. These fibers, when placed in a dye solution, have a positively charged group at the end of each polymer chain. Wool, for example, is a protein called *keratin*. (Proteins will be discussed in Chapter 15.) It consists of extremely large molecules having a carboxylate group at one end and an amino group, $—NH_2$, at the other end. In water solution the amino group gains a hydrogen ion to become $-NH_3^+$, and that part of the polymer chain has a positive charge, as shown schematically in Figure 14.15(a). Closely associated with the $—NH_3^+$ group is a negative ion such as Cl^- or NO_3^-, shown as X^- in Figure 14.15(a). This ion maintains electrical neutrality.

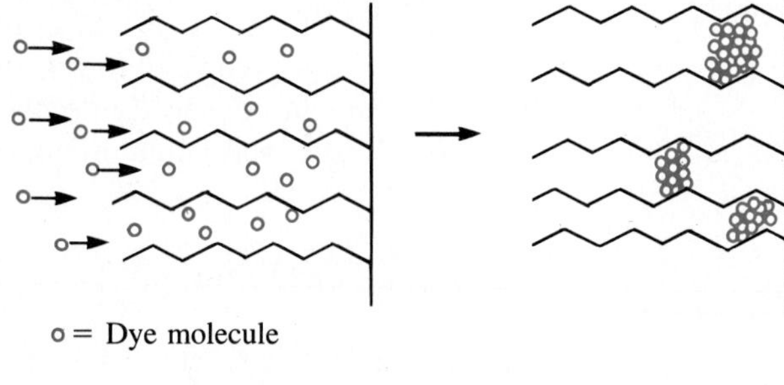

Figure 14.14. A dye that forms insoluble aggregates. (a) Dye molecules, which have been converted into an ionic, water-soluble form, penetrate the pores of the fabric. (b) After oxidation by air, the dye forms aggregates that do not wash out.

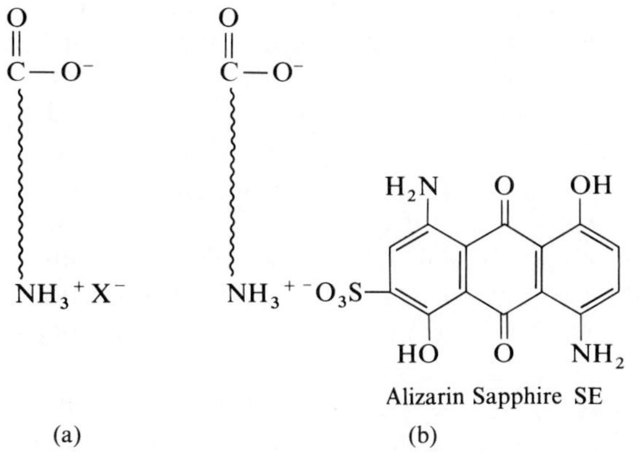

$NH_3^+ X^-$

(a)

$NH_3^+ \, ^-O_3S-$

Alizarin Sapphire SE

(b)

Figure 14.15. The dyeing of wool. (a) A simplified sketch of a protein in wool with its —COO^- and —NH_3^+ end groups. (b) In the presence of a sodium sulfonate dye, the dye replaces the counter-ion, X^-, and an ionic attraction holds the dye to the wool.

The dyes that fasten to wool best have a negative group, often a sulfonate group, —SO_3^-, somewhere in the molecule. A good example is "Alizarin Sapphire SE," a blue dye shown in part (b) of Figure 14.15. When a dye worker puts wool into a bath containing Alizarin Sapphire SE, the process shown in part (b) of Figure 14.15 takes place. The —SO_3^- group of the dye replaces the negative ion that originally was close to an —NH_3^+ group in the wool. Later dry cleaning or washing does not remove the dye because of the strong attraction between the —SO_3^- of the dye and the —NH_3^+ of the wool.

Sometimes the use of a mordant, usually an inorganic salt or hydroxide, increases the fastness of a dye. A mordant added to a fabric before dyeing alters the fabric so as to increase the attraction between dye and fabric. When mordants such as salts of chromium ion, Cr^{3+}, are mixed with the dye, they form complexes with it. The complex adheres more strongly to the fabric than the original dye does and may have a different color.

Colors are as important in the food industry as they are in the clothing industry, because one of the criteria on which consumers base their selection of food is appearance. The principles governing the colors of the dyes added to foods are the same as those governing fabric dyes. The one extra and critical consideration with respect to food additives is whether they are safe to eat. The number of food dyes permitted by the U.S. Food and Drug Administration (FDA) has been decreasing slowly.

14.7 COSMETICS

Personal appearance is at the heart of the cosmetics industry, which sells over $10 billion worth of products each year in the United States. Our interest in looking and smelling good goes back thousands of years. Women have long employed powders and creams to change their complexions and perfumes to create arresting aromas. In modern times men have also become interested in cosmetics; they consume products such as aftershave lotions and deodorants in great quantity.

Most consumers of cosmetics are concerned about two aspects—effectiveness and safety. Except for studies by a few organizations such as Consumers Union, the buyer's chief source of information about effectiveness is advertising. Often the major ingredient of a particular kind of cosmetic is the same for many brands. All have about the same effectiveness, and a buyer's choice depends on secondary considerations, such as fragrance and the appeal of a particular advertisement. Since 1977 U.S. manufacturers have been required to list any ingredient that comprises over 1% of the product, beginning with the predominant ingredient and continuing in descending order. Anyone who is motivated to read this list on competing products can judge the differences. However, if a manufacturer can convince the FDA that a certain ingredient is a trade secret, it can be listed simply as "fragrance" or "color."

Even though the Food and Drug Administration is not required by law to rule on each new cosmetic (as it is on each new drug), manufacturers make great efforts to evaluate the safety of their products. For cosmetics, all of which are applied to the skin, the chief concern is allergic reactions. A generally accepted procedure is for a producer to try a new preparation on many human subjects and certify that fewer than 1 in 10,000 people are sensitized by it. A person is considered to be sensitized to a foreign chemical if the body's defense system overreacts, creating an inflammation or rash. To further forestall sensitization, a manufacturer may alter some ingredients in a particular cosmetic every few years so that regular users will not have long exposure.

Some cosmetics—deodorants and antiperspirants—are designed to reduce body odor. Because perspiration is the source of the problem, a look at the sweat gland (see Figure 14.16) is a natural place to start. Two kinds of glands, eccrine and the less numerous apocrine, deliver a solution of water containing small amounts of sodium chloride, some sulfates, and organic compounds to the surface of the skin. This perspiration has very little odor, but bacteria that are naturally present on everyone's skin metabolize the organic compounds and produce substances with objectionable odors. Because the apocrine glands produce more organic matter than the eccrine glands, they are responsible for most of the odor formation. Since this kind of gland secretes to a canal containing a hair (Figure 14.16) and is stimulated by emotional stress, we produce body odor in regions having hair, such as the armpits, when we are nervous or under tension.

Deodorants and antiperspirants control the problem of body odor in two different ways. A **deodorant** kills skin bacteria. As long as the bacterial level is low, organic compounds in perspiration are not changed into smelly

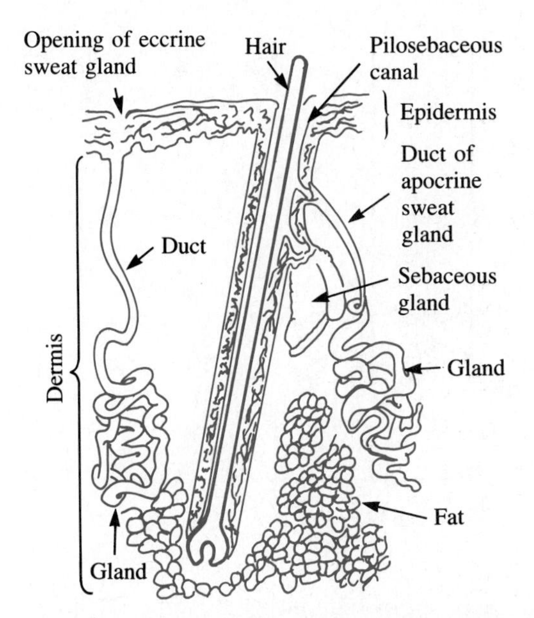

Figure 14.16. *A diagram of human skin, showing both eccrine and apocrine glands.*

products. An **antiperspirant,** on the other hand, slows down the operation of the sweat glands themselves, lowering the amount of perspiration. It is interesting to note that the FDA considers antiperspirants to be drugs, because they control sweating, a natural body function. Deodorants, however, are classified as cosmetics.

Deodorants employ a variety of specialized compounds to kill bacteria. Two of these are benzethonium chloride and zinc phenolsulfonate, the structures of which are shown in Figure 14.17. A few products have less esoteric substances, such as baking soda ($NaHCO_3$) and talcum powder, which have very mild antibacterial activity. Usually a fragrance is added. No doubt some buying decisions depend on the appeal of the fragrance.

The working ingredient in nearly all antiperspirants is an aluminum salt, most often a compound called *aluminum chlorohydrate* or *aluminum hydroxychloride,* with the simple formula $Al(OH)_2Cl$. Occasionally, either hydrated aluminum sulfate, $Al_2(SO_4)_3 \cdot 18\ H_2O$, or hydrated aluminum chloride, $AlCl_3 \cdot 6\ H_2O$, is used. The key constituent is the aluminum ion, Al^{3+}, which somehow closes pores through which perspiration escapes or otherwise hinders the flow.

Even though an aluminum salt is the active ingredient in virtually all antiperspirants, variations do exist in the performance of different brands. An FDA panel found that products containing aluminum chloride are more effective in inhibiting perspiration than preparations containing other aluminum salts. However, aluminum chloride is slightly more irritating to the skin than other aluminum salts and may be inadvisable for some people. This

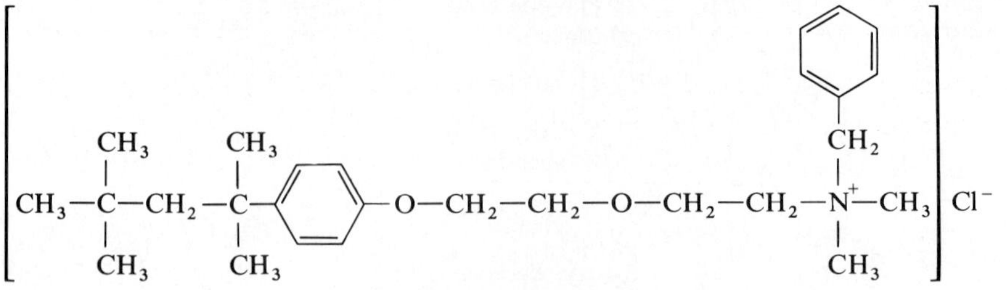

$$\left[CH_3-\underset{CH_3}{\overset{CH_3}{C}}-CH_2-\underset{CH_3}{\overset{CH_3}{C}}-C_6H_4-O-CH_2-CH_2-O-CH_2-CH_2-\underset{CH_3}{\overset{CH_2C_6H_5}{N^+}}-CH_3 \right] Cl^-$$

benzethonium chloride

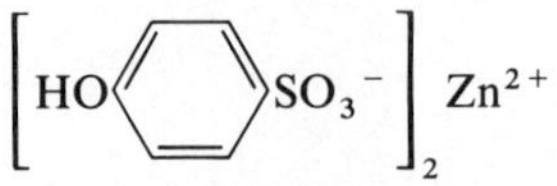

$$\left[HO-C_6H_4-SO_3^- \right]_2 Zn^{2+}$$

zinc phenolsulfonate

Figure 14.17. *Some active ingredients in deodorants.*

finding prompted the warning found on antiperspirant labels. The FDA also discovered that aerosol sprays and roll-on sticks tend to be less effective than lotions. In addition, there may be some hazard in inhaling aerosol sprays. In other words, the formulation in which the aluminum salt occurs does make a difference.

Sunscreens (see Figure 14.18) may not be regarded as cosmetics by most people, but they are applied to the skin. They provide vital protection for anyone working or playing in the sun. In addition to causing sunburn, overexposure to the sun can increase the probability of contracting certain kinds of skin cancer later in life. How do sunscreens work?

The harmful effect of the sun comes from the highly energetic, invisible ultraviolet rays. This high-energy light causes sunburn in skin that has little *melanin,* a dark pigment produced in the body. Melanin defends the body against ultraviolet light, because it absorbs these rays. Typically, a fair-skinned person is unable to produce much melanin and is easily sunburned. A dark-complexioned individual produces more melanin in response to sun exposure and thus is more likely to tan than to burn.

The function of a sunscreen is to filter out the ultraviolet rays. It absorbs the light and converts it into harmless heat. Because many organic compounds, particularly aromatic compounds, absorb ultraviolet light, there would seem to be a great many to choose from. Some, however, don't sufficiently absorb the kind of ultraviolet light that causes burning. Others may cause allergic skin reactions. And finally, still others wash off easily when a person goes swimming or perspires heavily. A once-popular ingredient, *p*-aminobenzoic acid,

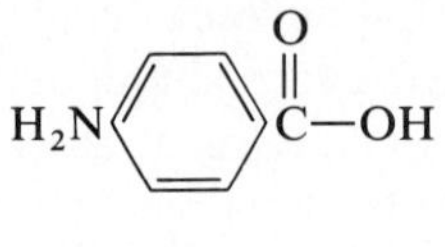

$$H_2N-C_6H_4-\overset{O}{\overset{\|}{C}}-OH$$

GOVERNMENT REGULATION OF SUNSCREENS?

Although *p*-aminobenzoic acid (PABA), an active ingredient in sunscreens, is probably harmless in sunscreen products because of the small amount present, it can cause nausea, vomiting, and skin rashes when taken in large amounts by mouth. Studies with dogs showed that an oral dose of 1.0 gram of the compound per kilogram of body weight caused 50% of the dogs to die.

It has also been reported that chromosomal damage occurs in mice when a dose of 1.0 gram per kilogram is injected into their abdominal cavity. The lowest oral dose that has been shown to affect normal birth in pregnant rats is 2.5 grams per kilogram of body weight. A review of carcinogens in the *Registry of Toxic Effects of Chemical Substances*, published by the National Institute for Occupational Safety and Health (NIOSH), an agency of the U.S. federal government, shows that PABA has an indefinite status as a cancer-causing agent. In other words, more research needs to be done to determine if there is any danger.

Thus PABA is known to be toxic to animals, but only in large doses taken internally. If human beings have about the same dose response as do dogs, mice, and rats, a 150-pound person would have to eat 70 grams, or about half a cup, of PABA before suffering severe toxic effects. There are no data in the NIOSH *Registry* that suggest dangers from the small amounts of PABA in sunscreens rubbed on the skin. Yet some people think that any conceivable danger from toxic chemicals is too much. Here is a situation where a value judgment must be made concerning the known benefits of sunscreens compared to their possible dangers.

What policy should the government have on this issue? Since sunscreens reduce the risks of cancer and skin disorders caused by sunlight, and the toxic effects seem to be very small, the government could simply allow PABA sunscreens to be sold without regulation. Another alternative would be to require a warning label stating the small chance of a toxic effect. Still another would be to ban these products altogether. These decisions are not easy ones; responsible citizens often disagree on the extent of governmental action that is needed.

Overall, with our present knowledge, we think that the health advantages of PABA sunscreens outweigh the possible dangers by a wide margin. As long as PABA and its derivatives continue to be popular sunscreens, it would seem prudent to continue research on any possible cancer danger, even though previous data have indicated no obvious risk.

What do you think?

---

has the last-named disadvantage; it does an excellent job of filtering out the ultraviolet rays but comes off very easily in water. Most sunscreens now contain other active ingredients. A popular one is a derivative of *p*-aminobenzoic acid that is less soluble in water. The FDA has developed a list of 21 safe and effective ultraviolet absorbers for sunscreens.

At the suggestion of the FDA, all manufacturers of sunscreens now rate their products on a scale of 1 to 15 (Sun Protection Factor), with 1 indicating slight filtering of ultraviolet rays and 15 designating complete blockage. This scale is well standardized and reliable. Because of their uniformity in ability

Figure 14.18. *Some familiar sunscreens.*

to screen ultraviolet rays, sunscreens are also rated by the Consumers Union according to their retention on the skin after swimming.

As a final example of a cosmetic, let's look at lipstick. The most obvious component of colored lipstick is a dye, not unlike the ones discussed earlier. Other ingredients serve to protect lips from drying out. Hydrocarbons of high molecular weight provide a protective coating that prevents evaporation of water. Wax is also present to make the lipstick harder so it retains its shape in the dispensing tube. Many lipsticks contain lanolin, which is a mixture of high-molecular-weight esters and some alcohols; lanolin comes from wool and probably softens the skin. Table 14.3 gives the composition of a typical lipstick.

Table 14.3 Composition of a Typical Lipstick

INGREDIENT	FUNCTION	AMOUNT
dye	provides color	4–8%
castor oil, alkanes, or fats	prevents drying of the lips	50%
lanolin	skin softener	25%
wax	makes lipstick stiff	18%
perfume	adds pleasant odor	small

14.8 POLYMERS

The twentieth century has seen a profound change in the materials we use. In earlier times, buildings, machines, tires, appliances, toys, and clothes were made of wood, metal, rubber, cotton, wool, and silk. Now plastics, synthetic rubber, and synthetic fibers substitute for many of these materials. Sometimes, as when a new plastic toy breaks before Christmas day has ended, the new materials seem inferior. When properly designed, however, articles made from plastic are extremely durable. More important, if we didn't have the new materials, there might not be enough wood in the forests and metal ores in the ground to satisfy our needs, and prices of goods would rise. As is true of so many of the substances discussed in this chapter, the raw materials for plastics and other synthetics come from petroleum. Billions of pounds of synthetic polymers are produced in the United States each year. Polyethylene is the leader; about 15 billion pounds are produced.

The story of the new materials, called **polymers,** is the story of large molecules, or **macromolecules.** The word polymer means "many parts." Not all macromolecules are man-made. We have already mentioned cellulose, the main constituent of wood and cotton. All plants and animals contain macromolecules. In the human body, proteins and nucleic acids play vital roles; they are discussed in more depth in Chapter 15. In this chapter we concentrate on synthetic polymers, although the principles governing the properties of both kinds of polymers are similar.

All polymer molecules consist of chains, which may be linear, branched, or cross-linked in a three-dimensional, covalently bonded structure. Their molecular weights can be very large, reaching 1,000,000 or more. Because carbon atoms form stable covalent bonds with each other, the backbone of a polymer is often carbon, but other structures are possible. Sometimes oxygen or nitrogen atoms are interspersed in a carbon chain at regular intervals. Silicone polymers have no carbon atoms at all in the backbone, which instead consists of alternating silicon and oxygen atoms.

An easy-to-visualize polymer structure is linear polyethylene, which is made by joining many ethylene molecules end to end. The simple molecule from which the polymer is constructed (ethylene in this case) is called a **monomer** (meaning "one part"). The double bond in the ethylene monomer is changed into a single bond, and the two electrons thus released participate in bonds with neighboring atoms in the chain:

$$[n+2]\ \underset{\mathrm{H}\ \ \mathrm{H}}{\overset{\mathrm{H}\ \ \mathrm{H}}{\mathrm{C{=}C}}} \longrightarrow \mathrm{CH_3{-}CH_2{-}}\left(\underset{\mathrm{H}\ \ \mathrm{H}}{\overset{\mathrm{H}\ \ \mathrm{H}}{\mathrm{{-}C{-}C{-}}}}\right)_n\mathrm{{-}CH{=}CH_2}$$

Except for the end groups, the polymer consists of identical **repeating units** linked together, as indicated by the parentheses above. The value of n might be 2000, giving a molecular weight of 56,000 (2000 · 28). Since two adjacent CH_2 groups rotate easily with respect to each other about the single bond that joins them, the long chain can assume a variety of shapes. Under certain reaction conditions the chains form random coils (see Figure 14.19). Under other conditions, the molecules of polyethylene are lined up next to each other in a more orderly arrangement.

What are some of the special properties of polymers, and how are these properties related to the structures of the polymer chains? Nearly all polymers are solids. A few are somewhat fluid or gooey at room temperature, but none are gases. When you think about the large surface areas of polymer molecules, this fact is not hard to explain. Large surface areas mean that there are large dispersion forces holding the molecules together. We have already noted in Chapter 9 that melting and boiling points increase with the surface area of molecules.

A peculiar fact about polymers, though, is that many of them, including most plastics, don't have definite melting points. As you raise the temperature of a plastic, it becomes softer and changes shape more easily, but there is no temperature at which it suddenly changes into a liquid in the way that ice turns into water at 0°C. Chemists who have studied how x-ray beams interact with polymers have discovered that many plastics have no regular crystal structures. These plastics, called **amorphous** solids, have been compared with supercooled or highly viscous liquids. Ice, on the other hand, has a well-defined crystal structure, as do most solids composed of simple

Figure 14.19. *A model of a small section of a polyethylene molecule, showing how the chain can bend on itself.*

molecules. Only crystalline solids have definite melting points.

When you squeeze or stretch a polymer, it responds in interesting ways. These responses are called the *mechanical properties* of the polymer, and they depend on how rapidly you attempt to change the shape. If you put a slab of a plastic such as Plexiglas in a machine that stretches it for many hours, it will slowly lengthen in the direction of the stretch. Its shape changes permanently; in other words, it shows an ability to flow. Pounding on it with a hammer, however, only briefly changes its shape (as shown in a high-speed photograph); it immediately springs back to its original shape. The same behaviors are shown by the ''Silly Putty'' sold in novelty stores. By squeezing it slowly, you can mold it into any shape, but if you mold it into a ball and drop it on the floor to cause a rapid distortion, it bounces like a rubber ball. For many plastics, both behaviors are important. A manufacturer wants to be able to mold a plastic at a high temperature, but then wants the plastic to be elastic at room temperature so that it can withstand a sudden shock without fracturing or changing shape permanently.

A simple model can explain these mechanical properties. Because of their great flexibility, polymer chains contain many kinks and bends. In an amorphous plastic such as Plexiglas (see polymethyl methacrylate, Table 14.4), the chains are coiled in random fashion, as shown in part (a) of Figure 14.20. When a stretching force, or stress, is applied to the polymer, the chains uncoil and tend to line up, as in Figure 14.20(b). This process allows the polymer to elongate in the direction of the stretch. If the stress is quickly removed, the coils assume their original configurations and the plastic snaps back into its original shape. On the other hand, a longer application of the stress causes polymer chains to slip past one another, and the shape changes permanently, as in Figure 14.20(c).

It is important to realize that segments of the polymer chain are in constant motion, just as molecules of any liquid vibrate back and forth and occasionally jump past one another. This motion allows the chains to change configuration and to slide past neighboring chains. As in all substances, molecular motion increases when the temperature increases. So chains slip

Figure 14.20. *Changes in the configurations of polymer chains under a stress. (a) Under no stress, the chains are random coils. (b) When stress is applied, the chains tend to line up. (c) After a long time under stress, the chains slip past one another.*

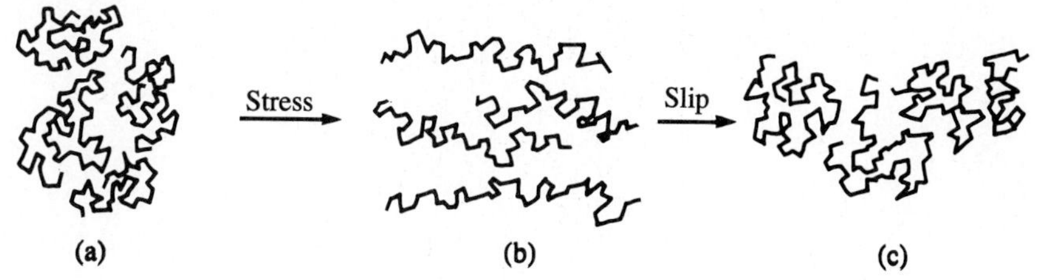

(a) Stress (b) Slip (c)

Table 14.4 Some Synthetic and Natural Polymers

REPEATING UNIT	CHEMICAL NAME	COMMON NAME
$-\!(\mathrm{CH_2{-}CH_2})\!-$	polyethylene	polyethylene
$-\!(\mathrm{CH_2{-}CH(CH_3)})\!-$	polypropylene	polyolefin
$-\!(\mathrm{CH_2{-}CHCl})\!-$	polyvinyl chloride	vinyl
$-\!(\mathrm{CF_2{-}CF_2})\!-$	polytetrafluoroethylene	Teflon
$-\!(\mathrm{CH_2{-}C(CH_3)(C{=}O)OCH_3})\!-$	polymethyl methacrylate	Plexiglas, Lucite
$-\!(\mathrm{CH_2{-}C(CH_3){=}CH{-}CH_2})\!-$ (cis)	poly-*cis*-1,4-isoprene	natural rubber
$-\!(\mathrm{CH_2{-}CH{=}CH{-}CH_2})\!-$ (trans)	poly-*trans*-1,4-butadiene	
$\mathrm{CH_2OH}$, O, H, H, O, OH, H, H, H, OH (glucose ring unit)	polycellulose	cotton, wood, viscose rayons, cellophane

Table 14.4 (cont.)

REPEATING UNIT	CHEMICAL NAME	COMMON NAME
$-\!\left(\overset{O}{\overset{\Vert}{C}}-(CH_2)_4-\overset{O}{\overset{\Vert}{C}}-\underset{H}{\underset{\vert}{N}}-(CH_2)_6-\underset{H}{\underset{\vert}{N}}\right)\!-$	polyhexamethylene adipamide	Nylon 66
$-\!\left(\underset{H}{\underset{\vert}{\overset{H}{\overset{\vert}{C}}}}-\underset{H}{\underset{\vert}{\overset{H}{\overset{\vert}{C}}}}-O-\underset{O}{\underset{\Vert}{C}}-C_6H_4-\underset{O}{\underset{\Vert}{C}}-O\right)\!-$	polyethylene terephthalate	polyester, Dacron
$-\!\left(\underset{H}{\underset{\vert}{\overset{H}{\overset{\vert}{C}}}}-\underset{CN}{\underset{\vert}{\overset{H}{\overset{\vert}{C}}}}\right)\!-$	polyacrylonitrile	acrylic, Orlon

past one another more easily at high temperatures, and the polymer is better able to flow but is less elastic. Thus high temperatures are best for molding plastics.

Not all polymers are amorphous. Under the right reaction conditions, ethylene molecules combine to give high-density polyethylene, in which there are many **crystalline** regions. The simply structured polyethylene chain, having only hydrogen atoms attached to the carbon backbone, can line up with adjacent chains in an ordered arrangement. These crystalline regions don't extend throughout the whole sample but instead are separated by amorphous, random regions, as sketched in Figure 14.21. The close packing

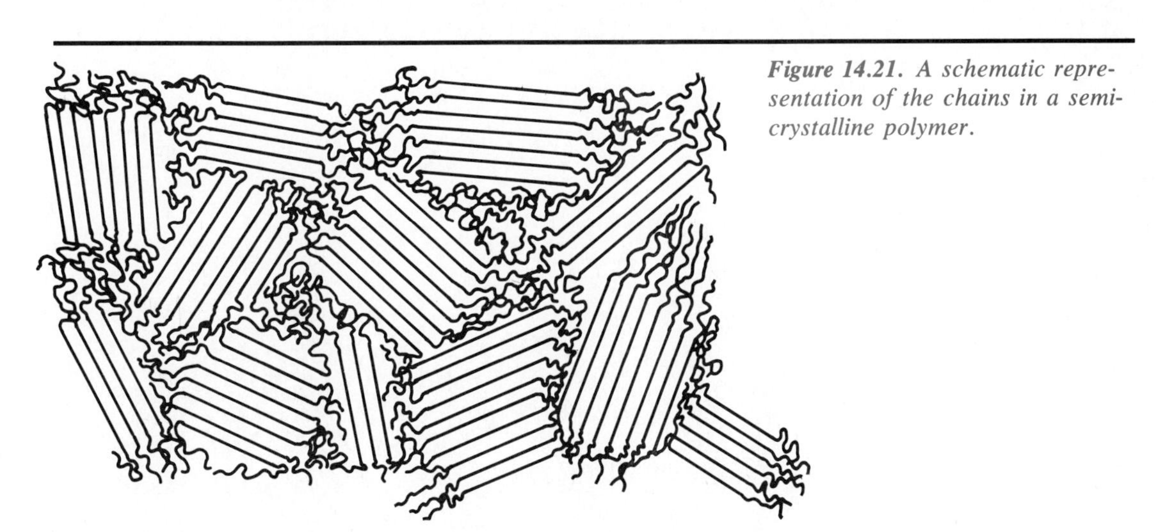

Figure 14.21. A schematic representation of the chains in a semicrystalline polymer.

in high-density polyethylene leads to large intermolecular forces and a hard, tough product.

Some polymers have still another characteristic: **cross-links** join neighboring chains to each other. We can show a cross-linked polymer in schematic fashion as follows:

```
            |                                         |
            X                                         X
            |                                         |
—R—R—R—R—R—R—R—R—R—R—R—R—R—R—R—R—R—
   |                     /                   /
   X                    X                   X
  /                     |                   |
—R—R—R—R—R—R—R—R—R—R—R—R—R—R—R—
              |                                   |
              X                                   X
               \                                   \
—R—R—R—R—R—R—R—R—R—R—R—R—R—R—R—R—R—R—
     |                          |
     X                          X
     |                          |
```

In this diagram R stands for the repeating monomer unit and X is an atom or group of atoms covalently bonded to adjacent polymer chains. Cross-linked polymers consist of one giant macromolecule, because covalent bonds extend throughout the entire system. Vulcanized rubber is an outstanding

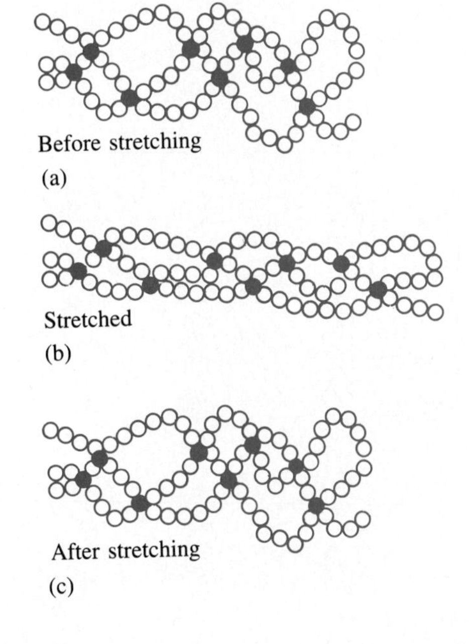

Figure 14.22. The effect of cross-links (solid circles) in vulcanized rubber.

example. Rubberbands and rubber tires can be stretched and compressed repeatedly without losing their shape. The cross-links prevent polymer chains from sliding past one another permanently. Instead, after the stress stops, the chains return to their original configurations, as suggested in Figure 14.22, in which the polymer chain is represented by a string of beads.

14.9 PLASTICS AND RUBBER

Both of the polyethylene bottles shown in Figure 14.23 were heated to 240°F in the same oven. The badly deformed bottle on the left was made from branched polyethylene, and the undamaged one on the right was made from linear polyethylene. A small section of the branched polymer is shown in Figure 14.24(a). When ethylene gas reacts at a moderately high temperature and pressure, the molecules do not all link together end to end. Instead, side chains occasionally form. These branches prevent the long chains from packing closely, reducing the intermolecular forces and producing a flexible, less heat-resistant plastic. This plastic is fine for the squeeze bottles used to hold distilled water in a chemical laboratory but not so good for high-temperature applications. Figure 14.24(b) shows a small portion of linear polyethylene, which is made by heating ethylene with a special catalyst. The closer packing that is possible with the unbranched chains yields a stiffer, heat-resistant material. Controlling the degree of branching is just one way a polymer chemist regulates the properties of plastics.

Closely related to polyethylene is polytetrafluoroethylene, normally called Teflon. Many people know it as the thin coating on no-stick frying pans.

Figure 14.23. *Two bottles made of different kinds of polyethylene.*

```
       H       H       CH3      H
    H  |    H  |    H   |    H  |
    |  C    |  C    |   C    |  C~~
 ~~C      C       C        C
    |  H    |  H    |  H     |  H
    H      CH2     CH2       H
            |         \
           CH3         CH2—CH2—CH3
                 (a)
```

```
       H       H       H       H
    H  |    H  |    H  |    H  |
    |  C    |  C    |  C    |  C~~
 ~~C      C       C       C
    |  H    |  H    |  H    |  H
    H       H       H       H
                 (b)
```

Figure 14.24. Small sections of (a) branched and (b) linear high-density polyethylene.

Teflon is also used in tubing, gaskets, bearings, and electrical insulation in coaxial cable, sockets, and plugs. Its chief asset is its remarkable resistance to heat and to chemicals of all kinds. Made from tetrafluoroethylene, $CF_2{=}CF_2$, Teflon has the repeating unit

```
   F  F
   |  |
 —C—C—
   |  |
   F  F
```

Thus it too is an **addition polymer,** a polymer formed by combining many molecules of a substituted alkene end to end. Teflon is one of the most stable organic materials on the face of the earth.

The principal reason for Teflon's toughness and resistance to heat is that it is 90% crystalline. A piece of Teflon has a white, opaque appearance because the millions of crystalline regions reflect light in many directions. Granulated salt or sugar does the same thing. Another reason why Teflon is not easily distorted is related to its high molecular weight of between 500,000 and 2,000,000. As molecular weight increases, so do intermolecular forces.

The complete replacement of hydrogen with fluorine in Teflon explains its chemical inertness. The C—F bond is strong, and fluorine holds onto its electrons so tightly that scarcely anything can alter it chemically. As a result, Teflon is unaffected by strong bases, strong acids, or any organic solvents. In a chemical laboratory, highly reactive reagents are often stored in Teflon bottles. The stability of Teflon makes it especially valuable in medical transplants.

Another widely used plastic is polyvinyl chloride (PVC); in the United States about 7 billion pounds are made each year. Like polyethylene and Teflon, polyvinyl chloride is an addition polymer made from a monomer in the ethylene family, in this case vinyl chloride, $CH_2{=}CHCl$. Unlike polyethylene and Teflon, however, vinyl chloride is asymmetric: the two ends of the molecule are different. When the polymer is formed, the chlorine end of one molecule may attach to the hydrogen end of the next molecule (head to tail) all along the chain, or other, less symmetrical arrangements may occur. Evidently, the pathway with the smallest energy barrier gives the symmetrical arrangement

$$
\begin{array}{ccccccccccc}
 & \mathrm{Cl} & & \mathrm{Cl} & & \mathrm{Cl} & & \mathrm{Cl} & & \mathrm{Cl} & \\
\mathrm{H} & | & \mathrm{H} & | & \mathrm{H} & | & \mathrm{H} & | & \mathrm{H} & | & \\
| & \mathrm{C} & | & \mathrm{C} & | & \mathrm{C} & | & \mathrm{C} & | & \mathrm{C} & \\
\diagdown\mathrm{C}\diagup & | & \diagdown\mathrm{C}\diagup & | & \diagdown\mathrm{C}\diagup & | & \diagdown\mathrm{C}\diagup & | & \diagdown\mathrm{C}\diagup & | & \diagdown \\
| & \mathrm{H} & | & \mathrm{H} & | & \mathrm{H} & | & \mathrm{H} & | & \mathrm{H} & \\
\mathrm{H} & & \mathrm{H} & & \mathrm{H} & & \mathrm{H} & & \mathrm{H} & &
\end{array}
$$

because this alignment occurs in the commercial product.

One of the more familiar uses of polyvinyl chloride is for plastic pipe, which now is used in most houses, replacing the steel or copper pipe of an earlier era. Other products made from this plastic are phonograph records, floor tile, and window sashes and siding for houses. For products requiring a softer, more flexible plastic, polyvinyl chloride is often mixed with a **plasticizer,** which is simply a low-molecular-weight organic liquid. The small molecules of the plasticizer mix uniformly with the polymer, separating the chains and weakening the forces between chains. Another way to make a softer polymer is to mix two monomers, such as vinyl chloride and vinyl acetate,

$$
\begin{array}{l}
\mathrm{CH_2{=}CH}\}\ \text{vinyl} \\
\qquad\quad | \\
\qquad\quad \left.\begin{array}{l}\mathrm{O{-}\underset{\underset{\displaystyle O}{\|}}{C}{-}CH_3}\end{array}\right\}\ \text{acetate}
\end{array}
$$

to create a **copolymer,** which contains both monomer units in every chain. Shower curtains, raincoats, dishpans, wire insulation, films, and other flexible products are often made from plasticized polyvinyl chloride or a copolymer of vinyl chloride.

Though very similar in structure to polyethylene and polyvinyl chloride, polypropylene has some special commercial uses. Like polyvinyl chloride, polypropylene has a head-to-tail linkage of its repeating unit,

$$
\begin{array}{ccc}
 & \mathrm{H} & \mathrm{CH_3} \\
 & | & | \\
- & \mathrm{C} - & \mathrm{C} - \\
 & | & | \\
 & \mathrm{H} & \mathrm{H}
\end{array}
$$

It can be made in a highly crystalline form that is even harder and stronger than polyethylene. Some of the applications for polypropylene include parts for appliances and automobiles, wire insulation, and pipe. Most often polypropylene is seen as the tightly fitting plastic wrap used to cover food, clothing, phonograph records, and many other items.

Long before the introduction of synthetic polymers, a natural polymer—

rubber—achieved economic importance. Known first as a gummy, elastic material obtained from the sap of rubber trees, natural rubber became much more useful in 1839, when Charles Goodyear, a U.S. inventor, discovered that heating rubber while mixing in a small amount of sulfur made it less sticky and more durable. His discovery of *vulcanization* led to the development of the automotive tire industry that we know today.

The monomer for natural rubber is isoprene, an alkene with two double bonds:

$$CH_2{=}\underset{}{\overset{CH_3}{\overset{|}{C}}}{-}CH{=}CH_2$$

Because rubber is an addition polymer, one double bond in the monomer is absent in the polymer. The other double bond in isoprene moves to the center of the repeating unit. In nature's hands each unit has a special geometry. Here the two CH_2 groups are on the same side of the carbon-carbon double bond, in what chemists call the ***cis*** arrangement, rather than on opposite sides of the double bond, in the ***trans*** arrangement. In other words, each unit looks like

```
—CH2        CH2—                      —CH2        H
    \      /                               \      /
     C=C             instead of             C=C
    /      \                               /      \
 CH3        H                           CH3        CH2—
       cis                                    trans
```

A two-dimensional formula cannot show the actual spatial relationships, but a section of polymer chain in natural rubber is approximately as follows:

```
—CH2        CH2—CH2         CH2—CH2         CH2—
    \      /       \       /       \       /
     C=C            C=C             C=C
    /      \       /       \       /       \
 CH3        H   CH3         H   CH3         H
```

There are other possible geometries: each repeating unit could be *trans*, or the polymer could consist of a random collection of *cis* and *trans* units. The actual *cis* arrangement is just one of countless examples of precise control of molecular geometry in living systems.

The vulcanization of rubber with sulfur altered the polymeric structure in a way that Charles Goodyear could not possibly appreciate because of the primitive state of chemistry in the nineteenth century. Sulfur atoms react with some of the double bonds in the polymer chains and simultaneously cross-link the chains to form an infinite network, as discussed earlier. The cross-links contain from one to half-a-dozen sulfur atoms in series:

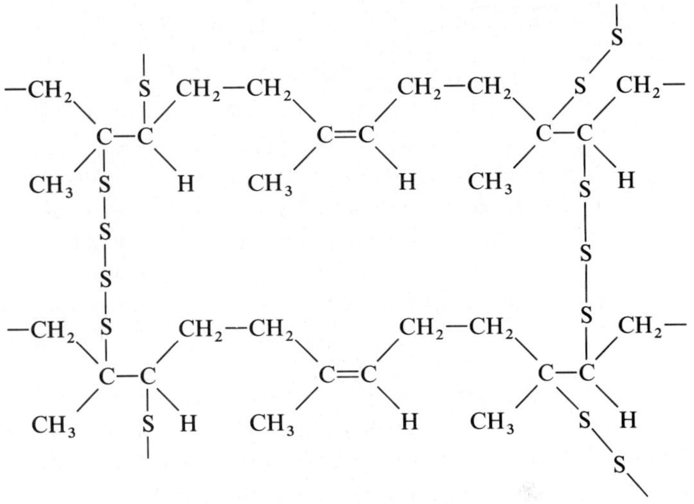

These cross-links give the rubber greater elasticity and prevent chains from slipping past one another (see again Figure 14.22). This is very important, because a bicycle, automobile, or truck tire must flex millions of times without losing its original shape.

During World War II, the United States faced a crisis when it was cut off from its supply of natural rubber in the South Pacific. Without rubber, all vehicles serving the armed forces were useless, not to mention the vehicles needed by the civilian population. Very quickly a synthetic rubber industry was developed, and the production of synthetic rubber went from none in 1940 to 670,000 tons in 1944. Borrowing a method developed in Germany a decade earlier, the U.S. producers used styrene and butadiene as the monomers:

$C_6H_5{-}CH{=}CH_2$ (styrene) $\qquad CH_2{=}CH{-}CH{=}CH_2$ (butadiene)

The synthetic rubber called SBR (styrene-butadiene rubber) is a copolymer in which there is considerably more butadiene than styrene. It also is an addition polymer in which one of the double bonds in butadiene and the double bond in the ethylenic part of styrene are converted into single bonds. It must be vulcanized to have the essential cross-links. SBR is still the most widely used synthetic rubber, employed primarily in tire treads. The second most popular synthetic rubber is made from butadiene alone.

The formulas and uses of the synthetic and natural polymers discussed in this chapter are given in Table 14.4. Also included are the fibers described in the next section.

14.10 FIBERS

As we have suggested before, the most important natural fiber in the world is cellulose, a major component of wood and plants. Cotton is composed almost entirely of cellulose. Cellulose from hemp is used to make rope, and cellulose from wood is used in construction and in making paper. Rayon, the first man-made fiber, is also composed of cellulose.

The repeating unit in cellulose is a glucose ring, as shown in Figure 14.25. Between 200 and 300 of these glucose units are linked together in a chain having a molecular weight of about 400,000. An important characteristic of the chain is the great number of —OH groups that are free to form hydrogen bonds with neighboring chains, preventing the chains from curling up as they do in many plastics. Instead, the chains are hydrogen-bonded together in extended form as tiny bundles. Many bundles make up a fiber, which has very great strength along its length. Breaking a fiber probably means breaking some chemical bonds. Fibers lend themselves to spinning and weaving into cloth—a necessity in all civilizations.

In the United States more than half of the fibers sold are synthetic. Just three fibers account for most of the market: nylon, polyesters, and acrylics. The production of nylon amounted to about 2.3 billion pounds in 1985, just behind the 3.3 billion pounds of polyester. Nylon is a condensation polymer rather than an addition polymer. When a **condensation polymer** forms, a small molecule such as water splits off each time one more monomer links to the polymer chain. Nylon 66, one of two kinds of nylon in common use, is made from adipic acid and hexamethylenediamine,

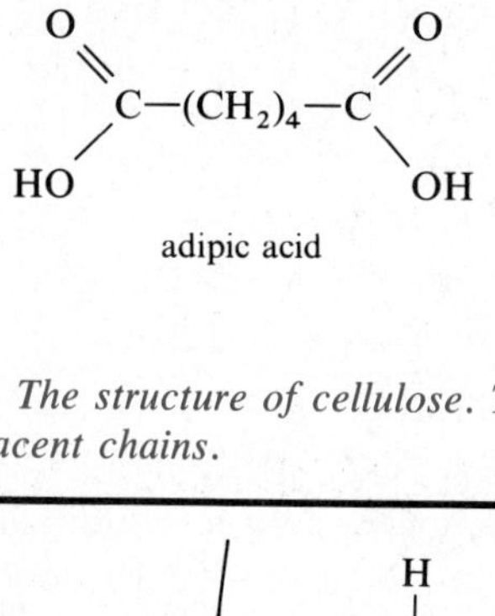

$$HO-C(=O)-(CH_2)_4-C(=O)-OH \qquad H_2N-(CH_2)_6-NH_2$$

adipic acid   hexamethylenediamine

Figure 14.25. The structure of cellulose. The many —OH *groups form hydrogen bonds to adjacent chains.*

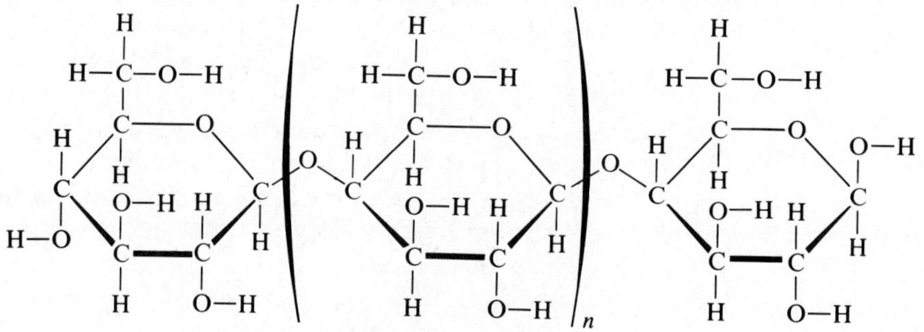

(n = 100 to 3,000)

These two compounds react in the same way as a simple acid and a simple amine (Table 11.4) combine to give an amide and water:

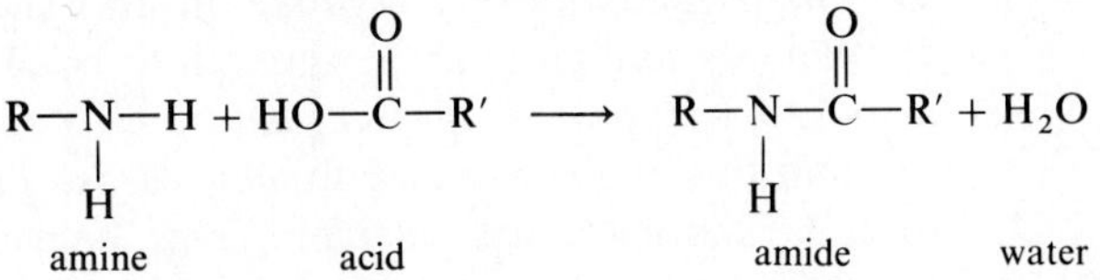

$$\underset{\text{amine}}{R-\underset{\underset{\displaystyle H}{|}}{N}-H} + \underset{\text{acid}}{HO-\overset{\overset{\displaystyle O}{\|}}{C}-R'} \longrightarrow \underset{\text{amide}}{R-\underset{\underset{\displaystyle H}{|}}{N}-\overset{\overset{\displaystyle O}{\|}}{C}-R'} + \underset{\text{water}}{H_2O}$$

Because hexamethylenediamine has two amine groups and adipic acid has two carboxyl groups, each molecule can react at both ends in alternating sequence to build up a polymer chain with the repeating unit

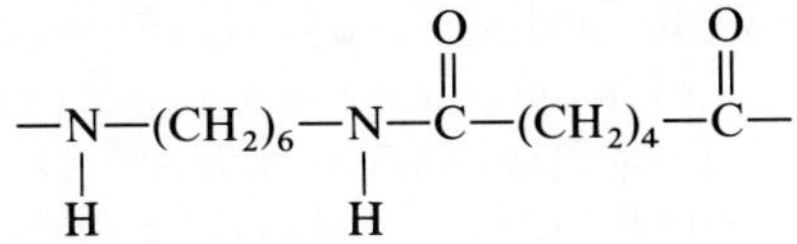

$$-\underset{\underset{\displaystyle H}{|}}{N}-(CH_2)_6-\underset{\underset{\displaystyle H}{|}}{N}-\overset{\overset{\displaystyle O}{\|}}{C}-(CH_2)_4-\overset{\overset{\displaystyle O}{\|}}{C}-$$

Figure 14.26. *A schematic representation of ordered polymer chains in nylon. Hydrogen bonds of the type* N—H ··· O *bind the chains together. The slanting lines pass through the groups that form hydrogen bonds.*

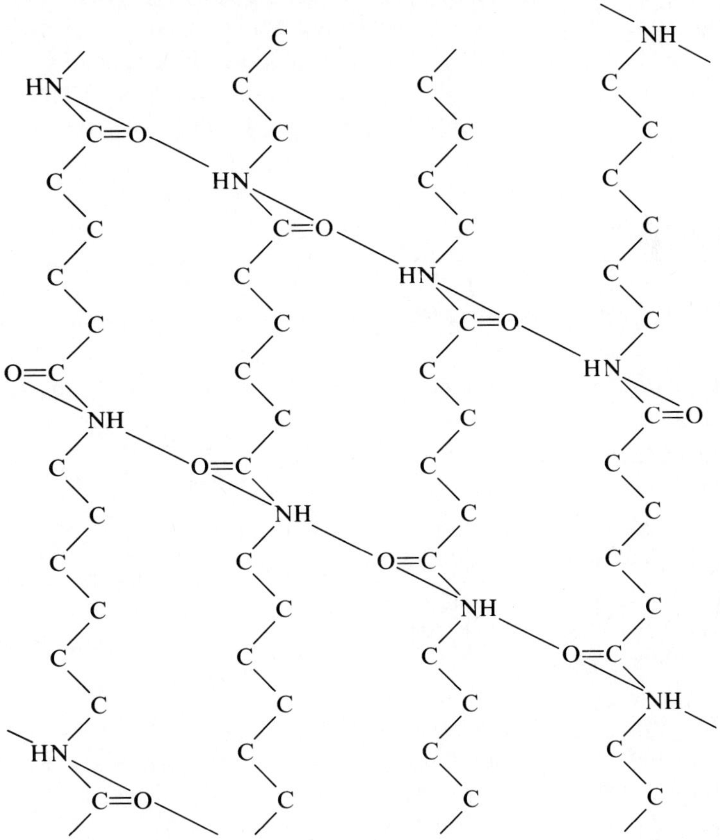

In nylon as in cellulose, hydrogen bonds form between adjacent chains, here between NH groups and the oxygen atoms of the CO groups. Again the chains pack closely together in an extended array that increases the crystallinity and gives great strength to nylon fibers (see Figure 14.26). The manufacturing process involves extruding the solution containing the reactants through tiny holes as the polymer forms. Thin filaments then harden into thread. Extrusion and stretching also help to align the polymer chains in one direction and to increase crystallinity.

Nylon was first prepared by the research group of Wallace Caruthers of the DuPont Company in 1935. It soon replaced silk as the favorite material for women's stockings, primarily because it is more durable and less expensive than silk. Although nylon is still the fiber of choice for stockings, over half of the nylon produced today goes into carpets for homes. Of the remainder, half is used in wearing apparel of various kinds, and half has industrial uses such as reinforcement for tires.

The other two synthetic fibers mentioned earlier, polyester and acrylic, also make significant contributions to our economy. Much of the wearing apparel sold these days consists of a blend of polyester with cotton or wool. Prices of such garments are considerably lower because of the inclusion of polyester. The repeating unit for polyester, a condensation polymer formed from an alcohol and an acid, is shown in Table 14.4. Acrylic, or polyacrylonitrile, is in the polyethylene family (Table 14.4). Although more expensive than polyester, it is popular in sweaters, socks, and other knitted clothing. Sometimes it is used in blankets as a substitute for wool.

IMPORTANT TERMS

soap Sodium salt or potassium salt of a fatty acid.

fat An ester of glycerol and three fatty acids; a triglyceride.

fatty acid A carboxylic acid with a hydrocarbon chain of 10 to 20 carbon atoms.

unsaturated fatty acid A fatty acid in which there is a double bond in the hydrocarbon chain.

saturated fatty acid A fatty acid containing no double bonds in the hydrocarbon chain. It is "saturated" with hydrogen.

micelle An aggregate of a large number (approximately 20–50) of long-chain molecules or ions in water solution. One portion of each molecule or ion avoids water and points toward the center of the micelle. The other, water-seeking end is at the exterior of the micelle.

detergent A substance that consists of molecules or ions with hydrocarbon chains attached to water-seeking groups. Calcium and magnesium salts of detergents are soluble in water.

optical brightener A component of some detergent formulations. It usually fluoresces blue and thus makes clothing seem whiter. (Fluorescence is light emitted by a substance when it is illuminated by light of a somewhat different wavelength.)

bleaching agent A reactive chemical that destroys stains by oxidizing them.

complementary colors A pair of colors that give white light when combined.

chromophore A group of atoms within a molecule that is primarily responsible for the color of a dye.

auxochrome A second group of atoms within a molecule that has some influence on the color of a dye.

deodorant A cosmetic that kills odor-forming bacteria on the skin.

antiperspirant A preparation that closes pores and reduces perspiration.

macromolecule Any very large molecule.

polymer A substance that is composed of very large molecules, made from many small molecules of a given kind linked together.

monomer A small molecule that reacts with many of its own kind to give a polymer.

repeating unit A group of atoms that occurs repeatedly along a polymer chain. It differs from the monomer because of changes in bonding during a polymerization reaction.

amorphous Lacking in order or regularity.

crystalline Possessing an ordered array of molecules.

cross-links Atoms that bind polymer chains together.

addition polymer A polymer resulting from the linking of monomers with no change in their chemical composition.

plasticizer A low-molecular-weight liquid that is mixed with a polymer to make it softer.

copolymer A polymer made up of two or more different monomers.

cis Describes a molecule having two identical or similar groups on the same side of a carbon-carbon double bond.

trans Describes a molecule with two identical or similar groups on opposite sides of a carbon-carbon double bond.

condensation polymer A polymer formed from monomers which may lose several atoms as the links between them are made. An example is Nylon 66.

QUESTIONS

1. Write the formula of a triglyceride that might be in beef tallow. Show specifically which fatty acids are incorporated in the triglyceride.

2. We said in this chapter that the longer the hydrocarbon chain is, the less soluble the soap is. Explain why sodium stearate, for example, ought to be less soluble in water than sodium laurate.

3. Give a characteristic of soap anions that makes them good cleaning agents in water, and explain how the cleansing action works.

4. Why does a detergent perform better than soap in hard water?

5. Explain how the functioning of an optical brightener differs from that of a dye.

6. What is the chemical role of most bleaches?

7. What is the chemical role of swimming-pool chlorine?

8. a. Why does the light coming through the glass in a green traffic light appear green to our eyes?

b. Why does the light coming from a house that has been painted green appear green to our eyes?

9. Explain why a dye that adheres well to wool probably wouldn't be suitable for cotton.

10. In what way does a deodorant function differently from an antiperspirant?

11. Compare the operation of a sunscreen and a dye at the molecular level. Are there similarities?

12. If you decide to seek the very best value the next time you purchase a sunscreen, what facts about the available sunscreens will you investigate?

13. Draw a sketch showing 12 to 14 atoms in the backbone of each of the following polymers. Omit all atoms attached to the backbone.

a. nylon **b.** Teflon
c. polyethylene **d.** natural rubber

14. Explain why the structure of the repeating unit in a polymer is not the same as that of the monomer from which the polymer is formed.

15. Many polymers have no definite melting point. Why?

16. Suppose that you have before you a chunk of sugar and a chunk of plastic. Give at least two significant differences between the properties of these two substances.

17. A certain sample of polyethylene is amorphous and has no cross-links. Explain what will happen to the polymer chains in each of the following cases.

a. It is held in a stretched position for a week.
b. It is quickly squeezed and released.

18. If you wanted to increase the elasticity of a polymer by changing the temperature, would you raise or lower the temperature? Explain.

19. How many repeating units per molecule are there in each of the following cases?

a. a sample of Teflon whose molecules have a molecular weight of 500,000

b. polyvinyl chloride with a molecular weight of 750,000

20. What is the distinction between an addition polymer and a condensation polymer? Include an example of each type.

21. Describe two ways in which you might alter the molecular structure of a polymer to make it softer and more flexible.

22. What does the vulcanization of rubber accomplish, and why is it important?

23. Explain the role of hydrogen bonding in fibers.

Public-Policy Discussion Questions

24. Is the world a better place because of synthetic materials? Present arguments for both points of view. Where do you stand?

25. Although manufacturers of cosmetics voluntarily test their products to determine whether allergic reactions occur in more than 1 in 10,000 people, the FDA is not required to certify each new cosmetic. Some people might argue that an allergic reaction in 1 out of 10,000 users of cosmetics is too many. Should the FDA be given more control over cosmetics?

26. The only reason for adding dyes to foods is to increase the attractiveness of the foods to consumers. The dyes add no nutritional value and may conceivably cause illness or disease in a few people. Should they be prohibited in foods? Present both sides of the issue.

Biochemistry

Biochemistry is the application of chemistry to biological structures and processes—that is, to life. All living organisms can respond to their environment, are able to grow and reproduce, and can use metabolism to get energy and stay healthy. Some organisms are very primitive and relatively simple; others are exceedingly complex. The human body may be the most complex and the most versatile chemical factory on earth. Biochemists have made great strides within the last 50 years in unraveling the chemical reactions that take place in our bodies and the chemicals that we depend on for life itself.

The fundamental structural unit of all organisms is the **cell** (see Figure 15.1), the site of biochemical metabolism. In brief, **metabolism** is the total complex of chemical and physical processes involved in the maintenance of life. There are many different kinds of cells: animal cells, plant cells, bacterial cells, liver cells, nerve cells, skin cells, and many others. Although they differ, most of them contain a number of similar compartments and chemicals. Most of the chemicals in living organisms are organic molecules, in which carbon atoms are covalently linked with hydrogen, oxygen, and nitrogen. These molecules conform to all the laws of chemistry.

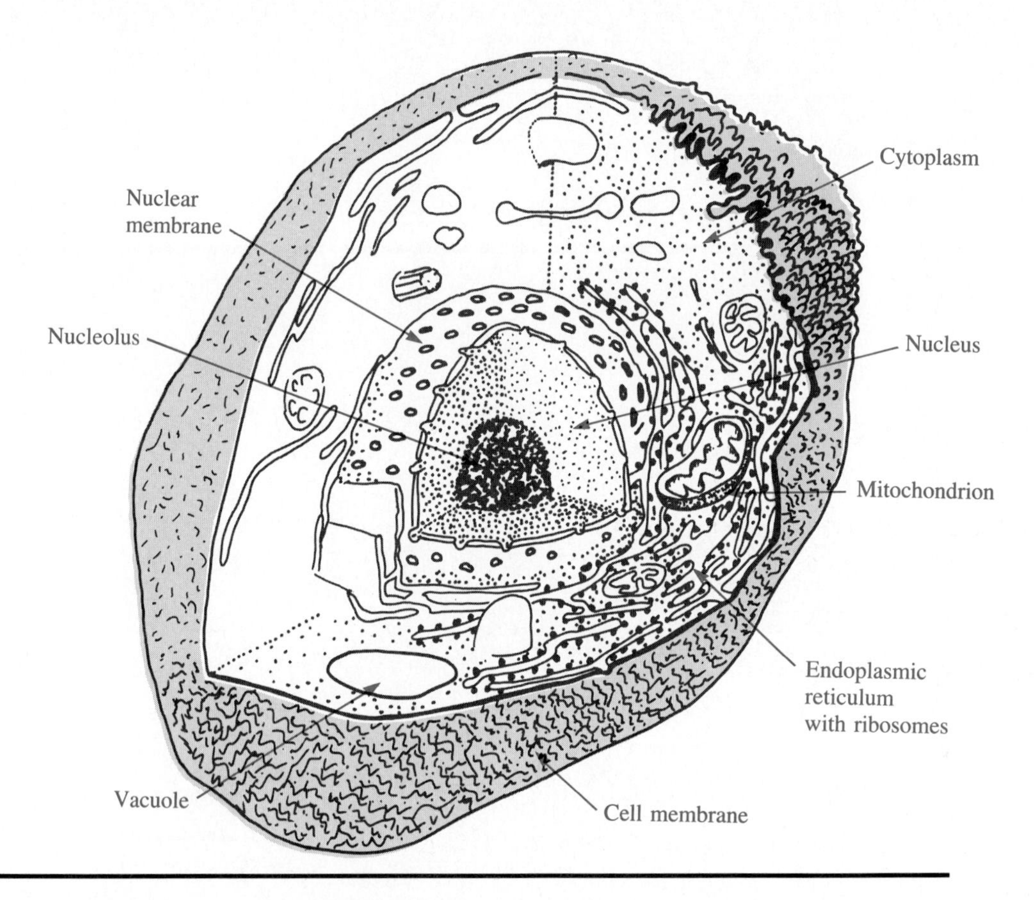

Figure 15.1. A simplified picture of a cell. Some of its important parts are labeled, but it is not necessary to learn the names.

There are four major groups of chemical compounds or biomolecules upon which all life depends. They are called (1) carbohydrates, (2) lipids, (3) proteins, and (4) nucleic acids. Many of these chemicals are macromolecules or polymers. They are made of repeating units, much like the condensation polymers mentioned in Chapter 14. The repeating units are monosaccharides in carbohydrates or polysaccharides, fatty acids in lipids, amino acids in proteins, and mononucleotides in nucleic acids. A single cell of a simple bacterium may contain 5000 different kinds of organic compounds, with perhaps 3000 different kinds of proteins and 1000 different kinds of nucleic acids.

Many structures and formulas of biomolecules are given in this chapter. Usually it is not necessary to memorize them; most are presented only as illustrations. Your instructor can provide guidance as to which structures and formulas you should learn.

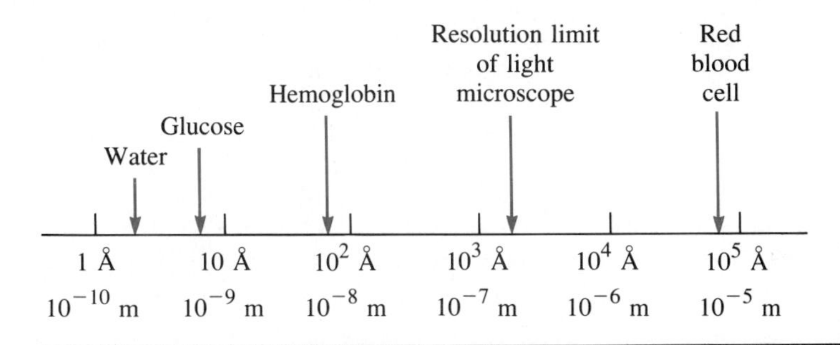

Figure 15.2. *The dimensions of some biomolecules and cells. The numbers are in Angstrom units (Å); 1 Å = 1 × 10^{-10} meter.*

The most abundant molecule in living cells is water, which may make up 70% of an average cell's weight. Protein is the next most abundant at about 15%. The remaining 15% is composed of nucleic acids, carbohydrates, lipids, and inorganic salts. The chemicals in plant and animal cells have a wide range of sizes, as shown in Figure 15.2.

15.1 CARBOHYDRATES—SUGARS

Carbohydrates, such as sugars and starches, provide much of the energy that we all run on. A well-balanced diet contains about 65% carbohydrates; grains, vegetables, and fruits are the main sources. The amount of carbohydrate in the biosphere probably exceeds the total amount of all other organic matter combined. Most plants contain large amounts of cellulose, which is a polysaccharide. The cell walls of many microorganisms are also made largely of polysaccharide compounds. The word *saccharide* comes from the Greek word for sugar. Some of the more common carbohydrates, as well as their classifications, are shown in Table 15.1.

Carbohydrates form from carbon dioxide and water in green plants, using the energy of the sun. The process is known as *photosynthesis,* without which there would be little food in the world. Carbohydrates are at a much higher energy level than is carbon dioxide, so energy must be used to drive their synthesis. On the average, plants all over the world use about 5 billion dietary calories of energy every second to produce carbohydrates. Plants carry out photosynthesis in specialized cellular structures where chlorophyll

Table 15.1 Different Examples of Carbohydrates

POLYSACCHARIDE	DISACCHARIDE	MONOSACCHARIDE
cellulose	sucrose	glucose
starch		fructose
		glyceraldehyde

molecules are assembled for the efficient trapping of sunlight. Chlorophyll gives plants their green color. The unbalanced chemical reaction of photosynthesis is

$$CO_2 + H_2O + \text{sunlight} \longrightarrow \text{carbohydrate} + O_2$$

Carbohydrates are so named because their simplest formula, CH_2O, seems to suggest that they are made of *carbon* and *water*. Since carbohydrates contain many CH_2O units, the balanced equation for photosynthesis is

$$x\,CO_2 + x\,H_2O + \text{sunlight} \longrightarrow (CH_2O)_x + x\,O_2$$

Chemists use the name **sugar** for any simple carbohydrate. The simplest type of sugar, called a **monosaccharide,** contains from three to seven carbon atoms. The two three-carbon monosaccharides have these structures:

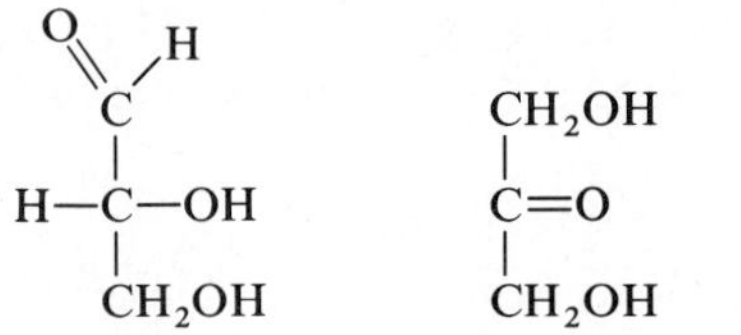

$$\begin{array}{ccc} \begin{array}{c} O\!\!\diagdown\!\!\diagup H \\ C \\ | \\ H\text{—}C\text{—}OH \\ | \\ CH_2OH \end{array} & \qquad & \begin{array}{c} CH_2OH \\ | \\ C{=}O \\ | \\ CH_2OH \end{array} \\ \text{glyceraldehyde} & & \text{dihydroxyacetone} \\ C_3H_6O_3 & & C_3H_6O_3 \end{array}$$

Notice that each of these sugars contains two alcohol (O—H), or hydroxyl, functional groups, as well as an aldehyde ($\text{—}\overset{O}{\overset{\|}{C}}\text{—}H$) or ketone group ($\text{—}\overset{O}{\overset{\|}{C}}\text{—}$) (see Table 11.4).

In modern use the term *carbohydrate* refers to substances that contain several hydroxyl groups and either an aldehyde or a ketone group. It also includes more complex chemicals that react with water to form such substances. Like other alcohols, sugars form hydrogen bonds easily and therefore tend to be soluble in water.

The most important monosaccharide sugars in our bodies are those that contain five or six carbon atoms. The most abundant monosaccharide is **glucose.** Natural sources of glucose are fruits, especially dates and grapes. Glucose is one of the fastest and most efficient sources of energy. No wonder a glucose solution is the food dripped into the veins of hospital patients who are unable to take nourishment by mouth. Glucose is sometimes called *blood sugar* because it is the main sugar transported in the bloodstream for use in the body. When the glucose concentration in the blood is too low (hypoglycemia) or too high (hyperglycemia), there are immediate symptoms. Going without breakfast can often produce a sluggish or dizzy feeling that accompanies hypoglycemia. People who suffer from diabetes mellitus can have too much glucose in their blood. In severe cases of hyperglycemia, diabetics can go into a coma. This condition can be relieved through the administration of insulin, a hormone that controls the metabolism of glucose.

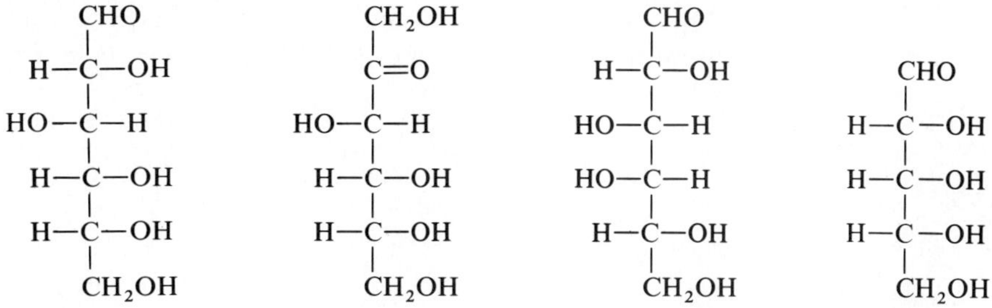

glucose	fructose	galactose	ribose
CHO	CH_2OH	CHO	
H—C—OH	C=O	H—C—OH	CHO
HO—C—H	HO—C—H	HO—C—H	H—C—OH
H—C—OH	H—C—OH	HO—C—H	H—C—OH
H—C—OH	H—C—OH	H—C—OH	H—C—OH
CH_2OH	CH_2OH	CH_2OH	CH_2OH
glucose	fructose	galactose	ribose
$C_6H_{12}O_6$	$C_6H_{12}O_6$	$C_6H_{12}O_6$	$C_5H_{10}O_5$

Figure 15.3. *Fischer structures of four important monosaccharides. The aldehyde group,* —C(=O)—H, *is written* —CHO.

If too much insulin is given, though, insulin shock can occur as a result of too little blood glucose.

Another six-carbon sugar, fructose, is the sweetest of the sugars. Fructose is very water soluble and is often used in liquid sweeteners. Diabetics can metabolize fructose much more readily than glucose or starches. The structures of glucose, fructose, and two other monosaccharides are shown in Figure 15.3.

The structures drawn in Figure 15.3 are called *Fischer structures*. Fischer structures provide a way to show the three-dimensional structures of the sugars, although they contain more information than we will use. Interchanging the hydrogen and hydroxyl groups connected by horizontal bonds to any of the tetrahedral carbon atoms will give the Fischer structure of a different sugar.

In water solution most simple sugars are found as cyclic compounds. One of the hydroxyl groups of the sugar molecule reacts with the carbon-oxygen double bond to form a ring. In its cyclic form, glucose has a ring of six atoms. There are two different cyclic forms of glucose that differ only in the arrangement of their atoms at one carbon atom. In one cyclic form, the hydroxyl group attached to the carbon atom at the far right is up; in the second one, it is down. All of the other atoms in glucose have the same positions.

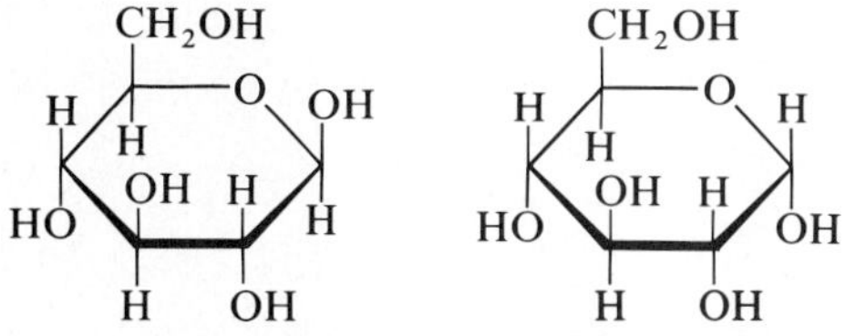

CH_2OH O OH H H OH H HO H H OH

CH_2OH O H H H OH H HO OH H OH

the cyclic forms of glucose

Ribose, a five-carbon sugar, has a ring of five atoms. We will discuss ribose later in conjunction with ribonucleic acid (RNA).

$HOCH_2$ O OH
H H H H
HO OH

ribose in the cyclic form

The structures of glucose and other monosaccharides are far simpler than those of proteins and nucleic acids, not to mention cell membranes and other cell structures. However, a hundred years ago some scientists predicted that glucose was so complex that we might never learn its detailed structure. A few years later the structure was solved in a brilliant series of chemical experiments performed by a German chemist, Emil Fischer (see Figure 15.4). The history of science is full of such events.

Figure 15.4. Emil Fischer, who determined the complete structure of glucose in 1891. He won the Nobel Prize in Chemistry in 1902.

15.2 SUGARS AND STARCHES

A sugar that contains two simple sugar units is called a **disaccharide.** Disaccharides are made of two monosaccharides joined together by an ether linkage (C—O—C), which results from the elimination of a molecule of water. Probably the most common disaccharide is **sucrose,** or table sugar. When sucrose reacts with water in the presence of acids, the monosaccharides glucose and fructose are formed. From this reaction we can infer that sucrose contains a glucose unit and a fructose unit.

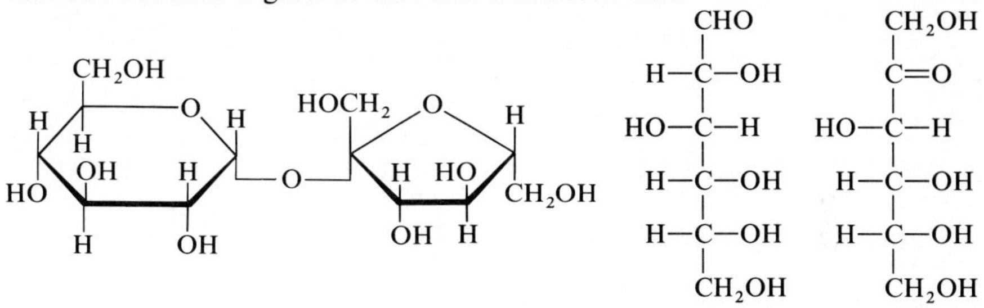

$$\underset{\text{sucrose}}{C_{12}H_{22}O_{11}} + H_2O \longrightarrow \underset{\text{glucose}}{C_6H_{12}O_6} + \underset{\text{fructose}}{C_6H_{12}O_6}$$

(shown as Fischer structures)

Sucrose is used to sweeten many foods, from candy to cereal. Occurring universally throughout the plant kingdom in fruits, seeds, flowers, and roots, sucrose is isolated commercially from sugarcane and sugar beets. One reason for its popularity as a sweetener is that it can easily be crystallized or granulated from water solutions, so it is easily purified or refined.

Of the many other disaccharides, we will mention only lactose, or milk sugar. The milk of mammals contains a fair amount of lactose, from 2 to 8%. In sour milk some of the lactose has been converted by bacterial action to the sour-tasting lactic acid. Adults in many cultures do not regularly drink milk, and so they lose part of their ability to convert lactose into the monosaccharides glucose and galactose. When lactose remains undigested in their intestines, it causes cramps and diarrhea.

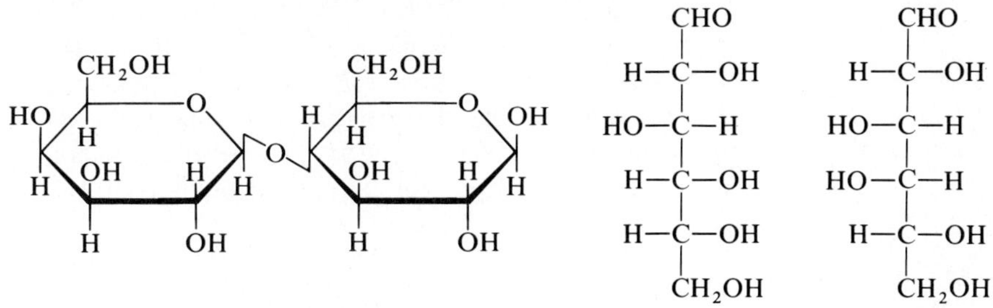

$$\underset{\text{lactose}}{C_{12}H_{22}O_{11}} + H_2O \longrightarrow \underset{\text{glucose}}{C_6H_{12}O_6} + \underset{\text{galactose}}{C_6H_{12}O_6}$$

(shown as Fischer structures)

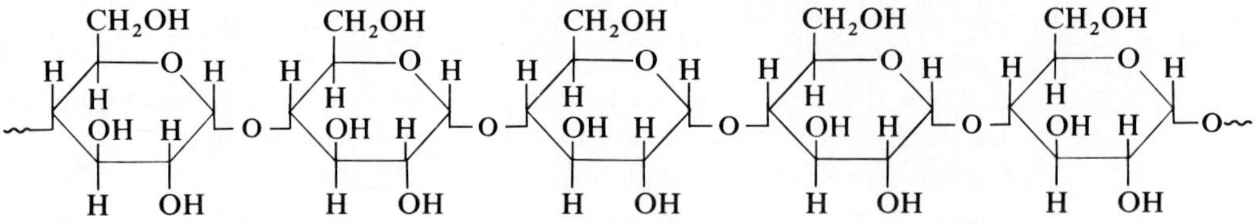

Figure 15.5. A partial structure of amylose, a soluble starch.

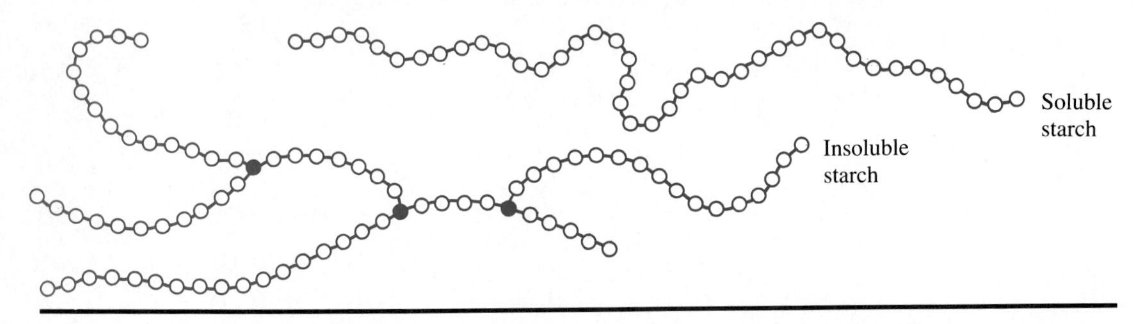

Figure 15.6. Partial structures of soluble and insoluble starches. The circles represent glucose units; the darkened circles are branching points.

If a large number of monosaccharide units are linked together, a **polysaccharide,** or sugar polymer, results. Glucose is far and away the most common monosaccharide unit in polysaccharides. When hundreds of glucose rings link together end to end, cellulose (Figure 14.25) or **starch** (see Figure 15.5) results. The major difference between starch and cellulose is that the cyclic glucose units are connected differently.

When we think of starch, we usually think of food. Most of the carbohydrates we eat are in the form of starches. In plants, the starches are stored as large granules that are insoluble in cold water. In warm water the granules release a soluble starch; the rest of the starch is insoluble. The soluble starch is a linear polymer; the insoluble form has branched chains of glucose units (see Figure 15.6).

In animals, glucose is stored as part of the polysaccharide glycogen, which is especially abundant in liver and muscle tissue. Glycogen has a structure much like that of insoluble starch except that glycogen is even more highly branched. When you need energy, glycogen from the liver is converted to glucose, which in turn is metabolized to produce the necessary energy.

15.3 LIPIDS—FATS AND OILS

A **lipid** is any oily or greasy water-insoluble chemical in cells that dissolves in nonpolar organic solvents. Unlike carbohydrates, all of which have strong structural similarities, the lipids are rather diverse in their structures. The one characteristic that all lipids have in common is their nonpolar nature. Many of them are hydrocarbons or esters (see Section 11.4); a few of them have functional groups such as hydroxyl groups that can take part in hydrogen bonding. Fats and oils are the most abundant class of lipids in nature, and fatty acids are their building blocks. They provide a means for plants and animals to store excess food energy. Other lipids include the steroids, such as cholesterol and the sex hormones.

Many people in the United States consume 40 to 50% of their calories in the form of fats or oils. Natural fats and oils are triglycerides (Figure 14.2). One of the common fats in butter has the structure

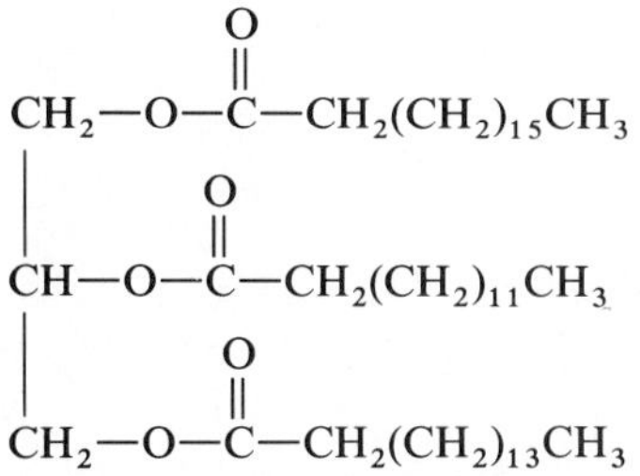

$$
\begin{array}{l}
CH_2-O-\overset{\overset{\displaystyle O}{\|}}{C}-CH_2(CH_2)_{15}CH_3 \\
| \\
CH-O-\overset{\overset{\displaystyle O}{\|}}{C}-CH_2(CH_2)_{11}CH_3 \\
| \\
CH_2-O-\overset{\overset{\displaystyle O}{\|}}{C}-CH_2(CH_2)_{13}CH_3
\end{array}
$$

Notice that this triglyceride has three different fatty-acid groups in it, each attached to an oxygen atom of glycerol.

In most animal and plant cells, fats and oils occur as tiny, finely dispersed droplets. Animals also have specialized fat cells in their connective tissues. Fat droplets fill almost the entire volume of fat cells, as shown in Figure 15.7. Fat cells exist in large numbers under the skin, in the abdominal cavity, and in the mammary glands. Many kilograms of fat are present in the fat cells of obese people, enough to supply minimal energy needs of the body for several months. By contrast, the body can only store about a day's energy supply in the form of glycogen. When fats are metabolized, they not only provide a great deal of energy, but also release a large amount of water. A camel's hump is an example of fatty tissues used as an internal water reservoir.

The **steroids** are an important group of lipids. Cholesterol (see Figure 15.8), vitamin D, and many of the sex hormones are steroids.

Cholesterol is present in all normal animal tissues, but it is concentrated in the brain and in the spinal cord. It is the major component of human gallstones. Deposits of cholesterol on the inner linings of arteries cause their hardening. As these deposits build up, they can clog the arteries and may lead to a blood clot and a heart attack. There is considerable controversy over the connection between a high cholesterol level in the blood and a diet rich in saturated fats, but it is not uncommon these days to see margarine and vegetable oils labeled with the words "low cholesterol" and "polyunsaturated."

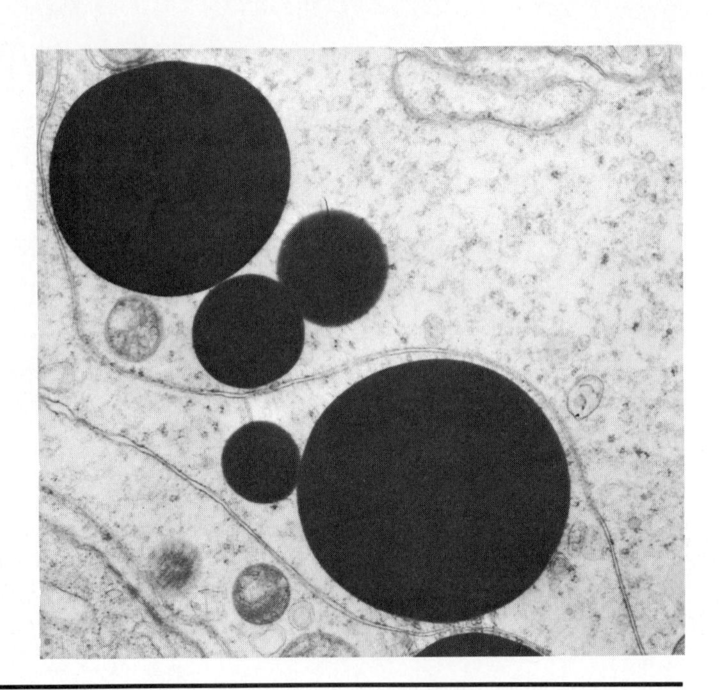

Figure 15.7. Fat cells. The huge fat droplets are spherical and fill most of the cell volume.

Some cholesterol is synthesized in the liver no matter what we eat. Normal body function requires it, in part because it is the molecule from which male and female sex hormones are produced. The androgens, male sex hormones, and the estrogens, female sex hormones, are steroids. In principle, maleness and femaleness are a matter of the balance between our secretions of androgens and estrogens (see Figure 15.9).

Androgens stimulate the growth and maintenance of the male reproductive system and accessory sex tissues, and estrogens regulate the female reproductive system. Androgens and androgen derivatives are also called *anabolic steroids*. They stimulate muscle growth and thus have been used

Figure 15.8. The structure of cholesterol (the basic steroid carbon skeleton is shown in gray).

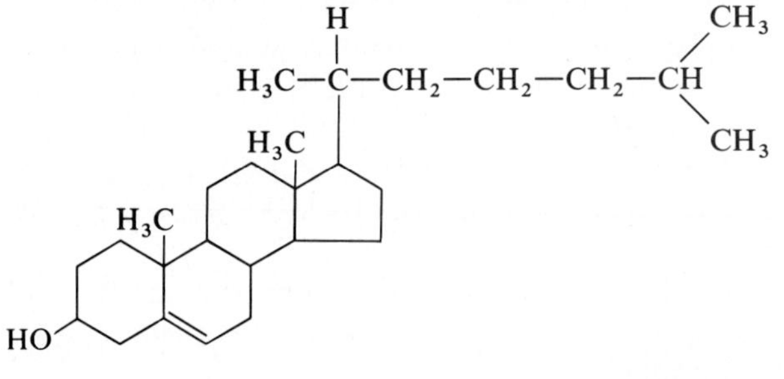

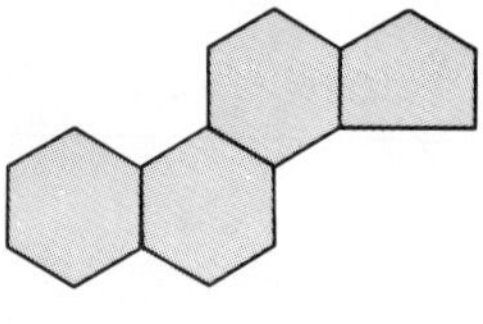

H

$H_3C-C-CH_2-CH_2-CH_2-CH(CH_3)_2$

H_3C

H_3C

HO

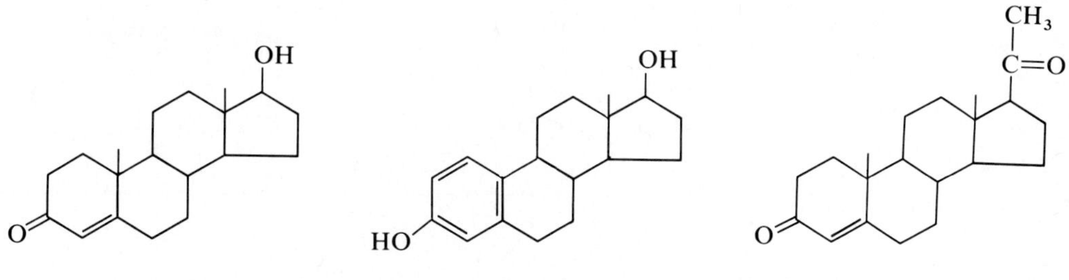

(a) (b) (c)

Figure 15.9. The three most important sex hormones. (a) Testosterone is the major androgen. (b) beta-Estradiol is the major estrogen. (c) Progesterone is the precursor of both testosterone and beta-estradiol in the female.

by athletes to increase muscle mass and strength. However, many people oppose their use because of their side effects and their questionable contributions to superior athletic performance. The controversy over the use of anabolic steroids often appears in the press, especially at the time of the Olympic Games.

The last important class of lipids that we will discuss is associated with cell membranes. A **membrane** plays a vital role in keeping biological functions going smoothly. Membranes separate one cell from another and each compartment within a cell from the other subcellular structures. They regulate the passage of chemicals in and out of cells and keep unwanted chemicals outside. Understanding the molecular architecture of cell membranes is an important objective of biochemistry today.

The most abundant membrane lipids are called *phospholipids,* because each one contains phosphorus within a phosphate group at the end or head of the lipid molecule (see Figure 15.10). The phosphate group carries a negative charge at body pH. Because of the ionic sites, these lipids are somewhat polar. The other end of the molecule, however, has the normal ester bonds to nonpolar fatty-acid groups. One can say that the phospholipids have polar heads, which are attracted to water molecules, and nonpolar tails, which prefer a nonpolar environment.

Figure 15.10. A representative membrane phospholipid.

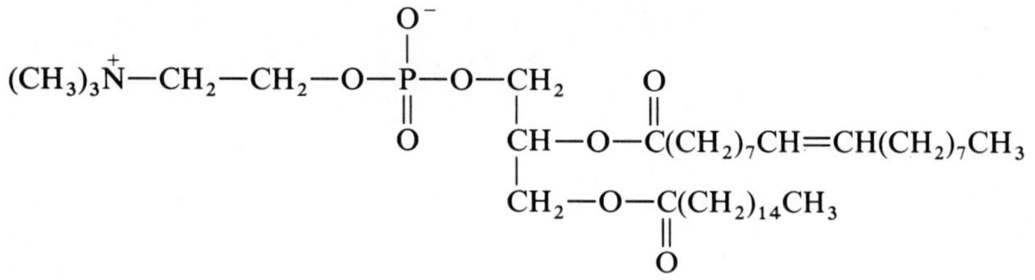

$$(CH_3)_3\overset{+}{N}{-}CH_2{-}CH_2{-}O{-}\underset{\underset{O}{\|}}{\overset{\overset{O^-}{|}}{P}}{-}O{-}CH_2$$

$$CH{-}O{-}\overset{\overset{O}{\|}}{C}(CH_2)_7CH{=}CH(CH_2)_7CH_3$$

$$CH_2{-}O{-}\underset{\underset{O}{\|}}{C}(CH_2)_{14}CH_3$$

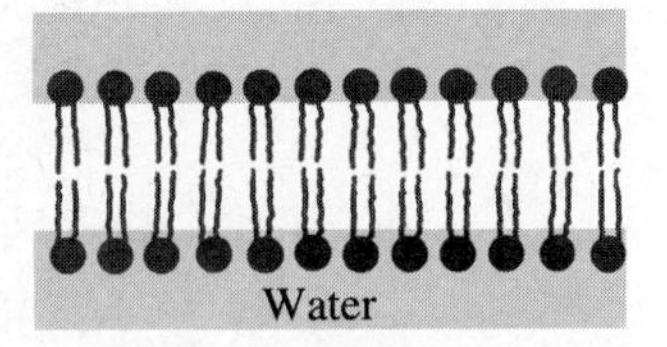

Figure 15.11. An example of a lipid bilayer. The solid circles represent the polar heads of the lipid molecules.

Just like the soaps discussed in Chapter 14, membrane lipids have polar heads and nonpolar tails. They can form micelles in the same way soaps and detergents do. They can also form bilayers. In a **bilayer,** the nonpolar hydrocarbon tails of the lipid molecules extend inward from the two polar surfaces, forming a continuous inner hydrocarbon core. The polar heads face outward into the surrounding water solution (see Figure 15.11).

Membrane bilayers are quite thin, rather like ribbons. Their width depends on which fatty acids are in the membrane lipids. They are flexible and even fluid-like in nature. Membrane bilayers allow water to pass through, but not cations or anions. Phospholipids spontaneously form bilayers in water through the operation of the chemical principle that similar molecular structures attract one another. Micelles and lipid bilayers are examples of a supermolecular structure that is vital in the maintenance of life.

Proteins are often embedded in natural lipid bilayer membranes. The fluid-mosaic model, a unifying theory of membrane structure proposed in 1972, is popular today. In Figure 15.12 you can see that many lipid bilayers are stacked together to create the membrane. On the two surfaces the polar heads are attracted to each other and to polar water molecules. In the interior of the membrane the long hydrocarbon tails attract each other. The bilayer is considered fluid because the unsaturated and saturated hydrocarbon tails in the lipid bilayer may be fluid at the normal temperature of the cells.

Figure 15.12. The fluid-mosaic model of membrane structure.

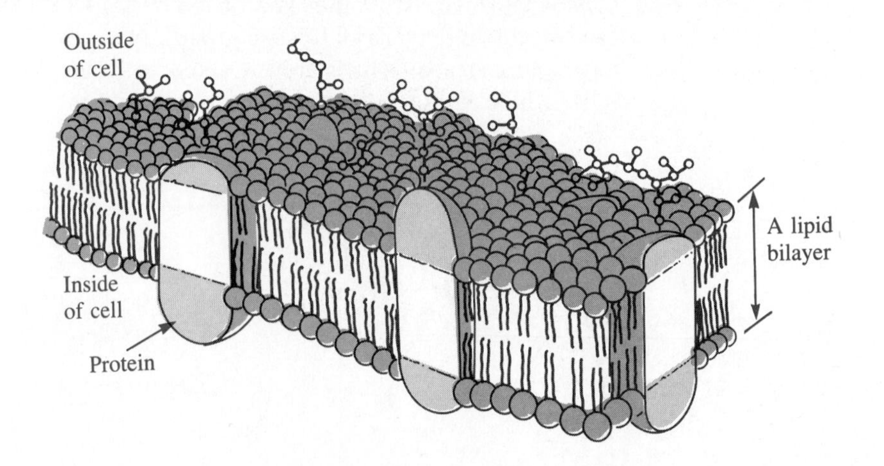

Some proteins are integrated into the membrane and serve to assist transport across the membrane. In this model it is possible for lipids and proteins to move in lateral directions—to float about in the bilayer "sea."

15.4 AMINO ACIDS—BUILDING BLOCKS FOR PROTEINS

Proteins occur in every major part of living cells. They are polymers, much larger and more complex than lipids and simple carbohydrates, comparable in size to starches and cellulose. But whereas starch is built up from only one kind of monosaccharide unit (glucose), proteins are made from about 20 kinds of amino acid units.

The number of different arrangements of 20 amino acid units in a protein chain having, say, 200 monomer units is so large that it is hard to imagine. If you were able to make one protein molecule with each possible arrangement, their combined weight would be greater than the mass of the whole earth. Having 20 amino acid units in the biochemical construction set is rather like having an alphabet of 26 letters which can be arranged into a huge number of words, sentences, paragraphs, chapters, and books. From these building blocks, organisms build proteins as diverse as enzymes, hormones, antibodies, and venoms. Figure 15.13 shows some protein crystals from a bacterium.

Although many proteins function in ways common to all species, each species has its own set of proteins. A human being may have over 50,000 kinds of proteins, all of which are different from those of any other species. In a way, our proteins define what we are. When we eat protein, it is broken down into its constituent amino acids, which are then built into our own set of proteins. Table 15.2 lists some important roles for proteins.

Figure 15.13. *Protein crystals formed by* Bacillus thuringiensis; *these crystals are produced on a commercial scale as an insecticide.*

Table 15.2 Classes of Proteins

TYPE	EXAMPLES
enzymes	proteins that catalyze biochemical reactions
transport proteins	hemoglobin in the bloodstream
contractile proteins	muscle proteins
defense proteins	antibodies, blood-clotting proteins, and venoms
hormones	insulin and growth hormones
structural proteins	proteins in ligaments, skin, and fingernails

Before we discuss the marvelous structures of proteins, we need to consider their monomer units, the amino acids. **Amino acids,** as their name suggests, have two functional groups—an amino group and a carboxylic acid group (Table 11.4). Since the amino group (NH_2) contains the element nitrogen, proteins contain four common elements: carbon, hydrogen, oxygen, and nitrogen. Two of the common amino acids, and therefore most proteins, also contain the element sulfur.

The formal structure of an amino acid is

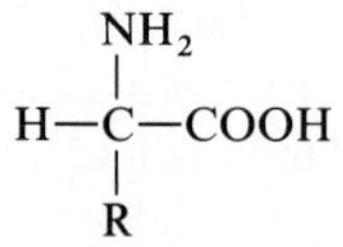

$$\begin{array}{c} NH_2 \\ | \\ H{-}C{-}COOH \\ | \\ R \end{array}$$

The COOH group is a way of showing $\overset{O}{\overset{\|}{C}}{-}OH$, the carboxylic acid group. The symbol R in the structure simply means that this group of atoms, or *radical*, can vary from one amino acid to another. Because the carboxylic acid group is a moderately strong acid, it is able to transfer a hydrogen ion to the rather basic amino group. Thus amino acids in water solution exist mainly as dipolar ions rather than as unionized molecules (see Figure 15.14).

The side chains (R) of the standard 20 amino acids range from a simple hydrogen atom to groups of over 25 atoms. A few examples are shown in Table 15.3.

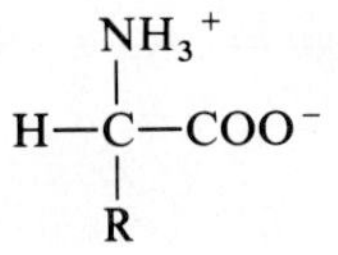

Figure 15.14. The structure of an amino acid in water.

$$\begin{array}{c} NH_3^{\ +} \\ | \\ H{-}C{-}COO^- \\ | \\ R \end{array}$$

Table 15.3 The Structures of Some Common Amino Acids as They Exist in Water Solution at pH 7 (The COO^- group represents $\overset{O}{\overset{\|}{C}}—O^-$)

Structure	Name	Structure	Name
COO^- | $^+H_3N—C—H$ | H	glycine (gly)	COO^- | $^+H_3N—C—H$ | $H—C—OH$ | H	serine (ser)
COO^- | $^+H_3N—C—H$ | CH_3	alanine (ala)	COO^- | $^+H_3N—C—H$ | CH_2 | SH	cysteine (cys)
COO^- | $^+H_3N—C—H$ | CH / \\ H_3C CH_3	valine (val)	COO^- | $^+H_2N——C—H$ | | H_2C CH_2 \\ / CH_2	proline (pro)
COO^- | $^+H_3N—C—H$ | CH_2 | (benzene ring)	phenylalanine (phe)	COO^- | $^+H_3N—C—H$ | CH_2 | CH_2 | C // \\ O O	glutamic acid or glutamate (glu)
COO^- | $^+H_3N—C—H$ | CH_2 | CH_2 | S | CH_3	methionine (met)	COO^- | $^+H_3N—C—H$ | CH_2 | CH_2 | CH_2 | CH_2 | $^+NH_3$	lysine (lys)

Glycine is the simplest of these ten amino acids, with only a hydrogen atom as its side chain. Glycine is used as a noncarbohydrate sweetener. Two of the amino acids have alkyl side chains (methyl and isopropyl groups). Phenylalanine has a benzene ring attached to its side chain; it is called an *aromatic amino acid.* Glutamic acid has on its side chain a second carboxylic acid group, which is ionized at neutral pH; glutamic acid is classified as an *acidic amino acid.* Similarly, lysine is a *basic amino acid* because it has an extra amino group on its side chain.

15.5 STRUCTURES OF PROTEINS

In proteins, amino acids link together through amide bonds. Take, for example, the dimer of glycine (see Figure 15.15). Glycylglycine is an example of a dipeptide. The amide bond that links the two amino acid units is also called a **peptide bond.**

The amino acid sequence in a peptide or protein is known as the **primary structure.** The tetrapeptide *gly-ser-met-ala* is shown in Figure 15.16. If the sequence of amino acid units in the tetrapeptide were given as *ser-met-ala-gly*, the peptide would be a different molecule with different properties. The two molecules would be as different as the words *stop* and *tops*.

Sickle-cell anemia is a hereditary disease in which the protein hemoglobin, which transports oxygen in the blood, has one "wrong" amino acid. In sickle-cell hemoglobin, valine is found in place of the normal glutamic acid unit at position 6 in the primary structure. This mutation gives the hemoglobin

Figure 15.15. (a) Glycylglycine and (b) the amide (peptide) bond.

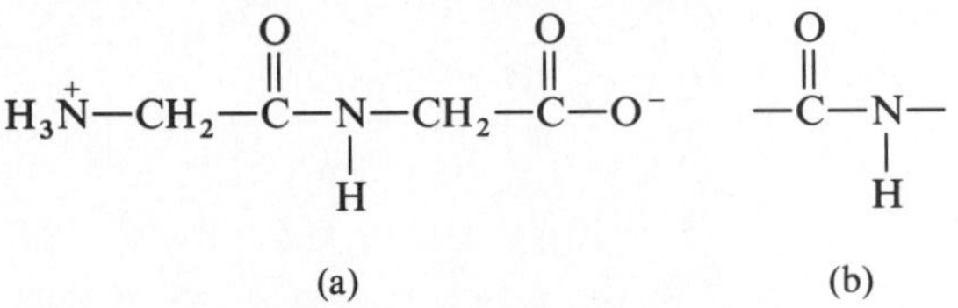

$$H_3\overset{+}{N}-CH_2-\overset{O}{\overset{\|}{C}}-\underset{H}{\underset{|}{N}}-CH_2-\overset{O}{\overset{\|}{C}}-O^-$$

(a)

$$-\overset{O}{\overset{\|}{C}}-\underset{H}{\underset{|}{N}}-$$

(b)

Figure 15.16. The structure of a tetrapeptide with the peptide bonds in boxes.

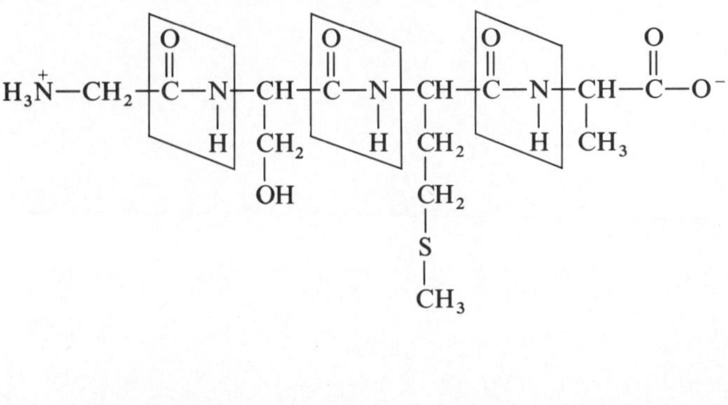

$$H_3\overset{+}{N}-CH_2-\overset{O}{\overset{\|}{C}}-\underset{H}{\underset{|}{N}}-\underset{\underset{OH}{|}}{\underset{CH_2}{\underset{|}{CH}}}-\overset{O}{\overset{\|}{C}}-\underset{H}{\underset{|}{N}}-\underset{\underset{\underset{\underset{CH_3}{|}}{S}}{\underset{|}{CH_2}}}{\underset{\underset{|}{CH_2}}{\underset{|}{CH}}}-\overset{O}{\overset{\|}{C}}-\underset{H}{\underset{|}{N}}-\underset{CH_3}{\underset{|}{CH}}-\overset{O}{\overset{\|}{C}}-O^-$$

molecule a different shape, causing low hemoglobin concentrations in the bloodstream and, consequently, trauma from physical exertion.

Proteins are made of long chains of amino acid units, each connected to the next by a peptide bond. They are condensation polymers (Section 14.10). The general term for an amino acid polymer is **polypeptide.** Proteins are the polypeptides important to living organisms. They are seldom found in nature as simple linear chains of amino acid units: many are tightly folded into spherical shapes, and others are arranged in long strands or sheets. Figure 15.17 shows a sheet-like structure in which several chains of a protein are linked by hydrogen bonding (see Section 9.10). In Figure 15.17 the hydrogen bonds, shown by the dotted lines, link the N—H groups to oxygen atoms nearby in space but some distance away in the primary structure. In other words, the protein chain folds back on itself (not shown in the drawing).

Probably the most famous type of hydrogen-bonded structure in protein molecules is the *alpha*-helix. Here the protein chain is tightly coiled around the chain's axis, as shown in Figure 15.18. Hydrogen bonding between N—H groups and oxygen atoms in adjacent turns of the spiral maintains the rigid, rodlike helix.

American chemists Linus Pauling and Robert Corey uncovered these sheet-like and helical structures during the late 1940s and early 1950s. Their research on the x-ray diffraction patterns of crystals of amino acids and

Figure 15.17. *A sheet-like structure in a protein. The hydrogen bonds are shown by dotted lines.*

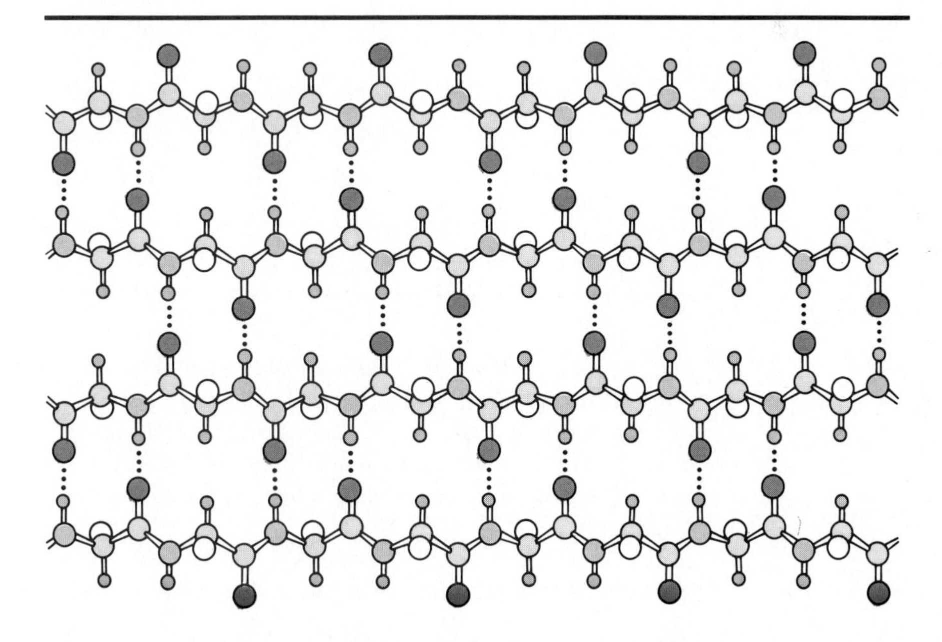

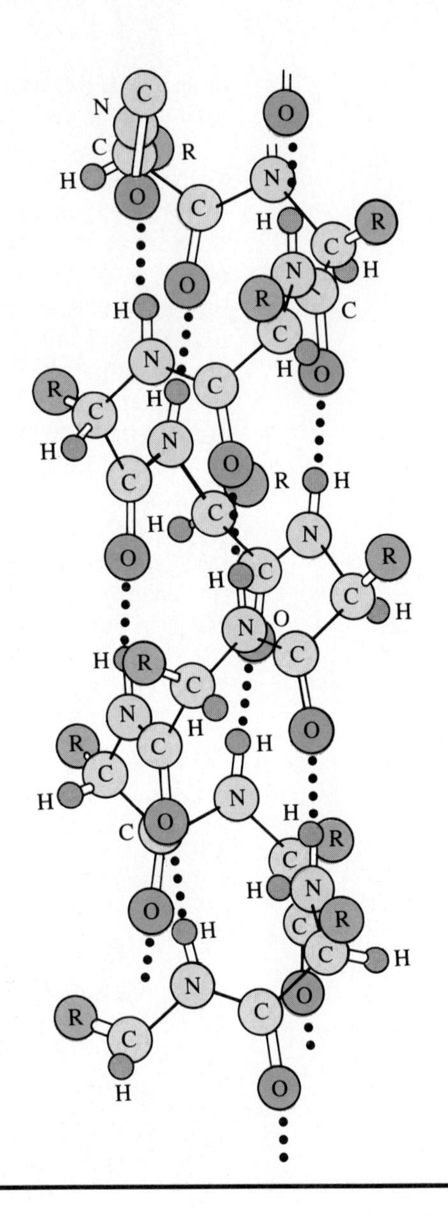

Figure 15.18. The structure of an alpha-helix. The hydrogen bonds are shown by dotted lines.

peptides was followed by careful study of molecular models derived from their diffraction data. Linus Pauling received the Nobel Prize in Chemistry in 1954 in recognition of his studies on chemical bonds and the structures of complex molecules (see Figure 15.19). In 1963 he was awarded the Nobel Peace Prize for his successful fight against the atmospheric testing of nuclear weapons.

Our hair is composed of protein, and the major protein in hair is *alpha*-keratin. Its basic structure is the *alpha*-helix. In straight hair the helical

Figure 15.19. Linus Pauling, a chemist of accomplishment in many fields.

coils of keratin are held in a parallel position by cross-linking sulfur-sulfur bonds from cysteine side chains (see Figure 15.20).

When the hair is curled in a "permanent," the sulfur-sulfur bonds are broken by mild reducing agents. The loss of these links allows the hydrogen bonds of the *alpha*-helixes of keratin to be broken and the polypeptide chains to be stretched out. After a time the reducing solution is removed. Then, while the hair is held by curlers, new and different sulfur-sulfur bonds are formed with an oxidizing agent. When the hair is rinsed, the keratin reforms the *alpha*-helix structures. The hair is now curly because the new sulfur-sulfur covalent bonds exert a twisting force on the keratin bundles in the hair fibers.

Proteins that are folded into spherical shapes have polar amino acid side chains on the outside near the water solvent and nonpolar amino acid units largely in the interior. Figure 15.21 shows the structure of myoglobin, with

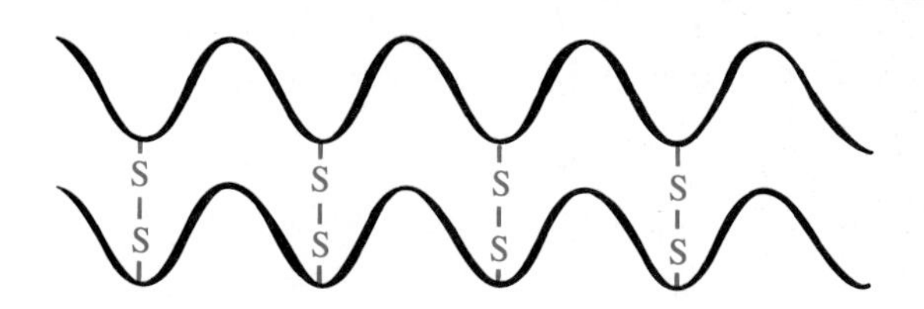

Figure 15.20. Sulfur-sulfur bonds cross-linking parallel keratin chains in hair.

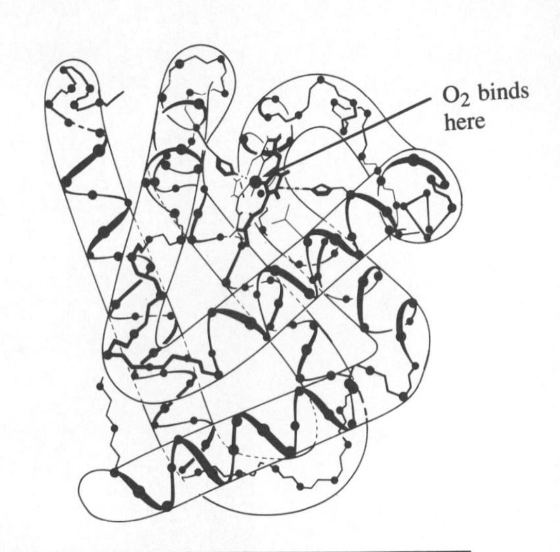

Figure 15.21. The 3-D structure of myoglobin.

its polypeptide chain folded into a complex three-dimensional shape. Myoglobin is a protein used to transport O_2. Within its structure you can see straight, rigid portions of the *alpha*-helix. In the upper part of the drawing, you can see the small disk-like group where the O_2 molecule attaches.

15.6 THE SHAPE OF MOLECULES AND BIOCHEMICAL RECOGNITION

The proteins, so important to life, are made up of amino acid units of a special shape. In fact, many biomolecules are recognizable by their special shapes. Our senses of smell and taste, for example, depend on the shapes of molecules and their complementary receptors in the nose and tongue.

How different the world would seem if we could not smell! The odors of a rose, a banana, a skunk, and perfume all conjure up memories. Much remains to be learned about the sense of smell, but we believe that the nose sorts out the basic chemical entities of an odor on the basis of their respective molecular shapes and then passes signals on to the brain. Our noses are incredibly sensitive organs, able to detect as little as 10^{-12} gram of a pungent, volatile substance.

Molecular shape is also a key part of vision. The major visual pigment in the retina of the eye is called *retinal*. It exists in two isomers that differ in the geometry about the carbon-carbon double bond between carbons 11 and 12, as shown in Figure 15.22.

As in ethylene (Chapter 11, page 250), the atoms attached to the C=C are held flat in a plane. In Figure 15.22(a), note the difference between the positions of the atoms around the C=C at carbons 7 and 8 and the positions of those around the C=C at carbons 11 and 12. Carbon atoms 10 and 13 are held on the same side of the flat alkene functional group. This arrangement is called the *cis* geometry (Section 14.9). In these simplified structural formulas, not all of the hydrogen atoms are shown. However, two hydrogen atoms are attached to carbons 11 and 12, also in the *cis* geometry.

All of the other C=C's outside the ring are in what is called the *trans* geometry, where the chain carbon atoms are on opposite sides of the double bonds. Usually this *trans* arrangement is more stable. Because the atoms attached to the C=C are held rigidly in a flat geometry, 11-*cis*-retinal [Figure 15.22(a)] has a very different shape from its isomer, the all-*trans*-retinal [Figure 15.22(b)].

In the eye the 11-*cis*-retinal is bound to a protein; the complex is called *rhodopsin*. Rhodopsin has a red-purple color, which means that it absorbs the complementary colors, blue and green, most efficiently. When visible light reaches the photosensitive retina, it is absorbed by the rhodopsin, causing the 11-*cis*-retinal to rearrange into the more stable all-*trans*-retinal. The all-*trans*-retinal, with its substantially different shape, no longer fits at the receptor site in the protein, and the rhodopsin falls apart into all-*trans*-retinal and the protein, opsin. This in turn activates the optic nerve, and we see the light. So the shapes of these biomolecules play major roles in vision.

Like many other biomolecules, amino acids come in a right-hand form and a left-hand form. This handedness, or **chirality,** is responsible for the most important kind of shape recognition in biochemistry. What is chirality, and how does it come about?

Figure 15.22. *Molecules that allow us to see: (a)* 11-*cis-retinal and (b) all-trans-retinal.*

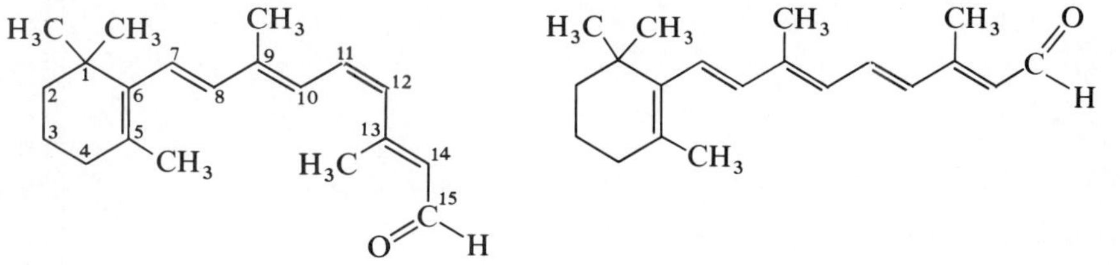

(a) (b)

If you examine your right and left hands, palms up, you will notice that your left-hand thumb points to the left and your right-hand thumb points to the right. Placing one hand on the other, again with palms up, you'll see that the fingers simply do not line up properly. You cannot superimpose the two hands. But if you look at your left hand in a mirror, palm facing the mirror, and compare its mirror image with your right hand, palm facing you, all of the fingers line up perfectly. Try it. The thumb of your right hand points to the right as before, and in the mirror image of your left hand the thumb also points to the right. Right and left hands are mirror images of each other (see Figure 15.23).

Speaking of mirrors, you may have seen a medical vehicle with these large letters printed above its windshield:

AMBULANCE

It almost looks like some foreign language unless the viewer is in a car ahead of an ambulance and reads the letters in the rear-view mirror. This is another case of nonsuperimposable mirror images.

At the molecular level such mirror-image relationships demonstrate the handedness of molecules. In Figure 15.24(a) we see the formula of the amino acid *alanine*, in which four different groups are bonded to the central carbon atom. Figure 15.24(b) shows two 3-D models of alanine, with the correct tetrahedral geometry around the central carbon atom (see Chapter 9, page 204). The two 3-D models are not the same; one is a mirror image of the other. The central carbon atom is called a *chiral center* because of the existence of these two chiral, or right- and left-handed, isomers that

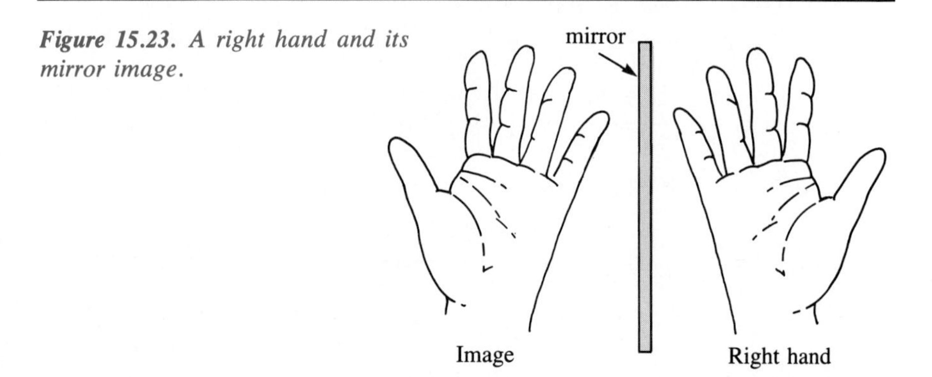

Figure 15.23. *A right hand and its mirror image.*

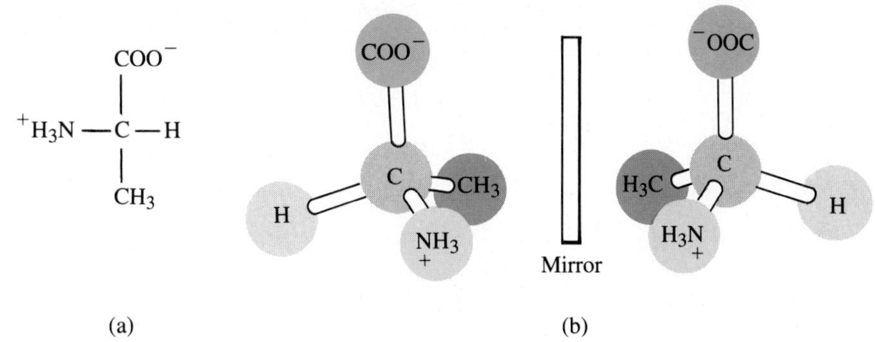

Figure 15.24. *(a) 2-D drawing of alanine and (b) 3-D models of right- and left-handed alanine molecules.*

differ only in the arrangements or configurations of the groups bonded to the central carbon atom.

In Figure 15.24(b) the 3-D model of alanine on the left is called (*R*)-alanine and the model on the right is called (*S*)-alanine. (*R*) comes from the Latin word *rectus*, or right, and (*S*) comes from the Latin word *sinister*, or left. (*R*)-alanine and (*S*)-alanine are nonsuperimposable mirror images, or enantiomers. Since they cannot be superimposed on each other, they are different isomers, but in most respects they are very much alike. They differ in two ways. The first is the way they interact physically with polarized light—light that has passed through polarizing plastic or certain minerals.

The second difference between (*R*)-alanine and (*S*)-alanine, and the one that is crucial to living organisms, is that (*R*)-alanine and (*S*)-alanine interact differently with other chiral molecules. Since all biochemical catalysts, or **enzymes,** are polymers of chiral amino acids, all enzymes are chiral. (*R*)-alanine just does not fit in the catalytic site, or **active site,** of the protein the same way (*S*)-alanine does. The enzyme complements the structure of (*S*)-alanine but not that of (*R*)-alanine. So these amino acid isomers are metabolized quite differently in our bodies. Chirality is a crucial factor in much of our metabolic body chemistry.

In order for molecules to be chiral or handed, it is usually necessary that they contain chiral atoms—atoms that have bonds to four different kinds of groups. Only one common amino acid found in proteins is not chiral—glycine (Table 15.3), the simplest one. It does not contain a chiral atom and its mirror images do superimpose on one another—in fact, they are identical.

The distinctly different smells of caraway seeds and spearmint leaves come from two isomers that differ only in their chirality. The major compound in both caraway oil and spearmint oil is carvone, an unsaturated cyclic compound with the formula $C_{10}H_{14}O$. One oil contains the (*R*)-isomer and the other the (*S*)-isomer (see Figure 15.25). As far as we know, chiral

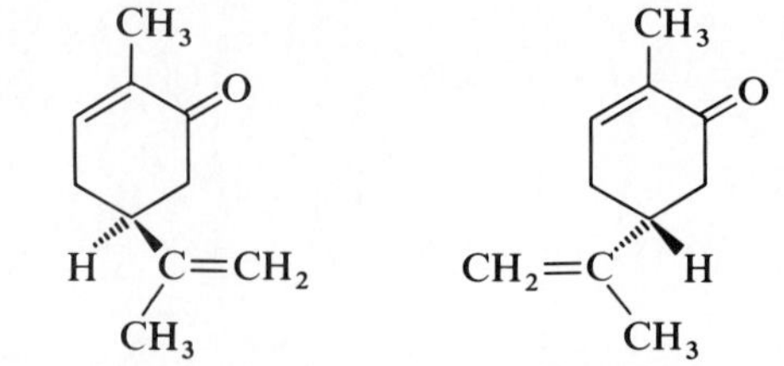

CH_3 ... O ... H C=CH_2 ... CH_3

(S)-carvone
the major component of caraway and dill seed oils

CH_3 ... O ... CH_2=C H ... CH_3

(R)-carvone
the major component of spearmint oil

Figure 15.25. Carvone from caraway seeds and spearmint.

organic compounds in nature exist only in living tissues or in matter that was once part of living tissue. It is a mystery why caraway plants produce (*S*)-carvone and spearmint plants produce its mirror image, (*R*)-carvone. Curiously, it has been reported that about one person in ten cannot tell the difference between the odors of caraway and spearmint.

15.7 NUCLEIC ACIDS AND GENES

The hereditary, or genetic, information in cells is contained in its **nucleic acids.** Passed down from generation to generation, specific giant molecules of **DNA,** or **deoxyribonucleic acids,** determine our heredity and maintain the continuity of life. DNA ensures that a mosquito begets mosquitos and an elephant begets another elephant. A second kind of nucleic acid, called **RNA,** or **ribonucleic acid,** carries the instructions from the genes to the building sites in the cell where proteins are made. The proteins, in turn, determine our biochemical metabolism.

In higher organisms, most of the DNA is found in the cell's nucleus (Figure 15.1). It is stored there in the **chromosomes.** A chromosome contains many **genes,** which are the physical basis for transmission of hereditary characteristics from one generation to the next. The detailed structure of a chromosome is still controversial. The cells of microorganisms often have no separate nuclei, and the genetic material is encoded in simple strands of DNA molecules. The molecular structure of the DNA of each gene contains the information for the amino acid sequence of one protein. Since 1952, when James Watson and Francis Crick discovered the helical structure of nucleic acids, along with the key to how nucleic acids reproduce their exact molecular structures, nucleic acid research has moved at a rapid pace.

Phrases such as "information processing," "genetic engineering," and "cloning" are much in the news today. Indeed, the nucleic acids are our information processors, and research on molecular genetics is an exciting, fast-moving field of science. The dominant technology of the twentieth

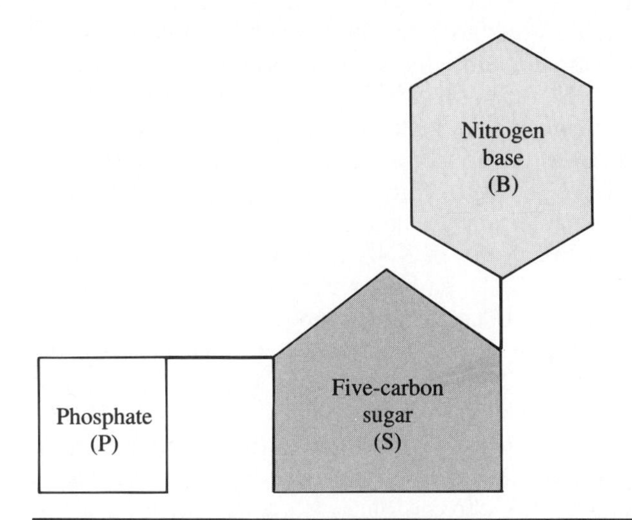

Figure 15.26. *Schematic drawing of a nucleotide unit.*

century is the computer (such as the one on which this book was written), and it is fashionable to see the human being as an information processor. Interestingly, back in the nineteenth century, when the dominant technology was the steam engine, the human being was seen as a heat engine, taking in food and converting it into energy.

All nucleic acids are polymers made up of **nucleotide** units, so a nucleic acid is also called a **polynucleotide.** Nucleotides are composed of three parts—a five-carbon sugar, a cyclic nitrogen base, and a phosphate group (see Figure 15.26).

The five-carbon sugar units are different in RNA and DNA. RNA contains ribose, shown in Figure 15.27(a), whereas DNA contains deoxyribose, shown in Figure 15.27(b). Deoxyribose is different from ribose only in that it has a second hydrogen atom instead of a hydroxyl group at carbon-2.

There are also two classes of cyclic nitrogen-containing bases: one set for DNA and a slightly different set for RNA. They are called bases because at least one nitrogen atom in each can bond to the H^+ from an acid. But

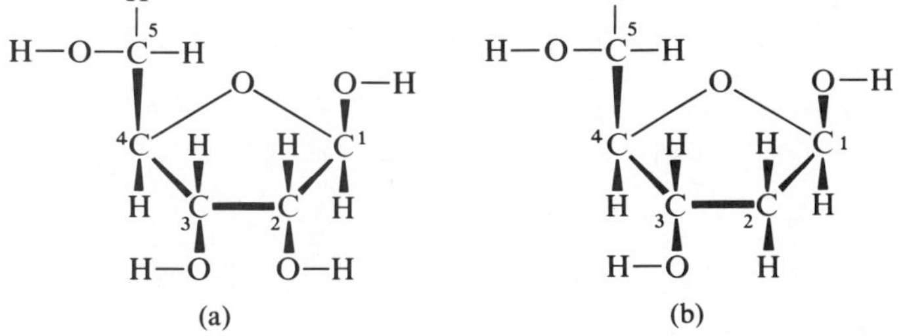

Figure 15.27. *(a) Ribose, the sugar unit in RNA, and (b) deoxyribose, the sugar unit in DNA.*

each set of bases commonly contains only four members, very different from the 20 amino acid units commonly found in our proteins. The nitrogen bases found in DNA and RNA are drawn in Figure 15.28. Those with one ring are called *pyrimidines*, and those with two rings are called *purines*.

Three of the bases in DNA and RNA are the same, but thymine occurs only in DNA and uracil only in RNA. Each nucleotide gets its identity from the base in it. The remainder of the unit is always the same: in DNA it is deoxyribose and phosphate, and in RNA it is ribose and phosphate. Biochemists use the letters A, G, C, and T as shorthand for the bases in DNA. The letters A, G, C, and U are used for RNA.

DNA can have a molecular weight of over 4 billion, which makes it the world's largest organic molecule. The polymer is held together by phosphate groups that link one sugar to the next in the long chain (see Figure 15.29).

Genetic engineering became possible when scientists learned how to determine the nucleotide sequences of DNA chains. Figure 15.30 shows the nucleotide order of the DNA chromosome of a small bacterial virus, as it was worked out in 1977 by Frederick Sanger and his colleagues in Cambridge, England. This chromosome contains nine genes. Sanger received his second Nobel Prize for this work; his first was for being the first to determine the primary structure of a protein, insulin, 25 years earlier.

Figure 15.28. *Structures of the major pyrimidine and purine building blocks of nucleic acids. (a) The set of cyclic nitrogen-containing bases found in DNA. (b) The set of cyclic nitrogen-containing bases found in RNA. The R group in each structure shows where the sugar is attached in the nucleotide unit.*

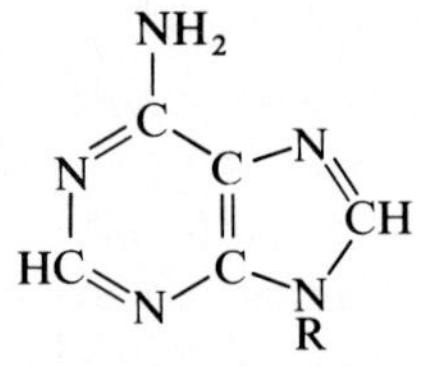

adenine (A) 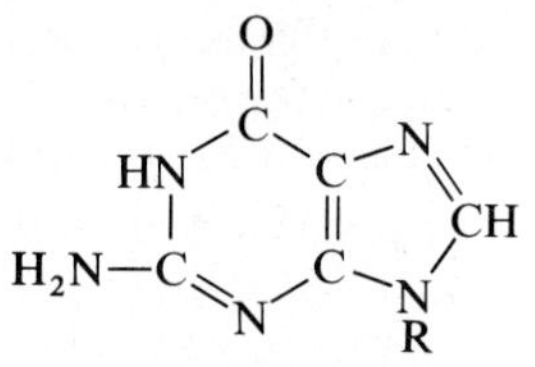 guanine (G) 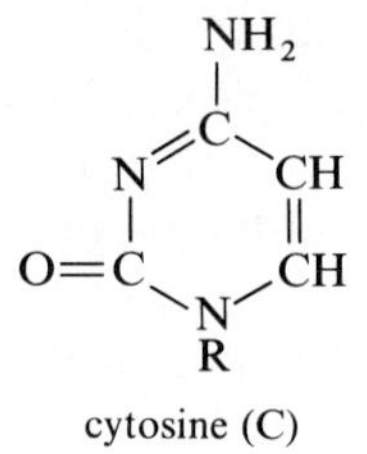 cytosine (C) 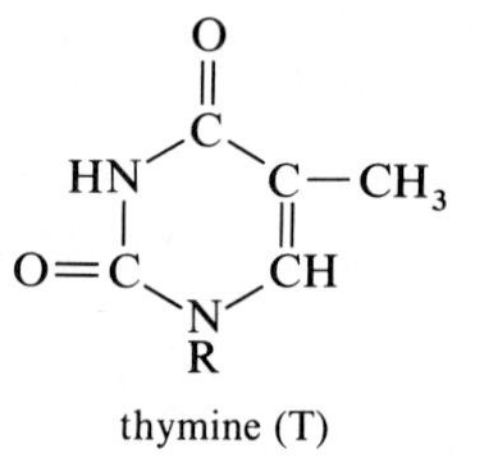 thymine (T)

(a)

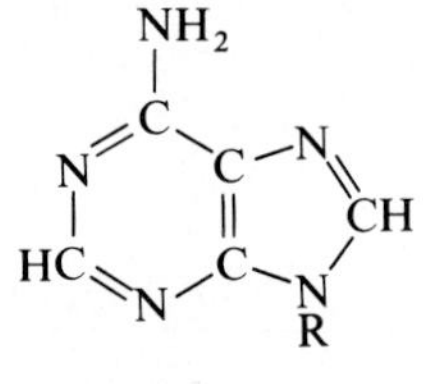

adenine (A) 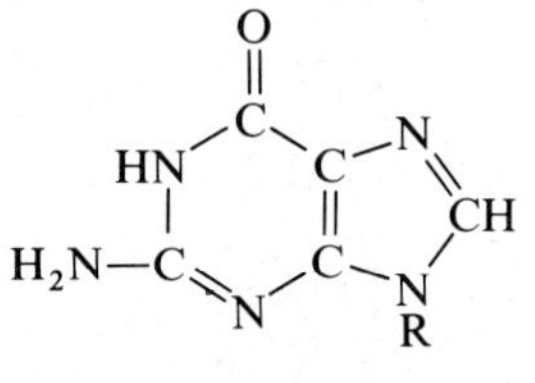 guanine (G) 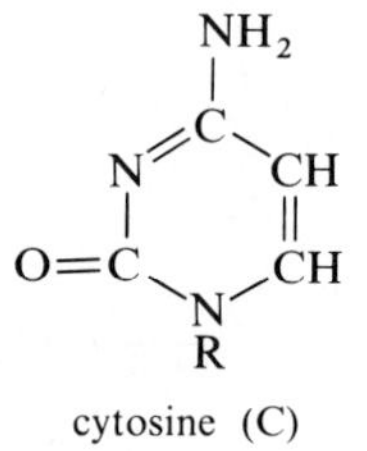 cytosine (C) 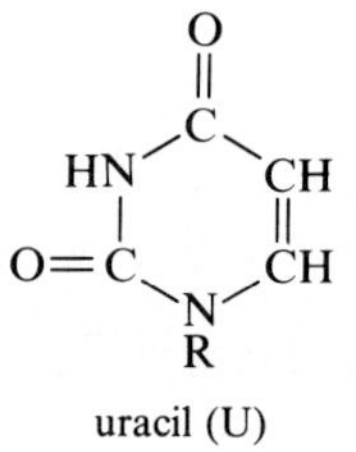 uracil (U)

(b)

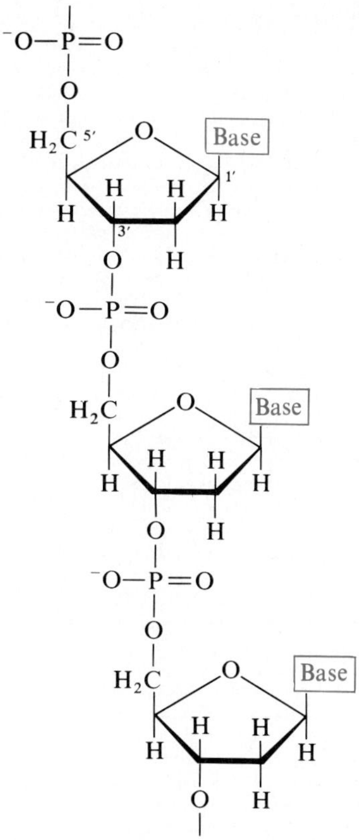

Figure 15.29. Structure of part of a DNA chain.

The fact that DNA comes in helical rods plays a major role in its biological function. The helix of DNA is reminiscent of the *alpha*-helix structures in proteins, but there is an important difference. DNA exists as a **double helix** (see Figure 15.31). Hydrogen bonding helps to hold the DNA double helix together, and the specific nature of this bonding plays the major role in the ability of DNA to transmit genetic information. Figure 15.31(a) shows how the nitrogen bases face each other in the double helix and are hydrogen-bonded to one another. *Watson and Crick deduced that in the double helix of DNA adenine must pair with thymine, and guanine with cytosine, because of shape and hydrogen-bonding factors*. An example of this kind of specific base pairing is shown in Figure 15.32 on page 388.

The most stable hydrogen bonding occurs between guanine (G) and cytosine (C) and between adenine (A) and thymine (T). The G-C pair forms three hydrogen bonds, whereas a G-T pair could only form one hydrogen bond. The A-T pair forms two hydrogen bonds. *The net result is that G always pairs with C and A always pairs with T*. The two DNA strands complement each other (see Figure 15.33 on page 389).

mRNA START → A START

CCGTCAGGATTGACACCCTCCCAATTGTATGTTTTCATGCCTCCAAATCTT CTTTTTTATGGTTCGTTCTTATTACCCTTCTGAA

TGTCACGCTGATTATTTTGACTTTGAGCGTATCGAGGCTCTTAAACCTGCTATTGAGGCTTGTGGCATTTCTACTCTTTCTCAATCCCCA

ATGCTTGGCTTCCATAAGCAGATGGATAACCGCATCAAGCTCTTGGAAGAGATTCTGTCTTTTCGTATGCAGGGCGTTGAGTTCGATAAT

GGTGATATGTATGTTGACGGCCATAAGGCTGCTTCTGACGTTCGTGATGAGTTTGTATCTGTTACTGAGAAGTTAATGGATGAATTGGCA

REGION OF ORIGIN OF DNA REPLICATION

CAATGCTACAATGTGCTCCCCCAACTTGATATTAATAACACTATAGACCACCGCCCCGAAGGGGACGAAAAATGGTTTTTAGAGAACGAG

AAGACGGTTACGCAGTTTTGCCGCAAGCTGGCTGCTGAACGCCCTCTTAAGGATATTCGCGATGAGTATAATTACCCCAAAAAGAAAGGT

ATTAAGGATGAGTGTTCAAGATTGCTGGAGGCCTCCACTAAGATATCGCGTAGAGGCTTTGCTATTCAGCGTTTGATGAATGCAATGCGA

CAGGCTCATGCTGATGGTTGGTTTATCGTTTTTGACACTCTCACGTTGGCTGACGACCGATTAGAGGCGTTTTATGATAATCCCAATGCT

TTGCGTGACTATTTTCGTGATATTGGTCGTATGGTTCTTGCTGCCGAGGGTCGCAAGGCTAATGATTCACACGCCGACTGCTATCAGTAT

TTTTGTGTGCCTGAGTATGGTACAGCTAATGGCCGTCTTCATTTCCATGCGGTGCACTTTATGCGGACACTTCCTACAGGTAGCGTTGAC

mRNA START →

CCTAATTTTGGTCGTCGGATACGCAATCGCCGCCAGTTAAATAGCTTGCAAAATACGTGGCCTTATGGTTACAGTATGCCCATCGCAGTT

CGCTACACGCAGGACGCTTTTTCACGTTCTGGTTGGTTGTGGCCTGTTGATGCTAAAGGTGAGCCGCTTAAAGCTACCAGTTATATGGCT

B START

GTTGGTTTCTATGTGGCTAAATACGTTAACAAAAAGTCAGATATGGACCTTGCTGCTAAAGGTCT CTAAAGAATGGAACAACTCA

CTAAAAACCAAGCTGTCGCTACTTCCCAAGAAGCTGTTCAGAATCAGAATGAGCCGCAACTTCGGGATGAAAATGCTCACAATGACAAAT

CTGTCCACGGAGTGCTTAATCCAACTTACCAAGCTGGGTTACGACGCGACGCCGTTCAACCAGATATTGAAGCAGAACGCAAAAAGAGAG

ATGAGATTGAGGCTGGGAAAAGTTACTGTAGCCGACGTTTTGGCGGCGCAACCTGTGACGACAAATCTGCTCAAATTTATGCGCGCTTCG

B END

ATAAAAATGATTGGCGTATCCAACCTGCAGAGTTTTATCGCTTCCATGACGCAGAAGTTAACACTTTCGGATATTTCTGATGAGTCGAAA

C START A END

AATTATCTTGATAAAGCAGGAATTACTACTGCTTGTTTACGAATTAAATCGAAGTGGACTGCTGGC AAATGAGAAAATTCGACCTAT

CCTTGCGCAGCTCGAGAAGCTCTTACTTTGCGACCTTTCGCCATCAACTAACGATTCTGTCAAAAACTGACGCGTTGGATGAGGAGAAGT

GGCTTAATATGCTTGGCACGTTCGTCAAGGACTGGTTTAGATATGAGTCACATTTTGTTCATGGTAGAGATTCTCTTGTTGACATTTTAA

mRNA START → D START C END

AAGAGCGTGGATTACTATCTGAGTCCGATGCTGTTCAACCACTAAT AAGAAATCATGAGTCAAGTTACTGAACAATCCGTACGTTT

CCAGACCGCTTTGGCCTCTATTAAGCTCATTCAGGCTTCTGCCGTTTTGGATTTAACCGAAGATGATTTCGATTTTCTGACGAGTAACAA

E START

AGTTTGGATTGCTACTGACCGCTCTCGTGCTCGTCGCTGCGTT CTTGCGTTTATGGTACGCTGGACTTTGTAGGATACCCTCGCTT

TCCTGCTCCTGTTGAGTTTATTGCTGCCGTCATTGCTTATTATGTTCATCCCGTCAACATTCAAACGGCCTGTCTCATCATGGAAGGCGC

TGAATTTACGGAAAACATTATTAATGGCGTCGAGCGTCCGGTTAAAGCCGCTGAATTGTTCGCGTTTACCTTGCGTGTACGCGCAGGAAA

E END D END J START

CACTGACGTTCTTACTGACGCAGAAGAAAACGTGCGTCAAAAATTACGTGCGG TGATGTAATGTCTAAAGGTAAAAAACGTTCT

GGCGCTCGCCCTGGTCGTCCGCAGCCGTTGCGAGGTACTAAAGGCAAGCGTAAAGGCGCTCGTCTTTGGTATGTAGGTGGTCAACAATTT

J END F START

TAATTGCAGGGGCTTCGGCCCCTTACTTG TAAATTATGTCTAATATTCAAACTGGCGCCGAGCGTATGCCGCATGACCTTTCCCAT

CTTGGCTTCCTTGCTGGTCAGATTGGTCGTCTTATTACCATTTCAACTACTCCGGTTATCGCTGGCGACTCCTTCGAGATGGACGCCGTT

GGCGCTCTCCGTCTTTCTCCATTGCGTCGTGGCCTTGCTATTGACTCTACTGTAGACATTTTTACTTTTTATGTCCCTCATCGTCACGTT

TATGGTGAACAGTGGATTAAGTTCATGAAGGATGGTGTTAATGCCACTCCTCTCCCGACTGTTAACACTACTGGTTATATTGACCATGCC

GCTTTTCTTGGCACGATTAACCCTGATACCAATAAAATCCCTAAGCATTTGTTTCAGGGTTATTTGAATATCTATAACAACTATTTTAAA

GCGCCGTGGATGCCTGACCGTACCGAGGCTAACCCTAATGAGCTTAATCAAGATGATGCTCGTTATGGTTTCCGTTGCTGCCATCTCAAA

AACATTTGGACTGCTCCGCTTCCTCCTGAGACTGAGCTTTCTCGCCAAATGACGACTTCTACCACATCTATTGACATTATGGGTCTGCAA

GCTGCTTATGCTAATTTGCATACTGACCAAGAACGTGATTACTTCATGCAGCGTTACCATGATGTTATTTCTTCATTTGGAGGTAAAACC

TCATATGACGCTGACAACCGTCCTTTACTTGTCATGCGCTCTAATCTCTGGGCATCTGGCTATGATGTTGATGGAACTGACCAAACGTCG

TTAGGCCAGTTTTCTGGTCGTGTTCAACAGACCTATAAACATTCTGTGCCGCGTTTCTTTGTTCCTGAGCATGGCACTATGTTTACTCTT

GCGCTTGTTCGTTTTCCGCCTACTGCGACTAAAGAGATTCAGTACCTTAACGCTAAAGGTGCTTTGACTTATACCGATATTGCTGGCGAC

CCTGTTTTGTATGGCAACTTGCCGCCGCGTGAAATTTCTATGAAGGATGTTTTCCGTTCTGGTGATTCGTCTAAGAAGTTTAAGATTGCT

GAGGGTCAGTGGTATCGTTATGCGCCTTCGTATGTTTCTCCTGCTTATCACCTTCTTGAAGGCTTCCCATTCATTCAGGAACCGCCTTCT

GGTGATTTGCAAGAACGCGTACTTATTCGCAACCATGATTATGACCAGTGTTTCAGTCGTTCAGTTGTTGCAGTGGATAGTCTTACCTCA

F END

TGTGACGTTTATCGCAATCTGCCGACCACTCGCGATTCAATCATGACTTCGTGATAAAAGATTGAGTGTGAGGTTATAACCGAAGCGGTA

G START

AAAATTTTAATTTTTGCCGCTGAGGGGTTGACCAAGCGAAGCGCGGTAGGTTTTCTGCTT TTTAATCATGTTTCAGACTTTTATT

TCTCGCCACAATTCAAACTTTTTTTCTGATAAGCTGGTTCTCACTTCTGTTACTCCAGCTTCTTCGGCACCTGTTTTACAGACACCTAAA

GCTACATCGTCAACGTTATATTTTGATAGTTTGACGGTTAATGCTGGTAATGGTGGTTTTCTTCATTGCATTCAGATGGATACATCTGTC

AACGCCGCTAATCAGGTTGTTTCAGTTGGTGCTGATATTGCTTTTGATGCCGACCCTAAATTTTTTGCCTGTTTGGTTCGCTTTGAGTCT

TCTTCGGTTCCGACTACCCTCCCGACTGCCTATGATGTTTATCCTTTGGATGGTCGCCATGATGGTGGTTATTATACCGTCAAGGACTGT

GTGACTATTGACGTCCTTCCCCGTACGCCGGGCAATAACGTCTACGTTGGTTTCATGGTTTGGTCTAACTTTACCGCTACTAAATGCCGC

G END H START

GGATTGGTTTCGCTGAATCAGGTTATTAAAGAGATTATTTGTCTCCAGCCACTTAAGT TTATGTTTGGTGCTATTGCTGGCG

GTATTGCTTCTGCTCTTGCTGGTGGCGCCATGTCTAAATTGTTTGGAGGCGGTCAAAAAGCCGCCTCCGGTGGCATTCAAGGTGATGTGC

TTGCTACCGATAACAATACTGTAGGCATGGGTGATGCTGGTATTAAATCTGCCATTCAAGGCTCTAATGTTCCTAACCCTGATGAGGCCG

CCCCTAGTTTTGTTTCTGGTGCTATGGCTAAAGCTGGTAAAGGACTTCTTGAAGGTACGTTGCAGGCTGGCACTTCTGCCGTTTCTGATA

AGTTGCTTGATTTGGTTGGACTTGGTGGCAAGTCTGCCGCTGATAAAGGAAAGGATACTCGTGATTATCTTGCTGCTGCATTTCCTGAGC

TTAATGCTTGGGAGCGTGCTGGTGCTGATGCTTCCTCTGCTGGTATGGTTGACGCCGGATTTGAGAATCAAAAAGAGCTTACTAAAATGC

AACTGGACAATCAGAAAGAGATTGCCGAGATGCAAAATGAGACTCAAAAAGAGATTGCTGGCATTCAGTCGGCGACTTCACGCCAGAATA

CGAAAGACCAGGTATATGCACAAAATGAGATGCTTGCTTATCAACAGAAGGAGTCTACTGCTCGCGTTGCGTCTATTATGGAAAACACCA

ATCTTTCCAAGCAACAGCAGGTTTCCGAGATTATGCGCCAAATGCTTACTCAAGCTCAAACGGCTGGTCAGTATTTTACCAATGACCAAA

TCAAAGAAATGACTCGCAAGGTTAGTGCTGAGGTTGACTTAGTTCATCAGCAAACGCAGAATCAGCGGTATGGCTCTTCTCATATTGGCG

CTACTGCAAAGGATATTTCTAATGTCGTCACTGATGCTGCTTCTGGTGTGGTTGATATTTTTCATGGTATTGATAAAGCTGTTGCCGATA

H END

CTTGGAACAATTTCTGGAAAGACGGTAAAGCTGATGGTATTGGCTCTAATTTGTCTAGGAAATAA

Figure 15.30. *The nucleotide sequence of the DNA chromosome of* ϕ*X174, as worked out in 1977. Each letter stands for an entire nucleotide unit of the DNA chromosome. The start and end notes in the figure can be ignored.*

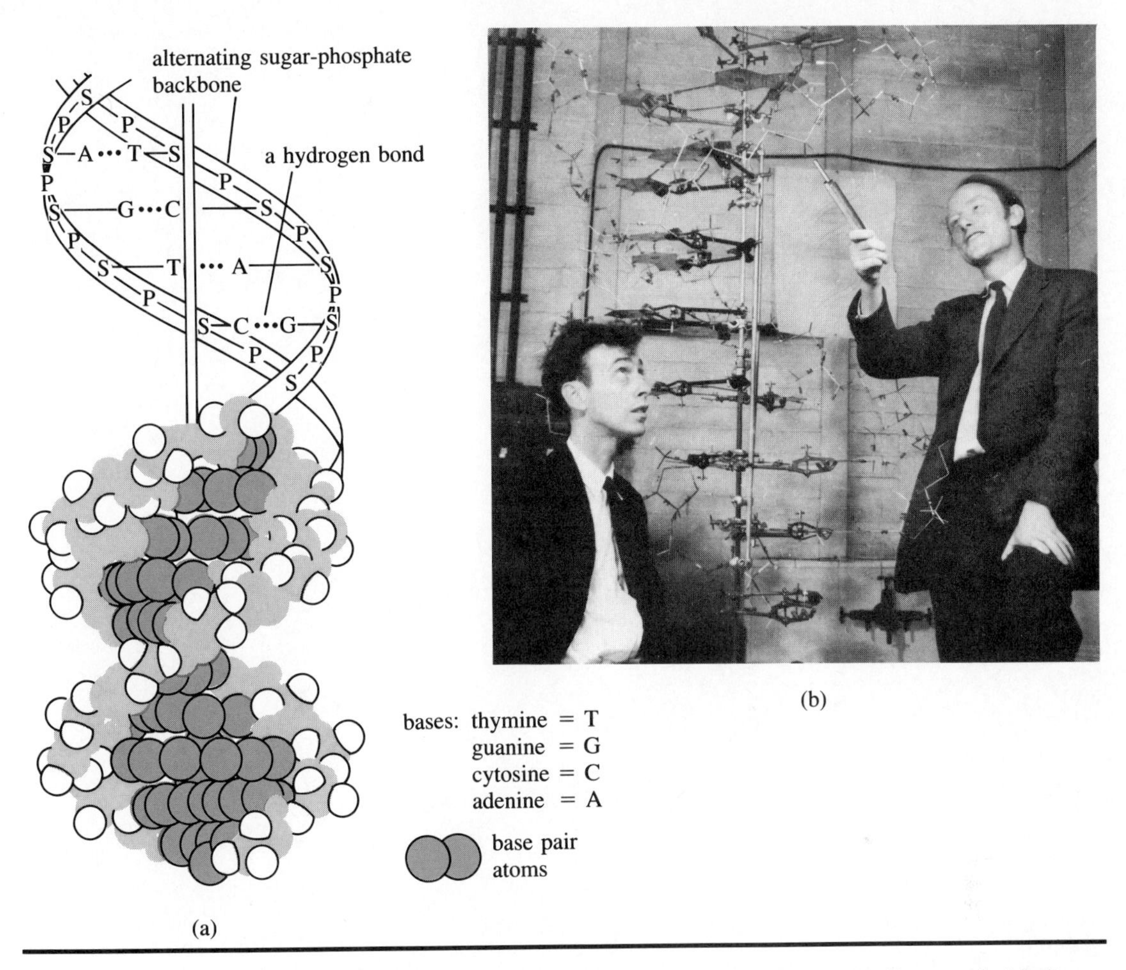

Figure 15.31. (a) The Watson-Crick model of DNA as a double-stranded helix. (b) Francis Crick and James Watson with one of their models of DNA in 1953.

It is through the specific sequence of the four bases A, T, G, and C in each strand of DNA that genetic information is encoded and our heredity determined. DNA replicates (reproduces) itself exactly when the two complementary strands of a double helix break apart and new complementary strands form. We can say that DNA provides a template for precise DNA replication (see Figure 15.34 on page 390). Each parental strand of the DNA directs the replication of a complementary daughter strand, and after replication there are two identical double-stranded helixes rather than one.

The genetic information contained in DNA is expressed through the directed synthesis of complementary messenger RNA molecules. Through a complex process, the RNA templates direct the synthesis of proteins. Overall, a sequence of three bases in the DNA chain determines which

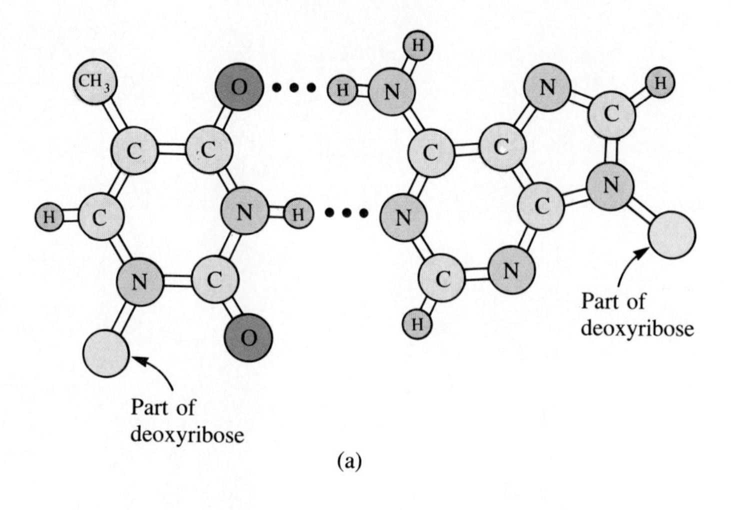

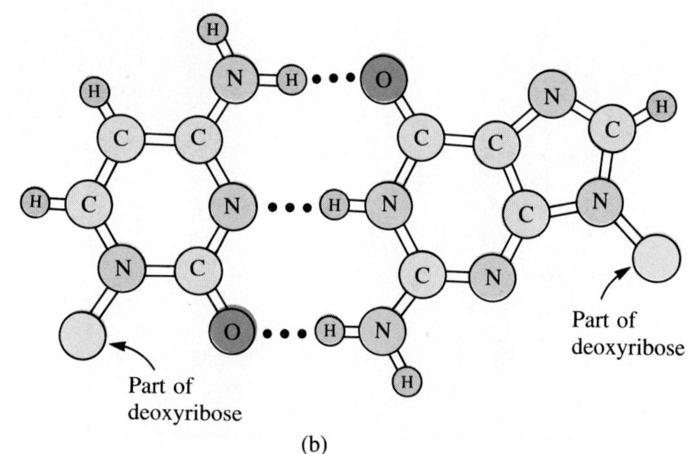

Figure 15.32. The specific base pairing in DNA; (a) thymine with adenine and (b) cytosine with guanine.

amino acid unit will come next in the primary structure of the protein being made. Cracking the genetic code was one of the great scientific achievements of the twentieth century.

A single human cell contains enough DNA to direct the synthesis of thousands upon thousands of proteins. Figure 15.30 shows the vast amount of information in the DNA base sequence of the small virus ϕX174. This DNA chromosome, with a molecular weight of 3.4×10^6, has 5386 base pairs, and its information requires almost a whole page of fine print in this book. If the entire base sequence of the 46 chromosomes of a human cell were printed in the same way, over 820,000 pages would be required—perhaps 2000 volumes the size of this book. The miniaturization of this information to the molecular level is an awesome accomplishment of nature, far more awesome than the most sophisticated computer.

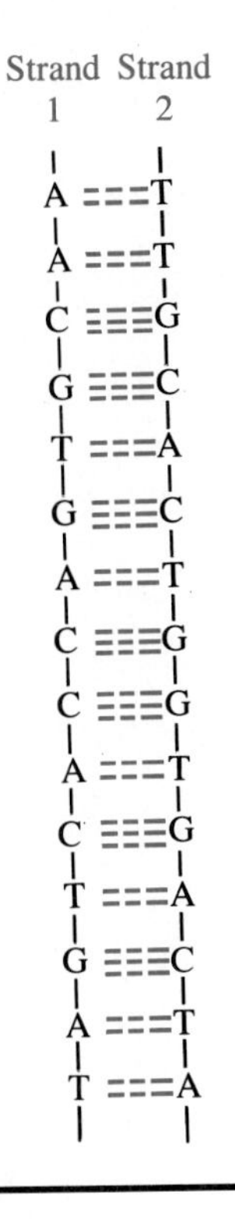

Figure 15.33. A drawing of the complementary strands of DNA proposed by Watson and Crick. The horizontal lines show hydrogen bonds; notice that there are always three between G and C and always two between A and T.

15.8 ENZYMES AND BIOCHEMICAL METABOLISM

The word *metabolism* means the entire complex of physical and chemical processes involved in the maintenance of life. All living organisms have a highly integrated network of chemical reactions for the extraction of energy from food and the synthesis of critical cell constituents. Our metabolic processes are regulated by the precise organization of our cells and by specific enzymes.

The enzymes that are produced so faithfully from the information stored in DNA are the **catalysts** for metabolic reactions. Like any catalyst, an enzyme increases the rate of a reaction without an increase in temperature. Enzymes provide tremendous rate increases in a controlled manner so that our bodies can function smoothly. They serve to control the balance among our many metabolic pathways, so that one set of reactions does not go too fast relative to another. This control is crucial to our lives, since to remain healthy we must maintain *homeostasis,* or physiological equilibrium. Increases and decreases in body temperature cannot be used to control metabolic reaction rates—everyone knows the discomfort of even a mild fever.

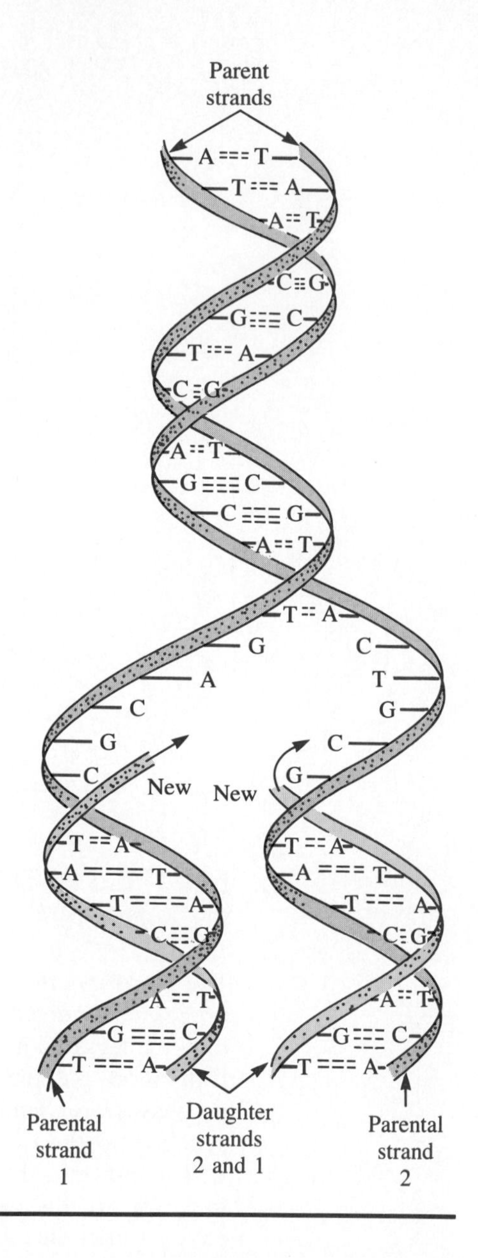

Figure 15.34. Replication of DNA as suggested by Watson and Crick.

Virtually every chemical reaction in the body is catalyzed by a specific enzyme. Consider the oxidation of glucose, which is one of our most efficient sources of energy. Burning a sample of glucose in the air would release a large amount of energy, but this would be a very painful way to obtain the necessary energy in our bodies.

$$C_6H_{12}O_6 + 6\,O_2 \longrightarrow 6\,CO_2 + 6\,H_2O + 2880\text{ kJ}$$

When molecules such as glucose and oxygen collide with one another, they can react to form new chemicals only if they hit each other with considerable force. Otherwise, they bounce apart unchanged. The minimum energy of motion (kinetic energy) for a reaction is called the *energy of activation* (E_a). Enzymes, like other catalysts, work by lowering the activation energies of chemical and biochemical reactions (see Figure 15.35) through the use of different, more efficient pathways. When the activation energy is low, many of the reactant molecules have enough energy to react and the reaction rate is faster.

Glucose is oxidized efficiently and rapidly at ordinary temperatures in the presence of the proper enzymes. It requires many enzymes and many steps, but the overall result is the same—each mole of glucose produces 2880 kilojoules of energy, along with carbon dioxide and water.

Even the liberation of CO_2 from our lungs is controlled by an enzyme; it is called *carbonic anhydrase*. This enzyme catalyzes the dehydration (loss of water) of carbonic acid.

$$H_2CO_3 \longrightarrow CO_2 + H_2O$$

Carbonic anhydrase has a zinc ion (Zn^{2+})in each protein molecule, and its molecular weight is about 29,000. It is an extremely efficient enzyme. In a second, a single molecule of carbonic anhydrase can convert 600,000 molecules of carbonic acid into carbon dioxide and water.

Enzymes are highly specific for a given reaction. They recognize the exact shapes of their **substrates,** or reactants, by the complementary shapes

Figure 15.35. *(a) The energy of activation for an uncatalyzed chemical reaction. (b) The smaller energy of activation for a catalyzed reaction using a different pathway.*

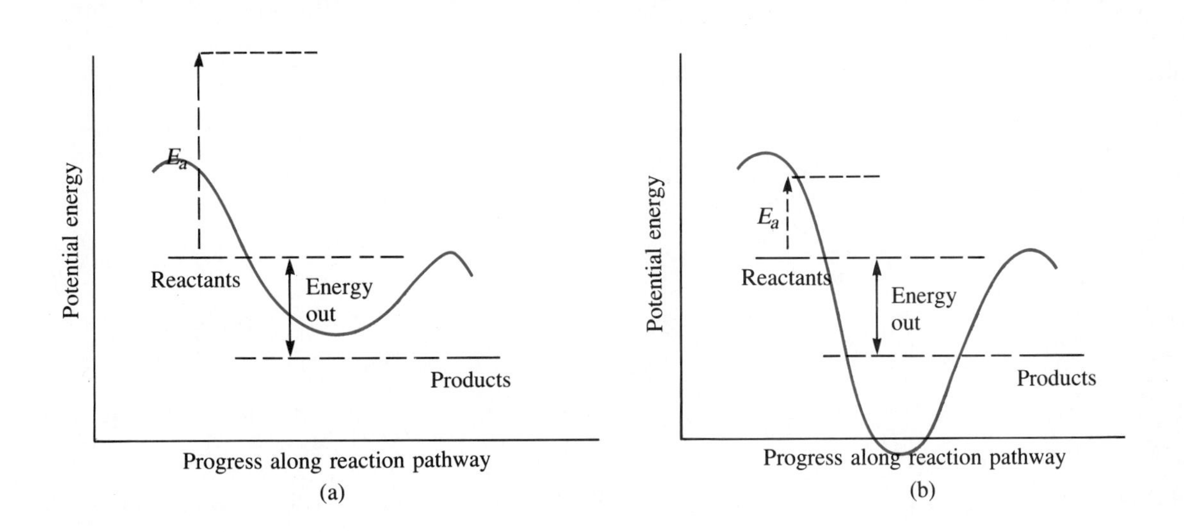

of their catalytic sites. The enzyme's active site fits together with its substrate in somewhat the same way as a key fits into a lock. Some enzymes catalyze their reactions simply by binding two or three substrates in close proximity so that unproductive collisions need not occur. In other cases, enzymes provide acids or bases or metal ions at just the right places to allow bonds to be broken or made with a minimum of activation energy. Our bodies require many vitamins (see Chapter 16), because we turn them into **coenzymes,** which serve as oxidizing, reducing, and transferring agents at the active sites of specific enzymes. Enzymes are usually named for the reactions that they catalyze, with the ending *-ase* added.

Oxidation of the carbohydrates and fats in the food we eat supplies much of the energy for our bodily functions. From our food comes the energy for chemical work such as the biosynthesis of proteins and nucleic acids, for the mechanical work of muscle contraction, for the electrical work of nerve conduction, and for the work of selective transport of molecules across membranes. Energy-yielding reactions are closely tied to specific energy-requiring processes.

Exactly how we turn our food into usable energy is a central question of biochemical metabolism. The oxidation of glucose in our cells, for example, involves an organized series of about 50 separate, complex reactions, each of which is catalyzed by a specific enzyme. In this series of reactions the released energy is effectively trapped in usable form by molecules of **adenosine triphosphate,** better known as **ATP** (see Figure 15.36).

The oxidation of a single molecule of glucose eventually produces 38 molecules of ATP. Molecular oxygen enters the process only at the end of a series of oxidation-reduction reactions. In the absence of O_2—say, in the active muscles of an athlete—some of the glucose is converted to lactate to produce ATP faster than it can be produced from other reactions (see Figure 15.37). The build-up of lactate produces the syndrome of sore muscles. Outside the human body, microorganisms (such as various yeasts) produce energy from glucose by the process of alcoholic fermentation (Figure 15.37),

Figure 15.36. *The structure of ATP, adenosine-5'-triphosphate, a nucleotide that is central to energy transfer in living organisms.*

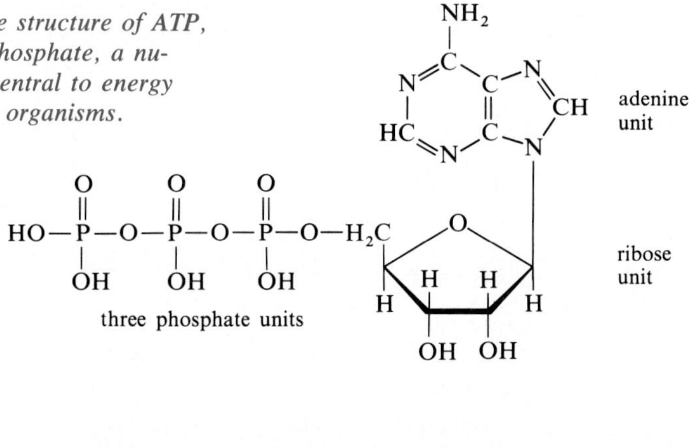

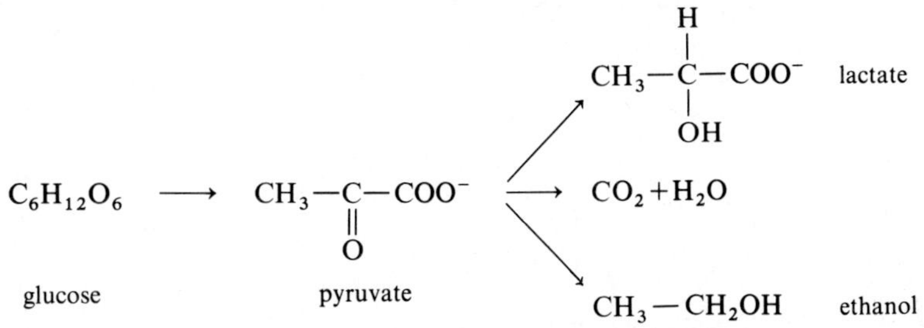

$$C_6H_{12}O_6 \longrightarrow CH_3{-}\overset{\displaystyle O}{\overset{\|}{C}}{-}COO^- \longrightarrow \begin{cases} CH_3{-}CH(OH){-}COO^- & \text{lactate} \\ CO_2 + H_2O & \\ CH_3{-}CH_2OH & \text{ethanol} \end{cases}$$

glucose — pyruvate

Figure 15.37. Some metabolic fates of glucose.

one of the oldest chemical technologies. Ethyl alcohol accumulates in reasonably high concentrations—up to 12% or so.

The energy trapped by ATP is transferred in our bodies to other metabolic intermediates, enabling them to undergo reactions that would otherwise be impossible (see Figure 15.38). ATP drives the movement of our muscles and other cellular motions, controls the transport of metabolites across membranes, and provides the energy to drive the synthesis of our biopolymers (proteins and nucleic acids).

The principle underlying the role of ATP in coupling energy-yielding reactions to those requiring energy can be shown by the following example. Consider a hypothetical, energy-requiring metabolic reaction that has an unfavorable chemical equilibrium:

$$X + Y \rightleftharpoons X{-}Y \qquad \text{(Very little product)}$$

With an unfavorable equilibrium, a very small percentage of reactants X and Y will form the product X—Y no matter how much time is allowed. However, a similar reaction can be accomplished through two favorable steps driven by the high reactivity of ATP. In the first step, ATP transfers a phosphate group to X; here, a phosphate group is shown by the letter P. *Adenosine diphosphate* (*ADP*) results from the loss of the phosphate

Figure 15.38. The formation and cleavage of ATP in energy-transfer processes.

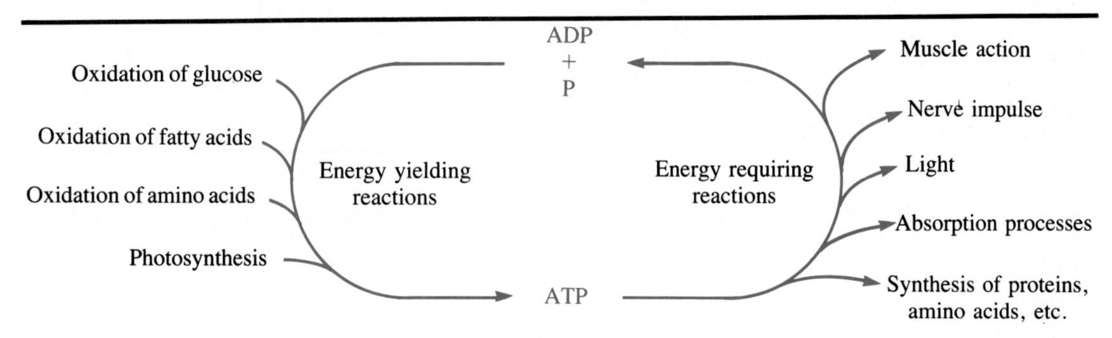

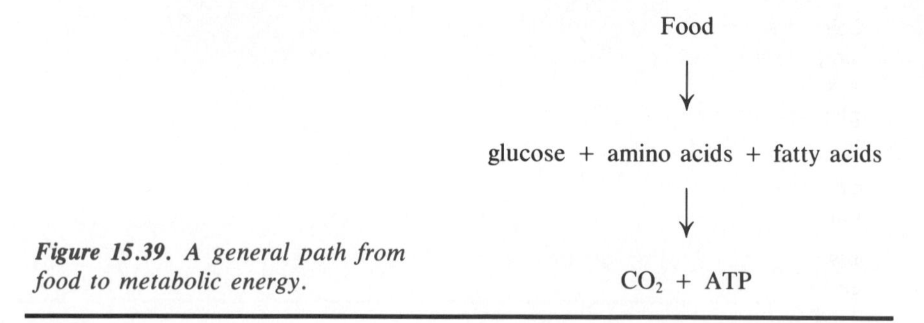

Figure 15.39. A general path from food to metabolic energy.

group. The molecule of X with a phosphate attached is much more reactive than X was by itself. So now the reaction of X—P with Y has a favorable chemical equilibrium, and X—Y can form in the necessary amount (assuming that there is no pathway by which X—Y can directly form X plus Y). A phosphate anion (HPO_4^{2-} at pH 7) is also formed.

$$\underset{\text{ATP}}{\text{adenosine—P—P—P}} + \text{X} \rightleftharpoons \underset{\text{ADP}}{\text{adenosine—P—P}} + \text{X—P} \qquad \text{Step 1}$$

$$\text{X—P} + \text{Y} \rightleftharpoons \text{X—Y} + \text{P} \qquad \text{Step 2}$$

The overall chemical process would be

$$\text{X} + \text{Y} + \text{ATP} \rightleftharpoons \text{X—Y} + \text{ADP} + \text{P} \qquad \text{(Favorable equilibrium)}$$

In addition to vitamins and minerals, our food is made up of a complex molecular mixture of proteins, carbohydrates, and fats. We must digest them—hydrolyze them by reaction with water—before they can be absorbed. The process begins in the mouth and continues through the stomach and small intestine. We break food down into amino acids, glucose and other sugars, and fatty acids and glycerol (see Figure 15.39).

We use the amino acids to build our own set of proteins or else convert them into carbohydrates for energy. We eventually excrete CO_2, H_2O, and excess nitrogen (for the most part as urea, H_2NCONH_2).

IMPORTANT TERMS

cell The highly structured, fundamental unit of all living organisms. It is surrounded by a cell membrane.

metabolism The entire complex of physical and chemical processes involved in the maintenance of life.

carbohydrate A chemical substance that contains several hydroxyl groups and either an aldehyde or a ketone group. Sugars and starches are examples of carbohydrates. Simple carbohydrates have the formula $(CH_2O)_x$.

sugar Any monosaccharide or disaccharide; a simple carbohydrate.

monosaccharide A carbohydrate containing from three to seven carbon atoms per molecule.

glucose A carbohydrate with the formula $C_6H_{12}O_6$. It is the monomeric unit of cellulose and starch. It is also called *dextrose* or *blood sugar*.

disaccharide A carbohydrate that contains two sugar units.

sucrose The common disaccharide that is used as table sugar.

polysaccharide A carbohydrate polymer with many repeating monosaccharide units.

starch The polysaccharide made of glucose units that is the major carbohydrate energy source in our food.

lipid A water-insoluble biochemical, such as a fat, oil, or steroid.

steroids The class of lipids that contains cholesterol and the sex hormones.

membrane A biochemical structure composed primarily of lipids which serves to separate cell compartments and cells from one another.

bilayer A supermolecular structure in which many lipid molecules associate together. Their polar heads face outwards on two sides, toward surrounding water molecules, and their nonpolar tails extend inward, forming a hydrocarbon core.

protein A biopolymer made of amino acid units.

amino acids The units from which proteins are built. They have two functional groups: an amino group (NH_2) and a carboxylic acid group (CO_2H).

peptide bond The special name for the amide bond that links any two amino acid groups in a protein.

primary structure The sequence or order of amino acid units in a polypeptide or protein. Also, the nucleotide sequence in a nucleic acid.

polypeptide The general name for any polymer of amino acids. Proteins are naturally occurring polypeptides.

chirality The handedness of molecules that gives rise to mirror-image isomers.

enzyme A protein that serves as a biochemical catalyst.

active site The small portion of an enzyme where the substrates bind and are converted to products; the catalytic site of an enzyme.

nucleic acids The DNA or RNA molecules in cells that contain genetic information and direct its transfer. They are polymers made of nucleotide units.

DNA, deoxyribonucleic acid The polymer that contains genetic information in its sequence of nucleotide bases and determines heredity. It is composed of nucleotide units based on the sugar deoxyribose.

RNA, ribonucleic acid The polymer that transfers genetic information from DNA for the synthesis of proteins within cells. It is composed of nucleotide units based on the sugar ribose.

chromosomes The constituents of a cell's nucleus that contain the cell's genetic material, the genes.

genes The DNA structures that are the physical basis for transmission of hereditary characteristics from one generation to the next. Each gene contains the information for the amino acid sequence of one protein.

nucleotide A repeating unit of an RNA or DNA polymer, composed of a sugar linking a cyclic nitrogen base and a phosphate group. The term is also used for linked sugar, base, and phosphate groups in smaller molecules.

polynucleotide Any polymer made of repeating nucleotide units.

double helix The structure of DNA, made of two helixes intertwined. It is important in the replication of DNA.

catalyst A substance that increases the rate of a chemical reaction but is not a reactant or product in the equation for the net reaction.

substrates Chemicals that are reactants in metabolic reactions.

coenzyme A nonprotein, organic molecule that is required by an enzyme to catalyze a biochemical reaction.

ATP, adenosine triphosphate A nucleotide phosphate of central importance in the transfer of biochemical energy.

QUESTIONS

1. In chemical terms, how do you differ from a cat? From a tree? From a rock?

2. Four elements account for more than 99% of the weight of living matter. What are they?

3. Is glucose a protein, a carbohydrate, or a lipid?

4. Proteins normally contain which five elements?

5. Distinguish between the following:

a. starches and sugars
b. amino acids and proteins
c. fats and oils
d. monosaccharides and polysaccharides

6. What are the basic differences between the structures of DNA and RNA?

7. a. Give an example of an amino acid, a monosaccharide, a lipid, and a nucleotide base.

b. Draw a simplified structure of each example you gave in part a.

8. Distinguish between fatty acids and fats.

9. Draw the structure of the peptide linkage that bonds amino acids together in protein chains.

10. In the lock-and-key analogy of enzyme activity, the enzyme is the ____ and the substrate molecule is the ____.

11. When a molecule of sucrose is hydrolyzed, which two monosaccharides are obtained?

12. Explain the concept of activation energy for a chemical reaction, and show how an enzyme can affect the activation energy when it is a catalyst for the reaction.

13. a. What is the source of energy for photosynthesis?

b. What are the reactants in the photosynthetic process?

14. Write a chemical equation for the digestion of the following:

a. a protein to amino acids
b. a disaccharide to a monosaccharide

15. Give the structure of ATP, and point out the region of the molecule containing the bonds that are broken when ATP transfers chemical energy.

16. a. What stabilizing bonds hold the double helix together in the DNA structure proposed by Watson and Crick? Are these bonds strong or weak compared to the bonds between nucleotides in each strand?

b. What are the implications of the difference in the strength of these bonds for life as we know it?

17. How many dipeptides could be formed from three different amino acids?

18. Since the properties and biological activities of biomolecules depend on their functional groups, it is important to be able to identify these groups. Identify and name the constituent functional groups in each of the following biomolecules.

a.

```
     H
     |
H—C—OH
     |
H—C—OH
     |
H—C—OH
     |
     H
```

glycerol

b.

```
        COOH
         |
  H₂N—C—H
         |
   H—C—OH
         |
        CH₃
```

threonine, an amino acid

c.

```
 ⁻O    O
   \  //
    C
    |
   CH₂
    |
   CH₂
    |
   NH
    |
   C=O
    |
  HC—OH
    |
CH₃—C—CH₃
    |
   CH₂OH
```

pantothenate, a vitamin

19. Although biomolecules can be enormously complex, their structures do contain a limited set of building blocks. Identify the constituent building blocks for each of the two important biomolecules below.

a.

```
                                   NH₂
                                   |
                                   C
                               N//   \C—N
                               |     ||   \\CH
                              HC     C—N/
                                \\N/     |
     O       O       O                   |
     ||      ||      ||                  |
HO—P—O—P—O—P—O—H₂C      O                |
     |       |       |    \_________/
     OH      OH      OH   H  H    H  H
                             OH   OH
```

adenosine triphosphate (ATP), an energy-carrying biomolecule

b.

```
                                   O⁻
                                   |
        +
(CH₃)₃N—CH₂—CH₂—O—P—O—CH₂          O
                   ||      |          ||
                   O       CH—O—C(CH₂)₇CH=CH(CH₂)₇CH₃
                           |
                           CH₂—O—C(CH₂)₁₄CH₃
                                 ||
                                 O
```

phosphatidylcholine, a major membrane component in cells of higher organisms

Public-Policy Discussion Questions

20. Our understanding of human genetics is moving rapidly, and we may eventually have the ability to bring about specified genetic changes in humans. Suppose it were possible to alter the brain chemistry to raise IQ substantially or to breed human beings who always acted out of love and respect for one another. Would you be in favor of making such changes for yourself? For everyone?

21. Cloning is a process by which almost exact genetic copies of a living organism can be produced asexually. In the not-too-distant future it may be possible to clone human beings either in a donor mother or in artificial wombs by using sperm from males.

a. Would you want to be a duplicate? Of whom?

b. Would you want to be duplicated?

c. Discuss some of the ethical and political implications of widespread cloning.

Chemistry and Your Health

Modern medicine—which is based on several sciences, including chemistry—contributes to the improving health that successive generations enjoy. Often forgotten when we think about health, however, are nutrition and food, the careful selection of which plays a vital role in our physical well-being. As Americans try to shorten the amount of time they spend cooking meals, the use of easily prepared foods is increasing. Because these foods require many additives, their long-term effects on health are of public concern.

In this chapter we deal primarily with foods and drugs—substances that are eaten. These substances are a mix of natural and synthetic chemicals. Since the early 1900s, nutritionists have discovered many of the vitamins and minerals necessary for good health and have learned about the proper balance among protein, carbohydrate, and fat in our diets. Synthetic chemicals that are exact duplicates of natural chemicals offer no new problems. In fact, they are often purer than their natural counterparts and therefore more predictable in their effects on the human body. On the other hand, new synthetic chemicals that people have never ingested before must be subjected to scrutiny. Scientists in the food and drug industries are constantly seeking to find better methods for evaluating the safety of these chemicals in a reasonable amount of time.

16.1 WHAT IS IN THE FOOD WE EAT?

We eat both for enjoyment and for survival. Tastes and preferences for foods are beyond the scope of scientific investigation, except in the area of marketing. Questions of nutrition, however, have been the subject of much research, and the molecular structures and biochemical functions of many essential nutrients are now understood.

Although the public focus is on food additives, over 99% by weight of our food intake is from natural sources. Most of the food additives that make up the remaining 1% are substances such as vitamin D (in fortified milk) and iron-containing compounds that also occur in nature but are thought to be lacking in some people's diets. Additives such as artificial coloring and preservatives account for a tiny percentage of our total food intake. In this section we will concentrate on the major components that come from natural sources and are essential to our diets. Additives are the subject of Section 16.4.

People must regularly consume three nutrients: proteins, carbohydrates, and fats (see Figure 16.1). These **macronutrients** provide the energy for all bodily activity. But they are not interchangeable; each serves some unique function. That is one reason why we are told to eat a well-balanced diet that includes choices from a number of food groups.

As mentioned in Chapter 15, proteins account for more of our body weight than any other biopolymer does. Because protein is slowly degraded and lost, it must be replaced by the food we consume so that body tissue and other nitrogen-containing structures can be rebuilt. During childhood, and after severe injury or surgery, when a large amount of blood has been lost, extra amounts of protein are needed for the growth of new tissue.

Once in the intestinal tract, proteins must be broken down by the digestive process into their constituent amino acids in order to be utilized in building tissue. The proteins in our food are digested with the help of enzymes made in the pancreas. Following absorption into the bloodstream, these amino acids are recombined into needed proteins in a process requiring ATP (Section 15.8). Although all 20 or so amino acids must be available in our bodies in order for the variety of essential proteins to be formed, not all must be present in our diets. Our bodies can synthesize some of them. Biochemists have identified eight amino acids, however, that must be present in the proteins we consume, because our bodies cannot produce them. These **essential amino acids** are listed in Table 16.1 (see page 402).

Five of these essential amino acids (methionine, tryptophan, isoleucine, threonine, and lysine) have been identified as lacking in certain diets, particularly those of children in developing countries. Children on near-starvation

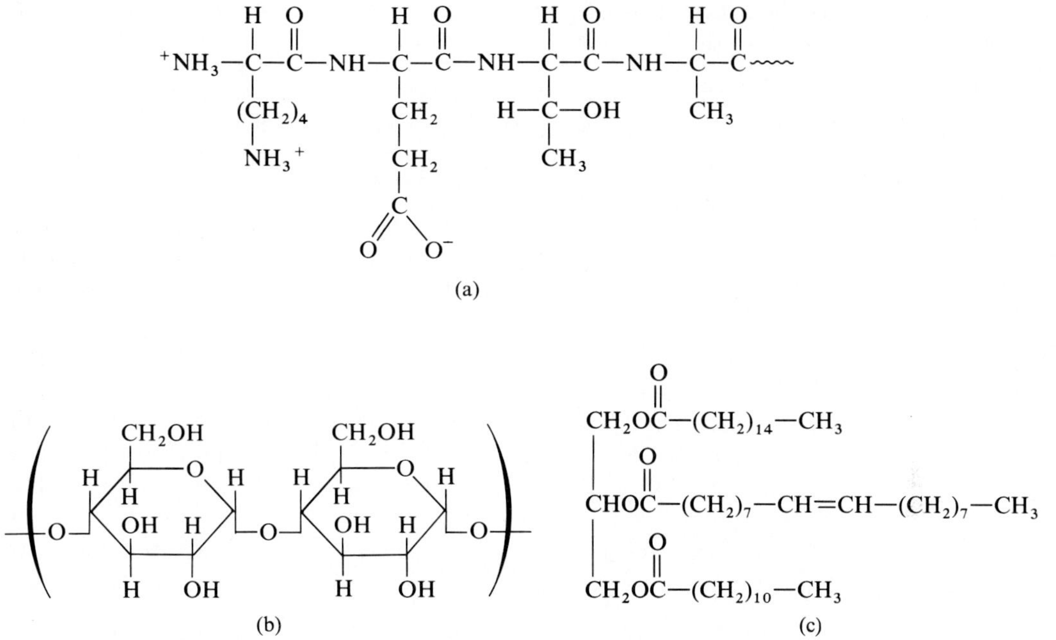

$$^{+}NH_3-\overset{\overset{\displaystyle H}{|}}{\underset{\underset{\underset{\displaystyle NH_3^{+}}{|}}{\displaystyle (CH_2)_4}}{\underset{|}{C}}}-\overset{\overset{\displaystyle O}{\|}}{C}-NH-\overset{\overset{\displaystyle H}{|}}{\underset{\underset{\underset{\underset{\underset{\displaystyle O\ \ O^{-}}{//\ \backslash}}{\displaystyle C}}{|}}{\displaystyle CH_2 - CH_2}}{\underset{|}{C}}}-\overset{\overset{\displaystyle O}{\|}}{C}-NH-\overset{\overset{\displaystyle H}{|}}{\underset{\underset{\underset{\displaystyle CH_3}{|}}{\displaystyle H-C-OH}}{\underset{|}{C}}}-\overset{\overset{\displaystyle O}{\|}}{C}-NH-\overset{\overset{\displaystyle H}{|}}{\underset{\underset{\displaystyle CH_3}{|}}{C}}-\overset{\overset{\displaystyle O}{\|}}{C}\sim$$

(a)

(b)

$$CH_2O\overset{\overset{\displaystyle O}{\|}}{C}-(CH_2)_{14}-CH_3$$
$$CHO\overset{\overset{\displaystyle O}{\|}}{C}-(CH_2)_7-CH{=}CH-(CH_2)_7-CH_3$$
$$CH_2O\overset{\overset{\displaystyle O}{\|}}{C}-(CH_2)_{10}-CH_3$$

(c)

Figure 16.1. The macronutrients in a well-balanced diet: (a) a very small section of a protein, (b) a section of starch, and (c) a typical fat.

diets usually develop Kwashiorkor disease, which results from severe protein deficiency. The most obvious effects are apathy, atrophied muscles, and a protuberant belly (see Figure 16.2). Before the food-shortage crisis in Africa during the early 1980s, the World Health Organization estimated that between 100 and 300 million children were afflicted with Kwashiorkor disease.

The Food and Drug Administration (FDA) rates proteins according to their quality. A **high-quality protein** contains all eight essential amino acids in about the proportions needed by the human body. High-quality protein most often comes from animal sources, such as meat, fish, poultry, eggs, milk, and cheese. In fact, the FDA has defined casein, the principal protein found in milk, as the standard. Any protein equal to or better than casein in nutritional value is labeled "high quality." The Food and Nutrition Board, under the sponsorship of two prestigious scientific societies, has determined that a well-balanced diet for adults and children 4 years of age and older should include 45 grams of high-quality protein per day. If the protein is of significantly lower quality than the standard, 65 grams is recommended for daily consumption. Either of these amounts constitutes the **U.S. Recommended Daily Allowance** under the regulations of the Food and Drug Administration. Protein of lower quality is typically found in

Table 16.1 The Essential Amino Acids

NAME	STRUCTURE
isoleucine	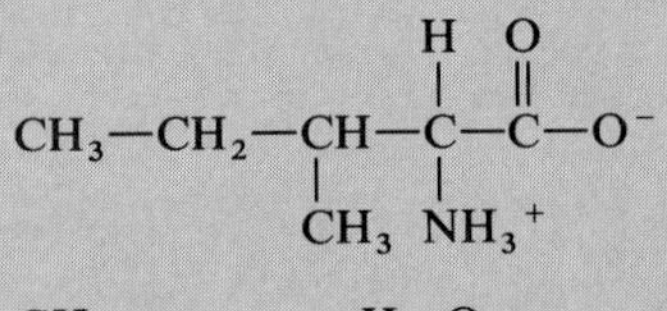 $CH_3-CH_2-CH(CH_3)-CH(NH_3^+)-C(=O)-O^-$
leucine	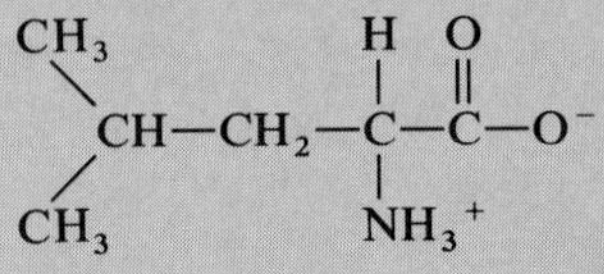 $(CH_3)_2CH-CH_2-CH(NH_3^+)-C(=O)-O^-$
lysine	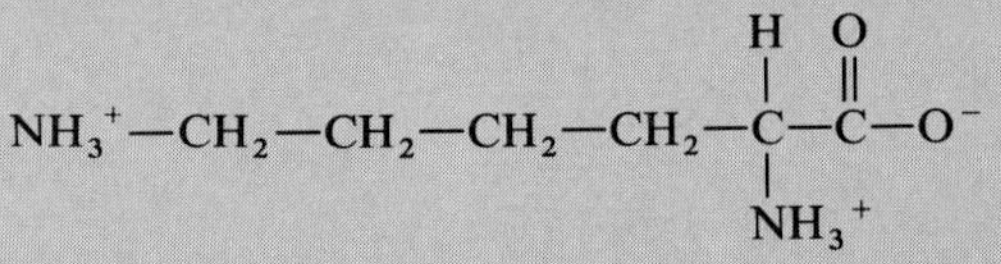 $NH_3^+-CH_2-CH_2-CH_2-CH_2-CH(NH_3^+)-C(=O)-O^-$
methionine	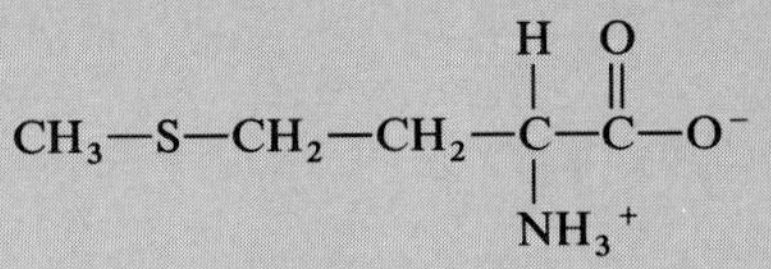 $CH_3-S-CH_2-CH_2-CH(NH_3^+)-C(=O)-O^-$
phenylalanine	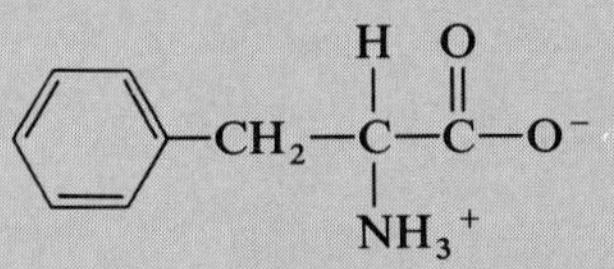 $C_6H_5-CH_2-CH(NH_3^+)-C(=O)-O^-$
threonine	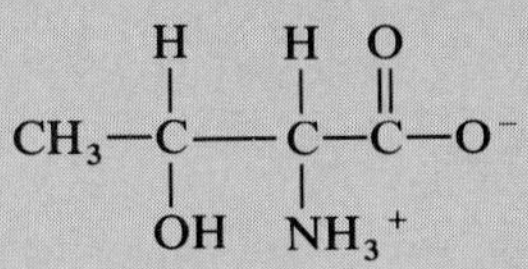 $CH_3-CH(OH)-CH(NH_3^+)-C(=O)-O^-$
tryptophan	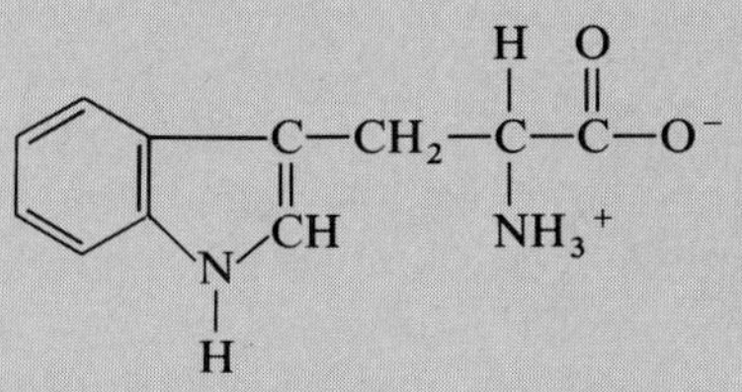 $C_8H_6N-CH_2-CH(NH_3^+)-C(=O)-O^-$ (indole ring: benzene fused to $C=CH-N(H)$)
valine	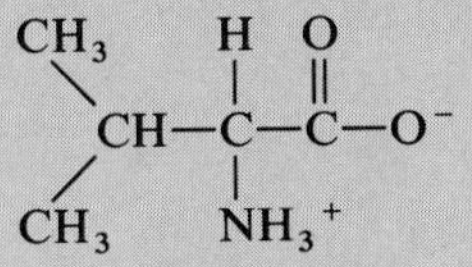 $(CH_3)_2CH-CH(NH_3^+)-C(=O)-O^-$

Figure 16.2. Starving children in Africa.

vegetables and grains. By combining several vegetables or grains, a person can often obtain all the essential amino acids, because an amino acid that is missing from one source is likely to be present in another. Peas and beans have more of the essential amino acids than most other vegetables do.

A strict vegetarian must be careful to eat foods from many plant sources in order to have enough protein of the right kind. Adding even small amounts of milk, cheese, or eggs to the diet makes it much easier for a vegetarian to get the essential amino acids (see Figure 16.3).

The chief role of the second macronutrient, carbohydrate, is to provide us with energy for all our physical and mental activities. We obtain carbohydrate mainly in the form of starch (Section 15.2), which is converted into glucose in the digestive process. Cereals, breads, rice, pasta, fruits, and potatoes are among the best sources of starch.

Although cellulose and amylose (one of the macromolecules in starch) have very similar structures (see Figure 16.4), human beings can digest only starch. None of the enzymes in the human body can catalyze the breakdown of cellulose into glucose. Thus, foods such as lettuce and celery, which contain cellulose, provide necessary roughage but they provide few

Figure 16.3. (a) Eating only vegetables and fruits makes it difficult to obtain enough of the essential amino acids. (b) Adding milk, cheese, or eggs minimizes this problem.

calories. It is the bacteria in the rumens of cows and sheep that digest the cellulose in the grasses they eat.

Like people in many other affluent societies, Americans tend to consume more carbohydrate than their energy needs require. The per capita consumption of carbohydrate in the United States is about 130 pounds per year. The excess carbohydrate ultimately becomes fat and can contribute to weight problems. Another problem with a diet heavy in carbohydrates is that other nutrients may be lacking. Pure sucrose has been called a source of "empty calories," for it has no other nutrients. There is no U.S. Recommended Daily Allowance for carbohydrates.

Figure 16.4. Structures of (a) cellulose and (b) amylose, a soluble starch.

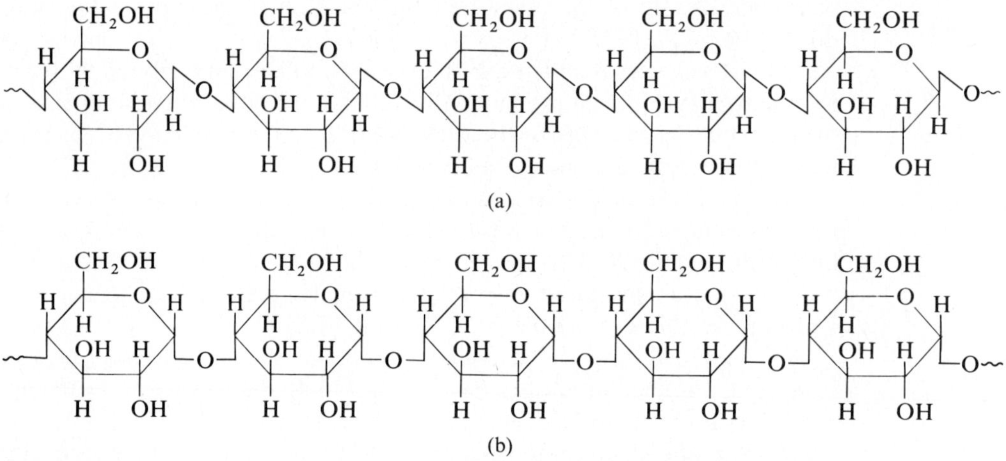

(a)

(b)

The third macronutrient, fat, is the most concentrated source of energy for our bodies. As discussed in the preceding chapters, fats are esters of glycerol and carboxylic acids having from 10 to 20 carbon atoms. Some of the fatty-acid components are unsaturated (that is, they have one or more carbon-carbon double bonds); others are saturated. Fats are digested by enzymes into fatty acids and glycerol in our intestines and eventually become oxidized to CO_2.

The benefits of unsaturated as opposed to saturated fats in our diets have received much publicity in recent years. Several **polyunsaturated fatty acids** (fatty acids containing at least two carbon-carbon double bonds) are thought to be essential, because the body cannot synthesize them either at all or in sufficient quantity. For example, linoleic acid,

$$CH_3(CH_2)_4CH{=}CHCH_2CH{=}CH(CH_2)_7COOH$$

linoleic acid, $C_{18}H_{32}O_2$

is not manufactured by the body but is essential for proper growth and a healthy skin. Eczema can result from a deficiency of this fatty acid. Nutrition specialists usually recommend inclusion of vegetable as well as animal fats in a diet, because the former have more polyunsaturated fats (see Figure 16.5). The amount of polyunsaturated fats needed is not great—just a few percent of our total fatty-acid intake. Corn, peanut, and soybean oils have above-average amounts of linoleic acid, whereas beef fat, coconut oil, and chocolate have high percentages of saturated fat. Hydrogenated oils, such as margarine, have been treated with hydrogen gas to convert double bonds into single bonds:

$$\underset{/\quad\;\;\backslash}{\overset{H\;\;\;\;\;\;H}{\overset{\backslash\quad\;\;/}{C{=}C}}} + H_2 \longrightarrow \underset{/\quad\;\;\backslash}{\overset{H\;\;\;\;\;\;H}{\overset{\backslash\quad\;\;/}{H{-}C{-}C{-}H}}}$$

The purpose is to make the product thicker and less fluid, but at the same time the amount of essential polyunsaturated fat is decreased. Whether this poses a problem in the average person's diet is a matter of debate.

Figure 16.5. Unsaturated fat in salad oil. Olive, corn, peanut, or soybean oil adds flavor to a salad and provides unsaturated fatty acids.

16.2 CALORIES AND FOODS

The idea that each food has a certain number of dietary calories is familiar to everyone. These calories come from the proteins, carbohydrates, and fats just discussed. Actually, calories, which are amounts of energy, are not *in* the foods; they represent the energy available when the foods are oxidized in metabolic reactions. The problem for food scientists is that it is difficult to measure the exact amounts of energy produced in the body for performance of vital functions. What then is the source of the values of dietary calories found in any book about nutrition or weight loss?

In brief, food scientists measure the amount of heat evolved when a known amount of food reacts with oxygen in an instrument called a *bomb calorimeter* (Figure 5.4). For fats and carbohydrates, the reaction products are carbon dioxide and water. The reactions are very similar to metabolic reactions, in which oxygen delivered to our lungs reacts with these same nutrients to yield carbon dioxide and water. When protein is burned in this manner, N_2 and nitrogen oxides are also formed.

In our bodies, part of the energy released drives processes such as muscle contraction and the synthesis of proteins or nucleic acids, and part provides the heat needed to maintain our body temperature. In the laboratory, all of the energy evolved is in the form of heat. However, the total amounts of energy are about equal in the two processes, and the reactants (food and oxygen) and the products are much the same. So, tables of food calories and the labels on cans of food and beverages give realistic and useful information to people concerned about their weight. Table 16.2 gives dietary calories for common foods. A **dietary calorie** is equivalent to 4.18 kilojoules. It is the energy needed to raise the temperature of 1 liter of water by 1°C and is equal to 1000 "small" calories.

Fats provide more energy per unit weight than either carbohydrates or proteins. Specifically, the combustion of 1 gram of fat yields about 9 dietary calories, whereas combustion of carbohydrates or proteins gives 4 dietary calories per gram. The difference between a fat and a carbohydrate is a result of the much greater amount of CO_2 and H_2O formed in the combustion of a fat. Let's compare glycerol tripalmitate [$(C_{15}H_{31}COO)_3C_3H_5$, molecular weight = 807] with sucrose [$C_{12}H_{22}O_{11}$, molecular weight = 342]. When the first compound, the fat, is oxidized to CO_2 and H_2O through our metabolism, the overall reaction is

$$(C_{15}H_{31}COO)_3C_3H_5 + \tfrac{145}{2}\,O_2 \longrightarrow 51\,CO_2 + 49\,H_2O$$

This equation says that 1 mole of glycerol tripalmitate, which weighs 807 grams, yields 51 moles of CO_2 and 49 moles of H_2O. The same weight of sucrose (807 grams) produces slightly over 28 moles of CO_2 and 26 moles of H_2O when it reacts with oxygen. So when equal weights of fat and carbohydrate are compared, the combustion of the fat gives much more CO_2 and H_2O.

Table 16.2 Dietary Calories for Foods

FOOD	SERVING	DIETARY CALORIES per serving	DIETARY CALORIES per gram
apple	1 (180 g)	93	0.5
bacon	2 cooked slices (15 g)	86	5.7
banana	1 (175 g)	101	0.6
beef	1 cooked hamburger (82 g), with 21% fat	235	2.9
beer	1 12-oz. can (360 g)	151	0.4
bread	1 slice, white (28 g)	76	2.7
butter	1 tablespoon (14.2 g)	102	7.2
carrot	1 cup, cooked (155 g)	48	0.3
chicken	2 pieces light meat (50 g), roasted	83	1.7
chocolate	1 oz. plain milk chocolate (28 g)	147	5.2
Coca-Cola	1 12-oz. can (369 g)	144	0.4
corn flakes	1 cup (25 g)	97	3.9
egg	1 fried (46 g)	99	2.2
lettuce	1 piece iceberg (20 g)	3	0.2
margarine	1 tablespoon (14.2 g)	102	7.2
peanut butter	1 tablespoon (16 g)	94	5.9
pizza	1 slice (67 g), with sausage topping	157	2.3
potato	10 french fries (50 g)	137	2.7
salad dressing	1 tablespoon mayonnaise (15 g)	65	4.3
spaghetti	1 cup, cooked (140 g)	155	1.1
sugar	1 tablespoon, granulated (12 g)	46	3.8

When a fat ''burns,'' strong C═O bonds and O—H bonds in the products replace the somewhat weaker C—C, C—H, C—O, and O═O bonds of the reactants. The formation of these strong bonds means that the potential energy of the products is less than that of the reactants. Such a reaction is highly exothermic (Chapter 4). Although the combustion of a carbohydrate is also exothermic, it is less so, because the compound contains much oxygen and is already partly ''burned.'' Figure 16.6 shows the energy relationships.

For years there has been a demand among consumers for foods that have a sweet taste without the considerable number of calories that accompany the ingestion of sucrose. Diabetics and people desiring to lose weight are

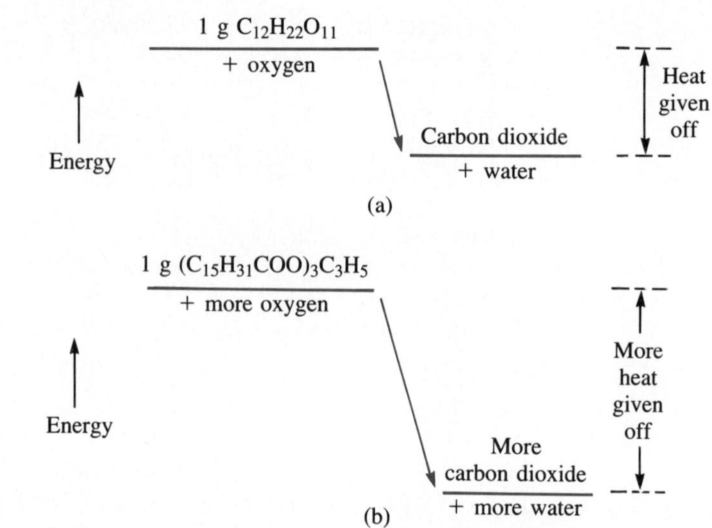

Figure 16.6. Energy diagrams. (a) The combustion of 1 gram of sucrose yields a moderate amount of energy. (b) The combustion of 1 gram of fat requires more oxygen and releases more energy.

particularly interested in low-calorie **artificial sweeteners.** The one with the longest history of use is saccharin,

saccharin   sodium salt of saccharin

which is normally marketed as the sodium salt or potassium salt. Saccharin is about 300 times sweeter than an equal weight of sucrose, and it passes through the body unchanged; therefore, it adds no calories to one's diet. However, some people find that it has a bitter aftertaste, and so some manufacturers add glycine, the simplest amino acid, to counteract this effect.

A more recent addition to the list of artificial sweeteners is aspartame,

$$H_3\overset{+}{N}-CH(CH_2COO^-)-C(=O)-NH-CH(CH_2C_6H_5)-C(=O)-OCH_3$$

which goes under the tradename NutraSweet. It is actually a peptide (Section 15.5) containing two amino acids. The left portion of the structure shown above is aspartic acid, and the right part is the methyl ester of phenylalanine. Although aspartame produces about the same number of calories as the same amount of sucrose, it is 180 times sweeter. Because such a small amount is needed to sweeten foods, it adds few calories. Like all other artificial sweeteners, NutraSweet has been tested extensively for toxic and carcinogenic effects. Currently aspartame is used in many products, including diet beverages, although it tends to break down when the beverages are stored for a long time.

16.3 VITAMINS AND MINERALS

Have you ever taken large doses of vitamin C to combat a cold? Do you know people who take multiple-vitamin supplements every day, or who take large amounts of vitamin A to prevent cancer or vitamin E to slow the aging process? There is no doubt that small amounts of 12 or 13 vitamins are essential for good health, but the benefits of taking megadoses of particular vitamins to prevent or cure certain ailments are open to considerable doubt.

Because knowledge of vitamins in our society is so commonplace, it is hard to believe that 80 years ago no one had heard of them. In the early part of this century Casimir Funk, an American biochemist, isolated a dietary growth factor from rice polishings (the outer covering of grains of rice). He found that it cured the disease beriberi when added to the food of those who had the disease. The factor was an organic compound of the class called *amines* (Table 11.4). Funk coined the term ''vitamine'' (life-giving amine) for this compound. We now call it *thiamine* or vitamin B_1. During the middle years of the twentieth century, chemists discovered many other simple substances that have to be present in our food if we are to enjoy good health. A number of them were not amines, and the word ''vitamine'' was shortened to the present-day **vitamin.**

A vitamin is sometimes called a **micronutrient,** because only small amounts, in the neighborhood of milligrams, are needed in the daily diet. By contrast, the proteins, carbohydrates, and fats that we consume each day are measured in grams. Because a vitamin is an organic compound, it is distinguished from a **mineral,** which is an inorganic salt. As more information about the chemical compositions and structures of vitamins has become available, general names such as vitamin B_1 and vitamin C have given way to chemical names—in this case thiamine and ascorbic acid, respectively.

Although vitamins are crucial to good health, we are not able to produce them in our own bodies in amounts large enough to keep us healthy. Probably vitamins have always been available in our food, so there was no need for our metabolism to produce them. Hormones are also organic compounds necessary in small amounts for normal health and growth, but most people make adequate amounts of hormones in their bodies and don't have to eat them. When someone cannot make enough of the hormone insulin, he or she suffers from the disease called diabetes.

Many vitamins are required by our bodies because we turn them into coenzymes, which serve as oxidizing, reducing, and transferring chemicals at the active sites of enzymes (Section 15.8). Two such vitamins are thiamine and niacin (shown in Figure 16.7), both of which are required by every living thing.

Thiamine, or vitamin B_1, plays a variety of roles in our bodies, from serving as a coenzyme in carbohydrate metabolism and energy production to affecting appetite, digestion, and nerve activity. Thiamine hydrochloride forms white crystals that melt at 244°C and dissolve easily in water. Wheat germ, rice bran, yeast, and soybean flour are excellent sources of thiamine. Many vegetables, nuts, and meats are also good dietary sources.

Niacin, or nicotinic acid, is another vitamin that was isolated from rice polishings by Funk. Medical research soon showed that the chronic disease pellagra, common in the southern United States until the 1940s, was caused by a niacin deficiency.

Another common vitamin is **vitamin C, or ascorbic acid:**

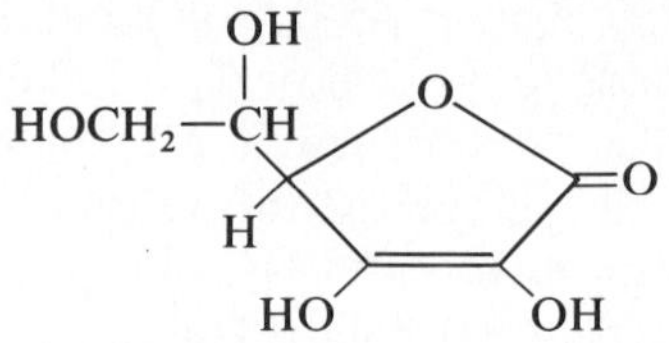

A structure as complicated as this one was not easy to make, but synthetic vitamin C was produced successfully in the 1930s. The best-known natural source of vitamin C is citrus fruit, although it is also present in many vegetables, as well as in rose hips. Several centuries ago British sailors who were long at sea without much fresh food often developed scurvy, a disease that causes bleeding at the gums and general weakness. Eventually the British navy realized that scurvy could be prevented by eating oranges, lemons, grapefruit, or limes. It is said that the name "limey" for a British sailor came from the requirement that he regularly eat limes. Today, synthetic ascorbic acid is cheaper than ascorbic acid isolated from natural sources, and so commercial vitamin C is generally synthetic.

The Food and Nutrition Board has determined our daily needs for 12 vitamins. The U.S. Food and Drug Administration has made these findings the basis for its Recommended Daily Allowances (RDA), given in Table 16.3. Because vitamins A, D, and E come in several forms, the recommended

Figure 16.7. Two vitamins: (a) thiamine hydrochloride, and (b) niacin.

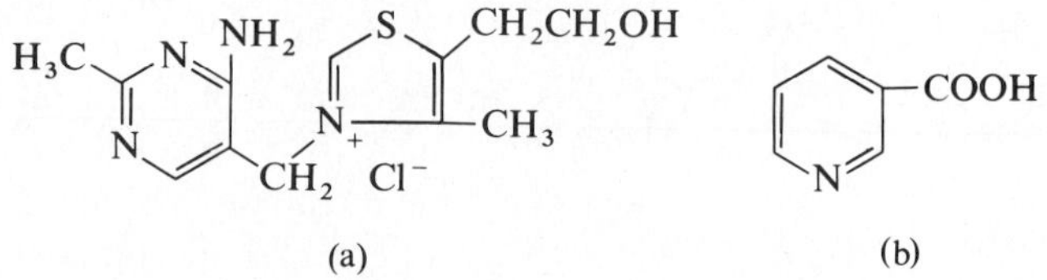

(a) (b)

Table 16.3 Characteristics of Vitamins

NAME	RDA FOR ADULTS	BENEFIT TO BODY	BEST FOOD SOURCES
Fat-soluble			
vitamin A (retinol)	5000 IU	new cell growth, night vision	green and yellow vegetables, eggs, milk
vitamin D (calciferol)	400 IU	incorporation of Ca and P in bone	fish, egg yolks; sunlight causes formation in skin
vitamin E (tocopherol)	30 IU	antioxidant; prevention of destruction of vital compounds	beans, eggs, whole grains, fruits, vegetables
Water-soluble			
vitamin B_1 (thiamine)	1.5 mg	digestion, growth, carbohydrate metabolism	pork, beans, peas, nuts, whole-grain breads and cereals
vitamin B_2 (riboflavin)	1.7 mg	obtaining energy from foods	leafy vegetables, whole-grain breads, dairy products
niacin	20 mg	healthy tissue cells	lean meats, peas, beans, whole-grain cereals, fish
pantothenic acid	10 mg	proper growth	liver, eggs, potatoes, peas, whole grains
folic acid	0.4 mg	growth of red blood cells, metabolism	liver, navy beans, green leafy vegetables
vitamin B_6 (pyridoxine)	2 mg	utilization of protein	liver, whole-grain cereals, red meats, green vegetables
vitamin B_{12} (cyanocobalamin)	6 mcg	growth of red blood cells, functioning of all cells	lean meats, fish, milk, eggs
biotin	0.3 mg	metabolism	eggs, milk, meats
vitamin C (ascorbic acid)	60 mg	tissue repair, tooth and bone formation	green peppers, broccoli, citrus fruits, tomatoes

Units: IU = International Unit, mg = milligram, mcg = microgram (10^{-6} gram)

amounts for them are specified in International Units (IU), which measure biological activity. Table 16.3 also gives the functions of each vitamin in the body, along with some examples of foods that are good sources. Since anyone who maintains a balanced diet automatically consumes at least the Recommended Daily Allowances of vitamins, vitamin deficiencies are not common in the United States. However, poverty, junk-food diets, and eating disorders can lead to vitamin deficiencies.

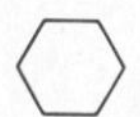

VITAMIN SUPPLEMENTS—HELPFUL OR NOT?

Evidently many Americans believe that they are not getting enough vitamins from the food they eat, despite strong evidence that a well-balanced diet provides ample vitamins of all kinds. An estimated 40 million people in the United States take supplemental vitamins; only about 7% of them do so at the suggestion of a physician. Often the amounts of vitamins ingested are large enough to constitute a **megadose:** ten or more times the Recommended Daily Allowance.

Research has established that for many vitamins there is a toxic dose—a level at which harmful effects occur. Such evidence raises some important public-policy questions. How do we balance the possible benefits of a large dose of a particular vitamin against the likelihood of toxic side effects? Is the toxicity at high dosage so well documented that the vitamin should be sold by prescription only? Information on vitamins A and C is presented below.

Vitamin A, popular as a supplemental vitamin, has been responsible for more cases of poisoning than any other vitamin. Harm can result from either a single very large dose or prolonged use at higher than recommended levels. The symptoms of vitamin A overdose include headache, nausea and vomiting, fatigue, swelling, hemorrhage, and pain in the arms and legs. If taken during pregnancy, large doses can cause birth defects. At the other extreme, however, a complete lack of vitamin A endangers health. According to estimates, 250,000 children in developing countries become blind each year because of a vitamin A deficiency.

Prolonged controversy has followed the assertion by a few leading figures, including Linus Pauling and Norman Cousins, that large doses of vitamin C have therapeutic value. In a 1970 book *Vitamin C and the Common Cold*, Pauling argued that daily doses of 200 milligrams or more reduce the frequency and severity of colds. More recently, Norman Cousins attributed his recovery from a serious illness partly to a large intake of vitamin C. Two questions arise: Is vitamin C effective in the treatment of colds, and are there harmful side effects from large doses of vitamin C?

Linus Pauling, twice a Nobel Prize winner, cited reports in the scientific literature and personal experience as evidence for the effectiveness of vitamin C in treating the common cold. Since his book was published, there

A thirteenth vitamin, vitamin K, is not listed in Table 16.3 because no Recommended Daily Allowance has been established for it. Scientists know that vitamin K must be present for the clotting of blood; a deficiency results in hemorrhage and liver damage. Egg yolks and vegetables such as spinach, lettuce, cabbage, and cauliflower are good sources of vitamin K.

The vitamins in Table 16.3 are grouped into two categories: fat-soluble and water-soluble. **Fat-soluble vitamins** (A, D, E, and K) migrate to fatty tissues in the body, where they can remain for a considerable period. So these vitamins can accumulate in potentially toxic levels if taken in large amounts. The remaining vitamins, known as **water-soluble vitamins,** pass through the body in a relatively short time. For example, two hours after have been many controlled experiments in which placebos (pills that look like vitamin C pills but contain inert ingredients) were given to a control group and real vitamin C pills to another group. No participant knew which kind of pill he or she was getting. Data on the number of colds and the length of each cold were recorded for everyone. These studies provided no evidence that vitamin C reduces the number of colds. There is evidence, however, that the intensity of cold symptoms in some people is decreased by large doses of vitamin C.

On the question of possible harmful effects, there is also much evidence. Although the water-soluble vitamin C passes very quickly through the system, it can have harmful effects on several groups of people. Megadoses of the vitamin can hasten the onset of gout or kidney stones in susceptible people. Diabetics who are taking large amounts of vitamin C may get misleading results from tests for sugar in their urine. And people who have been taking vitamin C for a long time may suffer the symptoms of scurvy if they suddenly stop their intake of the vitamin. It is advisable to decrease the dosage slowly. We need more evidence to develop a reliable picture of the side effects of vitamin C taken in megadoses.

Nine vitamins have been judged by an FDA panel to be safe and effective as over-the-counter drugs, but only for treatment of specific deficiencies under medical direction. In short, the FDA sees little validity in most of the claims of vitamins' curative effects, such as the allegations that vitamin E will help bald people grow hair, cure skin problems, ease arthritis pain, and prevent ulcers. Most people would do best to pay some attention to their diets and skip the vitamin supplements, unless a physician discovers a treatable deficiency.

Should the government take specific steps to discourage citizens from taking large amounts of vitamins? Some people may be harming themselves, but personal freedom is also an issue here. Vitamins could be classified as prescription drugs so that people could no longer use supplemental doses without a doctor's authorization. Another alternative would be to require warnings on the labels of all vitamin containers. A third possibility would be to conduct a public education campaign on the possible dangers of vitamin overdoses. We favor a mixture of the last two alternatives, depending on the specific vitamin involved. Where do you stand on this issue?

a 1000-milligram dose of vitamin C is taken, 950 milligrams has been excreted into the urine.

Significant structural differences account for the different solubilities of the two groups of vitamins (see Figure 16.8). The fat-soluble vitamins consist mainly of various hydrocarbon groupings and are structurally similar to fats. Water-soluble vitamins, on the other hand, have a number of groups such as —OH that form strong hydrogen bonds with water. Vitamin A, a fat-soluble vitamin, has only one —OH group and a long hydrocarbon chain, as shown in part (a) of Figure 16.8. Vitamin B_2, or riboflavin, has four —OH groups and several O and N atoms, all capable of forming hydrogen bonds with water. Its structure can be seen in part (b) of Figure 16.8.

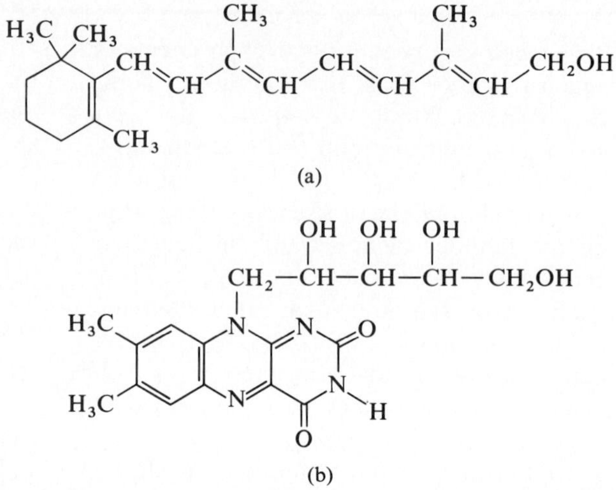

Figure 16.8. Structural differences between (a) a fat-soluble vitamin, vitamin A, and (b) a water-soluble vitamin, riboflavin.

A final class of essential micronutrients is the minerals, which are usually inorganic salts of particular metals. Nutritionists tend to speak of "minerals" such as calcium, phosphorus, iron, and iodine as though they were present in the body as pure elements. A moment's reflection shows this to be impossible. Calcium metal, for example, would react with water and other substances in the body with painful effect. In reality, all of these elements are combined with others in compounds, and in this combined form they serve indispensable functions in the body. So when the FDA states that the Recommended Daily Allowance for calcium is 1 gram, it means that we need calcium compounds containing 1 gram of calcium.

There are about a dozen-and-a-half elements known to play vital roles in our bodies, and others probably remain to be discovered. Most of these elements slowly pass through the body and must be replaced. The necessary daily intake of a few—such as calcium, phosphorus, and magnesium—is in the neighborhood of 1 gram. For others, the amounts needed are in the milligram range. Sufficient information is available to specify how much we need of only seven elements. See Table 16.4 for the RDA and other information about these seven minerals.

Calcium is probably the best known of the minerals. It is present in the body in a greater amount than any other mineral, primarily in bones and teeth. Small amounts of calcium are needed for normal functioning of the nervous system. At the other end of the scale, perhaps, is selenium, for which we have no Recommended Daily Allowance. Selenium has long been known as a poison; chemists working with selenium in the laboratory are careful not to get any in their mouths. Yet selenium at the trace level is considered essential for the proper functioning of vitamin E.

Table 16.4 Characteristics of Minerals

MINERAL	RDA FOR ADULTS	FUNCTION IN BODY	BEST FOOD SOURCES
calcium	1 g	bone formation, metabolic processes	milk, green leafy vegetables
copper	2 mg	functioning of hemoglobin	organ meats, shellfish, nuts, dried legumes
iodine	0.15 mg	thyroid function	seafood, iodized salt
iron	18 mg	carrying oxygen in hemoglobin	liver, egg yolk, green leafy vegetables, whole grains
magnesium	400 mg	carbohydrate and protein metabolism	many foods
phosphorus	1 g	bone formation, metabolic processes	meat, poultry, whole grains
zinc	15 mg	constituent of certain enzymes	meat, fish, egg yolk, milk

Units: g = grams, mg = milligrams

16.4 FOOD ADDITIVES

Even though well under 1% of the typical American diet consists of food additives, concern over these substances—particularly their possible cancer-producing properties—has run high in recent times. In the 1960s and 1970s public alarm over two artificial sweeteners, cyclamate and saccharin, caused a ban on the use of cyclamates in diet drinks and a temporary suspension of the use of saccharin. It is not easy to keep track of the advantages and potential hazards of the many food additives now used. A recent FDA listing of common food additives contained 127 substances, and the list was by no means exhaustive.

The introduction of food additives in large numbers is a result of some far-reaching changes that have taken place during the twentieth century. The percentage of family income spent on food has dropped, society has become more urban, and people have developed higher expectations for good health. The cost of food is relatively low today in part because each farmer produces much more per acre than before. But another important reason is that the rate of food spoilage has been greatly reduced. Foods that formerly had to be discarded after a few days on the grocery shelf can now be kept several times longer without loss of quality. Better food preservation techniques were developed when food had to be transported greater distances to feed the growing urban population. We now expect fruits and vegetables to be available throughout the year. We take for granted the absence of nutritional diseases (such as rickets, goiter, and pellegra), because foods are fortified with vitamins and minerals.

Food additives serve at least four functions. Some compensate for vitamin or mineral deficiencies that might otherwise occur in our diets. Others maintain freshness and prevent spoiling. Still others enhance the attractiveness of food. And finally, some additives facilitate the processing of food in the agricultural industry and its preparation at home. We will discuss the first three functions in more detail.

The history of **fortified foods**—foods enriched with nutrients (see Figure 16.9)—goes back to 1924, when potassium iodide (KI) was first added to table salt to prevent goiter, an ailment of the thyroid gland caused by iodine deficiency. Previously, goiter was common in the Great Lakes region and in the Pacific Northwest—areas with little iodine in the soil and in crops grown in that soil. Nowadays goiter is rare in the United States. Other diseases have also been eliminated by appropriate food additives. Most milk is fortified with vitamin D to prevent rickets, and some of the B vitamins are added to bread and cereals to prevent pellagra (skin eruptions resulting from a niacin deficiency) and beriberi (paralysis of the extremities from lack of thiamine). We have only to look at other countries that have deficiency-related diseases to recognize the benefits of fortified salt, bread, flour, cereal, milk, and other foods.

Sometimes food additives are included to replace nutrients lost during processing, as in white flour for white bread. During the milling of wheat for white flour, vitamins and minerals are lost along with the outer covering of the grain. As many as six vitamins and four minerals may be added later to restore the nutritional value of the bread. Another example is highly processed convenience foods, which increasingly are being eaten by a busy population. Although such foods may provide sufficient calories, they are often low in micronutrients unless they are fortified.

Why do so many lists of ingredients on food packages include sodium benzoate, sodium propionate, or BHA (butylated hydroxyanisole)? These are common examples of chemicals used to maintain freshness and prevent

Figure 16.9. Fortified foods.

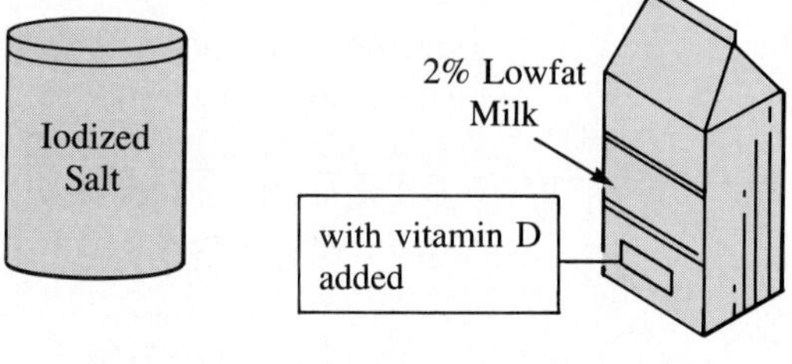

Iodized Salt

2% Lowfat Milk

with vitamin D added

spoiling. Sodium benzoate and sodium propionate are **preservatives,** additives designed to protect foods from the action of molds, bacteria, fungi, and yeasts. Molds are usually visible and signal a warning to throw out the spoiled food, but bacteria may leave no evidence of spoilage. Eating food containing harmful bacteria can cause severe stomach disorders and, in rare instances, death. Sodium benzoate,

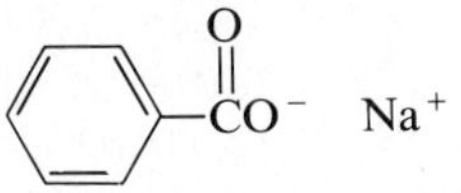

$$C_6H_5{-}\overset{\overset{\displaystyle O}{\|}}{C}O^- \quad Na^+$$

is the preservative used most often for bread and other baked goods. Sodium propionate and calcium propionate,

$$CH_3CH_2COO^-\ Na^+ \qquad \text{and} \qquad (CH_3CH_2COO^-)_2\ Ca^{2+}$$

respectively, are preservatives frequently used for fruit products and margarine.

Other chemicals used to prevent spoiling are the **antioxidants,** of which BHT and BHA are examples. BHT has the structure

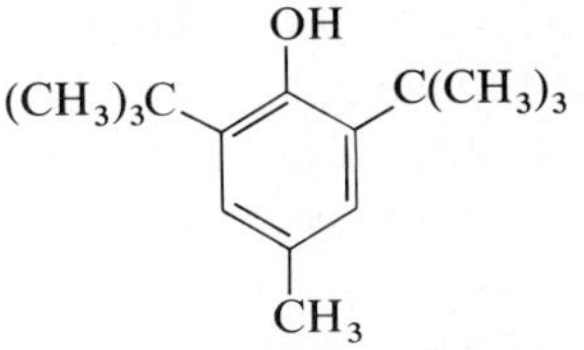

OH

$(CH_3)_3C$ $C(CH_3)_3$

CH_3

butylated hydroxytoluene (BHT)

BHA is actually a mixture of two compounds whose structures differ only in the location of a $C(CH_3)_3$ group:

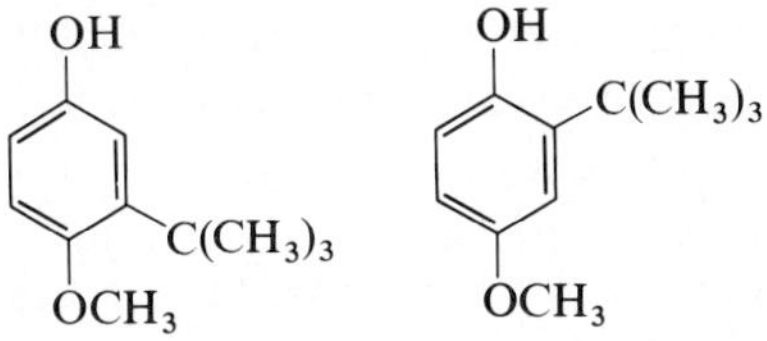

OH OH

$C(CH_3)_3$ $C(CH_3)_3$

OCH_3 OCH_3

two isomers of butylated hydroxyanisole (BHA)

BHT and BHA retard the reaction between oxygen and food. Oxidation is one reason foods discolor and spoil. The exposed surfaces of freshly cut apples or peaches, for example, soon turn brown as a result of oxidation, and butter becomes rancid as a result of oxidation. Because oxidation rarely causes serious illness, it is not so grave a problem as the growth of molds or harmful bacteria. Nevertheless, appearance and flavor are often affected, and consumers are not inclined to buy food that cannot be stored.

Sulfur dioxide can also be used to advantage as an antioxidant. In small amounts, either sulfur dioxide or sodium sulfite, Na_2SO_3 (which releases

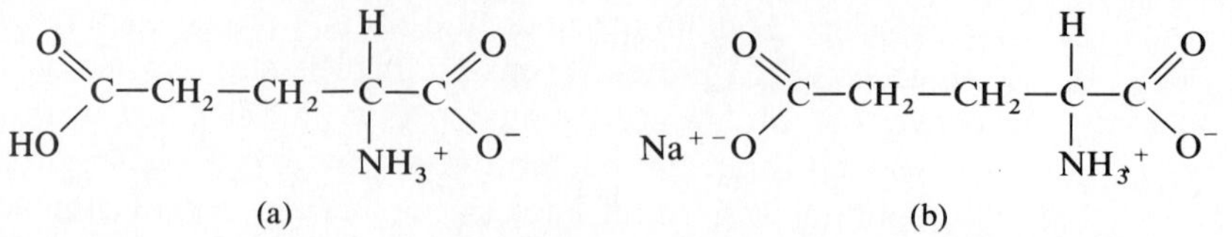

$$\mathrm{HO{-}C(=O){-}CH_2{-}CH_2{-}C(H)(NH_3^{+}){-}C(=O)O^{-}}$$

(a)

$$\mathrm{Na^{+}\,{}^{-}O{-}C(=O){-}CH_2{-}CH_2{-}C(H)(NH_3^{+}){-}C(=O)O^{-}}$$

(b)

Figure 16.10. *A comparison of (a) glutamic acid and (b) its salt, monosodium glutamate. Monosodium glutamate can be made from glutamic acid in the laboratory by adding sodium hydroxide.*

SO_2 in the presence of acid), keeps fruits and vegetables looking fresh by slowing oxidation and its attendant brown color and droopy appearance. Restaurants have used these additives in their increasingly popular salad bars and in seafoods and potatoes. Recently, however, the FDA has become aware that some people have allergic reactions to sulfites. People with asthma seem to be especially susceptible.

Monosodium glutamate is another chemical we often see on labels. A **flavor enhancer,** monosodium glutamate brings out the taste of other ingredients without contributing flavor itself. Restaurants use it, and it is in many prepared foods. Chinese restaurant dishes sometimes contain large quantities of this additive. Scientists aren't sure how monosodium glutamate works. It may increase the intensity of nerve impulses responsible for the perception of flavors.

Because monosodium glutamate is the salt of an amino acid (see Figure 16.10), we would expect that there would be little risk from ingesting it, and indeed few harmful effects have been found over the many years of its use. A small percentage of people, however, have allergic reactions to moderate amounts of monosodium glutamate. They suffer a burning sensation in the neck and arms, tightness in the chest, and headache. Others experience intestinal discomfort and diarrhea.

16.5 HOW TO READ A FOOD LABEL

Food labels contain much useful information for buyers who can interpret it. According to FDA regulations, all packaged foods must display the name of the product, the net weight of the contents, and the name and address of the manufacturer. In addition, nearly all labels must carry a list of ingredients, with the ingredient in greatest amount listed first, followed by the others in descending order. The only exceptions are "standardized" foods, such as mayonnaise or catsup, which by FDA regulation must contain particular ingredients in order to be called by those names. For these foods the ingredients need not be listed.

"Clam chowder—Ingredients: clams, potatoes, water, hydrolated plant protein, sodium phosphate, calcium carbonate, butylated hydroxytoluene. For external use only."

Increasingly we are seeing another section on food labels: **nutrition information per serving.** According to FDA regulation, "any food to which a nutrient has been added, or any food for which a nutritional claim is made, must have the nutritional content listed on the label." In other words, any product described as "enriched," "fortified," or "diet" must have this additional information. Many manufacturers whose products are exempt from this regulation are also including this useful nutritional information.

The best way to learn about a food label is to analyze one. The following list of ingredients was copied verbatim from the label of a popular brand of hot dogs:

> Ingredients: Beef and Pork, Water, Salt, Corn syrup, Dextrose, Flavoring, Sodium ascorbate (Vitamin C), Sodium nitrite.

As we would expect, the principal ingredients of these wieners are beef and pork. Next in line is water. Although the presence of water is not obvious, remember that all tissue contains water. Meats would be quite dry without it. Following water is salt, sodium chloride, which contributes to the flavor that most people expect in wieners. Two sweeteners, corn syrup and dextrose, are next in order of amount. Probably the manufacturer tried various combinations of sweeteners in order to discover one that gives the wieners an appealing taste. The major sweetener in corn syrup is fructose, and dextrose is another name for glucose. Apparently the taste was not yet optimal, because the next ingredient is "flavoring." The FDA does not require that the flavoring be specifically identified, allowing manufacturers to protect their trade secrets. Approximately 1700 natural and synthetic flavorings are now used in processed foods, to give them everything from smoky to fruity flavors. The single word *flavoring* on this label means that only natural flavors were used. If the wieners contained any synthetic flavors, the phrase *artificial flavor* would be required.

The next ingredient, sodium ascorbate (vitamin C), is the sodium salt of ascorbic acid. Vitamin C may have been added as a nutrient, but more likely it was added as a preservative and antioxidant. Even a very small amount of vitamin C can help retard spoilage.

The final ingredient, sodium nitrite, $NaNO_2$, is often seen on labels for processed meats and has been the subject of some controversy. Sodium nitrite is particularly effective in preventing botulism, a type of food poisoning caused by a dangerous toxin sometimes present in spoiled meats and other foods. Botulism can produce serious muscular paralysis and even death. According to an FDA bulletin, foods containing sodium nitrite account for about 7% of the food supply in the United States. Most processed meats—such as wieners, bacon, ham, sausage, bologna, and salami—contain sodium nitrite. Questions have arisen, however, over whether sodium nitrite in meats raises the risk of cancer. In the cooking process nitrites may be converted into nitrosamines (see Figure 16.11), which have been shown to cause cancer in laboratory animals. Although it is by no means certain that the use of nitrites poses dangers for humans, the controversy continues. Current research is directed toward finding a replacement for sodium nitrite.

Figure 16.12 shows another section of the label on a package of hot dogs: the nutritional information per serving. The first entries, involving serving size and servings per container, help a cost-conscious consumer choose the brand having the lowest cost per gram of hot dog. Then come data on the number of calories per wiener, along with the weights of protein, carbohydrate, and fat in one wiener. A person who is concerned about the

Figure 16.11. *(a) Sodium nitrite and (b) a nitrosamine.* (R_1 *and* R_2 *stand for organic groups.)*

(a) $Na^+NO_2^-$

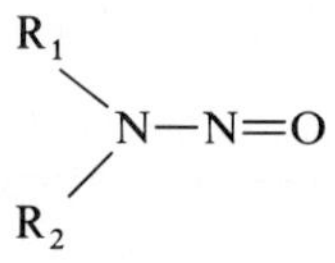

(b) R_1 and R_2 bonded to N: $(R_1)(R_2)N{-}N{=}O$

NUTRITION INFORMATION PER SERVING
Serving size—1 link (45 grams)
Servings per container—10

Calories	140	Cholesterol	.025 gram (25 mg)
Protein	5 grams		
Carbohydrate	1 gram	Sodium	.43 gram (430 mg)
Fat	13 grams		

Figure 16.12. Nutritional information on a label from a package of wieners.

amount of protein or other macronutrients in his or her daily diet can take this information into account when planning a meal.

Earlier we said that proteins and carbohydrates provide energy equal to about 4 dietary calories per gram, whereas fat can be metabolized to give about 9 calories per gram. The food label for wieners says that one hot dog yields 140 calories. Because the amounts of protein, carbohydrate, and fat are also given, we can calculate the total number of calories per serving for comparison.

$$
\begin{aligned}
\text{Protein:} \quad & (5\ \text{g})(4\ \text{cal/g}) = 20\ \text{cal} \\
\text{Carbohydrate:} \quad & (1\ \text{g})(4\ \text{cal/g}) = 4\ \text{cal} \\
\text{Fat:} \quad & (13\ \text{g})(9\ \text{cal/g}) = \underline{117\ \text{cal}} \\
& \text{Total food energy} = 141\ \text{cal}
\end{aligned}
$$

Given that these numbers are rounded off, the agreement is very good. To save the expense of measuring the heats of combustion of all their products, most manufacturers of processed foods calculate calories in this fashion. All other minor ingredients are assumed to contribute no calories.

Returning to the nutritional information per serving, we find two more items—the amounts of cholesterol and sodium. Most people are well aware that cholesterol in the bloodstream may contribute to hardening of the arteries and eventually heart disease.

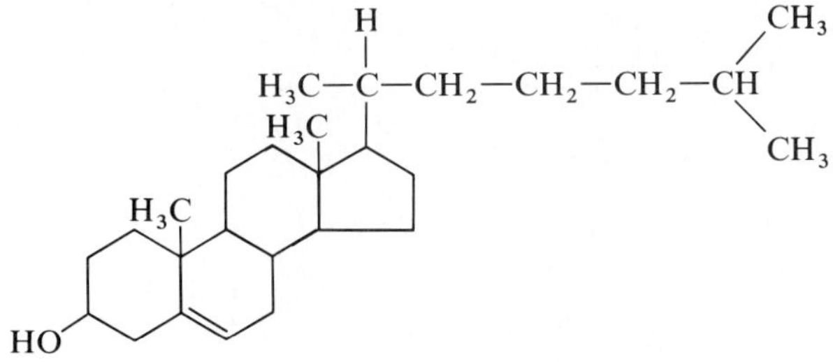

H, H_3C—C—CH_2—CH_2—CH_2—CH(CH_3)(CH_3), H_3C, H_3C, HO

cholesterol

Cholesterol is soluble in fats and is ingested whenever a person eats fatty meats. The human body also synthesizes this steroid. Although the connection between dietary cholesterol and heart attacks is still the subject of research, many people prefer not to take a chance. They try to minimize cholesterol in their diets. Anyone eating a wiener will take in 25 milligrams of cholesterol. For comparison, the yolk of an egg may contain 250 milligrams of cholesterol.

Finally, the label says that each serving contains 430 milligrams of sodium, most of which must come from the added salt. Very small amounts may come from the sodium ascorbate and sodium nitrite listed as ingredients. It is worth noting that salt is second only to sugar in amount added to food in the United States. The concern here is with high blood pressure, which may be aggravated by sodium in one's diet. People with high blood pressure are generally advised to reduce their intake of sodium. The connection between sodium and high blood pressure is still not thoroughly understood and is the subject of continuing study. The 430 milligrams of sodium in the wiener is a modest proportion of the 1100 to 3000 milligrams recommended by the Food and Nutrition Board as a safe and adequate daily intake. Most Americans consume much more sodium than is recommended.

Because most people like to eat wieners in a bun, we next examine the food label on a package of enriched hot dog buns. The word *enriched* means that the buns have been fortified with some vitamins and minerals, just as white bread usually is. Figure 16.13 gives some of the contents of a sample label. The nutrition information per serving is similar to that already discussed. Not surprisingly, there is less protein and fat and more carbohydrate in hot dog buns than in hot dogs. In addition, there is a list of ingredients, like the one for wieners shown earlier in this section.

A new section, **Percentage of U.S. RDA,** appears on the label for the buns, because the producer has used the word *enriched* to describe the buns and must show what vitamins and minerals have been added and give information about amounts. We have already listed the U.S. Recommended

Figure 16.13. *Information from a label on a package of enriched hot dog buns.*

NUTRITION INFORMATION PER SERVING	
Serving size	1 ounce
Servings per container	10
Calories	80
Protein	2 grams
Carbohydrate	14 grams
Fat	1 gram

PERCENTAGES OF U.S. RECOMMENDED DAILY ALLOWANCES (U.S. RDA)	
Protein	4
Vitamin A	*
Vitamin C	*
Thiamine	2
Riboflavin	2
Niacin	4
Calcium	6
Iron	8

*Contains less than 2 percent of the U.S. RDA of this nutrient.

Daily Allowances for both vitamins and minerals. The FDA-approved format is to state the percentage of the U.S. Recommended Daily Allowance of each nutrient provided by the food, including protein. Hot dog buns provide small percentages of our daily needs for protein, three vitamins, and the minerals calcium and iron. Few people take the time to add the percentages found on the labels of all the foods they eat daily, but this information can be important for anyone with a medical condition requiring careful attention to diet. People who rely heavily on processed foods to the exclusion of fresh vegetables, fruits, and meats may wish to pay some attention to the U.S. RDAs. Strict vegetarians need to seek sources of vitamin B_{12}, which is largely absent from plants.

Labels take some of the questions of food quality out of the realm of advertising and into the area of hard information. However, just how many consumers read the labels is unknown, and if few do, it is difficult to justify the additional expense to food producers (and therefore to consumers) of providing the information.

16.6 DRUGS

According to the important criterion of life expectancy at birth, citizens of the United States have enjoyed increasingly good health during the twentieth century. From 1900 to 1940 to 1980, life expectancy went from 47 to 63 to over 70 years, because of improved nutrition, sanitation, and medical care. One major advance is the development of a wide variety of medicines and drugs to prevent or cure diseases and to reduce pain and discomfort. Childhood diseases such as measles and poliomyelitis are now almost unknown in the United States because of vaccines. An ear infection, which 50 years ago often required surgery, now can be cured with antibiotics. High blood pressure can be controlled with drugs. Formerly untreatable mental illnesses such as schizophrenia have been alleviated through drugs, with a consequent decrease in the population of our mental hospitals. Other examples abound. Our discussion of drugs is largely confined to their use as medicines or curative agents.

Despite the great many nonprescription and prescription drugs now available, the need for new drugs continues. Some diseases have been arrested but not cured by drugs. Diabetics, for instance, can lead productive lives with the help of daily injections of insulin, but insulin does not cure the underlying causes. Drugs capable of a genuine cure are badly needed. As new strains of resistant bacteria develop, we require new antibiotics to cope with them. New surgical procedures, such as organ transplant and the implantation of artificial organs, require new drugs to diminish the body's natural rejection of foreign objects.

As the demand for new drugs grows, the difficulty of bringing them to market also increases. This is partly the result of stricter requirements for verifying the safety of new drugs and partly a consequence of our incomplete understanding of disease and drug action.

Because of the many mysteries surrounding drug action, the search for new drugs involves a screening process. Newly isolated compounds, either synthetic or natural, are subjected to standard tests to determine whether they are biologically active. The few that look promising ultimately receive clinical tests to determine their effectiveness. Many thousands of compounds must be tested for every one that proves useful.

Once an effective drug has been found, a series of molecular modifications may be employed to see if a more potent drug, with perhaps fewer unwanted side effects, can be synthesized. The structures of small parts of the original molecule are altered one at a time, and the resulting variants are tested. The ultimate goal of the drug scientist is rational drug design based on an understanding of the biochemical causes of a disease and how a drug might interfere with a reaction sequence critical to that disease.

The way in which a drug works can be simple. The main ingredient of an antacid is $Al(OH)_3$ or $Mg(OH)_2$, either of which combines with some of the hydrochloric acid (HCl) in the stomach through an acid-base reaction. Most drugs, however, act in a much more complex fashion. A drug can be an enzyme inhibitor, for example. When someone suffers an illness, an enzyme may promote an undesirable biochemical reaction. An effective drug can block the activity of the enzyme so as to slow down the damaging reaction. The research necessary to discover such a mechanism is difficult and slow. Designing a drug to do the job can take many years.

Aspirin is one of the most popular drugs in the United States. Americans consume over 20 billion tablets a year, which translates into about 40 million pounds of aspirin. A nonprescription drug, it is used by people to dampen the pain of a headache or muscle ache or to reduce a fever. In addition, it is usually recommended for decreasing the pain and swelling of joints caused by arthritis. Not only is aspirin sold under its own name; it is also a major ingredient in many cold remedies and "buffered" pain killers. Furthermore, it is effective. In a study more than two decades ago, aspirin was found to be more potent than any other drug in relieving the pain of patients suffering from inoperable cancer.

Because of undesirable side effects, however, the FDA warns people to be cautious about taking large doses of aspirin over a prolonged period. It can cause stomach upset, heartburn, nausea, and vomiting. There is a clear link between intake of aspirin and bleeding of the stomach lining, leading in extreme cases to ulcers. Recently, the use of aspirin for children with the "flu" has been discouraged, since it may play a role in Reyes Syndrome, a sometimes fatal disease.

The structure of aspirin, more properly called *acetylsalicylic acid,* is shown in Figure 16.14(a). As is true of most drugs, the relation between chemical structure and therapeutic action is difficult to understand. Research suggests that aspirin inhibits the synthesis of a class of hormones called

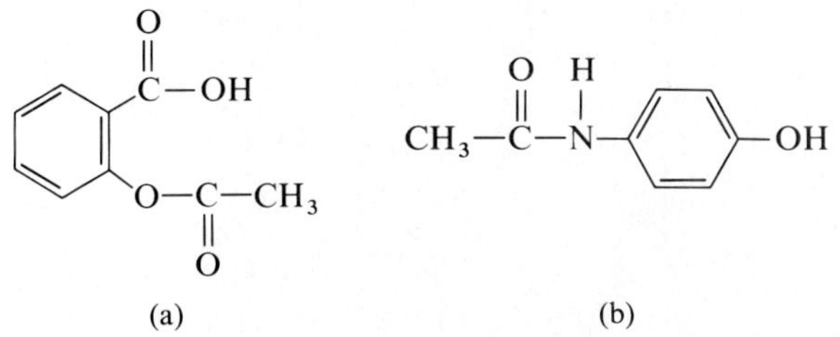

(a) (b)

Figure 16.14. The structures of two analgesics: (a) aspirin, or acetylsalicylic acid, and (b) acetaminophen.

prostaglandins. Although the chief role of prostaglandins is to stimulate muscle contraction, they may also cause pain, headache, and inflammation when present in higher than normal levels in the bloodstream. So aspirin, by inhibiting an enzyme crucial for the biosynthesis of prostaglandins, reduces pain.

A second **analgesic,** or painkiller, is acetaminophen, whose structure is shown in Figure 16.14(b). This compound goes under such trade names as Tylenol. Acetaminophen is preferred by some people because it offers effective relief from pain without most of the side effects of aspirin.

Of all the drugs, **antibiotics** have received the greatest public recognition because they are so effective in treating previously feared diseases such as pneumonia, tuberculosis, and syphilis. The story of antibiotics really began in 1928, when Alexander Fleming (see Figure 16.15), a British scientist, discovered penicillin as the result of a chance observation. After an absence from his laboratory, Fleming noticed that a Petri dish formerly covered with staphylococcal bacteria contained a blue mold. Around the mold was a region free of bacteria. Evidently the mold had produced a substance that killed the bacteria. Although Fleming gave the new agent the name *penicillin,* he turned his attention to other matters. It was not until the early 1940s that commercial production of penicillin began, stimulated by medical needs of World War II.

An effective antibiotic must be active against at least one microorganism in minute concentrations. Though produced by a microorganism, an antibiotic is a chemical in the usual sense; it has a well-defined molecular structure, which can be determined by the tools available to a chemist. The general structure of a penicillin is shown in Figure 16.16.

In the original penicillin, now called penicillin G, the part of the molecule designated by an R in Figure 16.16 is

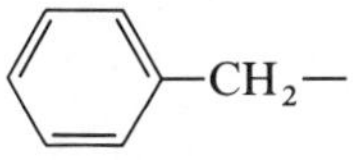

$$C_6H_5{-}CH_2{-}$$

Additional penicillins were made by reacting penicillin G with appropriate chemical reagents in the laboratory. So these new antibiotics were the result of a combination of biological and chemical procedures. After a microorganism

Figure 16.15. Alexander Fleming, discoverer of penicillin. He was awarded the Nobel Prize for Medicine in 1945.

has grown the original penicillin, a chemist "fine tunes" the drug for specific purposes. The R group of penicillin V, for example, is

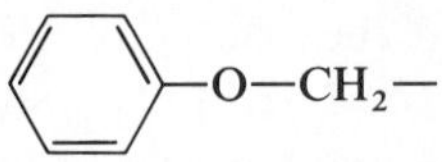

$$\text{(benzene ring)}-O-CH_2-$$

Penicillin V can be taken orally in small amounts because it is resistant to stomach acid. Many more units of penicillin G must be ingested when it is taken orally because it is readily decomposed by stomach acid.

Penicillins owe part of their activity to their ability to disrupt the formation of a bacterial cell wall by modifying an essential enzyme. If the cell walls of harmful bacteria cannot be made, the contents of their cells simply

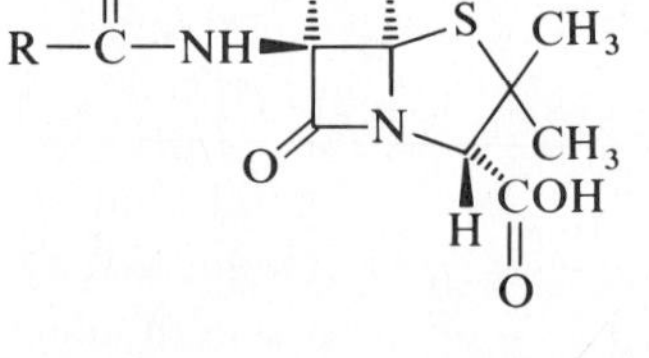

Figure 16.16. The general structure of a penicillin. The R group depends on the particular penicillin.

R—C(=O)—NH— (β-lactam ring fused to thiazolidine ring: H, H, S, CH_3, CH_3, N, O, H, COOH)

disperse and the bacteria die out. The key part of a penicillin molecule is the relatively unstable four-membered ring containing nitrogen (Figure 16.16). In the presence of an enzyme that catalyzes the biosynthesis of cell walls, the four-membered ring opens and the altered penicillin molecule bonds to the enzyme.

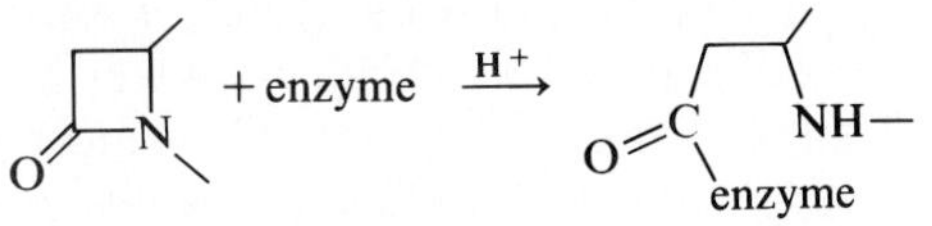

O N + enzyme $\xrightarrow{H^+}$ O=C NH— enzyme

Because the enzyme has been chemically modified, it is no longer able to catalyze bacterial cell-wall formation.

Since the 1940s a host of new antibiotics (tetracyclines and chloromycetin, among many others) have been made and tested. A small number have reached commercial production after approval by the FDA. These new drugs have great benefits; however, the problem of development of resistant strains of bacteria is becoming serious. With widespread use of an antibiotic against a particular infectious disease, a few mutants that are not attacked by the antibiotic survive and multiply. Gradually a new variation of bacteria that cannot be controlled by the antibiotic spreads through a population. The bacteria that cause malaria, typhoid, gonorrhea, and tuberculosis, for example, now have strains resistant to antibiotics that formerly were effective against these diseases. The quest for more new antibiotics is therefore accelerated.

Cocaine is an example of a drug having both uses and abuses. It was first isolated from coca leaves in 1860. Later it proved effective as an anesthetic for minor surgery when applied locally. The cocaine molecule contains several functional groups:

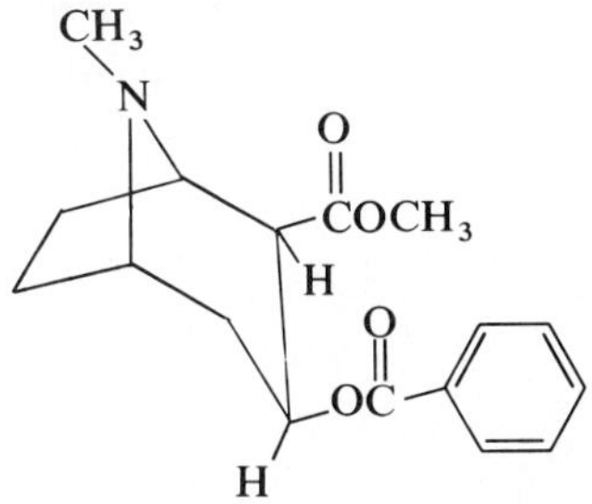

CH_3 N O ‖ $COCH_3$ H O ‖ OC H

Besides blocking pain, cocaine is a stimulant for the central nervous system. It increases heart rate and blood pressure and causes feelings of alertness and euphoria. Because of these pleasurable sensations, nonmedical uses of the drug also exist. Currently, cocaine rivals marijuana in popularity within the drug culture. Even though it may not be physically addictive, a psychological dependence results from long-term use of cocaine. An overdose has serious consequences: slowing of heart and respiration rates and even death.

16.7 CARCINOGENS

Because **carcinogens,** or external cancer-causing agents, are believed responsible for 60 to 90% of cancer cases, they are a source of great concern. Carcinogens can come from cigarette smoke, unsafe air, polluted water, chemicals at the work place, or even everyday food. Whereas pneumonia, influenza, and tuberculosis were the leading causes of death in 1900 and cancer ranked eighth, today cancer is second only to heart disease as a cause of death in the United States. In part this change has occurred because the most threatening diseases of the earlier period are now curable or preventable.

Cancer takes many forms, but a common characteristic of all of them is an uncontrolled multiplication of cells. In a healthy individual, cells divide under the control of DNA to replace damaged or worn-out cells. Imperfect DNA, however, can allow cells to replicate rapidly and produce a tumor, which causes bleeding or otherwise interferes with the function of the host organ (see Figure 16.17). A characteristic of cancer is that the growth is malignant; cells from the tumor spread through the bloodstream or lymph channels to other parts of the body, where they continue their abnormal growth. Healthy cells are displaced by the cancerous cells and are destroyed.

The causes of the changes in DNA structure are only partly understood. In a small percentage of the cases heredity is at fault. Some individuals

Figure 16.17. *Photographs showing a comparison between healthy and cancerous cells. (a) Normal bronchial tube cells from a patient with pneumonia have small, uniform nuclei. (b) Malignant cells from a bronchogenic carcinoma have much larger and more variable nuclei.*

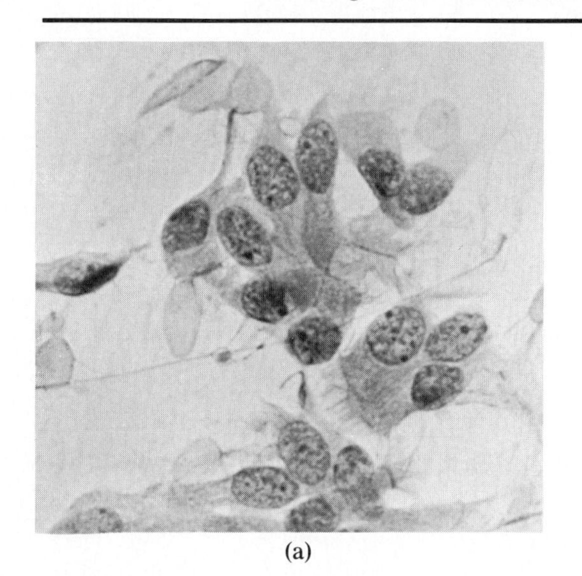

(a)

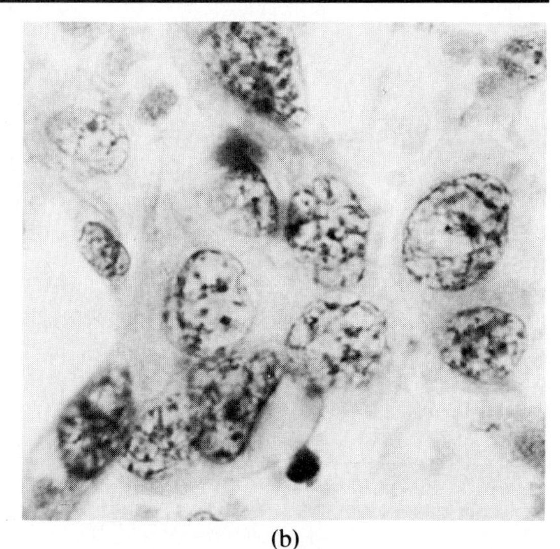

(b)

inherit weaknesses that make them prone to cancer. In another 10 to 40% of cancers, a cell mutation that is apparently unrelated to any source of radiation or foreign chemical develops into cancer. In the majority of cases an outside cause is at work. It may be long exposure to the sun (which results in skin cancer), an excessive dose of x-rays, nuclear radiation, or a virus. Or it may be a carcinogen: a chemical that is inhaled, eaten, or passed through the skin.

There is strong evidence that environmental causes of cancer are important. First of all, the prevalence of each kind of cancer varies from country to country. Stomach cancer, for example, is very common in Japan, Iceland, and Chile; it strikes much less frequently in Australia, Canada, and the United States. In the United States prostate and colon cancers are among the leading types. More important, when people emigrate from one country to another they take on the cancer pattern of their new country, and their children do also. People of Japanese descent living in the United States have a lower incidence of stomach cancer and a higher incidence of cancer of the colon than do Japanese living in Japan. The same is true of other nationalities. Differences in the environments must be responsible.

Because the delay between exposure to a carcinogen and development of cancer can be anywhere from 1 to 40 years, with the longer times predominating, identification of carcinogens depends on indirect evidence. Records of the incidence of lung cancer among smokers and nonsmokers over many years show a clear correlation between smoking and this form of cancer. In the past, workers in a particular industry have sometimes developed a certain cancer with a frequency much higher than that of the general population. For example, workers in dye industries were found to have an abnormally high rate of bladder cancer. This was eventually traced to benzidine and several other compounds used to synthesize dyes.

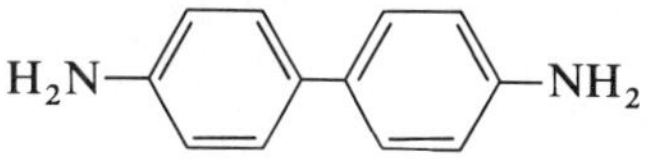

H_2N — — NH_2

benzidine

Benzidine, which is absorbed through the skin, is now recognized as a dangerous carcinogen. By law, workers cannot come into contact with any liquid or solid mixture containing more than 0.1% benzidine by weight.

One of the carcinogens in smoke, including cigarette smoke, is benzo[*a*]pyrene,

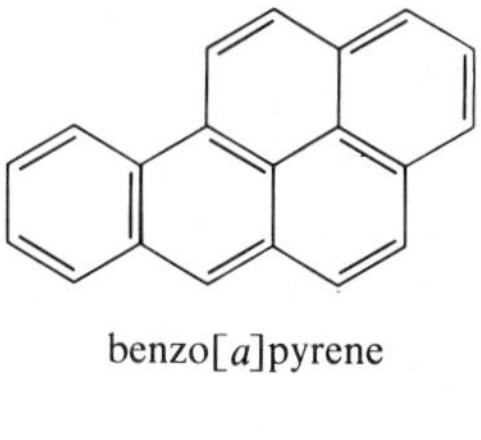

benzo[*a*]pyrene

This molecule, which consists of fused benzene rings, causes skin and lung cancer. A carcinogen with a totally different structure is vinyl chloride, $CH_2{=}CHCl$, the monomer used to make polyvinyl chloride, a polymer with important uses. Like some of the chemicals of the early dye industry, vinyl chloride was identified as a carcinogen through its effect on workers, in this case in the plastics industry. Vinyl chloride causes a variety of malignant tumors and now is carefully controlled.

Under some conditions the food we eat can contain carcinogens. It is especially important to avoid moldy food, such as moldy bread. One mold, for example, produces the carcinogen aflatoxin B_1,

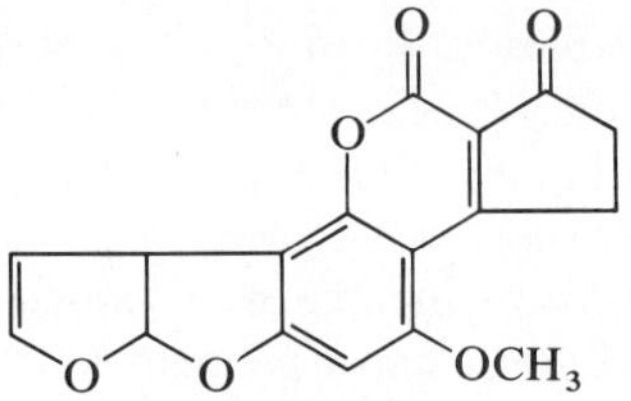

O O O O O OCH_3

aflatoxin B_1

which causes cancer of the liver. Other carcinogens have been found in food that has been highly browned or burnt in cooking. As research into the trace components of foods has progressed, very low concentrations of a number of carcinogens have been discovered. This is all the more reason for eating a variety of foods and not loading your diet with just a few kinds.

Besides close examination of the medical histories of groups of people, there are two ways to test for carcinogenicity of chemical compounds. One involves screening with animals, usually special breeds of rats or mice. A measured amount of a suspected carcinogen is applied to an area of the animal's skin, and the animal is monitored for the development of a tumor at that site. Although not conclusive, a positive result suggests that the compound is carcinogenic in humans, and additional testing is indicated. It takes from one to three years for tumors to appear on rats and mice; the delay time in humans is much longer. If quicker results are needed, short-term testing is employed. Cells are grown in laboratory cultures and exposed to a suspected carcinogen. If the growth becomes abnormally rapid or the cells mutate, it is possible that the substance may be carcinogenic for humans. Again, further testing is needed. The opposite result, normal growth, suggests but does not prove that the compound is safe.

As indicated earlier, a change in the structure of DNA in a cell may cause uncontrolled growth and cancer. It is possible that carcinogens react directly with the DNA to produce the change. Alternatively, they may affect the cell's growth control proteins, which normally prevent replication of DNA and cell division. Our understanding of what makes a molecule carcinogenic is limited. Often the structures of carcinogens seem to have little in common. And some carcinogens are highly potent in producing cancers, whereas others must be ingested in huge quantities to incur even a modest risk. Research on the quantitative dangers from various carcinogens will help to provide perspective in this controversial area.

IMPORTANT TERMS

macronutrients Proteins, fats, and carbohydrates, which energize our bodies. The body needs macronutrients in relatively large amounts.

essential amino acids Eight amino acids that must be in the proteins we consume, because they cannot be adequately synthesized by our bodies.

high-quality protein A protein that contains all eight essential amino acids in about the proportions needed by the human body.

U.S. Recommended Daily Allowance (RDA) The amounts of protein, vitamins, and minerals recommended by the Food and Drug Administration for daily intake to maintain good health. Percentages of U.S. RDA are included on some food labels.

polyunsaturated fatty acid A fatty acid with two or more double bonds in the hydrocarbon chain. Several polyunsaturated fatty acids are thought to be essential in the diet.

dietary calorie A unit of energy equal to 1000 calories, or 4.18 kilojoules (kJ).

artificial sweetener A sweetener that is not one of the monosaccharides or disaccharides found in nature. Examples are saccharin and aspartame, both of which are synthetic.

vitamin A constituent of food necessary for good health. Vitamins were originally thought to be amines, but they are now known to encompass a variety of organic compounds.

micronutrient A food component needed in an amount of 1 gram or less per day. The principal micronutrients are vitamins and minerals.

mineral An inorganic compound that provides a chemical element that is essential to our diets.

vitamin C, or **ascorbic acid** A common vitamin, found in citrus fruits and certain vegetables.

megadose Intake of a nutrient in an amount at least ten times as large as the Recommended Daily Allowance.

fat-soluble vitamin A vitamin that resides in fatty tissues because it is structurally similar to a hydrocarbon.

water-soluble vitamin A vitamin that is soluble in water because it contains OH and other groups capable of forming hydrogen bonds with water. Water-soluble vitamins pass through the body rather quickly.

fortified foods Foods to which nutrients have been added.

preservative A food additive intended to retard spoilage by molds, bacteria, fungi, and yeasts.

antioxidant An additive that slows the reaction between oxygen from the air and food.

flavor enhancer An additive, often monosodium glutamate, that enhances the perception of tastes.

nutrition information per serving A section of a food label that gives the serving size, number of servings per container, and number of dietary calories and weights of macronutrients per serving.

percentage of U.S. RDA The percentage of the Recommended Daily Allowance for a micronutrient in one serving of the food. The FDA requires that this information appear on the labels of certain foods.

analgesic A painkilling drug.

antibiotic A drug usually grown by a microorganism and capable of killing another harmful microorganism.

carcinogen A substance that causes cancer when it is inhaled, eaten, or passed through the skin.

QUESTIONS

1. The proteins we eat are not identical to the proteins that must be replaced in our bodies. Describe the chemical changes that ingested protein undergoes.

2. What essential nutrients may be lacking in a strictly vegetarian diet?

3. What is the principal kind of carbohydrate in our diets? What simple carbohydrate forms after digestion?

4. Suppose that the number of calories you get from your intake of protein, carbohydrate, and fat is consistently greater than your need. What side effects may occur?

5. Table 16.2 states that one fried egg yields 99 dietary calories. Describe the method by which this value was determined. What is wrong with saying, "There are 99 calories in one fried egg"?

6. Explain why the oxidation of a fat yields many more dietary calories than the oxidation of an equal weight of carbohydrate.

7. Several artificial sweeteners produce taste sensations at least 100 times stronger than the taste sensation of sucrose. Explain why this is important in the battle against calories.

8. Do you feel that vitamin C from natural sources is more effective than synthetic vitamin C? Discuss your answer.

9. Under what circumstances do you feel that it would be useful for someone to take daily vitamin supplements?

10. Suppose that your diet were to consist mainly of hamburgers and milkshakes. Might this diet have any harmful effect on your health? Discuss.

11. The U.S. Recommended Daily Allowance for phosphorus is 1 gram. Does this mean that your diet should regularly include some elemental phosphorus? Explain.

12. Give three functions that different food additives serve.

13. Give an example of a processed food that is fortified because nutrients are lost in processing. What nutrients are lost and later added?

14. Briefly describe the kinds of information that can be found on food labels.

15. Sodium benzoate, calcium propionate, butylated hydroxyanisole, and monosodium glutamate are among the chemicals often seen in the list of ingredients on food labels. Select one of these additives, and explain its function.

16. On a food label, the "Nutrition Information per Serving" lists the amounts per serving of a number of food components. Select one that is beneficial to health and another that may be harmful, and describe the effect of each.

17. Describe the benefits and harmful side effects of aspirin.

18. Describe the difficulties scientists may have in identifying carcinogens. What methods are used to determine whether a compound is likely to be carcinogenic?

Public-Policy Discussion Questions

19. Do you anticipate that in the future when you shop for groceries you will make use of any of the information on food labels? In your view, is the extra cost to the food manufacturers and to the FDA in its role of enforcement a waste of money? Explain your position.

20. There is still considerable controversy over whether or not vitamin C helps against the common cold. After doing some research on both sides of the issue, discuss which argument seems more persuasive.

21. Antispoilage agents such as sodium nitrite increase the shelf-life of foods, but some of these agents may be carcinogenic. Should there be a federal law banning these agents from foods?

22. In some states restaurants and other public places are required by law to provide a nonsmoking area for patrons. This law was enacted in the belief that breathing the air in a smoky room is unpleasant and may even be harmful to nonsmokers. In your view, is this a serious enough hazard that all smoking in public places should be made illegal?

SECTION TWO ⬣

Chemistry, Energy, and the Environment

Quantity and Quality of Energy

Ever since the oil embargo of 1973, when most oil-exporting nations suspended shipments of petroleum, people everywhere have been aware of how dependent industrial nations are on energy. In the United States about 37% of the energy consumption is in chemical and other industries. Examples of energy-consuming industries are the production of food, the conversion of raw materials to make iron, aluminum, glass, and concrete, and the production of synthetic fibers and plastics from petroleum.

The U.S. economy depends on the three fossil fuels—petroleum, coal, and natural gas—for 90% of its energy. Nuclear power and hydroelectric power contribute equally to the remaining 10%. Figure 17.1 shows the distribution by source of U.S. energy consumption for 1984. Although dependence on foreign oil has declined in the last decade, we still import close to one-third of the petroleum we use. The remainder of our energy is produced domestically, and we even export some coal. Fossil fuels are critical resources, and they are nonrenewable. U.S. coal reserves may last for 600 years, but petroleum and natural gas resources may run out in 100 years or less.

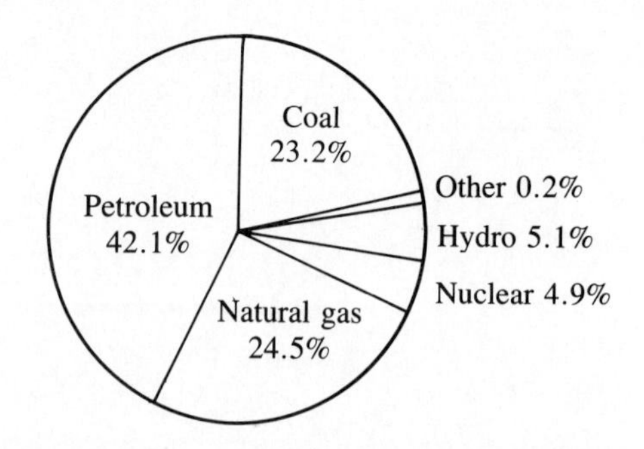

Figure 17.1. The distribution by source of the energy consumed in the United States during 1984.

17.1 THE SUN—AN INEXHAUSTIBLE ENERGY RESOURCE

The fossil fuels that the world consumes owe their existence ultimately to the sun. The vegetation that formed them grew as a result of sunlight. Over millions of years, the vegetation trapped underground decayed into coal, petroleum, and natural gas. The specific chemical reactions responsible for this conversion were favored by the lack of contact with oxygen from the air and by the high temperature and pressure deep underground.

Even hydroelectric power can be regarded as solar energy. Energy from the sun evaporates great quantities of water from oceans and lakes. The resulting water vapor eventually condenses as rain or snow and returns to the surface. Some of this water is found at high elevations, where it has a large potential energy. A portion of this potential energy is extracted at dams by water-driven turbines as the water falls to a lower elevation.

The sun constantly emits enormous amounts of energy in all directions through space (see Figure 17.2). Because of its great distance from the sun, the earth captures only 0.002% of this energy, and only a small part of the captured solar energy provides energy we can use directly. Even so, the sun accounts for about 99.98% of the earth's energy supply.

From the point of view of a traveler in space, the side of the earth facing the sun looks like a huge disk. Not all of the sunlight falling on this disk reaches the surface. At the top of the atmosphere, the energy arriving from the sun amounts to 81.1 kilojoules per square meter per minute. If through some magic all of this energy could reach the surface and that striking land could be completely converted into forms useful to society, all of our energy needs would be met 20,000 times over. Many facts, however, make this impossible. In the first place, about 30% of this solar energy is immediately reflected back into space by clouds and dust particles (see Figure 17.3). Another 47% of the light energy is absorbed directly by the atmosphere and the earth's surface and is changed into heat energy.

Only a tiny fraction of the sunlight reaching the earth's atmosphere, 0.02%, is captured by the chlorophyll in the leaves of green plants and

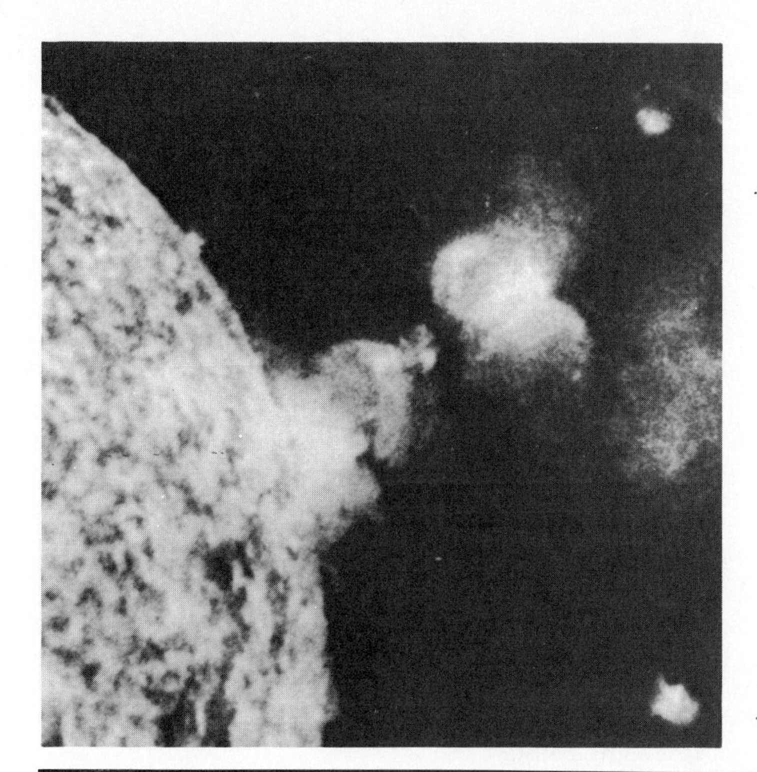

Figure 17.2. A portion of the surface of the sun. The active region is a solar flare.

Figure 17.3. The fate of sunlight reaching the earth.

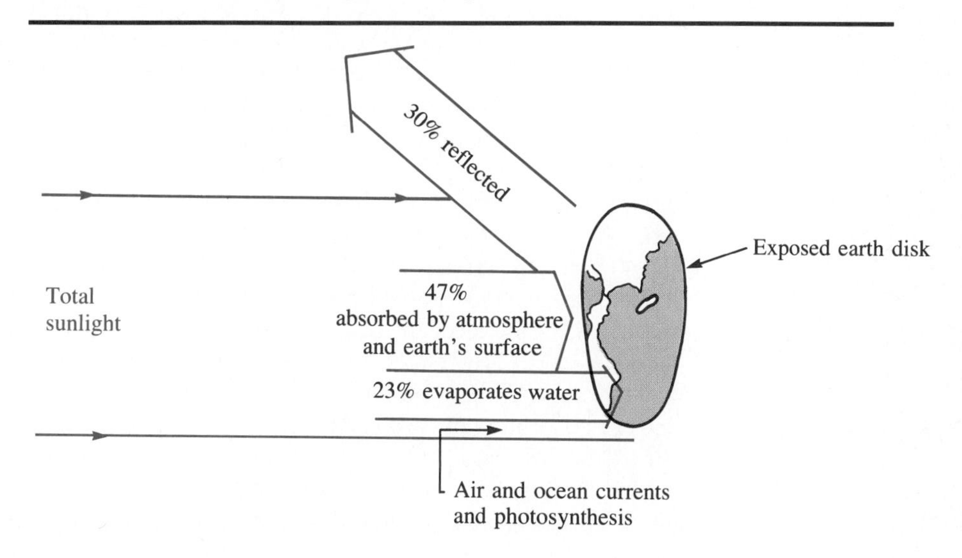

converted into chemical energy. This almost insignificant portion of the solar energy becomes the essential energy supply of the plant and animal kingdoms, replenishing oxygen in the air and thereby maintaining life.

17.2 WILL THE EARTH BECOME TOO HOT?

Because the surface of the earth is constantly bathed in sunlight, we might expect our surroundings to get hotter and hotter. Fortunately, another process counteracts this tendency. The earth not only absorbs energy from the sun but radiates energy into space at an equal rate. An analogy can be drawn between the earth and a lump of steel that is heated in a flame. The temperature of the steel rises until it glows with a brilliant white light. At this point its temperature reaches a plateau, as the energy it radiates exactly balances the energy it absorbs from the flame. The lump of steel has reached a **steady state.**

Of course the earth does not give off the characteristic light of a white-hot object. Nevertheless, it is constantly sending into space less powerful "infrared" radiation, which is not visible to our eyes. Most of the radiation produced by a heat lamp is infrared radiation. A balance is struck between energy input and outflow such that the earth becomes neither a cinder nor an icy wasteland. For thousands of years this balance has kept the temperature of the earth's surface fairly constant.

Concern over how human activities might affect the temperature of our planet has been mounting in recent decades. One possible problem is the increasing amount of carbon dioxide in the atmosphere, which comes from the burning of coal, gasoline and diesel fuel, and natural gas. By means of the **greenhouse effect,** the small concentration of carbon dioxide in the air allows visible light rays from the sun to pass through the atmosphere and warm the surface, but the CO_2 absorbs many of the infrared rays coming back from the surface (see Figure 17.4). This trapped energy helps to keep the earth warm. In this respect carbon dioxide is like the glass in a greenhouse, which allows sunlight in but helps to keep infrared radiation from escaping.

The problem, first recognized 100 years ago, is that too much carbon dioxide could cause our climate to become warmer. Already the carbon dioxide in the atmosphere has increased from 280 parts per million in 1860 to 340 parts per million in 1981. Carbon dioxide is the only substance whose global background concentration is known to be rising. Figure 17.5 shows the trend during a recent two-decade period. The annual variations evident in Figure 17.5 are related to the seasons of the year. Each spring, plants and trees grow a new set of leaves, renewing the process of photosynthesis and the absorption of carbon dioxide, so the level of carbon dioxide decreases. When the leaves drop in the fall, the process reverses.

So far no general temperature rise in the northern hemisphere has been discovered from our weather records of the last century. But a small trend

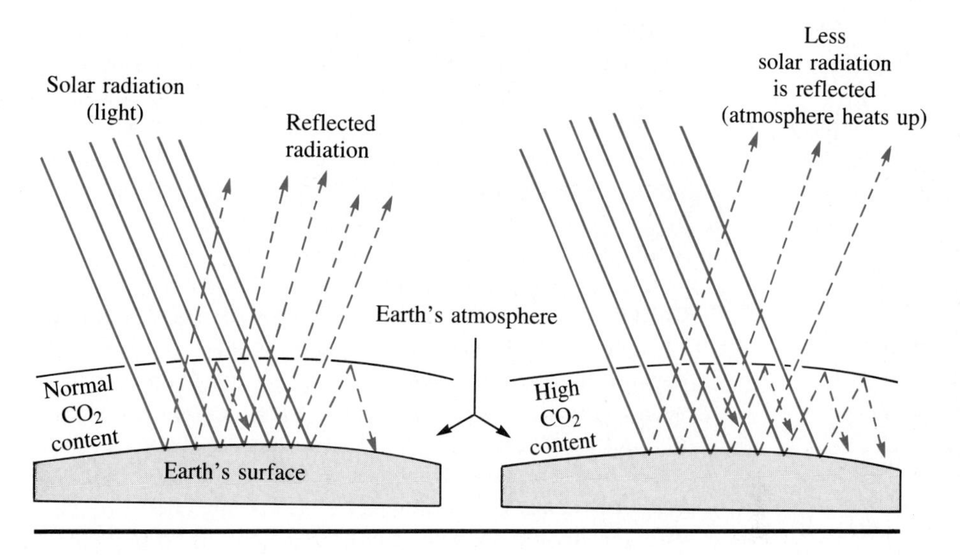

Figure 17.4. The greenhouse effect. Carbon dioxide lets sunlight pass through the atmosphere. Some of the infrared, or heat, radiation from the earth, however, is trapped.

would be difficult to see because of the considerable fluctuations in weather from year to year. Groups of scientists have developed complicated models to predict the temperature increase that may occur if carbon dioxide levels continue to increase. A recent report by a committee of the National Research Council predicted that the carbon dioxide concentration may double by late

Figure 17.5. Variations in CO_2 levels. These data were recorded at the observatory at Mauna Loa in Hawaii.

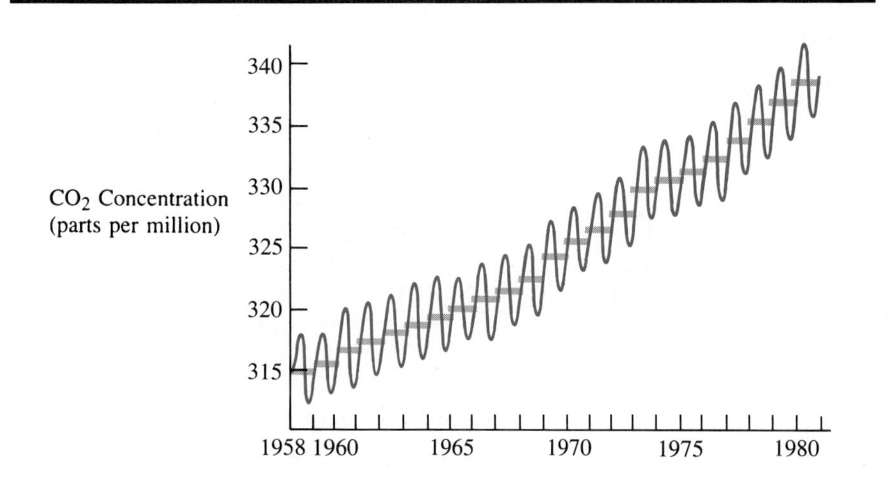

in the next century. This could cause an increase of from 1.5 to 4.5°C in the average temperature of the earth. The consequences could be far-reaching. In the northern hemisphere the agricultural belt would move northward. The plains states of the central United States would become arid and unproductive, while the climate of Canada would become much more favorable for agriculture. In this scenario, icebergs in the Arctic and Antarctic regions would melt, raising the sea level. This change would have a major effect on coastal cities.

These predictions are highly uncertain for several reasons. For one thing, no one can foretell what the world pattern of fossil fuel use will be several decades in the future. Also, it is difficult to say what fraction of the carbon dioxide increase may be absorbed by the oceans, which play an important role in the carbon dioxide cycle. Another uncertainty concerns the influence of dust and other particles in the air. Dust has an effect opposite to that of carbon dioxide. It reflects some of the incoming sunlight and sends it back into space. To what extent this cooling effect will balance the greenhouse effect remains unclear. For these and other reasons, study groups have differed greatly in their predictions of increases in world temperatures caused by increases in carbon dioxide concentration in the atmosphere.

17:3 FOSSIL FUELS

All three fossil fuels—coal, petroleum, and natural gas—contain carbon and hydrogen in varying proportions, along with small amounts of other elements. Of the three, coal has the highest carbon content and the smallest amount of hydrogen. After removal of moisture, most coals contain about 70–90% carbon, 5% hydrogen, and 5–15% oxygen by weight. The remainder consists of sulfur and nitrogen. The arrangement of atoms in coal is so complicated and variable that it is still not completely understood. The "molecules" are very large, and the atoms within them lie pretty much in one plane. In this respect coal is like graphite (Chapter 11, page 242); it is made up of sheets, or layers, which easily slide past one another. A simplified picture of the structure of one layer of coal is shown in Figure 17.6. Notice the great many benzene rings that are linked together.

Petroleum, or crude oil, has a higher proportion of hydrogen than coal does. It is a mixture of many kinds of hydrocarbon molecules, most of which contain from six to ten carbon atoms. The great majority of these compounds are normal alkanes, whose molecules have a single linear chain of carbon atoms. Mixed with these are very small amounts of branched-chain alkanes, which help raise the octane number of gasoline made from the crude oil. Also present in petroleum are some cyclic hydrocarbons, such as cyclohexane, and aromatic compounds, such as toluene and 1,2,4-trimethylbenzene.

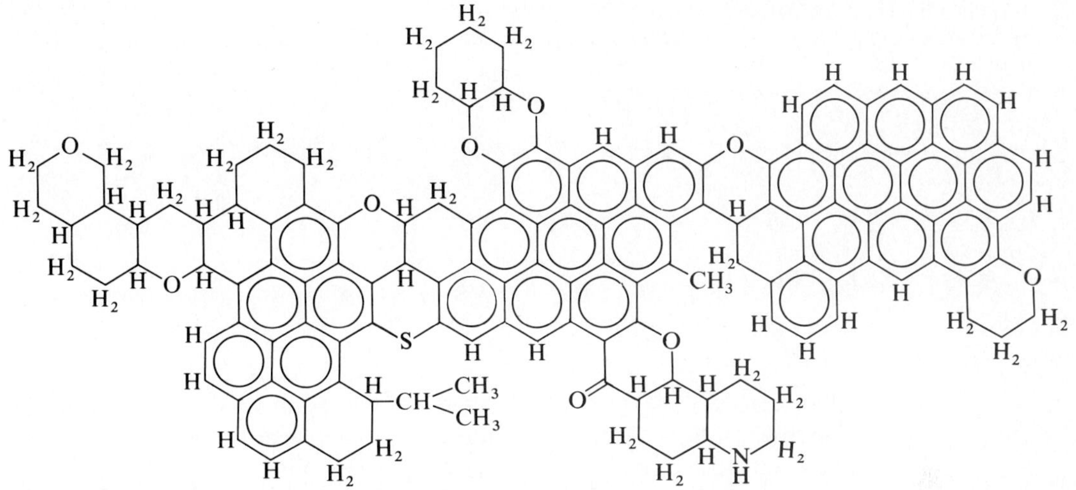

Figure 17.6. An approximate structure for one layer of coal.

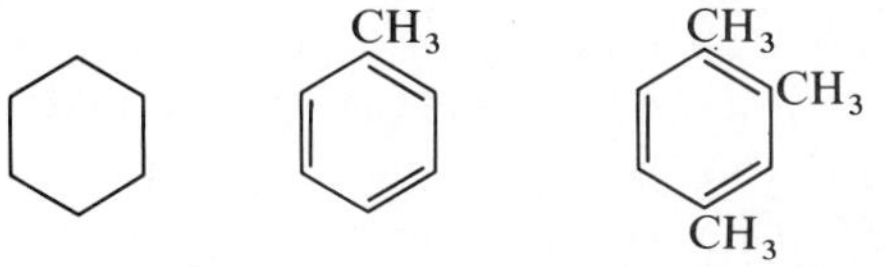

cyclohexane toluene 1,2,4-trimethylbenzene

Of the three fossil fuels, natural gas has the highest ratio of hydrogen to carbon. This is because it consists mainly of methane, CH_4, which has 75% carbon and 25% hydrogen by weight. Natural gas also contains small amounts of CO_2, N_2, H_2S, He, and H_2O. These gases are of no help in providing heat, because except for H_2S they do not burn.

There is an interesting correlation between the hydrogen content of a fossil fuel and the amount of heat given off when a unit amount of the fuel burns in air. The greater the percentage of hydrogen, the greater the amount of heat evolved. The "unit amount" can be either 1 gram of fuel or the amount of fuel containing 1 mole of carbon. This correlation is shown in Table 17.1. For simplicity, we have assumed coal to be pure carbon and have selected octane as a typical component of petroleum.

The combustion products formed are responsible for the increase in heat value with an increase in hydrogen content. Both water and carbon dioxide form when gasoline and natural gas are burned. Little water forms when coal is burned. The C═O bonds in CO_2 and the O—H bonds in H_2O are especially strong. Whenever the bonds in the products of a reaction are stronger than those in the reactants, heat is evolved. Since combustion of a fossil fuel always produces 1 mole of CO_2 per mole of carbon in the fuel, those reactions that form water as well produce extra heat.

Table 17.1 Heat Values for the Three Fossil Fuels

	COAL	GASOLINE	NATURAL GAS
typical % H	5	12	25
combustion reaction	$C + O_2 \longrightarrow CO_2$	$2\ C_8H_{18} + 25\ O_2 \longrightarrow 16\ CO_2 + 18\ H_2O$	$CH_4 + 2\ O_2 \longrightarrow CO_2 + 2\ H_2O$
heat evolved per mole of carbon (kJ/mole)	393	681	882
heat evolved per gram of fuel (kJ/gram)	33	48	55

Because petroleum contains so many compounds, with a multitude of properties, it must be refined before it becomes useful. Fuel oil for heating, motor oil and other lubricating oils, and gasoline are among the more important products of an oil refinery. Gasoline is particularly critical, because our transportation system strongly depends on its combustion in the internal-combustion engine.

A high compression ratio made the modern gasoline engine more efficient, but it also created a new problem. In high-compression engines a rapid explosion of the fuel-air mixture may occur in the cylinder rather than the desirable controlled burning. This causes spark **knocking** and a large loss of power. The problem of knocking was solved through careful selection of the hydrocarbons used as fuel and through the use of additives.

As long ago as the 1920s chemists knew that hydrocarbons with different structures have very different knocking tendencies. Since a compound called *iso*-octane has a high ability to resist knocking, they began rating each fuel by its **octane number.** *Iso*-octane and *n*-heptane, an alkane that knocks very badly, were used to define the octane scale (see Table 17.2). Mixtures of different percentages of these two alkanes give intermediate octane ratings when they are burned in test engines. For example, a mixture of 70% *iso*-octane and 30% *n*-heptane receives an octane number of 70.

Table 17.2 Reference Compounds for Octane Ratings

	CONDENSED FORMULA	OCTANE NUMBER
n-heptane	$CH_3CH_2CH_2CH_2CH_2CH_2CH_3$	0
iso-octane	$CH_3C(CH_3)_2CH_2CH(CH_3)CH_3$ (CH₃ groups above C and CH; CH₃ below C)	100

During the 1930s chemists realized that doing some chemistry on the gasoline before it left the refinery could radically improve its fuel value. They introduced processes by which hydrocarbons could be broken down into smaller molecules (cracking), changed into branched-chain compounds (isomerization), and turned into cyclic and unsaturated hydrocarbons (reforming). The remodeling of the complex hydrocarbons in oil by chemical reactions takes place in some of those complex masses of pipes in oil refineries. Table 17.3 gives the octane ratings of some additional hydrocarbons.

In the past the chemical additive most often used to reduce knocking was tetraethyllead.

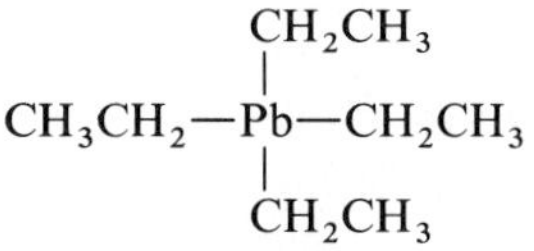

$$
\begin{array}{c}
CH_2CH_3 \\
| \\
CH_3CH_2-Pb-CH_2CH_3 \\
| \\
CH_2CH_3
\end{array}
$$

tetraethyllead

Addition of this compound to gasoline creates what is called leaded gasoline. Some years ago all premium gasoline was leaded. When smog became a major problem in cities in the United States, and people realized that exhaust from the engines of our automobiles was one of the causes, they called for a change. Converters were developed to clean up the exhaust. However, lead poisoned the converters and made them inactive. In order to allow the converters to do their job in helping clear up air pollution, all new cars in the United States are built to use only lead-free fuel. And to solve the knocking problem, oil companies now add more high-octane aromatic compounds and ethanol to gasoline and automobile companies have reduced the compression ratios of car engines. Decreasing the severity of the pollution problem required many compromises.

Table 17.3 Octane Numbers of Some Hydrocarbons

COMPOUND	NAME	OCTANE NUMBER
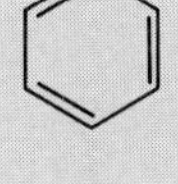	benzene	100
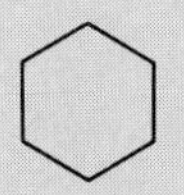	cyclohexane	83
$CH_3CH_2CH_2CH_2CH_2CH_3$	isohexane	73
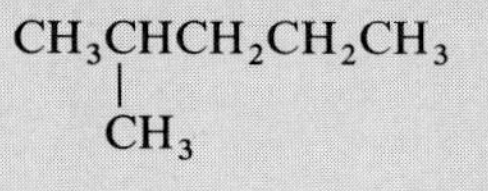 $CH_3CH(CH_3)CH_2CH_2CH_3$	*n*-hexane	25

17.4 HEAT AND WORK: TWO KINDS OF ENERGY TRANSFER

In Chapter 4 you encountered a very important law—the Law of Conservation of Energy. According to this law, energy is neither created nor destroyed in any physical or chemical change. When heat from a flame is transferred to a container of water at 100°C, boiling occurs. Some liquid water is converted into water vapor, as shown in Figure 17.7. The heat absorbed by the water in this endothermic process does not simply disappear. Instead, the potential energy of the molecules of water vapor is larger than that of the liquid by an amount exactly equal to the heat added. For every mole of liquid water vaporized, 40.6 kilojoules of heat is absorbed and the potential energy of the water is increased by 40.6 kilojoules (Figure 17.7). In Chapter 4 you saw similar examples involving endothermic and exothermic chemical reactions.

Because of the Law of Conservation of Energy, you might think that societies would not have to worry about how much energy they consumed. After all, energy never disappears. The energy used to transport goods, for example, should be recoverable. In principle this is true, but there are certain restrictions on converting one form of energy into another.

Heat is energy in transit. We don't speak of the amount of heat within an object; instead we talk about the quantity of heat transferred from a warm object to a cooler one. At the boundary between the two, rapidly moving molecules in the warm body collide with slower molecules in the cool body and transfer some of their kinetic energy. As this transfer of heat energy occurs, the temperatures of the objects become more nearly equal.

Most of our examples of energy transfer up to this point have involved heat. Perhaps even more important for society, however, is a second kind of energy transfer, called **work.** Whenever a force operates on a mass and

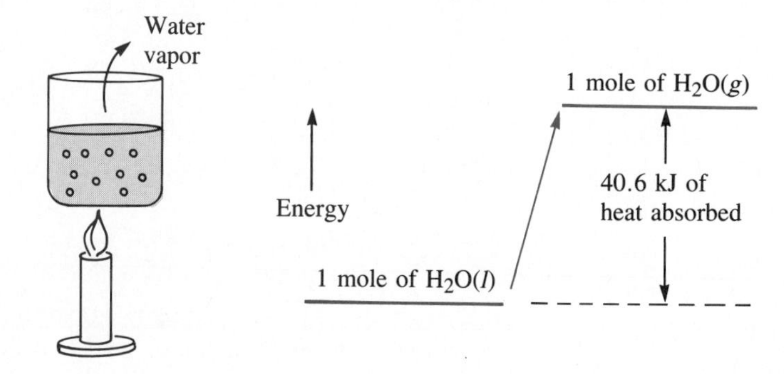

Figure 17.7. *The energy increase that accompanies the boiling of water. The heat absorbed equals the increase in the potential energy of water molecules.*

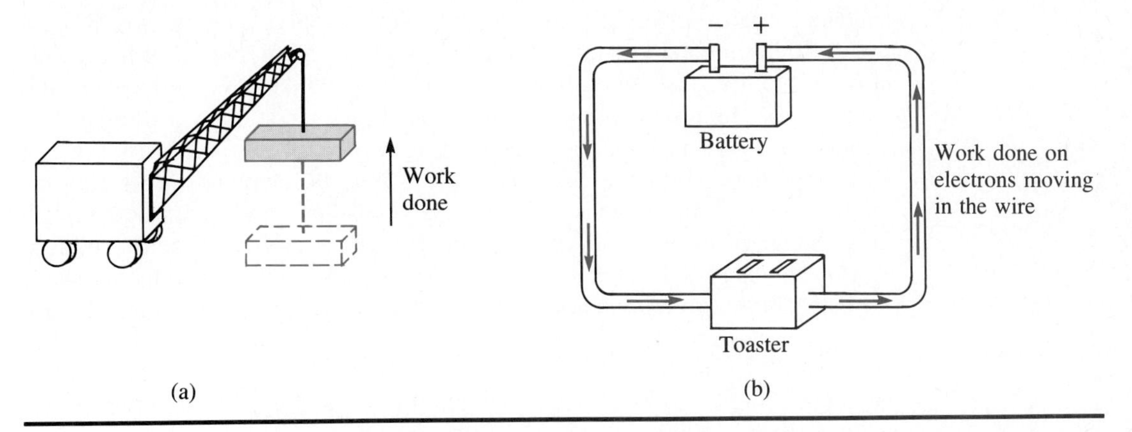

Figure 17.8. Two examples of work. (a) The upward force on a cable lifts a beam and does work on it. (b) An electrical force pulls electrons through wires and a toaster, doing work on the electrons.

causes it to move, work is done on that mass. When a derrick lifts a beam at a construction site, work is done on the beam and energy is transferred from the motor in the derrick to the beam, as shown in Figure 17.8(a). The motor receives its energy from the fuel, which burns and produces CO_2 and H_2O gases to drive the pistons. Similarly, work is done on a car or truck to overcome friction and keep the vehicle moving. Even a source of electrical energy, such as a battery, does work on the tiny electrons in a wire by forcing them to move past the stationary metal atoms in the wire, as shown in Figure 17.8(b).

We need heat to stay warm in the winter, and our industries need heat in order to make chemical reactions occur at high temperatures. The "cracking" of petroleum to increase the yield of gasoline is just one example.

Work is just as important as heat. Work is needed for all forms of transportation and to run the air conditioners that keep us cool in the summer. Work provides nearly all our electrical energy, because in power plants work turns the shafts of electrical generators. The Law of Conservation of Energy applies to work just as it does to heat. When a battery provides electrical work, a chemical reaction occurs within it. The potential energy of the products of this reaction is less than that of the reactants, and the difference equals the work supplied. The situation is analogous to the evolution of heat in an exothermic reaction.

17.5 INTERCONVERSION OF ENERGY

In complex energy systems transformations among several kinds of energy often occur. Think of how an automobile engine functions. The burning of gasoline does not directly produce the kinetic energy of motion along a road. The immediate result of the combustion is the giving off of heat. This flow of energy raises the temperature of the product gases, which expand because of their high pressure and thus drive a piston in a straight-line motion. So far, heat has been converted into work, but the work isn't of quite the right kind. Straight-line motion must be changed into rotational motion so that the wheels will turn. A crankshaft and gears accomplish this change from one kind of mechanical work to another. Finally, rotation of the wheels is converted into motion of the automobile down the highway.

In general, the conversion of work into heat is straightforward. Figure 17.8(b) shows the conversion of electrical work from a battery into heat in a toaster. According to the science of electricity, an electrical current through a metal produces heat. The quantity of heat produced in the toaster equals the amount of electrical work consumed. This is totally consistent with the Law of Conservation of Energy. In other similar experiments scientists have shown that electrical energy can always be converted into an equal amount of heat.

In the middle of the nineteenth century, James Joule, for whom the energy unit the *joule* is named, designed the paddle-wheel experiment shown in Figure 17.9. A weight is connected to a rope that is wound on a spool. The spool is on a shaft that has some paddles immersed in a container of water. Also dipping into the water is a thermometer. Joule first recorded the temperature of the water as registered on the thermometer. Then he allowed the weight to drop, permitting the force of gravity to act on the weight and do work on it. This work was transmitted to the paddle wheel, which stirred the water. After the weight reached the floor and the stirring stopped, Joule again measured the temperature of the water and found that it had risen slightly. The friction of the stirrer against the water had produced heat, which was distributed throughout the water and raised the water's temperature. Calculations showed that the work done equaled the thermal energy produced.

In a variety of experiments Joule and others showed that there is no

Figure 17.9. The Joule paddle-wheel experiment. Joule demonstrated the conversion of work into an equal amount of heat.

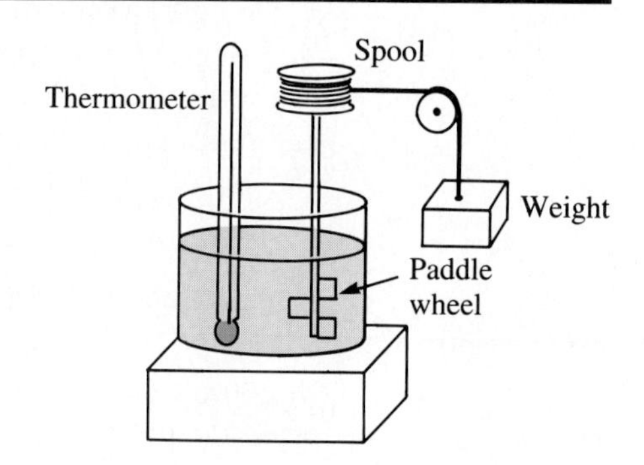

restriction on the conversion of work into heat; work in any form can be changed into an equal quantity of heat. By contrast, there is a serious limitation on the reverse process—conversion of heat into work. This principle is best illustrated by the generation of electricity in a power plant. In either a nuclear or a coal-burning power plant, heat is used to produce work in the form of electrical energy, as shown schematically in Figure 17.10.

In a coal-burning plant the first step is to convert into heat the chemical energy stored in coal. This thermal energy changes water into steam under high pressure at about 500°C, well above the normal boiling point. The efficiency of this process is about 90%—it is a very efficient energy conversion. Only 10% of the heat is shunted into the environment; the rest makes steam. Next, the superheated, high-pressure steam drives the blades of a turbine and does work in forcing the turbine to rotate. Thermal energy is converted into mechanical energy, which in turn is used to rotate an electrical generator. As the steam loses thermal energy by doing work, its temperature and pressure drop. The spent steam at the exit of the turbine has much less capability of doing work. In summary,

$$\text{fuel} \longrightarrow \text{heat} \longrightarrow \text{mechanical energy} \longrightarrow \text{electricity}$$

Imagine what would happen if there were simply a closed chamber at the exit of the turbine. The steam could not escape, and its pressure on the exit side would build up. The turbine, whose rotation depends on a

Figure 17.10. A simplified diagram of an electrical power plant.

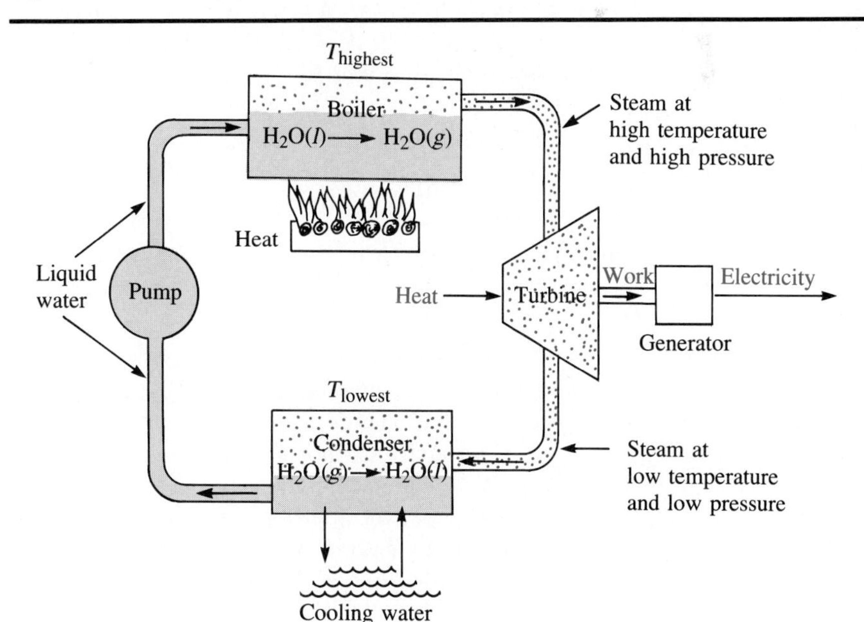

higher pressure of steam at the entrance than at the exit, would stop rotating, and the generation of electricity would stop. For maximum efficiency, then, we want the steam to be at a very high pressure on one side of the turbine in Figure 17.10 and the lowest feasible pressure on the other side. The best way to lower the pressure is to condense the steam by lowering its temperature to 100°C or below. Then vapor is continuously removed and replaced by a much smaller volume of liquid water. In most power plants, water from a nearby river is brought in to remove the heat that is released when the steam condenses. As a result, the water returned to the river is warmer than before.

The need to condense the low-pressure steam imposes a limit on the efficiency with which the thermal energy of the steam from the boiler can be changed into mechanical work in the turbine. The process involves one energy input and two energy outputs. The input is heat from the burning fuel, and the outputs are mechanical work on the turbine and heat evolved by the condensing steam. Some waste heat is always sent into the environment when electricity is generated. Because of the Law of Conservation of Energy, the work obtained from the turbine must be less than the heat energy put into the system. Although our discussion has applied to a power plant, a general principle is implied: *Heat can never be completely turned into an equal amount of work in an engine.* This statement applies to all ''heat'' engines: steam turbines, internal-combustion engines, and rocket engines.

Let us restate these ideas. Both heat and work are forms of energy that are transferred from one system to another. Heat flows from a warm object to a cool one. A system does work by exerting a force on a mass and causing the mass to move. Work can be completely converted into heat or other forms of work, but heat can never be completely converted into work in any engine. In other words,

$$\left.\begin{array}{l}\text{work} \longrightarrow \text{work} \\ \text{work} \longrightarrow \text{heat}\end{array}\right\} \text{with nearly 100\% efficiency}$$

but

$$\text{heat} \longrightarrow \text{work with much less than 100\% efficiency}$$

In 1824 French engineer Nicolas Carnot proposed a relationship for determining the maximum **efficiency of a heat engine.** In our example of the power plant, we had steam at 500°C at the turbine's entrance and 100°C at its exit. If we apply Carnot's time-tested relationship to this situation, we find that the greatest efficiency possible is 53%. In other words, only 53% of the heat from the fuel can become mechanical energy. The other 47% warms the river water flowing through the condenser. Moreover, this efficiency occurs only under ''ideal'' and unachievable conditions, which include operation of the turbine at a very slow speed. The sad truth is that any real steam turbine has an efficiency even lower than 53%.

In the real world of coal-fired and nuclear power plants, the overall efficiency is much less than the theoretical efficiency of the single stage described above. Failure to reach the ideal value for a heat engine and small losses in other stages bring the real efficiency down to about 35%. So 35% of the energy from the fuel is converted into electricity, and 65% is simply transferred to a river. A very large modern generating facility requires an enormous amount of water for cooling. About 10% of the total amount of freshwater flowing in the United States presently is used for cooling electrical generators. This amount will increase in the future as our population and economy expand. To be of any value, electrical energy must be transmitted from the power plant to users, who may be many miles away. For years metal wires have been used for this purpose, and inevitably some heat is generated as electrical current travels along the wires. Another 5% of the energy is lost in transmission, bringing the overall efficiency for delivered power to 30%.

Large amounts of energy are given in units called quads. A **quad** equals 10^{15} BTUs, and 1 BTU is the energy required to raise the temperature of 1 pound of water by 1°F. In a recent year 24 quads of energy was consumed by electric utilities in the United States. Of this energy 17 quads was dissipated as heat, and only 7 quads was transformed into useful electrical energy to provide light and run motors. It is not difficult to see why alternative ways of generating electricity are being explored.

17.6 ENTROPY AND EFFICIENCY

As discussed in Chapter 4, one of the driving forces for chemical and physical change is the tendency for molecular disorder to increase. The term **entropy** has been invented to describe the magnitude of this molecular disorder. It is not just chemical reactions and changes such as melting and boiling that are influenced by the natural tendency for entropy to increase. This principle is also behind the fact that much less than 100% of the heat put into a steam turbine or an automobile engine can be converted into work. So too is it consistent with the 100% efficiency for the opposite conversion, that of work into heat.

Let's think about the Joule paddle-wheel experiment again. Figure 17.11(a) shows the weight as it falls and does work. Of necessity, all the molecules and atoms in the weight are moving downward. Superimposed on this organized downward motion are the random, back-and-forth vibrational motions that are characteristic of all atoms and molecules in both stationary and moving objects.

After the weight has stopped falling, the water in the other part of the

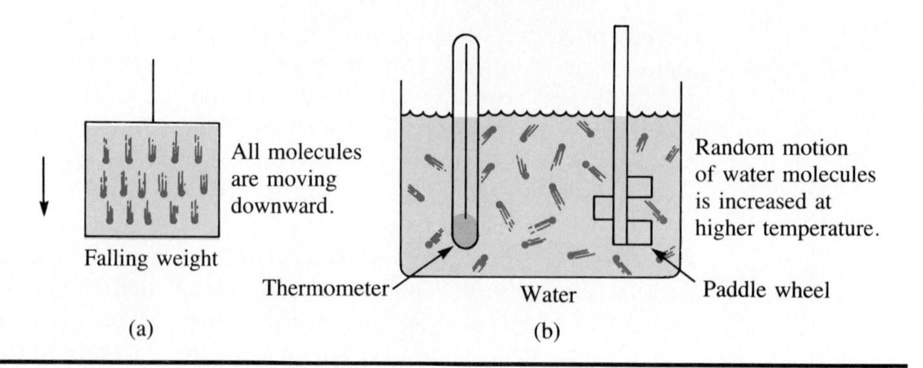

Figure 17.11. *The increase in molecular disorder in the Joule paddle-wheel experiment. (a) Organized molecular motion in the falling weight. (b) Increased disorganized molecular motion in the warmer water.*

apparatus, shown in Figure 17.11(b), is warmer than before. This rise in temperature is accompanied by an increase in the kinetic energy of the water molecules. The random motion of molecules in all directions is now faster, and the molecular disorder is greater.

The overall effect has been to change organized motion of molecules in the falling weight into increased random molecular motion in the water. The entropy of the whole system has increased. *This process is just another example of the universal tendency in nature for entropy, or molecular disorder, to increase*. We saw earlier that the net effect is to convert work into an equal amount of heat. The complete conversion of work into heat is in harmony with the concept of increasing entropy.

Now imagine the Joule experiment operating in reverse. We start with warm water in the container with the paddle wheel, and the weight on the floor. Now, magically, the water begins to cool as heat flows from the water to the paddle wheel. Much to our surprise, this heat changes into work, which turns the paddle wheel and lifts the weight off the floor. If this were to happen, our apparatus would have converted some heat into an equal amount of work, and random molecular motion would have been replaced spontaneously by organized molecular motion. The fact that we have never seen such an event testifies to the improbability of a spontaneous decrease in molecular disorder. Again we see the tendency for entropy to increase rather than decrease.

According to the **principle of increasing entropy,** then, no machine can convert heat completely into work. This is the fundamental reason why nuclear and coal-burning power plants are inherently inefficient. It's not that the engineers who design the power plants are incompetent. Nature simply conspires against efficient conversion of heat into electricity. Future changes in design will bring only marginal improvements. We are up against a law of nature that has nothing to do with conservation of energy. It is the tendency for entropy to increase that limits the amount of work available from heat.

17.7 QUALITY OF ENERGY

After the oil embargo of 1973 the U.S. public became much more conscious of our sources of energy and the efficiency with which we use energy. In the summer of 1974 some leading scientists and engineers met to work out criteria for evaluating the efficiency of everyday energy conversions. Their objective was to develop a theoretical basis for designing future energy systems. They came to two important conclusions. The first is that certain forms of energy are more valuable than others and that all forms can be ranked according to their quality. The second is that an energy source of high quality should not be wasted on an end-use that requires only low-quality energy. Unfortunately, many examples of this short-sighted practice still exist.

To evaluate the quality of different kinds of energy, we can begin with our previous analysis of a power plant. We saw that work can be converted quantitatively into heat, whereas only a fraction of heat can be converted into work. Because work is completely versatile and heat is not, work is more valuable to society than heat is. Electrical work can heat bread to toast in a toaster, and it can also drive the motor that turns the blade of a power saw. On the other hand, although heat from a gas stove can change bread into toast, it can't drive a motor. Only electricity can illuminate a fluorescent light; heat from a flame can't do it. As these examples demonstrate, electrical energy, a major form of work in our society, is more precious than an equal amount of heat produced by burning a fuel.

Whereas different forms of work are equally valuable because they can be converted into one another, the quality of a given amount of heat depends very much on the temperature at which the heat is provided. This is because heat only flows from a high to a low temperature. If we want to heat a house to 68°F, the air from the furnace must be at least several degrees above that temperature. Air at 60°F can slowly bring a house to 55°F if that is what is wanted, but it can't heat a house to 68°F. Furnace air at 75°F can heat a house to either 55° or 68°F. Therefore, heat provided at 75°F has a greater value. The higher the temperature, the greater the quality of the heat.

Another look at the operation of a heat engine also shows the greater value of high-temperature heat. As we have seen, any engine that converts heat into work must also discharge some relatively low-temperature heat into the environment. According to Carnot's mathematical analysis, the greater the difference between the temperature of the heat supplied and that of the heat discharged, the more efficient is the engine. The only way this difference can be large is if the upper temperature is very high, since

Table 17.4 The Quality of Energy

TYPE	QUALITY	EXAMPLES OF USES
work	high	electric motors, electronics, computers, transportation, aluminum production, lighting, air conditioning, refrigeration, electric space-heating
high-temperature heat	medium	transportation (internal-combustion and jet engines), production of electricity, heat for industrial processes, space-heating
low-temperature heat	low	space-heating

the exit temperature cannot be lower than that of the environment. So, high-temperature heat can drive a steam turbine or an automobile engine with reasonable efficiency. Heat at a temperature not much higher than that of our environment can only power an engine with low efficiency.

Table 17.4 gives the different **qualities of energy** and their uses.

To profit from this ranking of work and heat, we must introduce a second idea: *An energy source of high quality should only be coupled with an end-use requiring that quality*. To do otherwise is to waste energy just as surely as one does by designing a building with poor insulation or using a furnace that sends a lot of heat up the chimney. Table 17.4 shows that certain uses require high-quality energy. Some energy-using devices (such as computers) require high-quality electrical energy for their operation. Aluminum metal, Al, can be produced from aluminum ores only with high-quality energy. This is why the Hall process for making aluminum from bauxite, Al_2O_3, consumes large quantities of electrical energy. Other uses, however, appear two or three times in Table 17.4; space-heating, for example, can be accomplished by using high-, medium-, or low-quality energy.

In some parts of the United States, where natural gas has been in short supply, it has been popular to build new homes with electrical heating. Heating elements somewhat like those in a toaster are placed beneath the floor in every room, yielding a uniform, comfortable heat. This kind of heating has advantages, but it is inherently wasteful of energy resources. Remember that coal-burning and nuclear power plants consume 100 joules of heat to produce about 30 joules of electrical energy. In an electrically heated home, the 30 joules of electrical work is changed into 30 joules of heat. Thus, 100 joules of high-temperature heat at the power plant yields 30 joules of low-temperature heat in the home—not a very efficient process. The inefficiency is revealed in the cost; heating a home with electricity can be very expensive.

From this point of view, burning a fossil fuel such as natural gas directly in the home is preferable. To be sure, not all of the heat from a furnace is delivered to the rooms in a home. Roughly 30% of the thermal energy may escape up the chimney with the products of combustion (CO_2 and H_2O), resulting in a 70% efficiency for this step. New furnaces, however, are capable of efficiencies of 90% or better.

In the future other parts of our energy economy will benefit from efficient design of energy-using devices. The automobile is a clear candidate for improvement. Engineers have estimated that only 8% of the energy available from diesel fuel or gasoline is actually delivered to the drive wheels. In addition to attempting to improve engine design, engineers are looking for ways to utilize the energy wasted in the car brakes. In today's automobiles the kinetic energy of motion is converted into heat by the brakes whenever the driver wishes to slow down. An alternative would be to transfer this energy to a flywheel (a heavy wheel that possesses considerable kinetic energy when it rotates). As the car slowed down, the flywheel would speed up. The energy from the flywheel could then be used to aid the next acceleration of the car.

17.8 ENERGY CONSERVATION

Annual energy consumption in the United States during a recent three-decade period is shown in Figure 17.12. After a short-lived decrease in 1974 and 1975 following the oil embargo, the upward trend of many years continued. But since 1979 energy consumption in the United States has been lower than the 1979 peak. It appears that efforts at energy conservation are taking hold; however, the reduction could be the result of a decline in the U.S. economy that occurred at the same time. Many analysts maintain that our consumption of energy is closely tied to the ups and downs in the economy.

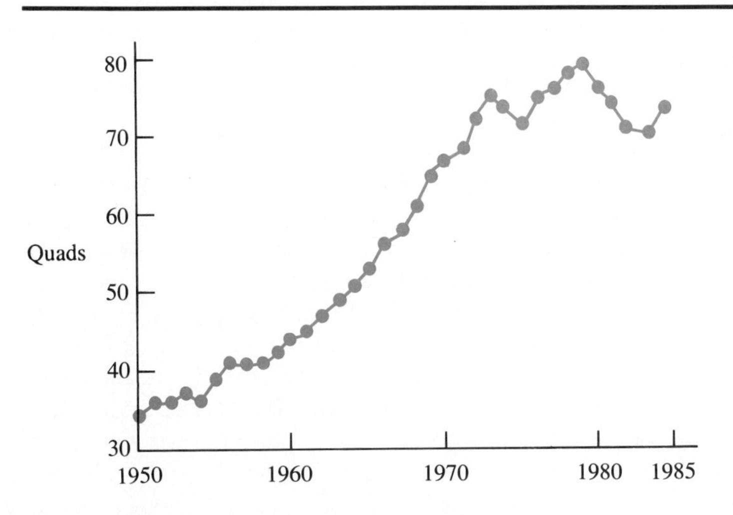

Figure 17.12. U.S. energy consumption, in quads per year (1 quad = 10^{15} BTUs).

A recent study done in the Energy Division of Oak Ridge National Laboratory attempted to assess the various causes of energy consumption. According to the study, energy use in 1981 was 28 quads less than it would have been had the pattern prior to 1973 continued. Of this decrease, 12 quads was attributed to downward swings in the economy and 11 quads to higher fuel prices, verifying that consumers do respond to traditional market forces. The other 5 quads was saved as a result of nonmarket forces, including conservation programs by the federal government, industry, and utility companies.

Contributing to the 11-quad saving were decisions to lower the house thermostat in winter, to raise the thermostat on the air conditioner in summer, and to travel fewer miles in the family cars. Contributing to the 5-quad reduction were technical improvements in automobiles, air conditioners, and appliances. Both kinds of savings resulted from conservation efforts. Over the short term, conservation will remain the most effective way of reducing reliance on imported petroleum and maintaining economic growth in the face of rising energy costs.

One way in which individuals have conserved energy is by choosing more efficient cars. The change toward smaller cars is now having a noticeable impact on gasoline consumption. Not only have car owners been motivated to decrease their costs; manufacturers have been motivated by government regulations to increase the fuel efficiency of the automobiles they produce.

A second step that individuals have taken is to minimize the loss of heat from their homes during the winter and the heat gain during the summer. In the northern part of the United States, as much as 70% of the home energy bill during the winter months is for space-heating. Because of the long lifetime of houses, it will be many years before a majority of homes meet the conservation standards used for new construction. In the meantime, it will pay to upgrade old homes.

In most old homes the primary need is for additional insulation. The main function of insulation is to slow the flow of heat energy through the walls and ceiling. This energy is transferred by means of collisions of rapidly moving "warm" molecules with slowly moving "cold" ones. If a house had perfectly insulated walls and no air leaks, there would be little need for a furnace or an air conditioner.

The most common kinds of insulation are rock wool, fiberglass, and cellulose. Rock wool is made from molten rock by passing steam through the liquid and then letting it solidify as a fibrous material. Fiberglass is a loose mat of thin glass filaments, and cellulose (Section 14.10) is a naturally occurring organic polymer. When placed between the interior and exterior walls, these materials prevent air molecules from traveling easily across the space between the walls and thereby transferring heat. The insulating value of a material is expressed as an **R value,** which measures the **resistance to heat flow.** The greater the R value, the greater the resistance to heat flow and the better the insulator.

Table 17.5 R Values for Insulating Materials

TYPE OF INSULATION	R VALUE/INCH
rock wool batts	3.1
fiberglass batts	3.1
loose-fill cellulose	3.7

Table 17.5 gives R values per inch for the three kinds of insulation mentioned. For each inch of thickness, rock wool and fiberglass insulation have a resistance of 3.1, and cellulose has a resistance of 3.7. So 2 inches of cellulose has a resistance of $2 \cdot 3.7 = 7.4$, and so on. As you can see from Table 17.5, cellulose is a better insulator than fiberglass.

The recommended R value for a house depends on the climate. For Texas and Georgia, for example, the Department of Energy has suggested enough attic insulation to give an R value of 26. For Minnesota, the recommendation is an R value of 38. To obtain an R value of 38, you would use $38/3.1 = 12$ inches of fiberglass but only $38/3.7 = 10$ inches of cellulose. A homeowner would have to consider the cost per inch of each material to determine the most economical alternative.

Escape of heat through the ceiling is a major source of heat loss in most poorly insulated homes, because warm air accumulates next to the ceiling. Adding attic insulation to homes has become a popular do-it-yourself project (see Figure 17.13). Because the walls of existing houses are difficult to

Figure 17.13. Adding insulation to an attic.

insulate, an R value of 11 is acceptable there. In new homes, however, the R value for wall insulation may be as high as 30.

If a house is even moderately well insulated, then the greatest heat loss is probably from air infiltration. Cold air leaks in and warm air leaks out through tiny holes and cracks. Although any single leak may be small, the entire collection may be equivalent to a hole 1 square foot in area. A common source of air filtration is loosely fitting doors and windows. Installing weatherstripping, a painstaking but inexpensive procedure, will seal most of these leaks. Houses with many windows and doors are vulnerable to significant heat loss across these openings. Unlike walls, doors and windows cannot be well insulated without destroying their function. The most common remedy is putting up storm doors and storm windows, which add a second layer and a dead air space between. Some new buildings in northern climates have triple glazing—three layers of glass separated by thin air spaces.

17.9 ENERGY AND INFORMATION—THE SILICON CHIP

The use of electricity to activate electronic components is critical to our technological society. Since the transistor was invented around 1950, the building blocks of the electronics industry have changed from bulky and power-hungry vacuum tubes to tiny silicon **chips** containing thousands of electronic components in a space a quarter inch square. The development of the microprocessor, a chip that processes information, has resulted in increasing automation of appliances and automobiles, as well as miniaturization of the circuits for radios and TV sets. We now have pocket calculators, digital watches, and programmed dishwashers, microwave ovens, and videocassette recorders. Because of improvements in methods of mass production and quality control, the prices of electronic components continue to fall, speeding the introduction of microprocessors into new areas. Soon the home computer may be commonplace. Offices will continue their trend toward automation, and robots will perform many manufacturing operations.

One of the key chemical elements in the production of microelectronic components is silicon, the second element in Group IV of the periodic table. (See the inside of the back cover for a periodic table.) Like diamond, the precious form of carbon, silicon is a solid in which the atoms are linked together by covalent bonds in three dimensions [review Figure 11.4(a)]. The melting point of silicon is very high, 1410°C, which indicates strong bonds holding the silicon crystal together.

Silicon belongs to the narrow band of elements between metals and nonmetals in the periodic table. Highly purified silicon does not conduct electricity very well and therefore is not useful for electronic components. The reason for the lack of conductivity is that all valence electrons in silicon crystals are used in covalent bonds (see Figure 17.14). Because each silicon

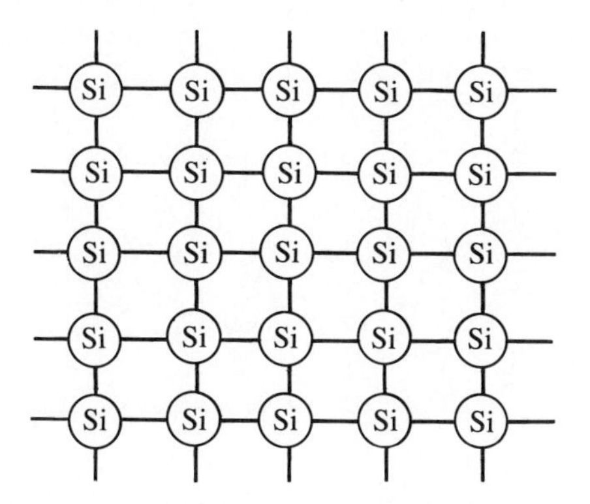

Figure 17.14. The bonding in a crystal of pure silicon. For simplicity, the true tetrahedral geometry around each silicon atom has been flattened into a two-dimensional drawing.

atom has four valence electrons, there are just enough electrons to complete four covalent bonds to each atom. No electrons are free to move through the crystal and carry an electrical current, as they do in metals.

When a small amount of a Group III or Group V element is added as an impurity, however, the silicon becomes a **semiconductor** and has a conductivity between that of an insulator and that of a metal. For example, 0.001% arsenic (As) in silicon causes a considerable increase in conductivity. Because arsenic and silicon have similar atomic sizes, the occasional arsenic atoms fit easily into the crystal structure of silicon. Each arsenic atom brings with it five valence electrons, one more than is needed to form four covalent bonds with neighboring silicon atoms, as shown in Figure 17.15(a). The fifth electron is not held in place and is free to migrate throughout the crystal. When it moves away from the arsenic atom, it leaves behind a +1 arsenic ion, which is fixed in position, as shown in Figure 17.15(b). So this semiconductor consists of a very small number of positive ions and an equal number of mobile electrons. It is called an ***n*-type semiconductor,** because the current carriers (electrons) are *negative*.

It is also possible to make a semiconductor from silicon by adding a very small amount of a Group III element, such as boron. Boron atoms also fit readily into the silicon structure but bring only three valence electrons per atom. Consequently, one of the four bonds from a boron atom contains only one electron, as shown in Figure 17.16(a). Because a stable covalent bond requires two electrons, an electron from a neighboring silicon atom can move into the vacancy. This extra electron in the valence shell of a boron atom converts the boron atom into a −1 boron ion. Simultaneously, the loss of an electron from the valence shell of the silicon atom leaves a positive charge and a vacancy there, as shown in Figure 17.16(b).

Electron vacancies—such as those shown in Figure 17.16(b)—and their associated positive charges are called **holes.** When an electron in a bond extending from one silicon atom moves to fill a hole in the valence shell

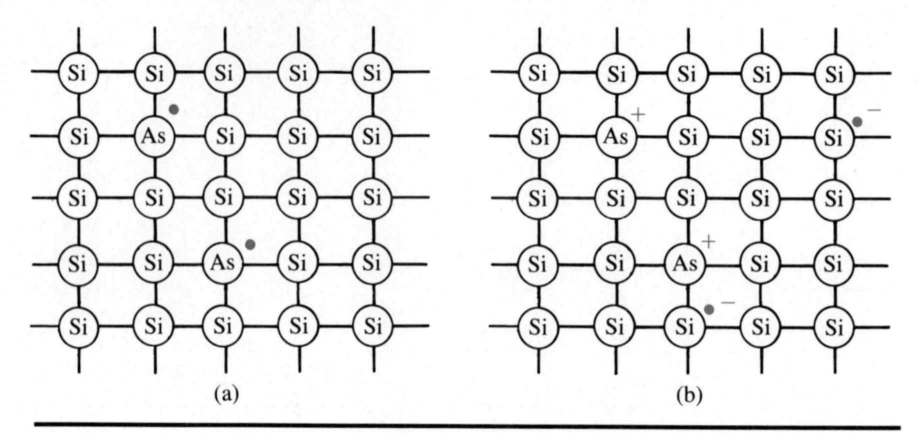

Figure 17.15. *An n-type semiconductor. (a) Each added arsenic atom has one electron (shown in color) that is not tied down by covalent bonding. (b) The mobile electrons (shown in color) have moved away from the arsenic atoms, converting them into positive ions. These mobile electrons have in turn converted silicon atoms into negative ions.*

of a second adjacent silicon atom, the hole jumps from the second atom to the first atom. Scientists in the field of semiconductors like to think of mobile, positively charged holes moving throughout the silicon crystal. These *positive* current carriers are responsible for the name *p*-type semiconductor. A ***p*-type semiconductor** consists of a small number of fixed, negative Group III ions and an equal number of mobile, positive holes.

Most modern electronic components are some combination of *n*- and *p*-type semiconductors. The simplest of all is an *n*-type semiconductor in

Figure 17.16. *A p-type semiconductor. (a) An electron vacancy, or hole, is shown in color next to each boron atom. (b) Because of electron jumps from silicon to boron, the vacancies are now near silicon atoms.*

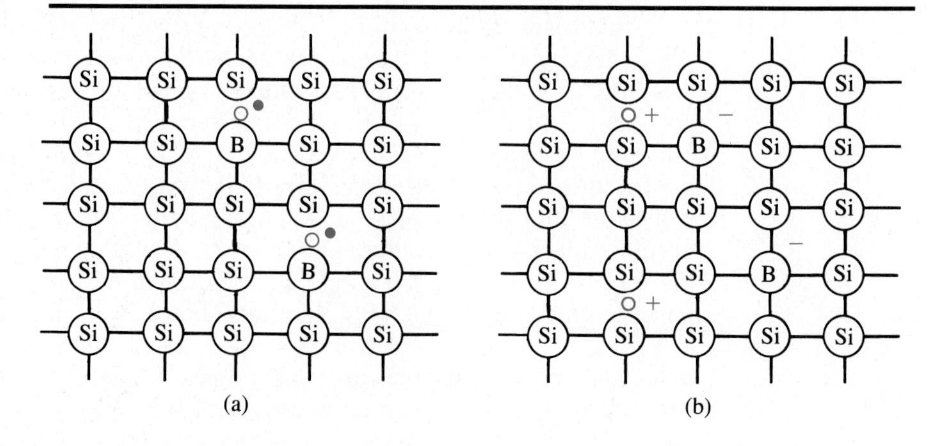

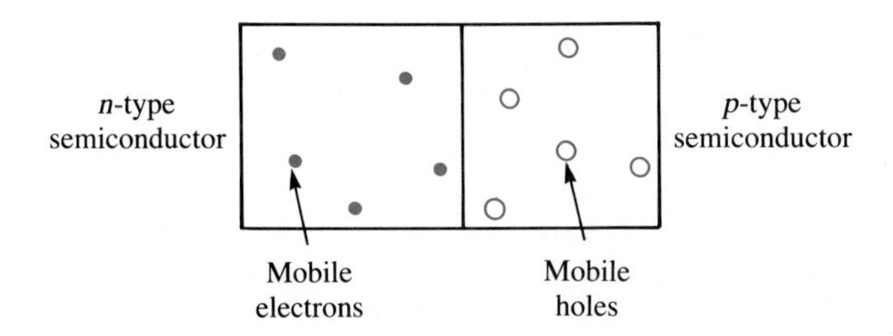

Figure 17.17. A simplified diagram of a rectifier, consisting of an n- and a p-type semiconductor. This device permits current to flow in only one direction.

contact with a *p*-type semiconductor (see Figure 17.17). Such a device, called a **rectifier,** permits electrical current to flow in one direction only. This restriction is often required in electronic equipment.

Suppose that the terminals of a battery are connected by wires to the two ends of the rectifier so that the left end is negative (has an excess of electrons) and the right end is positive (has a deficiency of electrons), as shown in Figure 17.18(a). Because electrons are repelled by a negative charge and are attracted to a positive charge, the electrons in the *n*-type semiconductor on the left move to the right. As they cross the boundary between the two semiconductors, they encounter and neutralize the positive holes. At the same time, more electrons move into this semiconductor from the wire at the left. A similar process occurs in the *p*-type semiconductor. Holes are propelled to the left. As some cross the boundary, they neutralize electrons, while more holes are created by the positive connection at the right end. This process of electrical conduction can continue indefinitely.

In Figure 17.18(b) we see what happens when the connections to the battery are reversed. Now electrons in the *n*-type semiconductor on the left flow toward the positive wire on the left, and there are no additional electrons to replace them. Simultaneously, holes move to the right, and no new holes are created. The only way the current can continue to flow is if electrons in the *p*-type semiconductor move left and holes in the *n*-type semiconductor move right. Because *p*-type semiconductors have very few

Figure 17.18. The operation of a rectifier. (a) When the left end is negative and the right end is positive, current readily flows. (b) When the polarity is reversed, scarcely any current flows.

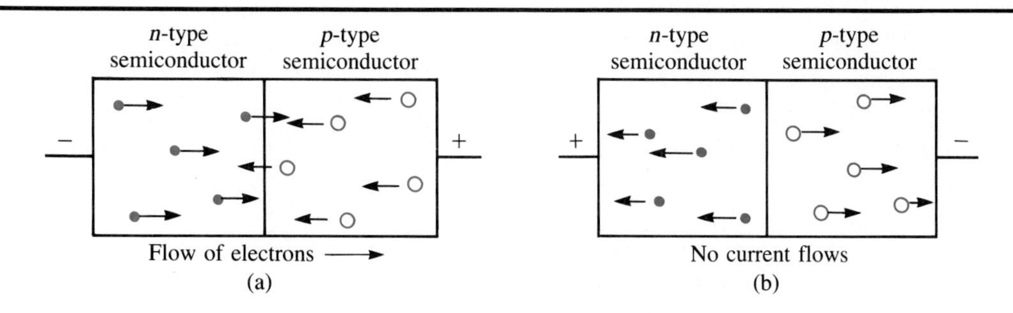

mobile electrons and *n*-type semiconductors have very few holes, this process meets with great resistance. Hardly any current flows. So when this device is in a circuit, electric current can flow in only one direction—namely, to the right, as shown in Figure 17.18(a).

Different combinations of *n*- and *p*-type semiconductors are found in transistors, which are semiconductor devices having three connections. Most of the transistors in a complex circuit amplify current or act as switches. Extremely sophisticated methods are used to design the circuits that are etched into tiny silicon chips (computers are often used to help in the design). Then a large-scale layout of the actual circuit is made. Finally, the surface of a silicon wafer is modified in such microscopic detail that the result can scarcely be seen without magnification. The manufacturing process occurs in rooms that are nearly dust-free, because particles of dust can cause flaws in chips, making them useless.

The development of microelectronics is rapidly changing our society. As factories and offices are automated, the amount of work required from each individual will probably decrease. In the future we may be as concerned with the productive use of leisure time as we now are with our careers and professions.

IMPORTANT TERMS

steady state A state in which there is no visible change in a property even though one agent tends to increase it while another tends to decrease it. The two flows, in and out, are equal in magnitude and therefore cancel each other. For example, if water from a spigot runs into a partly filled bathtub at the same rate as it leaves through the drain, the amount of water in the tub is constant, and the water level is at a steady state.

greenhouse effect The heating that occurs when a barrier (such as glass) prevents infrared rays from escaping from the object but allows visible rays to be admitted. The temperature rises because more energy enters than leaves.

knocking Pinging sounds caused by lack of smooth combustion in a gasoline engine.

octane number A measure of the antiknock properties of gasoline.

work The result of a force's acting on a mass and causing it to move. More technically, the amount of work equals the magnitude of the force times the distance through which the mass moves in the direction of the force.

efficiency of a heat engine Work output divided by heat input. If a gasoline engine delivers 15 joules of work for every 100 joules of heat from fuel combustion, the efficiency is 0.15, or 15%.

quad A unit of measure used for very large amounts of energy; 1 quad is equal to 10^{15} BTUs.

entropy A measure of molecular disorder.

principle of increasing entropy The idea that changes are accompanied by an increase in entropy. An offshoot of this principle is that no machine can convert heat completely into work.

quality of energy Versatility and usefulness of a given kind of energy. Work is most highly valued, followed by high-temperature heat. Low-temperature heat is least useful to society.

R value A unit of measure used to express the insulating value of a material.

resistance to heat flow The ability to prevent heat from flowing rapidly from a high-temperature region to a low-temperature region. A good insulator has a high R value and a high resistance to heat flow.

chip A tiny thin square of a semiconducting material containing thousands of electronic components.

semiconductor A piece of silicon to which a very small amount of a Group III or Group V element has been added. It has a small electrical conductivity.

***n*-type semiconductor** Silicon with a Group V impurity, such as arsenic. The charge carriers for electrical current are electrons, which are *negative*.

hole An electron vacancy. It has a positive charge.

***p*-type semiconductor** Silicon with a Group III impurity, such as boron. Because of a shortage of electrons, *positive* holes are the charge carriers.

rectifier A combination of different kinds of semiconductors that permits electrical current to flow in only one direction.

QUESTIONS

1. It is said that over 95% of the energy used by the United States originally was supplied by the sun. Give the justification for this statement.

2. Describe the steady state responsible for the nearly constant temperature of the earth's surface.

3. Suppose that the average amount of cloud cover and dust in the atmosphere should decrease. Would the temperature at the surface of the earth increase, decrease, or remain the same? Explain.

4. What special properties of carbon dioxide permit it to participate in a greenhouse effect involving the earth?

5. Neither oxygen nor nitrogen, which together make up 99% of the air, absorbs visible or infrared radiation. Are these two elements involved in a greenhouse effect? Explain.

6. As the hydrogen content of a fossil fuel increases, the heat of combustion per gram of fuel also increases. Explain why this is so.

7. Table 17.3 lists the octane number of cyclohexane as 83. Suppose you were given the task of determining the octane number of cyclohexane. What procedure would you follow?

8. In your own words define the terms *heat* and *work*.

9. In view of the fact that energy is neither created nor destroyed, explain why society must extract fossil fuels at considerable expense in order to meet its energy needs.

10. Figure 17.8(b) shows a toaster powered by a battery. Where does the heat given off by the toaster come from? Explain how the Law of Conservation of Energy applies to this process.

11. Explain why a steam turbine cannot convert into work all of the heat that drives it.

12. Give an example not used in this chapter of the complete conversion of a quantity of work into heat.

13. Explain why work is more useful to society than heat is.

14. Why is high-temperature heat more valuable than low-temperature heat?

15. Decide which of the two methods given in each part below is a better match of the quality of energy with the stated task, and justify each answer.

a. heating a home with hot water at 200°F or with hot water at 100°F

b. heating a home electrically or with a gas furnace

c. powering an automobile with rechargeable batteries or with a gasoline engine

d. browning toast in an electric toaster or over a gas flame

16. Both insulation A and insulation B are sold as square batts 8 inches thick and 18 inches on a side. Insulation A has an R value per inch of 2 and costs $1.10 per batt. Insulation B has an R value per inch of 4 and costs $1.90 per batt. If you wanted to insulate your attic to a total R value of 32, which insulation would you use? Why?

17. Explain why the kind of impurity added to silicon determines whether the material is a *p*-type or *n*-type semiconductor.

18. Give an example not used in this chapter of the principle of increasing entropy.

Public-Policy Discussion Questions

19. If you were a policy maker in the U.S. Department of Energy, would you promote intensive exploration for petroleum within the country to increase our petroleum production, or would you emphasize additional energy conservation measures? Discuss the pros and cons of each position.

20. Should restrictions be placed on the use of electricity for space-heating?

Nuclear Reactions and Nuclear Power

By the 1920s, Rutherford's model of the nuclear atom (Chapter 6) was generally accepted. According to this model, shown in Figure 18.1, the protons and neutrons, which account for nearly all of the atomic mass, are concentrated in the tiny nucleus of an atom, and the electrons surround the positively charged nucleus. It had also become clear to scientists that radioactive disintegrations involve changes in the nuclei of atoms. Alpha (α) and beta (β) particles and gamma (γ) radiation come from the nuclei of unstable isotopes. The electrons in the outer reaches of atoms rarely participate in these radioactive processes.

With so many protons so close together, shouldn't the nucleus simply fall apart? What holds a nucleus together? Although many physicists are actively seeking the answer to this question, the nature of the forces that overcome the repulsions between protons and hold protons and neutrons together is still mysterious, at least to those of us who lack powerful backgrounds in physics. Physicists have postulated the existence of a unique kind of force called the **strong force** or simply the **nuclear force.** To be consistent with the known characteristics of nuclei, this force must act only

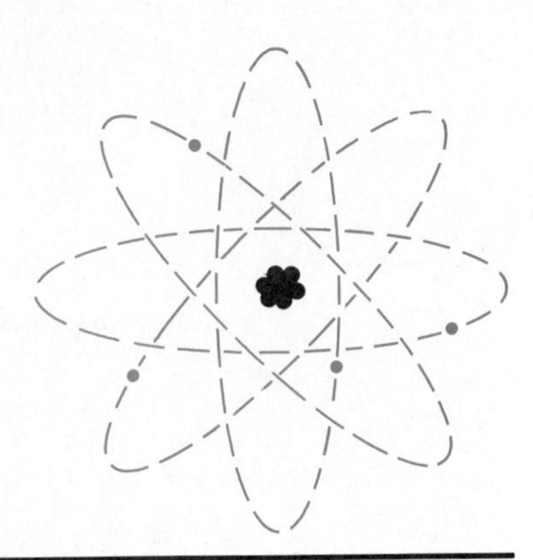

Figure 18.1. Simulated view of a nuclear atom.

over the very short distances typical of nuclear diameters (10^{-13}–10^{-12} centimeters), and it must be 100 times stronger than the ordinary electrical force between charged particles. In the language of modern physics, the strong force acts between *quarks,* which are thought to be the building blocks of protons and neutrons. Because of the complex nature of this theory, our discussion will emphasize the experimental side of nuclear science.

Fission is one kind of nuclear reaction known to almost everyone, since it occurs both in an atomic bomb and in a nuclear power plant. After we have covered some of the fundamental aspects of nuclear reactions, we shall investigate fission and how it is harnessed to produce electricity.

18.1 NATURAL RADIOACTIVITY

Since the time the earth was formed, long before the appearance of humans, elements in the crust and interior have been changing into other elements through radioactive decay. In Chapter 6 we saw how Henri Becquerel first discovered radioactivity in uranium compounds. There are many radioactive elements found in nature. Some were formed in ancient stars, others came from radioactive decay processes in minerals, and still others are continually being produced in our upper atmosphere. As discussed earlier, we now know that three kinds of radioactive emissions—α, β, and γ—account for most radioactivity. For the purpose of review, the properties of these emissions are summarized in Table 18.1.

The kinetic energies of the α and β particles (helium nuclei and electrons, respectively) are very large, in the neighborhood of 1 million electron-volts (1 Mev). For comparison, the electron in a hydrogen atom has an average kinetic energy of only 13.6 electron-volts. Despite these large energies, α and β particles come to a complete stop after passing through just a few centimeters of air, because they lose energy by knocking electrons from

Table 18.1 Particles and Rays Given Off by Radioactive Atoms

PARTICLE OR RAY	CHARGE	MASS (on atomic weight scale)
alpha (α)	+2	4
beta (β)	−1	0.00054
gamma (γ)	0	0

air molecules. By contrast, γ rays can travel through air for great distances, even though their energies may be considerably less than 1 million electron-volts. Gamma rays give up only a small fraction of their energy on each encounter with a molecule of nitrogen or oxygen in the air. Consequently, the chief hazard to a person standing some distance from a source of radioactivity is from the γ rays. The distances traveled by α, β, and γ emissions in various media are compared in Table 18.2.

On the basis of our simple model for the nucleus, it is not difficult to understand how an α particle might be produced from a nucleus: Two protons and two neutrons could simply leave as a single unit if that nucleus was unstable. A crude representation of how a particular isotope of beryllium emits an α particle is shown in Figure 18.2.

Less understandable, however, is the appearance of a β particle from a nucleus, because according to our model electrons do not reside there. For bookkeeping purposes it is simplest to think of a neutron as dissociating into an electron and a proton. The electron is spewed forth, and the proton remains in the nucleus. Whether a neutron actually consists of a proton and an electron is a question not easily answered, and it need not concern us here. It is an experimental fact that after β emission the new nucleus contains one more proton and one less neutron than did the parent nucleus.

Table 18.2 Ranges of 1-Mev α and β Particles and γ Rays in Several Substances

TYPE OF EMISSION	ABSORBING SUBSTANCES	RANGE (in centimeters)
α	air	0.5
	biological tissue	negligible
β	air	300
	biological tissue	0.5
	aluminum	0.2
γ	air	very large
	aluminum	large[a]
	lead	large[b]

[a]A thickness of 17 centimeters is required to reduce the γ ray intensity to one-sixteenth of its original value.
[b]A thickness of 3.5 centimeters is required to reduce the γ ray intensity to one-sixteenth of its original value.

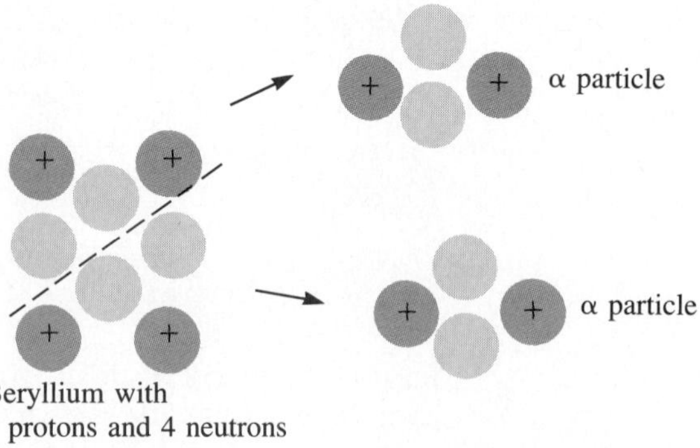

α particle

α particle

Beryllium with
4 protons and 4 neutrons

Figure 18.2. A crude representation of how an α particle is emitted by an unstable isotope of Be *having 4 protons and 4 neutrons. In this case the nucleus that remains is also an α particle.*

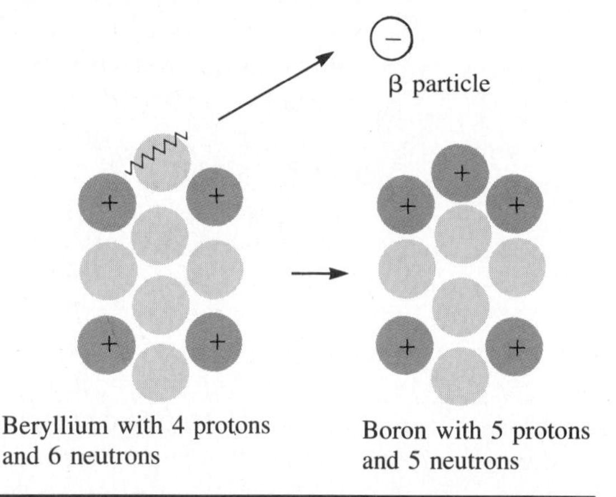

Figure 18.3. An idealized picture of the emission of a β particle from an isotope of beryllium containing 4 protons and 6 neutrons. A neutron becomes a proton, creating an isotope of boron (B).

For example, a thorium nucleus contains 90 protons and 144 neutrons. When thorium emits a β particle, protactinium, with 91 protons and 143 neutrons, is formed. Figure 18.3 illustrates a simpler case—the loss of a β particle from another isotope of beryllium.

18.2 RADIOACTIVE URANIUM

Essential to the writing of any nuclear reaction is the notion of **mass number.** We have seen that a proton and a neutron each have a mass of almost exactly 1 on the atomic weight scale. An electron has a mass equal to 0.00054 of that value. Because of its very small relative mass, an electron is arbitrarily assigned a mass number of zero. The approximate mass of any nucleus (on the atomic weight scale) is therefore equal to the sum of the number of its protons and neutrons. This sum defines the mass number of the nucleus.

$$\text{Mass number of the nucleus} = \text{number of protons} + \text{number of neutrons}$$

For example, the element bromine (Br) consists of almost equal parts of two isotopes, which have atomic weights of 78.918 and 80.916. Each isotope owes its mass almost entirely to the protons and neutrons in the nucleus. So the atomic weight, rounded off to a whole number, is the mass number; we have Br-79 and Br-81. Referring to a periodic table, we find the atomic number of bromine to be 35. The first isotope, then, has 35 protons and 79 − 35, or 44, neutrons. The second isotope has 35 protons and 46 neutrons.

Nuclear scientists use a standard notation for writing specific isotopes of elements, diagrammed in Figure 18.4. The mass number of the isotope appears as a superscript before the symbol of the element. The atomic number is a subscript, also to the left of the symbol. In this system, the two isotopes of bromine are written as ${}^{79}_{35}\mathrm{Br}$ and ${}^{81}_{35}\mathrm{Br}$.

In any radioactive decay the mass number of the parent (original) nucleus must equal the sum of the mass numbers of the daughter (product) nucleus and the particle (α or β) emitted. When an α particle is emitted, the mass number of the radioactive nucleus decreases by 4; when a β particle is emitted, the mass numbers of the original and the new element are the same. A second condition for any radioactive disintegration is that nuclear charge must be conserved. The sum of the charges of the final nucleus and the ejected α or β particle must equal the original nuclear charge.

These two principles can be illustrated by some examples involving radioisotopes found in nature. Consider first the emission of an α particle by the isotope of uranium having a mass number of 238:

$${}^{238}_{92}\mathrm{U} \longrightarrow {}^{234}_{90}\mathrm{Th} + {}^{4}_{2}\mathrm{He}$$

It is clear that the sum of the mass numbers after the transformation (234 + 4) equals the original mass number (238). Similarly, in agreement with the conservation of charge, the total charge on the right side of the equation (90 + 2) equals the original charge (92).

Generalizing, we can say that in the equation for any nuclear reaction the sum of the superscripts on each side is the same, as is the sum of the subscripts. Because the atomic number of any element determines the element's identity, we can deduce the symbol of the new element shown on the right by using a periodic table to find the symbol of element 90 (Th). Should we want to know the number of neutrons in any isotope, we can subtract the atomic number from the mass number. Thus, ${}^{238}_{92}\mathrm{U}$ and ${}^{234}_{90}\mathrm{Th}$ have 146 and 144 neutrons, respectively.

$$\begin{array}{r} \text{Mass number} \longrightarrow \\ \text{Atomic number} \longrightarrow \end{array} {}^{y}_{z}X$$

Figure 18.4. *Notation for isotopes. X is the symbol for the element, y is the mass number, and z is the atomic number.*

$^{234}_{90}Th$ is itself radioactive and decays by β emission. This process can be written as

$$^{234}_{90}Th \longrightarrow {}^{234}_{91}Pa + {}^{0}_{-1}e$$

Conventionally, the β particle (a highly energetic electron) is written as $^{0}_{-1}e$ to show its zero mass number and its minus one charge. Observe again that the total charge and the total mass number are constant. It is evident from these examples that α decay results in a new element with an atomic number that is 2 less than the original number, whereas β decay produces an element with an atomic number that is 1 greater than the original number.

The decay of $^{238}_{92}U$ into $^{234}_{90}Th$ and the decay of $^{234}_{90}Th$ into $^{234}_{91}Pa$ are the first two steps in a long series of decays which result finally in a stable, nonradioactive isotope of lead, $^{206}_{82}Pb$. This sequence, called the *uranium series,* is one of three that occur in nature. The others, the thorium and actinium series, also produce isotopes of lead as stable end products.

18.3 HALF-LIVES

The rates at which different radioactive isotopes emit subatomic particles cover a huge range. For instance, the decay of ^{238}U described above is so slow that it takes over 4 billion years for the number of ^{238}U atoms in any sample to decrease to one-half the original number. On the other hand, the daughter element, ^{234}Th, requires only 24 days to decay to half of its original amount. The halving time, or **half-life,** is commonly used to describe how fast an isotope decays. It doesn't depend on the amount of radioactive isotope present. After 24 days, a 10-gram sample of ^{234}Th becomes 5 grams of ^{234}Th plus 5 grams of decay products. The same amount of time is required for an 8-gram sample to decay to 4 grams. After another 24 days, only 2 grams of ^{234}Th (of the original 8 grams) is left; this is $\frac{1}{2} \cdot \frac{1}{2} = \frac{1}{4}$ of the original amount. Figure 18.5 depicts this behavior graphically. The curve is typical of exponential decay, which also occurs in some chemical reactions.

Table 18.3 gives half-lives for a few radioactive isotopes. The rates of decay range from very rapid (extremely short half-life) to very slow. ^{131}I and ^{90}Sr are examples of isotopes that remain after the fission of ^{235}U in either an atomic bomb or a nuclear reactor. If ^{131}I were in the environment, it would be a severe radiation hazard, because it emits many β particles (and γ rays) in a short time. It would last only a month or so, however. Plutonium, ^{239}Pu, a by-product in the operation of a nuclear power plant, has a half-life of 24,440 years. The handling and disposal of such a long-lived, dangerous isotope present serious problems.

Archeologists have made clever use of the half-life of carbon-14 to determine the ages of ancient carbon-containing materials. The method was developed by Willard F. Libby (see Figure 18.6) in 1946 at the University of Chicago.

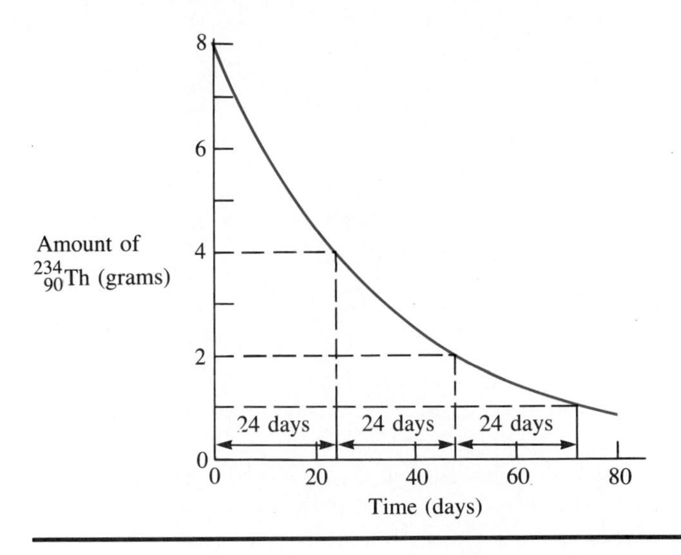

Figure 18.5. *Radioactive decay of* $^{234}_{90}Th$ *with a half-life of 24 days.*

Libby received a Nobel Prize in 1960 for his work. In the upper atmosphere a nuclear reaction between nitrogen atoms and neutrons from cosmic rays produces carbon-14:

$$^{14}_{7}N + ^{1}_{0}n \longrightarrow ^{14}_{6}C + ^{1}_{1}H$$

This carbon-14 eventually finds its way into the carbon dioxide in the lower atmosphere, so a small fraction of $^{14}CO_2$ becomes mixed with the much more common $^{12}CO_2$. This fraction has remained roughly constant over many thousands of years, because the formation of carbon-14 described above is just balanced by radioactive decay (half-life = 5730 years).

Since CO_2 is taken up by plants in photosynthesis, plants contain carbon with the same fraction of carbon-14. Animals also contain carbon-14 as a result of eating plants. Once a plant or animal dies, the intake of carbon-

Table 18.3 Half-Lives of Some Radioactive Isotopes

ISOTOPE	TYPE OF EMISSION	HALF-LIFE
$^{212}_{84}Po$	α	3.04×10^{-7} second
$^{42}_{19}K$	β	12.4 hours
$^{131}_{53}I$	β	8 days
$^{90}_{38}Sr$	β	28 years
$^{14}_{6}C$	β	5730 years
$^{239}_{94}Pu$	α	24,440 years
$^{238}_{92}U$	α	4.5×10^{9} years

Figure 18.6. *Willard F. Libby.*

14 stops, and the amount of this isotope slowly decreases because of radioactive decay. The sequence of events is pictured in Figure 18.7. It is possible to determine the amount of carbon-14 in a sample by measuring with a Geiger counter the number of β particles per minute coming from the ^{14}C. The greater the number of counts, the greater the amount of ^{14}C.

To determine the age of the Dead Sea scrolls (that is, the time since the papyrus plants were harvested to make the paper), archeologists found the number of counts per minute coming from a small specimen. They also determined the total amount of carbon in that specimen. By comparing the counts per minute from a living organism having the same amount of carbon with the counts per minute from a specimen, one can determine the age of the specimen. For example, if the number of counts from the specimen is equal to one-half the number of counts from a living sample, the amount of ^{14}C in the specimen is one-half the amount present when the specimen was produced and the age of the specimen is 1 half-life, or 5730 years. The number of counts per minute from the Dead Sea scroll was more than half the number from the corresponding living sample, and the age of the scroll was calculated as 1900 $\pm$ 200 years.

By this method, archeologists have determined the age of charcoal at Stonehenge, England, of mummies in Egyptian tombs, and of rope sandals in a cave in Oregon. The maximum age that can be determined from carbon-14 is about 50,000 years, because older objects have such a small amount of carbon-14 that it cannot be measured. Other isotopes with longer half-lives, such as potassium-40 and uranium-238, have been used in a similar way to date much older objects, including meteorites and even the earth itself.

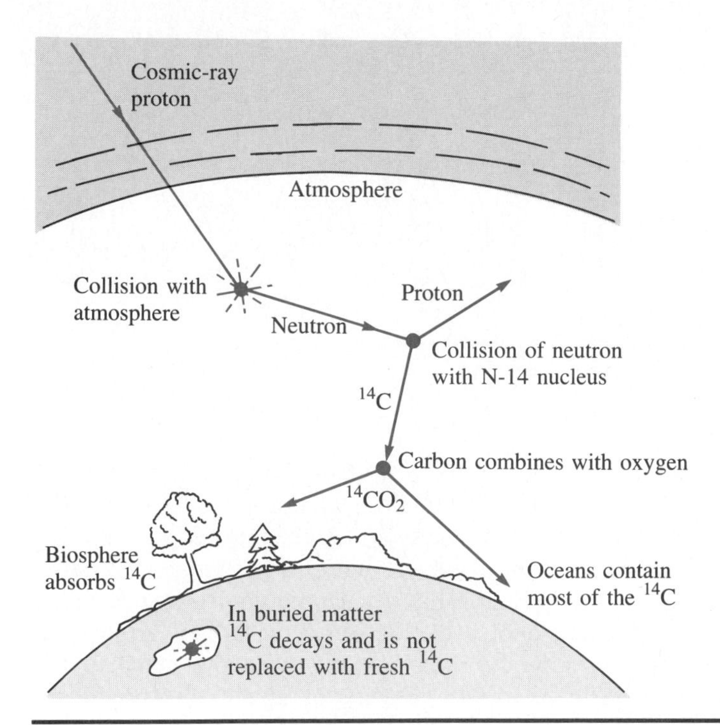

Figure 18.7. *The process by which $^{14}_{6}C$ forms in the upper atmosphere and eventually decays underground.*

The carbon-14 dating method as described above is slightly inaccurate because of changes in the rate of carbon-14 formation in the upper atmosphere. It has been possible to correct for this inaccuracy by painstakingly counting the rings in very old trees (one ring per year) and determining the carbon-14 content in these trees.

18.4 CONVERSION OF MASS INTO ENERGY—RADIOACTIVITY, FISSION, AND FUSION

The very large energies of the α and β particles emitted by radioactive nuclei are truly phenomenal when compared with the energies associated with chemical reactions. The amount of heat absorbed or evolved by a chemical reaction is just a few electron-volts per reacting molecule; the kinetic energy of an α or β particle that has been expelled from a nucleus is several million times that amount. A comparison is given in Table 18.4.

Evidently the changes occurring within nuclei are on a very different scale from those involving rearrangements of atoms and electrons during chemical reactions. Reasons for the difference began to unfold at the beginning of this century.

Table 18.4 Energies Released in Radioactive Decay and in Chemical Reactions

REACTION	ENERGY EVOLVED (in electron-volts per atom)	(in kilojoules per mole)
Nuclear		
$^{14}_{6}C \longrightarrow ^{14}_{7}N + ^{0}_{-1}e$	156,000	15,100,000
$^{239}_{94}Pu \longrightarrow ^{235}_{92}U + ^{4}_{2}He$	5,240,000	506,000,000
Chemical		
$C + O_2 \longrightarrow CO_2$	4.1	393
$C_6H_{12}O_6 + 6\ O_2 \longrightarrow 6\ CO_2 + 6\ H_2O$ glucose	29.2	2,820

In 1905, not long after the discovery of radioactivity, Albert Einstein, then a clerk in the Swiss Patent Office at Bern, published a series of very important papers on the theory of relativity. Among his ideas was the proposal that it should be possible to convert matter into energy and energy into matter. The quantitative relationship giving the **equivalence between mass and energy** is that often quoted expression $E = mc^2$. If an amount of mass, m, disappears in a process, an amount of energy equal to the mass multiplied by the square of the velocity of light, c, is produced. Einstein maintained that the old Law of Conservation of Mass had to be replaced by the Law of Conservation of Mass plus Energy.

Einstein suggested that it might be possible to test this equation by studying nuclear reactions, because the large energies involved should correspond to measurable changes in mass. It was not until 1932, however, that experimental techniques became sufficiently sophisticated. Prior to that time, nuclear physicists had produced many artificial nuclear reactions by bombarding various nuclei with high-energy α or β particles coming from radioactive isotopes, but the energies of the particles were not accurately known. In 1932, hydrogen nuclei (that is, protons) of precisely controlled energy were produced in an accelerator and directed toward a target of lithium metal. The transformation

$$^{7}_{3}Li + ^{1}_{1}H \longrightarrow ^{4}_{2}He + ^{4}_{2}He$$

occurred, and the energies of the ^4He nuclei formed were accurately determined. By subtracting the kinetic energy of the reacting proton (energy lost) from the total kinetic energy of the two helium nuclei (energy produced), researchers found the net energy released to be 17.2 million electron-volts (Mev).

In order to test the Einstein hypothesis, we need to know the masses of the reactant and product nuclei so that we can look for a change in mass. For the nuclei involved in the 1932 experiment, highly accurate mass values, based on the atomic weight scale, are as follows:

$$\begin{array}{lccccccc} & {}^{7}_{3}\text{Li} & + & {}^{1}_{1}\text{H} & \longrightarrow & {}^{4}_{2}\text{He} & + & {}^{4}_{2}\text{He} \\ \text{atomic mass units (a.m.u.):} & 7.01600 & & 1.00782 & & 4.00260 & & 4.00260 \end{array}$$

Then

$$\begin{aligned} \text{change in mass} &= \text{masses of product nuclei} - \text{masses of reactant nuclei} \\ &= (2)(4.00260) - (7.01600 + 1.00782) \\ &= -0.01862 \text{ atomic mass unit (a.m.u.)} \end{aligned}$$

There is definitely a small decrease in mass not revealed by the mass numbers we used previously because they were rounded off.

How much energy should be created by this loss of mass? We need a mass-energy conversion factor for this calculation. If you were to calculate the actual mass in kilograms corresponding to 1 atomic mass unit and substitute that value into the equation $E = mc^2$, you would find an energy of 931 million electron-volts after changing the energy units. That is,

$$1 \text{ a.m.u.} = 931 \text{ Mev}$$

The energy equivalent to the above change in mass of 0.01862 atomic mass unit can now be calculated as

$$(0.01862 \text{ a.m.u.}) \left(\frac{931 \text{ Mev}}{1 \text{ a.m.u.}}\right) = 17.3 \text{ Mev}$$

which agrees with the experimental value of 17.2 million electron-volts after allowance is made for experimental uncertainties. Einstein's equation seems to be justified!

It now appears that all spontaneous nuclear transformations are associated with a decrease in mass and a release of energy. This is true both for radioactive decay and for **nuclear fission,** which is the splitting of a very heavy nucleus into several nuclei of roughly equal mass. Both involve the fragmentation of nuclei. **Nuclear fusion,** on the other hand, consists of combining very light nuclei, such as hydrogen or helium, into nuclei of only slightly greater mass number. Such processes occur in the hydrogen bomb and under more controlled conditions in the laboratory. Paradoxically, fusion also causes a release of energy, which can only mean that the total mass decreases. We need to explore the reason for mass decreases in both fission and fusion.

18.5 THE BINDING ENERGY OF THE NUCLEUS

The mass of any nucleus is slightly less than the sum of the masses of the protons and neutrons that go into it. For example, the mass of ${}^{4}_{2}\text{He}$ is

4.00260 atomic mass units, whereas the sum of the masses of two protons and two neutrons is 4.03298 atomic mass units. If, with the help of some magic apparatus, one could combine two separate protons and two separate neutrons to make a $^{4}_{2}He$ nucleus, the mass would decrease by 0.03038 atomic mass unit. Simultaneously, an energy equal to

$$(0.03038 \text{ a.m.u.})(931 \text{ Mev}/1 \text{ a.m.u.}) = 28.3 \text{ Mev}$$

would be released. This energy is called the **binding energy** of the $^{4}_{2}He$ nucleus. In the reverse process—the complete separation of the $^{4}_{2}He$ nucleus—the binding energy of 28.3 million electron-volts must be supplied from an outside source. The potential energy of $^{4}_{2}He$ is less than the potential energy of two protons and two neutrons by this amount.

Binding energy is similar to the decrease in potential energy of an apple as it falls from a branch to the ground. Just before striking the ground, the apple has a kinetic energy equal to its decrease in potential energy. Similarly, if a helium nucleus were formed from two protons and two neutrons at rest, it would have a kinetic energy of motion equal to its binding energy, 28.3 million electron-volts. The release of this energy in the form of kinetic energy accompanies the decrease in potential energy; $^{4}_{2}He$ is more stable than its component particles.

Every nucleus has a binding energy, which can be calculated in the manner just shown for $^{4}_{2}He$. The binding energies of $^{56}_{26}Fe$ and $^{235}_{92}U$ are 492 and 1783 million electron-volts, respectively. The binding energy increases as the number of protons and neutrons in the nuclei increases. A better idea of the stability of a nucleus, however, is given by the binding energy per nucleon, or the total binding energy divided by the total number of nuclear particles (protons plus neutrons). This number stands for the average energy holding the nuclear particles together. Values for the nuclei discussed above are given in Table 18.5. Of the three nuclei in Table 18.5, $^{56}_{26}Fe$ has the greatest binding energy per nucleon and is the most stable.

Figure 18.8 gives a more complete view. It shows how binding energy per nucleon varies with mass number through a full range of isotopes. Beginning with the smallest mass number, the binding energy values increase sharply toward a maximum of 8.8 million electron-volts per nucleon at a

Table 18.5 Total Binding Energy and Binding Energy per Nucleon (nuclear particle)

ISOTOPE	TOTAL BINDING ENERGY (in Mev)	BINDING ENERGY PER NUCLEON (in Mev/nucleon)
$^{4}_{2}He$	28.3	7.07
$^{56}_{26}Fe$	492	8.79
$^{235}_{92}U$	1783	7.59

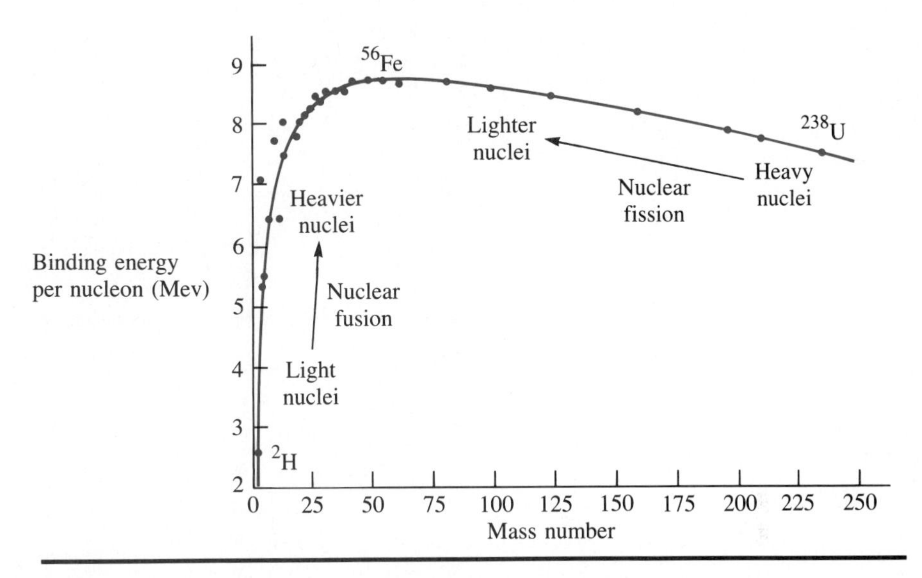

Figure 18.8. *Binding energy per nucleon as a function of mass number.*

mass number of about 56. Then they decrease slowly with increasing mass number.

All nuclear transformations, including natural radioactivity, fission, and fusion, occur with a release of energy as the original nuclei change into more stable nuclei. The most stable nuclei have intermediate mass numbers. The reason why both the fission of heavy nuclei and the fusion of light nuclei occur with the release of energy now becomes clear. A nucleus at the right end of the binding-energy curve (Figure 18.8) splits into nuclei that are closer to the central region of maximum stability. The splitting of a nucleus of mass 235, for example, into two nuclei of mass 100–150 results in increased binding energy and the output of a large amount of energy. Similarly, the fusion of isotopes of hydrogen or helium, at the left end of the binding-energy curve, produces nuclei that lie closer to the center and higher on the curve. Fusion causes an even greater production of energy than does fission.

18.6 NUCLEAR FISSION AND NUCLEAR REACTORS

The story of controlled nuclear fission began in 1942, when Enrico Fermi and his colleagues achieved a nuclear chain reaction in the athletic stadium at the University of Chicago (see Figure 18.9). Two-and-one-half years later, the atomic bomb had been developed and was used by the United States to bring World War II to an end. Partly as a reaction to this destructive

(a)

(b)

Figure 18.9. *(a) Enrico Fermi. (b) The first nuclear reactor. At the University of Chicago on December 2, 1942, a controlled nuclear chain reaction was achieved.*

use of atomic energy, a number of scientists began working toward the development of peaceful uses of atomic energy, principally nuclear reactors that would generate electricity for use by society. In the years since World War II the governments of the United States and European countries have given considerable financial support to this enterprise.

In April 1986 there were 101 nuclear reactors operating in the United States; they provided about 17% of the nation's electricity. Since 12% of the useful energy consumed in this country is in the form of electricity, nuclear reactors supplied about 2% of our total energy needs. Some people say that there will have to be many more nuclear power plants in the United States in the future; others advocate the elimination of nuclear power plants altogether. Although nuclear power is more a matter of physics and engineering than of chemistry, nuclear power is such an important and controversial technology that we have chosen to discuss it here.

To understand how a nuclear power plant works, we must first look at the fission process itself. A few of the very heaviest isotopes known, such as ^{233}U, ^{235}U, and ^{239}Pu, undergo fission. When they are bombarded by neutrons of moderate energy, they disintegrate in a manner very different from α or β emission. The fission produces two nuclei having mass numbers one-third to two-thirds the mass of the original nucleus, along with two or three neutrons. Of greatest importance is the large amount of energy released, 200 million electron-volts. This energy is produced because nuclei more stable than the original uranium or plutonium are formed, as shown in Figure 18.8.

One of the many ways in which ^{235}U can experience fission is

$$\underset{\text{uranium}}{{}^{235}_{92}U} + {}^{1}_{0}n \longrightarrow {}^{236}_{92}U \longrightarrow \underset{\text{krypton}}{{}^{90}_{36}Kr} + \underset{\text{barium}}{{}^{143}_{56}Ba} + 3\,{}^{1}_{0}n$$

Simultaneously, many other similar fission reactions take place, producing a wide variety of new isotopes, many of which are themselves radioactive and decay over a long time. Whereas the above fission yields three neutrons, others produce two. For ^{235}U the average number of neutrons produced per fission is 2.5. Note that a neutron must collide with a ^{235}U nucleus to trigger the process; spontaneous fission is rare.

Some of the neutrons formed in one fission collide with other ^{235}U nuclei to cause more fissions, which then produce more neutrons, and so on, in a **chain reaction.** Something like a chain reaction can occur in everyday life. On a crowded freeway, a sudden stop by one car can cause dozens of collisions among trailing cars in several lanes. A single event can produce a build-up of accidents.

Whether the chain reaction builds up explosively [see Figure 18.10(a)], continues at a constant rate [see Figure 18.10(b)], or dies out depends on the number of neutrons produced. In a nuclear reactor, control rods made of boron, an element that absorbs neutrons, are located between fuel rods

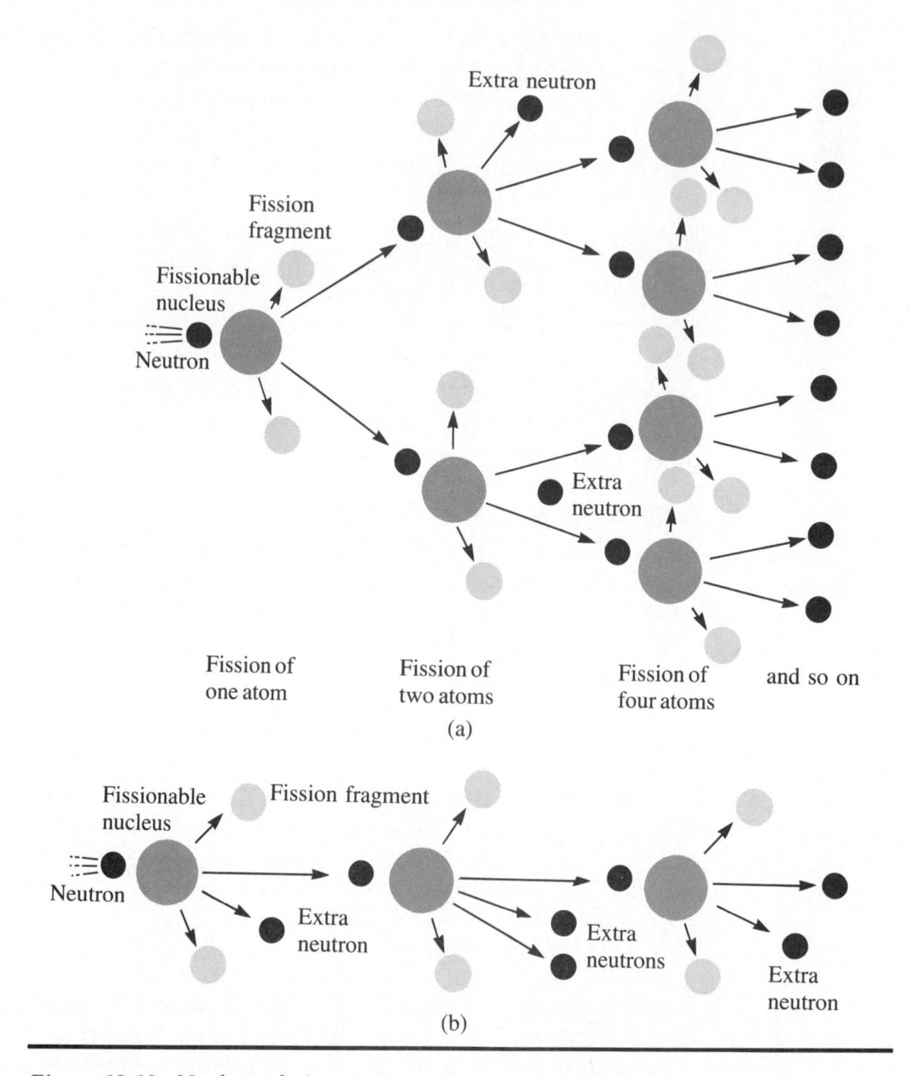

Figure 18.10. Nuclear chain reactions. (a) An uncontrolled chain reaction. Nearly all the neutrons produced in fissions cause additional fissions. (b) A controlled chain reaction. Many of the newly formed neutrons are absorbed by control rods.

containing ^{235}U. The length of the control rod inserted between fuel rods is controlled so that on the average only one of the neutrons produced in each fission causes a new fission. Fissions therefore continue at a steady rate.

The very large kinetic energies of the fission products generate a great deal of heat, which must be removed by a coolant. Otherwise the reactor would undergo a meltdown—the fuel rods and other parts would melt at the very high temperatures reached. As ^{235}U is used up over a period of months, the chance that a neutron will strike a ^{235}U nucleus before being lost decreases, and the control rods must be partially withdrawn in order to sustain the chain reaction. Eventually fission stops if new fuel is not provided.

18.7 NUCLEAR POWER

In an electrical power plant (see Figure 18.11) the function of the nuclear chain reaction is to provide heat to boil water. The resulting steam then turns a turbine exactly as described in Chapter 17 and Figure 17.10, where a power plant using fossil fuel was discussed. This steam must finally be condensed and waste heat released to the surroundings. One difference between nuclear and conventional power plants is that the temperature of a nuclear reactor must be kept lower than that of a coal or oil fired boiler in order to minimize deterioration of the reactor. A lower steam temperature means that less of the heat from the nuclear fuel becomes electrical energy. Therefore, the efficiency of a nuclear power plant is somewhat lower than that of a conventional plant, and more heat must be absorbed by the river or air near the plant.

Reactor design has become a highly developed technology because of the requirements for safety and for precise control of the amount of heat produced by the reactor. In the United States nearly all reactors have ordinary water circulating around the fuel rods. Thus, they are called *light-water reactors*. By contrast, another design uses *heavy water*, ^2H_2O, containing ^2H, or deuterium. Of the light water reactors, two types exist: the pressurized-water reactor and the boiling-water reactor.

Figure 18.11. *A modern nulcear power plant.*

A simplified drawing of a **pressurized-water reactor** appears in Figure 18.12. Water is pumped through the reactor pressure vessel, where it is heated to 300°C. Although at a temperature far above the normal boiling point, the water remains a liquid, because the high pressure raises the boiling temperature well above 300°C. The water then flows to the steam generator to provide heat for the boiling of other water, which is maintained at a lower pressure. The water in the reactor vessel also serves as a moderator, slowing the neutrons produced in the core to a speed at which they are more readily absorbed by ^{235}U nuclei.

Within the core of the reactor are assemblies of fuel rods containing uranium dioxide, UO_2. The uranium in this compound has been enriched in ^{235}U (from the 0.7% present in uranium ore to 3%) so that the chain reaction can be sustained. Control rods are moved into the core to decrease the concentration of neutrons or out of the core to increase the concentration. The neutron level, in turn, determines how fast the fissions occur and hence the amount of heat generated by the reactor. If for some reason the rate of nuclear fission should suddenly speed up, the reactor would immediately "scram." The control rods would automatically move completely into the core and stop the chain reaction. Otherwise, the reactor core could become hot enough to melt. Figure 18.13 gives some idea of the size of a reactor core.

Perhaps because of an association of nuclear power with nuclear weapons, some people believe that in an accident a nuclear power plant might explode

Figure 18.12. Pressurized-water reactor.

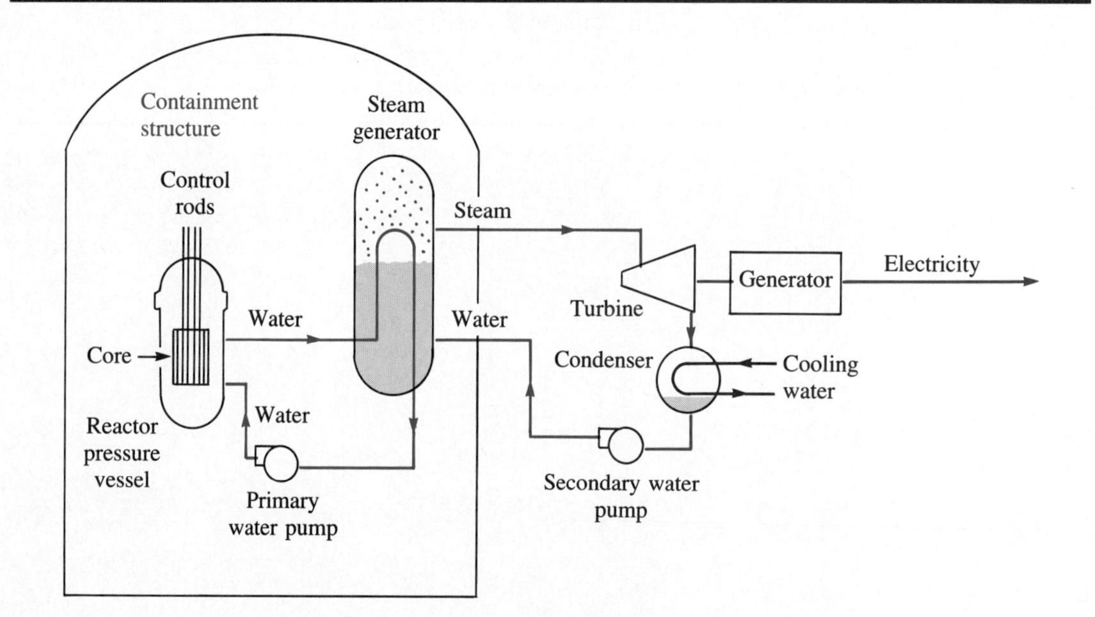

Figure 18.13. *The core of a nuclear reactor.*

with bomb-like force. The concentration of ^{235}U in the fuel, however, is much too low for this to happen. For a nuclear bomb to explode, a critical mass of uranium consisting almost entirely of the ^{235}U isotope must be assembled in a small volume during a very short time. Such a high concentration of ^{235}U could not exist within a reactor even if the fuel rods were to melt. Rupture of the containment vessel is conceivable, of course, but would result from the build-up of excessive heat and pressure and would not constitute a nuclear explosion.

The combination of many parts—such as control rods, reactor coolant, steam generator, and emergency cooling—makes for a complex system requiring sophisticated controls. Some idea of the complexity of a nuclear power plant is given by a view of a control room, shown in Figure 18.14.

18.8 THE NUCLEAR FUEL CYCLE

The **nuclear fuel cycle** is a long and demanding series of steps that occurs between the mining of uranium ore and the disposal of spent nuclear fuel. Uranium that is extracted from ores is first converted into a yellow compound, U_3O_8, called *yellow cake*. This compound contains only 0.7% ^{235}U. The other 99.3% is ^{238}U, which is not subject to fission. So, after U_3O_8 has been

Figure 18.14. The control room at a typical nuclear plant.

isolated and purified, the concentration of ^{235}U must be increased in order to get a fuel with enough ^{235}U nuclei to sustain a chain reaction. This is not a trivial task, because ^{235}U and ^{238}U are nearly identical in their chemical and physical properties.

Before long, more efficient methods may be used, but now most enrichment is done by the gaseous-diffusion method. First, U_3O_8 is converted into uranium hexafluoride, UF_6, which is a gas. The gas is then passed many times through a porous membrane (like a sausage casing). Because the lighter molecules of $^{235}UF_6$ diffuse through the membrane a bit more rapidly than $^{238}UF_6$, the concentration of ^{235}U in the uranium hexafluoride increases slightly after each pass. Many passes are necessary before the concentration of ^{235}U reaches 3%. This process of pressurizing the gas in the diffusion separators is expensive and consumes much energy. It has been estimated that it takes almost two years for the electrical output of a new nuclear power plant to equal all the energy consumed in building and fueling the plant. Most of that input energy is used for ^{235}U enrichment.

After enrichment, a chemical reaction converts the UF_6 into uranium dioxide, UO_2, which is stable at the high temperatures in the core of the reactor. Then the uranium dioxide is formed into pellets, which are placed inside hollow metal cylinders made of a zirconium alloy. These are the **fuel**

rods, the design of which has been given careful attention because they must withstand the rigors of high temperature and radiation damage. Premature failure of the fuel rods probably means a shutdown of the entire plant for repairs. Finally, the rods are collected into bundles, or "assemblies," before installation in the reactor.

The lifetime of a fuel rod is about three years. A typical schedule calls for replacing one-third of the fuel assemblies each year so that the power of the reactor does not drop too low. During the lifetime of a fuel rod, the ^{235}U concentration falls to about 1%, and radioactive products of fission accumulate. At the same time, some of the ^{238}U in the fuel is transformed into ^{239}Pu by the sequence shown in Figure 18.15. Plutonium-239 is also radioactive and decays slowly by α emission. Its half-life is 24,440 years. The fission products, on the other hand, have half-lives of less than 100 years. Both kinds of isotopes must be considered in the treatment of spent fuel.

After removal from the core, spent fuel assemblies are placed in large pools of water (see Figure 18.16) at the power plant to allow the intense, short-lived radioactivity to decay over a period of several months or more. At the present time some pools are nearly filled, because the next stage in the nuclear fuel cycle has not yet been implemented in the United States.

In the early days of the nuclear power industry, the plan was to ship the used fuel in heavily shielded containers to a **reprocessing plant.** Here, long-lived isotopes such as ^{239}Pu, unused ^{235}U, and ^{238}U were to be separated from short-lived radioactive species by various chemical and physical techniques. The difficulty of conducting these operations by remote control behind thick walls is not hard to imagine. The scene resembles some of those in science fiction. Because ^{239}Pu, much like ^{235}U, is capable of undergoing fission, it could be recycled to nuclear power plants to serve as new fuel. It is also a critical component of certain kinds of nuclear bombs.

A reprocessing plant at West Valley, New York, operated from 1966 until 1972, but since then no spent fuel from power plants has been treated. In 1977, President Carter banned all reprocessing activity, fearing that plutonium from a reprocessing plant might fall into the hands of terrorists set on making a bomb. President Reagan lifted this ban in 1981 but opposed the use of government funds for construction of reprocessing plants. A privately funded, partially completed plant at Barnwell, South Carolina, remains idle.

$$^{238}_{92}U + ^{1}_{0}n \longrightarrow ^{239}_{92}U$$

$$^{239}_{92}U \longrightarrow ^{239}_{93}Np + ^{0}_{-1}e$$

$$^{239}_{93}Np \longrightarrow ^{239}_{94}Pu + ^{0}_{-1}e$$

Figure 18.15. Equations for the formation of ^{239}Pu *from* ^{238}U *in a nuclear reactor.*

Figure 18.16. Underwater storage of spent fuel rods at a power plant.

The Nuclear Waste Policy Act, passed in 1982, sets up a timetable for developing a system of **nuclear waste disposal.** Whether the waste will be processed to remove plutonium-239 and the uranium isotopes is uncertain. With these two elements removed, it would take about 1000 years for the radioactivity in the waste to decay to a low level; without their removal, it would take at least 250,000 years for the waste to become safe.

In either case, the last step in the nuclear fuel cycle is long-term disposal, probably underground, of the highly radioactive material. According to the 1982 law, the selection of a site for the first repository will take a number of years. By 1998 the first shipment of spent nuclear fuel is to be sent to the chosen site. The chief problems that must be solved are preventing the spread of the radioactive isotopes from the original burial site, contamination of groundwater supplies, and accidental mining of these dangerous substances in the future by people seeking new underground resources.

Present research indicates that the waste should be encapsulated in a special glass or ceramic to retard attack by underground water. The burial site should be as free from underground water as possible. One possibility is a Nevada site composed of rock made of volcanic ash. Another is a salt bed, such as those in some of the southwestern states. The presence of salt, which is soluble in water, indicates that very little water has been in the vicinity for a long, long time. Safe, long-term storage of radioactive wastes is a problem still awaiting a solution.

Although estimates of future reserves of any natural resource are subject to considerable uncertainty, experts are fairly well agreed that uranium for conventional nuclear power plants will be in short supply by the end of the twentieth century. Many proponents of nuclear power argue that a natural outgrowth of the light-water reactor is the **breeder reactor,** which creates new fuel while it provides heat for generating steam and electricity. The most likely raw material for making new fuel is ^{238}U, which forms fissionable plutonium (^{239}Pu) under neutron bombardment, as shown in Figure 18.15. Recall that ^{238}U is not itself a fuel. It is a by-product of the enrichment process, and roughly 250,000 tons of leftover ^{238}U is stockpiled around the United States. So the breeder reactor would actually "breed" new nuclear fuel while providing electrical power.

In the most common design, the coolant in the breeder reactor is liquid sodium, which transfers heat from the reactor core to a water boiler. At the center of the core are fuel assemblies containing ^{235}U or ^{239}Pu; these isotopes undergo fission in a chain reaction. Surrounding the fuel is a blanket of "fertile" isotope (for example, ^{238}U). Excess neutrons from the central core strike the nuclei in the blanket, creating new fuel faster than the original fuel is consumed.

Many people are fearful of the "plutonium economy" that would develop if breeder reactors were employed on a large scale. Huge quantities of plutonium would need to be shipped and reprocessed. Since plutonium is a raw material for nuclear weapons, it is easy to imagine a group of terrorists stealing some. On the other hand, proponents of the breeder reactor argue that the least attractive way to make a bomb is by using material from a breeder reactor. The mixture of plutonium isotopes in this material, they say, is not favorable for the construction of a bomb. European countries have been more receptive to the breeder reactor than has the United States.

18.9 DOES THE UNITED STATES HAVE A NUCLEAR FUTURE?

Such a controversial question as the desirability of nuclear power can only be touched upon lightly in this short space, but it is too important an issue to ignore.

The future of nuclear power depends in part on the total amount of energy needed in the future. Predictions of future energy demand in the United States are highly uncertain and vary widely. Currently we consume somewhat over 70 quads of energy annually. A quad is about the amount of energy used by a city of 3 million inhabitants in a year.

Simple extension into the future of the average rate of growth over the past 35 years gives a figure of over 200 quads for U.S. energy consumption in the year 2000. The amount of energy presently being consumed by the entire world is not much more than 200 quads. Many people agree that we simply can't go on using more and more energy. After allowance is made for a reduced rate of population and economic growth, the projection comes to about 170 quads in the United States for the year 2000. That projection, however, is based on the assumption that the fraction of our national annual expenditures devoted to energy will be the same in the future as it has been in the past. This assumption is being questioned increasingly as writers argue that quality of life need not be tied to high energy consumption. Recently a committee of the National Research Council released several energy scenarios, one of which suggested a possible consumption of 58 quads in the year 2010. This would be a significant decrease from present usage and would involve some changes in the way we live.

Predictions of the demand for electrical energy are just as uncertain as those of future total energy consumption. Between 1960 and 1972 the consumption of electricity in the United States increased an average of 7.1% per year. During the decade from 1973 to 1982, however, the average rate of increase fell to 2.6% per year. If the rate of increase continues to drop, there will be little reason to build new power plants of any kind in the near future, because the plants currently under construction will serve our needs. On the other hand, if the demand for electricity increases by 4% per year, which is quite possible, there will be pressure to construct more nuclear power plants. Therefore the question concerning nuclear power may not be "whether" but rather "how much." Should it be de-emphasized or expanded?

Over the next several decades the alternative to nuclear power will be increased burning of coal, because the supplies of oil and natural gas are clearly inadequate. The advantages and disadvantages of these two options need to be compared instead of discussed in isolation from one another.

One issue is the environmental and health effects associated with power plant operation. According to most studies, nuclear plants cause fewer harmful effects during normal operation than do coal-fired plants. Radioactive emissions from nuclear plants are extremely low, and the safety records for workers within the plants have been good. Accidents associated with the mining of uranium ore have been relatively rare. By contrast, illnesses and premature deaths caused by emissions from the stacks of coal-burning plants cannot be ignored. Another consequence is acidic lake water and stunted tree growth from acid rain (Chapter 12). Although risks to workers in coal-burning plants are low, the hazards to eastern coalminers from underground explosions and cave-ins and from black lung disease have been well publicized. These hazards are minimal in the strip mining of western surface coal, but adverse environmental effects accompany this form of mining. Even after restoration, the land lacks its original contours and vegetation, and the soil can be quite acidic.

Another environmental effect of burning coal is a possible change in world climate from the increased concentration of carbon dioxide in the atmosphere. The greenhouse effect and its implications were discussed in Chapter 17. Nuclear fission, of course, presents no problem in this regard.

On the other hand, there is concern about the possibility of an accident at a nuclear power plant. This concern escalated in 1979, when a combination of poor maintenance, faulty equipment, and operator errors led to overheating and severe damage to the core of a reactor at the Three Mile Island plant in Pennsylvania (see Figure 18.17). No significant amount of radiation escaped, but people living nearby were alarmed and some families left the area for several days. The reactor was beyond repair, at least in economic terms, and the cleanup process will require years and will cost over a billion dollars.

Early in the morning of April 26, 1986, the worst accident in the history of nuclear power occurred at a 1,000-megawatt plant near Chernobyl, in

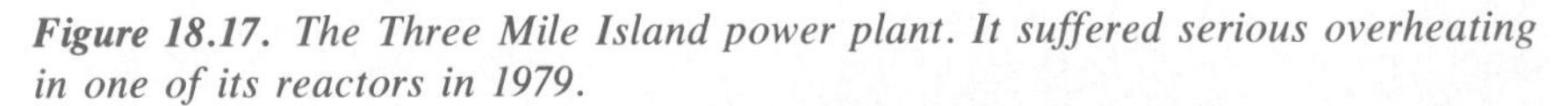

Figure 18.17. *The Three Mile Island power plant. It suffered serious overheating in one of its reactors in 1979.*

Figure 18.18. An aerial view of the damaged reactor at the Chernobyl nuclear power station. Damage to the reactor can be seen below the chimney.

the western part of the U.S.S.R. (see Figure 18.18). During the 24 hours preceding the accident, technicians prepared for an experiment by disabling most of the safety systems of the reactor, leaving it in an extremely unstable condition. After a final, and fateful, adjustment, the temperature shot upward, and two violent explosions occurred. The first, caused by a high pressure of steam, ruptured the reactor. The second, probably an explosion of hydrogen gas formed from a reaction between steam and the zirconium metal that covered the fuel rods, blew off the roof of the reactor building. Several workers were killed immediately, and perhaps eight tons of highly radioactive products of fission were released into the air and eventually spread to neighboring European countries.

The danger was compounded by a fire in the 1,700 tons of graphite (carbon) used in many Soviet reactors as a moderator to slow the neutrons. At least part of the reactor core melted, and the emission of radioactive material continued for several weeks. Many thousands of people were evacuated from the area within a 19-mile radius around the plant. Although some inhabitants of Kiev, a city of 2.5 million 80 miles south of the plant, left for safer areas, most remained and appeared not to be in serious danger. Four months after the accident, 31 people who had been at the site had died, and more than 200 had contracted radiation sickness. Medical experts estimate that 5,000 people may eventually die from cancer caused by exposure to radiation released from the Chernobyl nuclear reactor.

In an effort to smother the graphite fire and diminish the escape of radioactivity, Soviet helicopters dropped 5,000 tons of sand, clay, lead, and boron on the damaged plant. Later, a tunnel under the site was filled with concrete, and the remains of the reactor above ground were also covered with concrete in order to seal the still highly radioactive source from the environment. For many years people in the region will be concerned with the presence of radioactivity in the ground water and in agricultural products.

Because of the accidents at Three Mile Island and Chernobyl, public confidence in nuclear power has declined. During the several years before the Three Mile Island accident, polls showed that over half of the American public favored continued construction of nuclear plants; only 28% opposed additional construction. Currently, about 50% of those polled oppose an increase in the number of nuclear power plants. Some people everywhere will oppose the construction of nuclear power plants near their homes. Without increased public support, the future of plans for expansion of nuclear power in this country is in doubt.

The nuclear power industry has also experienced other difficulties. Cost overruns in construction, construction times of over ten years, and malfunctioning equipment have made the economics of nuclear power less favorable than had been expected in the early days of nuclear power. The Nuclear Regulatory Commission has periodically issued new regulations, which have caused costly changes in modes of operation for existing reactors and in plans for new reactors. Because of the new technology involved, managing a nuclear plant has been more demanding than managing a conventional plant, and not all electric power companies have met the challenge. Lack of a standard design for reactors means that many of them are experimental and unproven even though the industry is two decades old.

Another issue related to nuclear power is the danger of proliferation of nuclear weapons. From our vantage point at least, it seems desirable to limit the number of countries possessing nuclear weapons to the present six (plus the several others that may have developed them secretly). It is likely that the greater the number of nuclear nations, the greater the probability of nuclear war. The vulnerability of weapons stockpiles to theft by terrorist groups also increases. The link between nuclear power and nuclear weapons is in the recovery of plutonium by reprocessing from spent fuel or from the blanket of uranium surrounding the core of a breeder reactor. As already suggested, such procedures would necessitate transportation of large amounts of plutonium, from which a bomb might be made. If plutonium and the facilities for recovering it become widespread, the temptation for other countries to undertake production of nuclear weapons will certainly increase.

It has been argued by some that proliferation of nuclear weapons will occur in the future whether or not the United States begins recycling plutonium. Other countries will supply plutonium, the argument goes, even if we do not. An opposing point of view, however, is that the influence of the United States should not be underestimated. It is our role to convince other countries suffering from a shortage of fossil fuel that plutonium recycling and the breeder reactor are not economic necessities. Light-water reactors can operate for a number of decades if the United States and other countries continue to supply slightly enriched uranium to the world market. This argument will be convincing only if the United States defers plutonium recycling and the development of breeder reactors. The position of the U.S. government on this issue has changed several times.

The level of uncertainty associated with predicting future risks and environmental impact is so great that a clearcut case for the superiority of either coal-fired or nuclear power plants cannot be made. It appears that both will be developed until the evidence is more conclusive. A task facing scientists and governments is to provide information to the public as objectively as possible so that emotional responses do not dominate public opinion.

IMPORTANT TERMS

strong force or **nuclear force** A force that holds together the protons and neutrons in an atomic nucleus. It is 100 times stronger than the force between charged particles outside of the nucleus.

mass number The sum of the number of protons and the number of neutrons in a nucleus.

half-life The time required for the number of radioactive nuclei in a sample to decay to one-half the original number.

equivalence between mass and energy The Einstein relationship, $E = mc^2$, which gives the energy released, E, when the decrease in mass accompanying a nuclear reaction is m. The symbol c stands for the velocity of light.

nuclear fission The splitting of a heavy nucleus into two or more lighter nuclei. It is accompanied by the release of energy.

nuclear fusion The combining of two light nuclei into one nucleus. It is accompanied by the release of a large amount of energy.

binding energy The energy equivalent to the difference between the mass of a nucleus and the sum of the masses of the separate protons and neutrons that make up that nucleus. The binding energy would have to be added to a nucleus to separate it into protons and neutrons.

chain reaction A sequence of reactions in which a product from one reaction is a reactant in the next reaction. The sequence repeats many times.

pressurized-water reactor A nuclear reactor in which water under high pressure is heated by nuclear fission to a temperature well above its usual boiling point.

nuclear fuel cycle The complete set of events from the mining of an ore containing a fissionable isotope to the disposal of spent nuclear fuel.

fuel rod A cylinder of fissionable material inside a special metal container.

reprocessing plant A plant designed to separate used nuclear fuel into radioactive isotopes with short half-lives and those with long half-lives.

nuclear waste disposal The procedure for storing radioactive waste, probably underground, so that it will not harm future generations.

breeder reactor A type of nuclear reactor that produces new fuel while it provides heat for generating electricity.

QUESTIONS

1. a. When an unstable nucleus emits an α particle, what is the relationship between the mass number of the original nucleus and the mass number of the new nucleus? What is the relationship between the atomic numbers?

b. In β emission what is the relationship between the mass numbers of the parent and daughter isotopes? What is the relationship between their atomic numbers?

2. For each of the following, determine the identity of the missing nucleus or particle, and write the complete nuclear reaction.

a. $^{239}_{94}\text{Pu} \longrightarrow\ ? + {}^{4}_{2}\text{He}$
b. $^{210}_{83}\text{Bi} \longrightarrow\ ? + {}^{0}_{-1}\text{e}$
c. $^{14}_{6}\text{C} \longrightarrow {}^{14}_{7}\text{N} + ?$

3. Complete the following equations for nuclear reactions.

a. $^{16}_{8}\text{O} + ? \longrightarrow {}^{17}_{8}\text{O} + {}^{1}_{1}\text{H}$
b. $^{226}_{88}\text{Ra} \longrightarrow\ ? + {}^{4}_{2}\text{He}$
c. $^{227}_{89}\text{Ac} \longrightarrow\ ? + {}^{0}_{-1}\text{e}$
d. $^{235}_{92}\text{U} + {}^{1}_{0}\text{n} \longrightarrow {}^{139}_{54}\text{Xe} + ? + 2\,{}^{1}_{0}\text{n}$

4. By means of a sequence of two radioactive decays, the unstable nucleus $^{218}_{84}\text{Po}$ becomes $^{214}_{83}\text{Bi}$. Are two alphas, one alpha and one beta, or two beta particles emitted? With the aid of the periodic table, write the equations for the two nuclear reactions that would account for the information given.

5. a. In the involved process that turns $^{238}_{92}\text{U}$ into $^{206}_{82}\text{Pb}$, the first nuclear reaction is the loss of an α particle. Write a balanced nuclear reaction for this process, and include the symbol of the new element produced.

$$^{238}_{92}\text{U} \longrightarrow\ ? + \alpha \text{ particle}$$

b. In the conversion of $^{238}_{92}\text{U}$ to $^{206}_{82}\text{Pb}$, how many α particles and how many β particles are given off?

6. $^{131}_{53}\text{I}$ is radioactive with a half-life of 8 days. This isotope is a product of the fission of $^{235}_{92}\text{U}$ and could be introduced into the environment in an accidental release from a nuclear power plant. How long would it take for three-quarters of a given amount of $^{131}_{53}\text{I}$ to disappear?

7. a. Potassium-40 is a naturally occurring radioactive isotope with a half-life of 1.4 billion years. If the amount of potassium-40 now present on the earth is one-eighth of the amount originally present and potassium-40 has not been produced since the earth was formed, what is the estimated age of the earth?

b. Potassium-40 emits a β particle to become the stable isotope calcium-40. If you had a sample of rock containing potassium-40, how might you determine the amount of potassium-40 present in that rock when it was formed? What important assumption is necessary for this determination?

8. a. Explain why the amount of $^{14}_{6}\text{C}$ in a living tree remains constant.

b. Explain why the amount of $^{14}_{6}\text{C}$ in a piece of lumber used in construction decreases steadily.

9. Could the fusion of $^{56}_{26}\text{Fe}$ with $^{4}_{2}\text{He}$ result in the release of energy? Explain.

10. The masses of a proton and a neutron are 1.007825 and 1.008665 atomic mass units, respectively. Calculate the binding energy in $^{11}_{5}\text{B}$, which has a mass of 11.00931 atomic mass units.

11. Calculate the binding energy per nucleon in $^{9}_{4}\text{Be}$, which has a mass of 9.01218 atomic mass units. The masses of the proton and neutron are 1.007825 and 1.008665 atomic mass units, respectively.

12. Use Figure 18.8 to explain why nuclear fusion produces much more energy per event than does nuclear fission.

13. One gram of ^{235}U is the energy equivalent of 2.7 metric tons of coal or 13.7 barrels of crude oil. Where does the heat energy come from when uranium is used as a fuel? Be explicit.

14. What factors determine the number of free neutrons in the core of a nuclear reactor? Explain how each affects the number of neutrons.

15. What is the difference between the chain reaction that occurs in a nuclear power plant and the chain reaction that occurs in a nuclear weapon?

16. Why is the presence of $^{239}_{94}Pu$ in radioactive waste much more of a problem than the presence of $^{131}_{53}I$?

Public-Policy Discussion Questions

17. Several European countries plan to develop breeder reactors for the commercial production of electricity; the United States does not. In your view, should the United States change its policy and introduce the breeder reactor? Explain your position.

18. Do you favor a moratorium on all new construction of nuclear power plants until we have solved the problem of disposal of spent fuel rods? Should we go so far as to shut down currently operating nuclear power plants until this problem is solved? Explain your position.

19. On balance, do coal-burning or nuclear power plants pose the greater health hazard to society? Find other sources of information that go beyond the brief discussion in this chapter.

19 Chemistry, Resources, and the Environment

Chemistry is one of the natural sciences, but it has close connections to technology. Since chemistry spans the two areas, this book has dealt not only with scientific principles but also with some of the many practical applications of chemistry in society. Science, which comes from the Latin word for knowledge or learning, is the search for understanding of ourselves and our world. It may also be thought of as the knowledge of natural phenomena. In other words, science is a process—the investigation of nature—as well as the knowledge that results from the search. Technology, on the other hand, is concerned with the application of knowledge to practical uses. Technology involves the creation of designs, machines, products, and techniques. It has been said that the goal of science is to understand the natural world, whereas technology aims to change the world.

The relationships between science and technology are complex and not fully understood. The production and use of antibiotics has given physicians a powerful technology for combatting human disease. It is clear that the development of these antibiotics depended on science (the knowledge of chemistry and biology). There are many such examples of technological advances that have come from the application of scientific learning. Yet the process of scientific investigation depends on technology as well. Without telescopes, microscopes, and the many other products of technology used by scientists, we would know far less about the world around us.

In the United States it is probable that more chemists are pursuing technological than scientific objectives. Chemical industries produce many billions of dollars' worth of products each year. Thousands of chemists work in chemical and chemical-related industries. They have economic as well as humanitarian objectives, practical as well as intellectual goals. This is nothing new. For centuries, people have used chemistry to meet human needs. The production of iron, glass, concrete, and plastics involves chemistry. The production of adequate food for the world's population depends on the use of chemical fertilizers. Our fuels are chemical resources formed long ago.

Human technology has provided tremendous benefits to us all, but during the last 200 years advanced societies have depended more and more on resources that are not being replaced by natural processes. Consumption of resources does not mean that key elements disappear from the earth. All scientists believe in the Law of Conservation of Mass, which means that every atom in an original resource is still around somewhere after that resource has been converted by chemical reactions into a consumer product or has been used to provide energy. The carbon atoms in petroleum end up in carbon dioxide molecules after the petroleum has been distilled into gasoline and then burned in an automobile engine. However, to change the carbon dioxide back into gasoline requires much more energy than was gained from the gasoline in the first place, so it is not at all practical. **Recycling** of any scarce material is always possible in principle, but the amount of energy required to recover essential resources may be substantial.

There is no known way to recycle energy. Once useful energy has been degraded into low-temperature heat, there is no economical way to change it back into electrical energy or high-temperature heat. For this reason, future sources of energy are a major concern. The United States is particularly vulnerable on this issue, because its per-capita consumption of energy is about twice that of other developed countries. The quantity of energy used every year in the United States is equivalent to the amount that would be generated by burning 12 tons of coal for each resident. In countries such as the Netherlands, West Germany, and Norway, the figure is 6 tons per person, and in Japan the per-capita consumption of energy is even less. Yet the standard of living in all of these countries is not so very different.

A serious concern connected with our abundance of material goods is the gradual fouling of the air we breathe and of the water we drink. Many people are now aware of the connection between automobile exhaust and smog, between the burning of coal and acid rain, and between trace amounts of industrial chemicals and hazardous water supplies. During the last several years the environmental danger that has received the most attention is the great quantity of hazardous chemicals stored at over 14,000 dump sites around the United States. Many of the stored chemicals are by-products of industrial processes for which few practical recycling procedures exist. Years ago, when there were few environmental regulations, companies often kept their costs competitive by disposing of hazardous chemicals in unregulated landfills. The public has not been entirely innocent in this matter, because it too has taken a short-term view, demanding manufactured goods at low prices.

19.1 RESOURCES AND MATERIALS

Our material resources are closely linked to the availability of fossil fuels. In the first place, energy is required to refine metals from their ores, as in the reduction of iron ore in a blast furnace and especially in the electrolysis of aluminum ores. Other industrial processes also require a high temperature, which means the input of much heat. Fossil fuels are the major source of this energy. Second, most of the organic chemicals for society, including synthetic polymers, are derived from petroleum. Propylene, for instance, is a raw material in the production of solvents and polymers, such as polypropylene, acrylic fiber, and polyurethane. The only practical source of propylene at present is petroleum.

Some people argue that in the future our petroleum should be used entirely for **petrochemicals,** because our economy is heavily dependent on them. According to this point of view, petroleum should not be "wasted" in providing energy, because alternative sources of energy can be developed.

Along with propylene, another petrochemical of prime importance is ethylene, the first member of the olefin, or alkene, family. It ranks about fifth among industrial chemicals with respect to annual production in the United States. It is part of the gas that sometimes issues from oil wells. A more important source of ethylene is the breaking apart of high-molecular-weight hydrocarbons into smaller molecules, done by "catalytic cracking" at a petroleum refinery. This process results in more gasoline from a given amount of crude oil. A by-product is a gas containing ethylene.

Ethylene is the parent compound for many useful products, including low-molecular-weight substances as well as polymers (some of these were discussed in Chapters 11 and 14). Often a sequence of chemical reactions yields the final product. Figure 19.1 summarizes the varied uses of ethylene.

Petroleum is very important, but it is not the only material resource needed by the societies of the world. Many inorganic minerals are essential sources of metals and nonmetals. For these we are limited to the crust of the earth, which extends several miles below the surface. The deepest mines go down about 6 miles.

As we saw in Table 3.1, only a few of the essential elements, such as iron and aluminum, are abundant. Some other metals, such as copper and

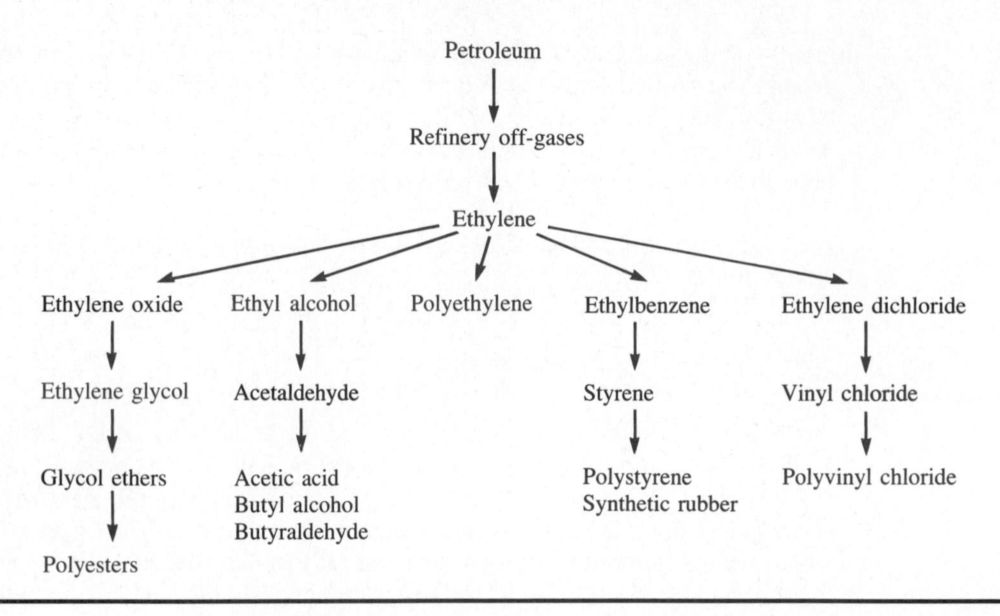

Figure 19.1. Final products and intermediates that can be derived from ethylene.

lead, appear to be in ample supply, even though their average abundance in the earth's crust is only 0.001–0.01% by weight. Many other industrially important elements, however, are present in average concentrations of only a few parts per million. Fortunately, in different regions of the world there are huge variations from the averages.

During and after the formation of the earth, different solid compounds concentrated at different locations, giving the surface of the earth tremendous variety. When a considerable amount of a compound of a desirable element is found at a certain place, it can be mined economically. The common ores of iron, for example, are usually located in what geologists call *sedimentary deposits,* which precipitated from oceans several billion years ago. Before that time the iron dissolved in seawater was in the form of Fe^{2+} ions. As the oxygen content of the earth's atmosphere increased, this oxygen removed electrons from the Fe^{2+} ions and oxidized them to Fe^{3+} ions. Then, because compounds of Fe^{3+} are much less soluble in seawater than are compounds of Fe^{2+}, the Fe^{3+} ions deposited on the ocean floor. Some of these deposits are now on land, and through subsequent reactions they have been converted into the common iron ores.

In the United States we are fortunate to have ample reserves of many key minerals, containing elements such as sulfur (see Figure 19.2), chlorine, magnesium, vanadium, copper, and phosphorus. A significant segment of our chemical industry is involved in treating these minerals to obtain the elements in useful forms. Because metallic elements are present as positive ions in their ores, electrons must be added to reduce them to the neutral

Figure 19.2. The mining of sulfur. Deposits of uncombined sulfur occur in the southern United States.

atoms of the metals themselves. Chemical agents with a strong tendency to donate electrons are adequate to reduce iron ores, but aluminum ions are reduced with such difficulty that electrical energy must be fed to an electrolysis cell, as described in Chapter 6.

Despite the bountiful supply of natural resources with which the United States has been blessed, there are about 16 elements in short supply here. They are contained in minerals that are abundant in Canada or on other continents. In normal times importing these elements is not difficult, since the United States has agricultural produce and manufactured products to export in payment. In times of international tension, however, some critical materials may be cut off, causing serious dislocations in our economy. The disruption caused by the oil embargo of 1973 is the most recent illustration of how reliance on foreign sources for critical materials can backfire.

We are almost totally dependent on foreign sources for three critical metals—chromium, manganese, and nickel. Each of these metals is an essential ingredient in a particular kind of steel. A small amount of chromium, for example, is alloyed with iron to make stainless steel and other steels that can withstand high temperature. These steels are used in the chemical and petroleum industries for high-temperature equipment and in the aerospace industry. Although this country has small reserves of low-grade chromium ore, we currently import all our chromium from the Union of South Africa and Zimbabwe, because it is cheaper. To protect against disruption of foreign supplies of critical materials, the federal government has been building small stockpiles of them.

19.2 NONRENEWABLE RESOURCES AND THE FUTURE

How likely is it that severe worldwide shortages of resources will develop in the future? At issue here are **nonrenewable resources,** because renewable resources such as wood and other plant products are partially subject to our control. During the past several decades, teams of researchers have attempted to project current trends into the future. All have used computer models to varying extents and have dealt with the broad questions of food supplies, population, energy, and resources. Their conclusions have run the gamut from pessimistic to optimistic.

An early study, which received much attention when it was published in 1972, was *Limits to Growth.* This study relied heavily on a computer model which incorporated a large amount of data on world resources and population. The model was not intended to predict the ultimate fate of world civilizations but instead to show what problems could occur if current trends in population growth, energy consumption, and resource utilization continued.

The first major conclusion of *Limits to Growth* was that the carrying capacity of the earth is finite. In other words, there is a limit to the number of people, roads, buildings, and machines that can be supported. Sophisticated technology can alter but cannot eliminate this limit. The second conclusion was that both the number of people and the number of physical objects are increasing exponentially. That is, each number doubles after a certain amount of time, then doubles again after an equal amount of time, and so on. Finally, it was pointed out that unless people recognize the existence of limits and adjust their behavior accordingly, there will be a collision between growth, with its increased consumption of resources and output of pollution, and the carrying capacity of the earth (see Figure 19.3).

Limits to Growth engendered much controversy. One camp accepted its major conclusions and stressed personal fulfillment in the future through cultural pursuits rather than increased material well-being. The other side claimed that the study was flawed, since it didn't recognize either the great adaptability of societies or the possibility of future technology that would greatly alter our ways of living. In this view, a society is capable of rapid changes in its practices when it becomes aware of serious problems. Recall the great increase in energy conservation in the United States after the oil embargo of 1973. Furthermore, substances in the crust of the earth are regarded as resources only when we have economical means for converting

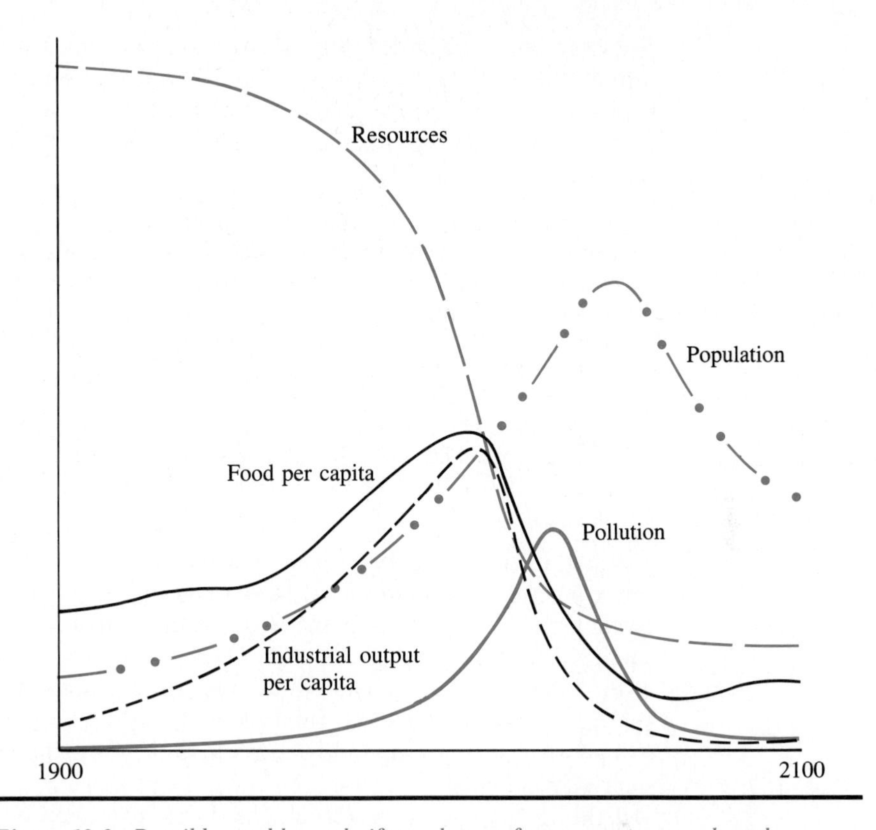

Figure 19.3. Possible world trends if no change from current growth and consumption rates occurs. Sometime in the twenty-first century, population could peak and then decline because of inadequate resources and a high level of pollution.

them into useful forms. Aluminum has been plentiful in the crust for billions of years, but only during the last century have we developed methods for cheaply converting the ore into the metal.

Since *Limits to Growth* appeared, other projections of the future state of the world have been undertaken. In 1980 the "Global 2000 Report to the President" was completed. It was pessimistic in tone, predicting that

> If present trends continue, the world in 2000 will be more crowded, more polluted, less stable ecologically, and more vulnerable to disruption than the world we live in now. Serious stresses involving population, resources, and environment are clearly visible ahead. Despite greater material output, the world's people will be poorer in many ways than they are today.

Questioning the conclusions of this report, the Heritage Foundation (a public-policy research institute in Washington, D.C.) sponsored another study, which appeared in 1983 under the title "Global 2000 Revised."

Although they conceded that some serious problems will appear in the future, the authors found a number of hopeful trends. For example, life expectancy has been increasing in nearly all parts of the world, indicating demographic and economic improvements. ''Global 2000 Revised'' concluded that

> Global problems due to physical conditions (as distinguished from those caused by institutional and political conditions) are always possible, but are likely to be less pressing in the future than in the past. Environmental, resource, and population stresses are diminishing, and with the passage of time will have less influence than now upon the quality of human life on our planet. . . . Because of increases in knowledge, the Earth's ''carrying capacity'' has been increasing throughout the decades and centuries and millennia to such an extent that the term ''carrying capacity'' has by now no useful meaning. These trends strongly suggest a progressive improvement and enrichment of the Earth's natural resource base, and of mankind's lot on Earth.

What is the ordinary citizen to make of these conflicting predictions about our future resource base? Does a world of plenty or a world of want await us and our children, or, as is so often the case, does the truth lie somewhere in between?

Pessimists and optimists agree that some of our minerals and energy sources are being depleted. Improvements in methods for processing ores will allow us to draw upon lower-grade ores, but problems will accompany such a development, both with respect to the processing of the ores themselves and the impact on the environment. First, treating huge volumes of low-grade ore will require a large amount of energy and a larger-scale mining operation. Increased cost is inevitable. Then the exhausted ore must be transported and disposed of. At best this will create unsightly piles of material, and at worst it may add harmful and polluting substances to the environment. Also, since the metallurgical industries, like many other industries, are no longer allowed to emit gaseous pollutants (such as sulfur dioxide) from their smokestacks or to put liquid wastes into nearby rivers, the companies must now bear the costs of removing these pollutants from their waste streams. Processing a large volume of ore means extra costs to control these effluents. The question about which there is so much disagreement is whether or not new and as yet undiscovered technologies and construction materials will allow a smooth transition to a better life.

19.3 RECYCLING OF SCARCE MATERIALS

If in the future our metallurgical industry will be processing ores that contain only very small percentages of the desired element, then shouldn't we pay more attention to recycling discarded products that have those scarce elements? Certainly the iron content of a junked car and the aluminum content of a beverage can are greater than the concentrations of these metals in low-grade ores. On a limited scale, these two metals have been involved in recycling for many years. Approximately 30% of iron production comes from scrap. The recycling of aluminum runs at about 5% of production. These figures have fluctuated in both directions over the years, and no upward trend is evident.

Glass and paper are also commonly recycled. The recycling of glass containers depends primarily on the willingness of consumers to save bottles, sort them by color, and periodically take them to a pick-up point for a glass factory. Old glass is then crushed and mixed with virgin glass to form new bottles. Recycled glass that is off-color or contains impurities may be used as an extender in cement or blacktop, for glass-wool insulation, or in bricks and tiles, since color and purity are not factors in these applications. Recycling of paper is most often done when the quality of the product need not be high. During the repulping of waste paper, the fibers are disrupted, resulting in a weaker paper. Coatings, glue, dyes, pigments, and inks from the original paper further weaken the recycled product and give it an off-white color. For some uses, however, these properties are not a disadvantage. Unfortunately, the market for waste paper fluctuates, and sometimes there is little demand for it. Therefore, it is difficult for consumers to develop a regular habit of saving old paper.

The kinds of recycling we have been discussing all require consumers to act voluntarily to save used materials. Sometimes there is a small payment for old bottles or cans, but it is not much of an incentive. An alternative is to process municipal waste so as to isolate all substances that have recycling value. This practice is rarely followed in the United States, since the methods for separating the useful components are expensive. Today the universal practice is to bury municipal waste in landfills. These landfills may be regarded as "mines" containing many valuable elements, which are simply part of discarded products rather than ores. Some day it may be profitable to work these mines.

One of the factors that determines whether or not recycling a metal is economical is the extent of the difference between the energy required for recycling and the energy used in the original production. Generally, much less energy is consumed in recycling. The contrast is especially great for aluminum because of the tremendous expenditure of energy in its production. Figure 19.4 shows ratios of energy used in original production to energy used in recycling for iron, copper, and aluminum. These ratios range from almost 2:1 to greater than 10:1 in favor of recycling.

The energy used in the recycling process is not the only cost, however; other demands on energy and materials also exist. Trucks must be used to pick up aluminum cans and scrap steel. Construction of these trucks requires more steel, aluminum, rubber, and glass, along with energy to process these materials. New activities, also requiring energy and materials, are put into motion by the recycling industry. A certain amount of pollution accompanies

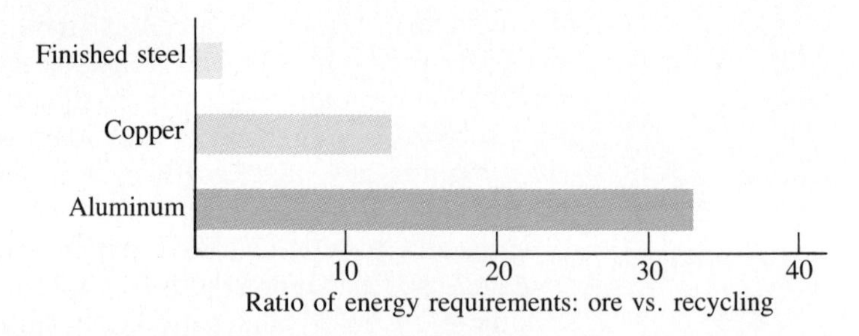

Figure 19.4. Ratio of energy consumed in forming a metal from its ore to energy consumed in recycling the same amount of metal.

these new activities, too. Therefore, a more complete ledger sheet must be drawn up in order to compare recycling with original production.

People conscious of the need to preserve the environment may look upon recycling of material resources as a moral obligation. Economists, on the other hand, regard the mechanism of the marketplace as quite adequate to determine the appropriate mix of recycling and production from raw materials. As we turn to lower-grade ores for new supplies of a particular metal, the price of the new metal will increase and may exceed the price of the recycled metal. The inevitable result is more reliance on the latter. If the price of the metal from both sources increases greatly, then the incentive to develop and produce substitutes will become strong. As long as the price of the metal doesn't increase so rapidly that there is insufficient time to react to it, entrepreneurs will analyze the trends and develop substitute materials in time to prevent a shortage.

19.4 AIR POLLUTION

More and more people have come to demand that the air and water not be fouled by U.S. agricultural and manufacturing industries. A ''safe'' level for any pollutant is a subject of continuing debate, but the principle that the cost of pollution control must be included in the price of a product is now generally accepted. As the methods of chemical analysis become more powerful and our knowledge of medicine increases, the list of potentially harmful substances increases. It is not uncommon to measure the levels of potentially toxic substances in the parts-per-billion (ppb) range or lower these days. Just a few years ago most substances could not be measured below the parts-per-million range because the technology to measure smaller concentrations had not been developed. So in the last few years chemists have extended the range of their analytical methods more than a thousandfold. As our knowledge and awareness increase, the problem of pollution may seem more severe in the short run. It may be, however, that the problem is actually less severe, and only our knowledge and awareness have increased.

Air pollution has affected societies at least since the beginnings of the industrial revolution about 200 years ago. Although the technology to control emissions has greatly improved in recent decades, the quality of the air in some places is still far from acceptable. At the end of Chapter 12 we discussed two classes of pollutants: sulfur dioxide, SO_2, and two oxides of nitrogen, NO and NO_2, all of which play a role in the formation of acid rain. The bulk of the SO_2 comes from coal-burning power plants, whereas the most significant source of oxides of nitrogen is automobile exhaust.

Automobile exhaust is responsible for another serious irritant, **photochemical smog,** which remains a problem in large cities such as Los Angeles and Denver despite years of corrective effort. Smog consists of a haze (see Figure 19.5) and harmful chemicals that irritate the eyes, throat, and lungs. The adjective *photochemical* is applied to this kind of smog because some of the crucial reactions in the formation of the smog take place only under the action of sunlight. Besides NO and NO_2, other poisonous components of smog are carbon monoxide (CO), ozone (O_3), and organic compounds such as aldehydes and peroxy compounds (containing the —O—O— linkage). Some of these substances are more toxic than hydrogen cyanide (HCN), which is known to be highly dangerous.

Extended research on photochemical smog has shown that many chemical reactions are involved in the formation of the toxic compounds; only a small part of the picture can be given here. One component, CO, undoubtedly owes its existence to the incomplete combustion of the hydrocarbons of gasoline in the automobile engine (see Chapter 11). A second component, NO, is formed from N_2 and O_2 in the air used to bring about the combustion in the engine:

$$N_2 + O_2 \longrightarrow 2\ NO$$

Figure 19.5. *These photographs, taken almost 30 years ago, show downtown Los Angeles (a) on a clear day and (b) with heavy smog.*

(a)

(b)

At 25°C, NO is almost nonexistent but at the high temperature of the automobile engine and exhaust pipe, its concentration reaches a few percent, and the reaction rate is fast.

The remaining pollutants in smog are formed outside the automobile, and here the story becomes more complicated. Several conditions must exist in order for photochemical smog to form. First, the air must be very still, so that the key intermediate compounds in the chemical reactions are not blown away before they have a chance to react together. This stable air mass is often the result of a temperature inversion, in which a layer of hot air lies above the cooler air next to the ground, contrary to the usual situation. In the Los Angeles basin a temperature inversion occurs about 80% of the time during the summer months. Second, the air must contain a compound that absorbs the sunlight and becomes very reactive as a result of its extra energy.

This compound is NO_2, a brown gas that is the most highly colored constituent of smog. Any substance that has a color absorbs some wavelengths of visible light. Because most of the light coming from the sun is in the visible range, NO_2 is the compound in smog most likely to absorb sunlight.

How is NO_2 formed? Very little NO_2 is present in the exhaust gas of a car or truck. Rather, it results from reactions between NO in the polluted atmosphere near a crowded highway and O_2 or some other oxygen-containing compound. The simplest reaction is

$$2\,NO + O_2 \longrightarrow 2\,NO_2$$

Given enough time, this reaction goes almost to completion at the usual temperature range of the atmosphere. However, for the very low concentrations of NO found in polluted air, the rate of the reaction is known to be too slow to account for the concentrations of NO_2 that have been detected. Other compounds containing oxygen must also help convert NO to NO_2.

Once formed, NO_2 absorbs energy from sunlight, as mentioned earlier. In this process a photon of light gives up its energy to the NO_2 molecule. This energy causes NO_2 to become very reactive, and therefore to participate in a number of chemical reactions.

One of the many substances resulting from reactions of NO_2 is ozone, O_3. Ozone has a greater ability to oxidize or remove electrons from other substances than does almost any other agent. Consequently, it can inflict great damage on animals and plants. Another product is acetaldehyde,

$$CH_3\overset{\overset{\large O}{\|}}{C}H$$

which is the simplest member of a class of compounds called aldehydes. Most aldehydes are highly toxic. Another particularly bad actor is peroxyacetyl nitrate (PAN),

$$CH_3\overset{\overset{\large O}{\|}}{C}OONO_2$$

which can be traced back to incompletely burned hydrocarbons in automobile exhaust. PAN is one of the strongest eye irritants in smog. It causes tears to flow endlessly, and it oxidizes human tissues readily.

In a smog-filled area the concentrations of the different components of the smog change in a regular fashion as the day progresses. This variation is shown in Figure 19.6. Early in the morning most of the pollutants are at a low level, since automobile traffic is infrequent. As traffic increases, the compounds emitted directly by the exhaust, hydrocarbons and NO, build up. Somewhat later NO_2 increases, and after that concentrations of O_3 and the aldehydes build, in agreement with the sequence of reactions that is understood to occur in smog. By midday, those highly reactive species arising from NO_2 react with hydrocarbons and aldehydes, decreasing their concentrations.

The principal means of dealing with the problem of photochemical smog has been to require **catalytic converters** in the exhaust systems of cars. The active agent in a converter is a mixture of metals that catalyze the complete oxidation of CO and hydrocarbons in the exhaust gas. If, as a result, the

Figure 19.6. *Typical variation in concentrations of components in photochemical smog during a day of intense smog.*

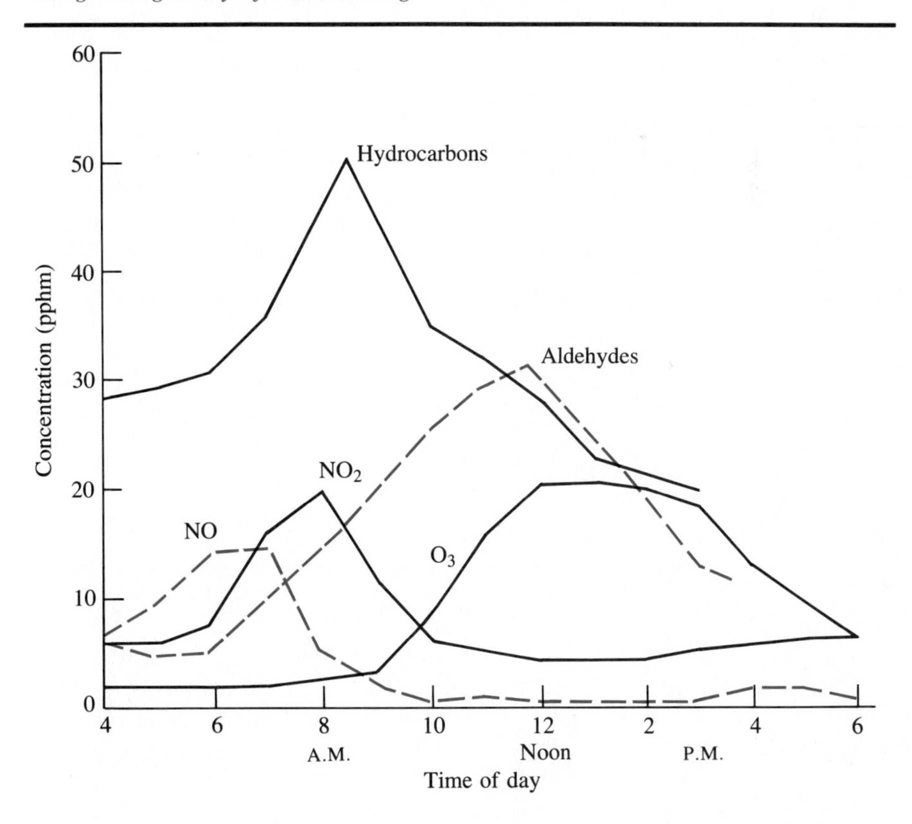

concentration of hydrocarbons is reduced, then the formation of aldehydes and PAN is inhibited. To get rid of NO in the exhaust, the converter recirculates a portion of the exhaust gas through the combustion chamber. The effect is to lower the temperature of the combustion, thereby slowing the rate of formation of NO from N_2 and O_2. So far, this method has not been completely effective. Some cities are so subject to smog that the best efforts to date have not been good enough. New methods must be found for the control of photochemical smog.

19.5 WATER POLLUTION

For centuries, contamination of drinking water by human and animal waste was the major kind of water pollution. The harmful bacteria present caused much disease. This problem may still exist in some communities, but the widespread municipal treatment of sewage has largely conquered it in the United States.

With the industrialization of society and the modernization of agriculture, other pollutants have appeared. Among them are pesticides and heavy metals. Farmers are strongly dependent on synthetic chemical **pesticides** to combat insects that can literally destroy entire crops. The high productivity of farms in the United States would be greatly diminished without pesticides. Yet when rainwater from fields runs into streams, pesticide residues find their way into drinking water.

The earliest pesticides were chlorinated hydrocarbons, such as DDT. Without any doubt the use of DDT to control pests, such as malaria-spreading mosquitos, has saved millions of human lives around the world. After much controversy, these pesticides were banned in the United States around 1970 because they did not break down in the environment and were seen as dangerous to birds and as potentially dangerous to human beings. Organic phosphates were then developed as substitutes. Parathion is one example (see Figure 19.7). Although highly toxic (it is chemically related to a nerve gas), parathion is fairly rapidly attacked by bacteria and so does not remain in the environment very long. In one way this is a disadvantage to the farmer, since parathion must be applied to crops frequently in order to be effective.

Figure 19.7. Parathion, a biodegradable pesticide.

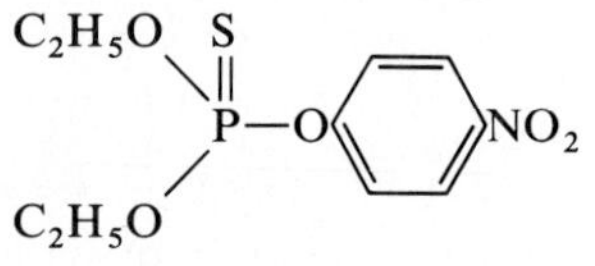

C_2H_5O S
P—O—(ring)—NO_2
C_2H_5O

O,O-diethyl-*O*-*p*-nitrophenyl phosphorothioate

The pollutants called **heavy metals** are the metals near the right-hand side of the bottom part of the periodic table. When present in a water supply, these metals are generally in the form of positive ions. A number of commercial products contain compounds of heavy metals. Examples are mercury compounds, used in paints to resist mildew and also in some batteries; compounds of arsenic; components of some pesticides; cadmium, sometimes present in water pipes, batteries, and tires; and lead, in leaded gasoline and in old plumbing. Heavy metals are also formed as by-products of some industrial processes.

Mercury nitrate was at one time used to treat the fur used to make felt hats. People who made these hats were exposed to mercury over many years, and from this long exposure they developed the symptoms of mercury poisoning—nervous disorders and loss of hair and teeth. Mercury poisoning may have given rise to the term "mad as a hatter" and the character the mad hatter, in *Alice in Wonderland*. For mercury to exert its toxic effect, it must enter the bloodstream, either from the digestive tract or from the lungs. Many mercury compounds are rather insoluble and are therefore less dangerous. Because of their insolubility, mercury amalgams (solutions of mercury and other metals) have long been used in dentistry without harmful effect.

Chlorine gas was formerly produced in large quantities through electrolysis of salt solutions with metallic mercury as one electrode. Although the mercury was recycled, a small fraction escaped and often was discharged into rivers and lakes. Since mercury compounds are mostly insoluble it was believed that they would sink harmlessly to the bottom of the rivers and lakes. This belief was dangerously incorrect; bacteria are now known to be able to convert insoluble mercury compounds into compounds such as methyl mercury, $CH_3—Hg^+$, which is soluble in water. Therefore, instead of lying harmlessly at the bottom, the mercury was transferred into the water. It accumulated in plants and then reached even higher concentrations in fish that ate the plants. In Minimata, Japan, industrial mercury wastes taken up by fish were responsible for the deaths of more than 50 people who ate the fish caught there. Other people suffered from mercury poisoning, and many children were born with serious birth defects. Procedures have now been changed, and mercury is no longer discharged as a by-product of chlorine production. Nevertheless, heavy metals are still a potential problem in drinking water.

19.6 AN INDUSTRIAL ACCIDENT—THE BHOPAL DISASTER

Ever present in a plant manager's mind is the possibility of a serious accident that would cause injury and death. The potential for disaster is certainly present in chemical and other manufacturing processes. The synthesis of useful chemicals ranging from pesticides to polymers involves many toxic substances, and concern for safe procedures has been a high priority in most companies. In the United States this concern has resulted in an enviable industrial safety record for workers: 2.5 workdays lost annually per 100 employees as a result of job-related injuries for the entire chemical industry. The comparable figure for workers in all kinds of manufacturing is 4.3 days lost. Yet, during the evening of December 2, 1984, an accident of unprecedented proportions occurred at a Union Carbide plant in Bhopal, India.

The plant produced a pesticide with the tradename Sevin, belonging to the class of pesticides called carbamates (see Figure 19.8). One of the chemical intermediates used to prepare Sevin is highly toxic methyl isocyanate, which was synthesized at the plant and stored in tanks. During the December evening, gas pressure within one tank increased to an abnormally high level, a safety valve opened, and a large quantity of methyl isocyanate was released into the atmosphere for about two hours. The gas, which stays close to the ground because it is twice as dense as air, spread to a heavily populated portion of the city. Especially hard hit were several densely populated slum areas, one of which was directly downwind of the plant.

The effects were immediate and devastating. Many people died in their sleep without knowing what killed them. Others ran choking and blinded into the streets, where they died within minutes. Some succeeded in reaching emergency treatment centers, only to perish after a few hours. A great many, exposed to smaller concentrations, survived with eye, lung, and other injuries yet to be fully assessed. The numbers are staggering: at least 2000 people killed, perhaps 10,000 with serious health impairments, and possibly 200,000 affected to some degree.

The structure of methyl isocyanate, $CH_3N{=}C{=}O$, explains in part its hazardous nature. The two adjacent double bonds make it fairly unstable and reactive toward a variety of compounds. Most of these reactions are exothermic and give off considerable heat. A liquid at room temperature, methyl isocyanate boils at 39°C. It readily evaporates at temperatures below the boiling point, and it has long been known to attack the respiratory system, eyes, and skin. By regulation of the U.S. Occupational Safety and Health Administration, the maximum exposure for workers is 0.02 parts per million averaged over eight hours.

Figure 19.8. Sevin, a carbamate insecticide used in agriculture.

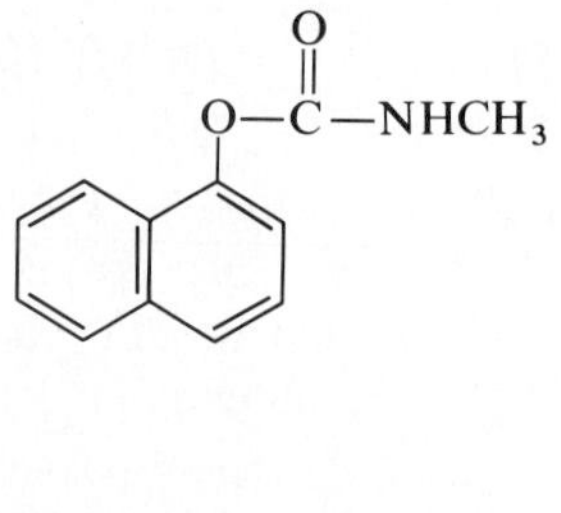

Presently available information indicates that the accident resulted from negligence and failure to follow standard safety procedures. The triggering event was the pumping, for unknown reasons, of about 1000 pounds of water into the tank containing the methyl isocyanate. This water reacted rapidly with the isocyanate, which should have been at 0°C but instead was at 15–20°C because a refrigeration unit had been inoperative for five months. The heat given off by the reaction caused the temperature and pressure to rise even higher, until a safety valve opened. The escaping gas then went to the first protective device, a scrubbing tank containing sodium hydroxide to react with the methyl isocyanate. This process could not handle the load, and the final safety feature, a flare to burn the gas, was out of service at the time.

A major question was the extent of responsibility borne by the parent company, Union Carbide, which owns 50.9% of the stock of the Indian subsidiary. Top company officials in the United States were apparently unaware of the extent to which safety regulations were being ignored in India. According to some critics, the U.S. company failed to monitor compliance with safety standards at the Indian plant. Admittedly, India had accepted the plant only on the condition that it be operated by Indian managers. And even though the scientific and technical backgrounds of the Indian managers were good, turnover among workers was high, and so their understanding of their jobs may have been limited.

Immediately after the Bhopal accident, a similar Union Carbide plant in Institute, West Virginia, was closed to determine the chance of methyl isocyanate leakage from that site. Investigation revealed very few exceptions to safety procedures, and the plant had experienced an extremely low accident rate. A few members of Congress expressed misgivings about some violations of EPA regulations at the plant, but most observers felt that there was little likelihood of a serious accident. Even so, as a result of the Bhopal experience, Union Carbide has added more safety measures at Institute.

A final question concerns the conditions under which industrial processes should be exported to less-developed countries. Until recently the main considerations have been economic ones, such as the potential for increased productive capacity, benefits from new products, and increased local employment. The Bhopal plant was important to India because pesticides improve agricultural yields and reduce the level of poverty and hunger. In the future, established chemical companies will make certain that a recipient country has adequate safety controls before exporting new technology. Although India at the time of the Bhopal incident had safety regulations and a system of inspection, enforcement was extremely lax. The Bhopal accident will probably be recorded in history as unique; it has heightened the awareness of people in all countries to the demands as well as the rewards of chemical technology.

19.7 HAZARDOUS WASTE

On August 2, 1978, the health commissioner of New York State ordered the evacuation of 240 families in the immediate vicinity of Love Canal (see Figure 19.9) in Niagara Falls. For some years these families had experienced illness and even death, which considerable evidence indicated was attributable to toxic chemicals dumped into the abandoned Love Canal from 1947 to 1952. It was not until August of 1980 that the remaining hundreds of residents in a wider area around Love Canal were evacuated by the U.S. Environmental Protection Agency (EPA). This incident shocked the American public into an awareness of the chemical dangers that may await us at many dump sites around the country.

During the 1880s an entrepreneur named William Love had planned at this location a model city whose industry was to be powered by cheap electricity from hydroelectric plants. Water for these plants was to flow through a new canal from the Niagara River. After the canal had been partially constructed, the plan was abandoned. All that remained of Love's dream was an incomplete canal, which filled with water and served as a swimming hole.

In 1905 a company then known as Hooker Electrochemical Corporation began manufacturing industrial chemicals at a plant in Niagara Falls. Like most chemical manufacturing operations, the operations at Hooker produced

Figure 19.9. *A section of a fence surrounding a contaminated dump site, formerly part of Love Canal.*

unwanted by-products. Moreover, occasionally there would be a malfunction in a plant operation, and an entire batch of product would be of unsatisfactory quality and therefore useless. These wastes had no economic value. To turn them into harmless substances by suitable chemical reactions would have been costly and would have placed the company at a disadvantage relative to its competitors.

Hooker Chemical purchased Love Canal in 1946 and a year later began dumping its unwanted chemicals there. According to an estimate made in 1979, 21,000 tons of waste were dumped there during the five years that Hooker used the site. Major components of this waste were hexachlorobenzene (a pesticide) and many other hydrocarbons containing chlorine. It is believed that long-term exposure to such organic chlorides increases the risk of contracting cancer. Such compounds are known as **carcinogens.**

It seemed as though nearly every family in the vicinity of Love Canal had some story to tell about a health problem that developed during the 1960s or 1970s. A number of children were born with birth defects. Others had frequent attacks of bronchitis, asthma, and skin rashes. A survey by the New York Department of Health showed that women between the ages of 30 and 34 living in the area had nearly four times as many miscarriages as did women in that age group in the population at large. Investigations revealed that drums of waste buried in the canal had corroded and leaked their contents. On certain days the stench of chemicals was overpowering. According to chemical analysis, the air, earth, and water were heavily contaminated with toxic and carcinogenic chemicals. These sometimes oozed out of the ground on homeowners' land or penetrated their basements.

The acute health problems of the residents of the Love Canal region are beyond dispute. It is more difficult to evaluate the long-term effects on their health, because much of the evidence is anecdotal and subjective. Scientists recognize that those personally involved in a serious problem may be unable to analyze that problem objectively. They try to solve this predicament by designing controlled studies, done by people with no vested interests. Several outside investigators did study the health problems of Love Canal residents, but because of the urgency of the situation they had to cut some corners. A pilot study in 1980 showed that residents exposed to the chemical waste may have experienced changes in their chromosomes. Yet, in 1983, on the basis of additional data, the U.S. Center for Disease Control concluded that people adjacent to the dump site were no more likely to undergo damage to chromosomes than were residents of other parts of Niagara Falls.

Evidence from dump sites all over the country suggests a widespread problem. At Times Beach, Missouri, unusual illnesses were traced to contaminated waste oil that had been spread on country roads in the early 1970s to keep down the dust. After several years of investigation, scientists found the toxic agent in the oil to be dioxin, a dangerous poison. Eventually, in 1983, the Environmental Protection Agency bought out the homes and

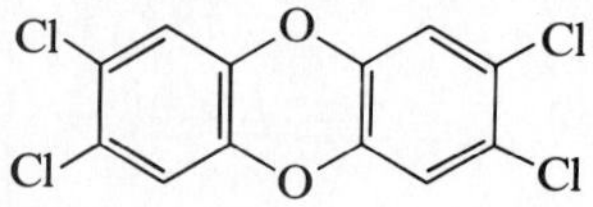

Figure 19.10. The structure of 2,3,7,8-tetrachlorodibenzo-p-dioxin.

businesses of the town at a cost of $24 million. Most of the 800 residents moved out.

Dioxin is a shorthand term for 2,3,7,8-tetrachlorodibenzo-*p*-dioxin, whose molecular structure is shown in Figure 19.10. Dioxin is formed in small amounts in the chemical synthesis of some herbicides and industrial chemicals. Its concentration in many contaminated areas is at the parts-per-billion level, an extremely low amount. The fact that dioxin is harmful to human beings at fairly high concentrations is well established. Much less is known about the long-range effects of tiny concentrations. Our knowledge rests primarily on data from laboratory animals. Dioxin is a carcinogen in animals, causing liver and other tumors. However, its toxicity differs considerably among different species of animals. It will be years before the full story is known, and in the meantime we must be cautious in dealing with dioxin.

The list of toxic wastes produced in the United States is long, and many of them come from industries other than the chemical industry. In addition to the chlorinated hydrocarbons already mentioned, there are polychlorinated biphenyls, or PCBs (see Figure 19.11), used for many years as heat-transfer fluids in high-voltage electrical transformers and as lubricants in pumps. When the transformers and pumps were discarded, the PCBs were thrown out with them, and some have leaked into the environment.

Congress has mandated that the Environmental Protection Agency identify the hazardous waste sites most in need of cleanup. The result was a National Priorities List, which by late 1984 had 786 names. The number of sites within each state is shown in Figure 19.12. The cleanup has been retarded by disagreements about fiscal priorities. In addition, the construction of modern and effective waste disposal sites meeting safety criteria has been slowed down by the objections of citizens who assume that having any disposal site nearby is undesirable. Almost everyone agrees that the sites are necessary for our well-being, but few are willing to have them nearby.

Figure 19.11. Structural formulas of two polychlorinated biphenyls.

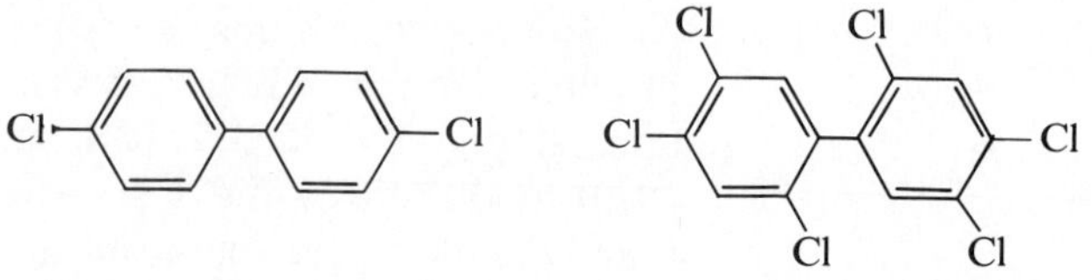

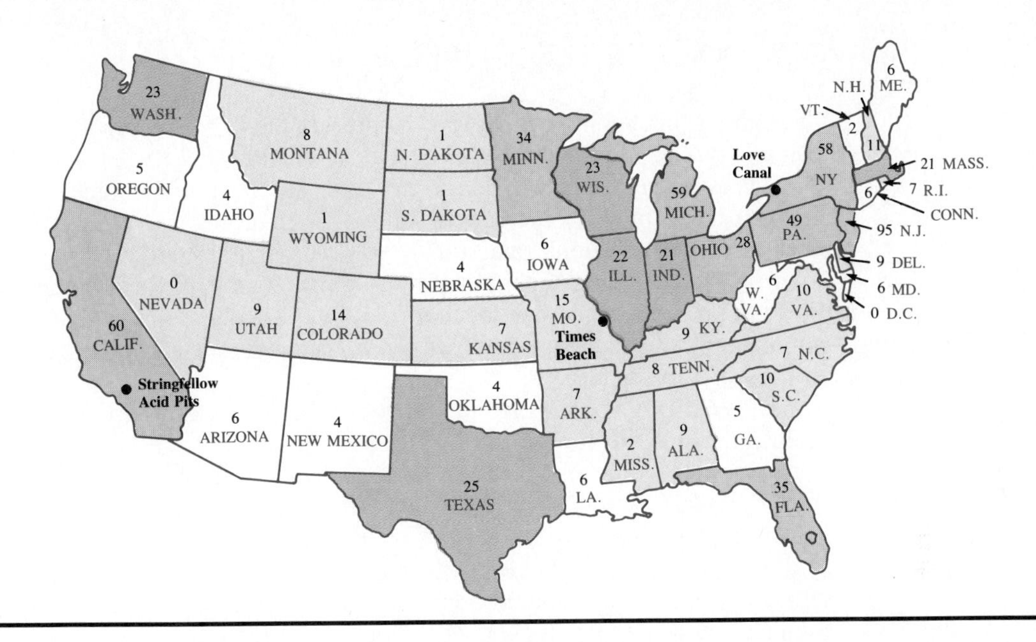

Figure 19.12. The number of high-priority hazardous waste sites in each state. Numbers are actual and proposed sites on the Environmental Protection Agency's national priorities list as of October 1984.

The federal government and many state governments in the United States now seem firmly committed to stopping the flow of hazardous waste to landfills and beginning the cleanup of contaminated areas. The Resource Conservation and Recovery Act, passed in 1976, controls the disposal of much newly-generated chemical waste. Enforcement of this law is difficult, however, and significant amounts of waste are still unregulated. In 1980 the first "Superfund" law provided \$1.6 billion for aid to states in cleaning up designated sites. The process will be slow, and the best we can hope for is removal and detoxification of the heaviest concentrations of pollutants.

The chief obstacle to combatting the release of hazardous waste to the environment is the lack of an economic incentive. Chemical reactions that will destroy most of these materials are known. For example, dioxin can be decomposed into nontoxic compounds by exposing it to ultraviolet light. High-temperature incineration converts many wastes into harmless substances. All such processes, however, are costly. In the long run, designing industrial processes to give only usable by-products is the best solution. This approach is superior to placing hazardous waste in even the best-designed and most secure landfill. It will certainly be much cheaper than cleaning up, at a later time, a dump site of the kind so often used in the past.

19.8 OUTLOOK FOR THE FUTURE

Throughout this book we have tried to present a balanced view of chemical technology. This technology has brought great benefits to the lives of us all in almost countless ways. As the earth's resources become scarcer, chemistry will play an increasingly important role in providing effective substitutes for traditional materials. However, it is important for society to ensure that chemical technology is used responsibly, in ways that present the least dangers to us and our world. This last chapter has dealt more with some of the challenges that we face. The economic and political trade-offs are often difficult.

We began this book with the statement that everything around us is a chemical. Some are simple, such as water and silicate minerals; others are complex, such as the enzymes that control chemical reactions in our bodies. Over the years, as our increasing knowledge of chemistry has allowed us to form metals from ores, to synthesize polymers and medicines, and even to design genes, there have been great benefits, as well as dangers and setbacks caused by a failure to take a broad view of chemical systems. If chemicals and chemical technology are used in the future with wisdom and resolve, they will continue to enrich our lives.

IMPORTANT TERMS

recycling The practice of reclaiming metals or other substances from discarded products.

petrochemicals Chemicals synthesized from petroleum.

nonrenewable resources Material resources that cannot be replaced by natural processes. Examples are petroleum and minerals.

photochemical smog An unpleasant and dangerous haze produced from automobile and industrial emissions by chemical reactions, some of which are catalyzed by sunlight.

catalytic converter An attachment to the exhaust system of an automobile that catalyzes the complete combustion of hydrocarbons in gasoline.

pesticides A broad term for all chemical agents used to kill harmful insects, weeds, fungi, or rodents.

heavy metal A metal located at the lower right portion of the periodic table. Heavy metals get their name from the fact that they are usually quite dense.

carcinogen A substance that causes cancer.

QUESTIONS

1. Describe the dual role played by petroleum as a natural resource.

2. The earth's crust contains only 0.6% by weight of titanium, which is used in certain steels. Explain why it is possible to recover titanium from the earth despite this very low average concentration.

3. A number of the chemicals in photochemical smog are mentioned in this chapter. Group them into three categories: those formed in an early stage of smog development, those formed in a middle stage, and those formed in a late stage.

4. Describe the operation of a catalytic converter.

5. What household item containing hazardous waste have you discarded during the last several years? How was it disposed of?

Public-Policy Discussion Questions

6. What kind of incentive do you feel would be effective in motivating consumers to recycle aluminum cans? Examples are payment for each used can returned to a local center, a reduction in garbage collection cost if the consumer separates cans from other refuse, and a law prohibiting the discarding of aluminum cans. Give the reasons for your answer.

7. *Limits to Growth* and "Global 2000 Revised" present very different views on whether or not we need to conserve carefully our material resources. Where do you stand on this issue? Why?

Projects

8. Because it is mostly carbon, coal could serve as a source of industrially important organic chemicals. Do some reading on this subject, and write a paper on whether this use of coal seems feasible to you.

9. Select a metal that is in extremely short supply in the United States, determine the sources, if any, of the metal in this country, and find where it may be found outside the United States. Write a short paper on your findings, and tell whether the shortage in this country has the potential to become a major problem.

10. Stringfellow Acid Pits, in southern California, is a well-publicized hazardous waste site. After doing some library research, write a paper describing the particular problems at this site.

11. Using library resources, write a paper describing a hazardous waste site in your state.

Appendix

SOME COMMON ENGLISH-METRIC AND METRIC-ENGLISH CONVERSIONS

Table A.1

ENGLISH-METRIC	METRIC-ENGLISH
1 inch = 2.54 centimeters (cm)	1 cm = 0.3937 inches
1 yard = 0.914 meters	1 meter = 1.094 yards
1 mile = 1.609 kilometers (km)	1 km = 0.621 miles
1 fluid ounce = 29.57 milliliters (mL)	1 mL = 0.03380 fluid ounces
1 quart = 0.946 liters (L)	1 L = 1.057 quarts
1 pound = 0.4536 kilograms (kg)	1 kg = 2.205 pounds
1 calorie = 4.184 Joules (J)	1 Joule = 0.2390 calories

Note: Table 2.2 summarizes the common prefixes.

FAHRENHEIT AND CELSIUS TEMPERATURE SCALES

These scales differ in two ways: the size of the degree and the zero of each scale. The Fahrenheit degree is smaller and equals $\frac{5}{9}$ of a Celsius degree. At 0°C, which is the freezing point of water, the Fahrenheit temperature is 32°F. The zero of the Fahrenheit scale corresponds to a colder temperature than the zero of the Celsius scale. Conversions between the two scales are accomplished by the equations

$$(°F) = \tfrac{9}{5}(°C) + 32$$

and

$$(°C) = \tfrac{5}{9}[(°F) - 32]$$

The table below presents selected equivalent temperatures.

Table A.2

°C	°F	°F	°C
−273.15	−459.67	−459.67	−273.15
−200	−328	−400	−240.0
−150	−238	−300	−184.4
−100	−148	−200	−128.9
− 50	− 58	−100	− 73.3
0	32	0	− 17.8
50	122	100	37.8
100	212	200	93.3
150	302	300	148.9
200	392	400	204.4
250	482	500	260.0
300	572		

INDEX

Bold page numbers refer to definitions of important terms.

Periodic Table of the Elements

GROUPS

TRANSITION METALS

PERIODS	I	II							
1	1 **H** Hydrogen 1.01								
2	3 **Li** Lithium 6.94	4 **Be** Beryllium 9.01							
3	11 **Na** Sodium 22.99	12 **Mg** Magnesium 24.31							
4	19 **K** Potassium 39.10	20 **Ca** Calcium 40.08	21 **Sc** Scandium 44.96	22 **Ti** Titanium 47.90	23 **V** Vanadium 50.94	24 **Cr** Chromium 52.00	25 **Mn** Manganese 54.94	26 **Fe** Iron 55.85	27 **Co** Cobalt 58.93
5	37 **Rb** Rubidium 85.47	38 **Sr** Strontium 87.62	39 **Y** Yttrium 88.91	40 **Zr** Zirconium 91.22	41 **Nb** Niobium 92.91	42 **Mo** Molyb- denum 95.94	43 **Tc** Technetium 99	44 **Ru** Ruthenium 101.07	45 **Rh** Rhodium 102.91
6	55 **Cs** Cesium 132.91	56 **Ba** Barium 137.33	57 ***La** Lanthanum 138.91	72 **Hf** Hafnium 178.49	73 **Ta** Tantalum 180.95	74 **W** Tungsten 183.85	75 **Re** Rhenium 186.2	76 **Os** Osmium 190.2	77 **Ir** Iridium 192.22
7	87 **Fr** Francium 223	88 **Ra** Radium 226.05	89 ****Ac** Actinium 227	104 **Unq** 261	105 **Unp** 262	106 **Unh** 263			

***Lanthanides**	58 **Ce** Cerium 140.12	59 **Pr** Praseody- mium 140.91	60 **Nd** Neodymium 144.24	61 **Pm** Promethium 145	62 **Sm** Samarium 150.35	63 **Eu** Europium 151.96
****Actinides**	90 **Th** Thorium 232.04	91 **Pa** Pro- tactinium 231.0	92 **U** Uranium 238.03	93 **Np** Neptunium 237.0	94 **Pu** Plutonium 244	95 **Am** Americium 243